Current Topics in Microbiology and Immunology

Volume 411

More information about this series at http://www.springer.com/series/82

Elke Mühlberger · Lisa L. Hensley
Jonathan S. Towner
Editors

Marburg- and Ebolaviruses

From Ecosystems to Molecules

Responsible series editor: Yoshihiro Kawaoka

Springer

Editors
Elke Mühlberger
Boston University School of Medicine
Boston, MA
USA

Jonathan S. Towner
Centers for Disease Control and Prevention
Atlanta, GA
USA

Lisa L. Hensley
NIAID
National Institutes of Health
Frederick, MD
USA

ISSN 0070-217X ISSN 2196-9965 (electronic)
Current Topics in Microbiology and Immunology
ISBN 978-3-319-68946-3 ISBN 978-3-319-68948-7 (eBook)
https://doi.org/10.1007/978-3-319-68948-7

Library of Congress Control Number: 2017960183

Printed on acid-free paper

This Springer imprint is published by Springer Nature
The registered company is Springer International Publishing AG
The registered company address is: Gewerbestrasse 11, 6330 Cham, Switzerland

Preface

Filoviruses have captivated the imagination of scientists and the public alike since Marburg virus was first isolated in Germany in 1967. Through the years, these viruses have gained much notoriety for the devastating nature of the outbreaks they cause and their repeated sensationalization by mainstream media and the entertainment industry. Over the last 50 years, we have seen tremendous advances in our understanding of these agents, and this book endeavors to capture the major areas of discovery but in no way expects to be an all-encompassing source. Perhaps it is appropriate that the completion of this book coincides with the 50-year anniversary of the discovery of the first filovirus.

The outline for this book was conceived in 2013, prior to the start of the West African Ebola virus outbreak. Many of the contributors to this book were among those who volunteered to respond to the outbreak, some for over a period of years, or set aside their normal work to help support outbreak response efforts.

A wide spectrum of renowned experts in the field worked together to make this book happen. They range from clinicians to virologists to biochemists, and we are incredibly grateful for their contributions. Some of the authors have worked on filoviruses since their discovery, while others are much newer to the field. Despite these differences, all of the authors have one common goal—to better understand how filoviruses work, and to use their knowledge to help prevent or mitigate the impact of future filovirus outbreaks.

We have separated this book into four parts. Part I covers filovirus ecology, outbreaks, and clinical management. It begins with a fascinating first-hand account of the challenges that faced researchers 50 years ago in a small German town when they encountered a highly virulent infectious agent of unknown origin. Chapter "Filovirus Research: How it Began" was written by one of the original filovirus discoverers and describes the first isolation of Marburg virus in 1967 during a time long before virologists had use of modern biocontainment facilities. We then move on to a global view of filovirus distribution and emergence in the chapter "Ecology of Filoviruses" in which we learn about the natural origins of some enigmatic viruses, how they persist long term in nature, and what drivers might promote their spillover to other animals including humans. One of these spillover events led to the largest Ebola virus outbreak on record and is described in the chapter "West Africa 2013 Ebola: From Virus Outbreak to Humanitarian Crisis". This comprehensive account describes the devastating epidemic that not only brought Ebola virus directly to Europe and the USA, but overwhelmed the long-neglected public health infrastructures in Guinea, Sierra Leone, and Liberia and the international response alike. The next two chapters, "Clinical Management of Ebola Virus Disease Patients in Low Resource Settings" and "Clinical Management of Patients with Ebola Virus Disease in High Resource Settings", describe the challenges and risks facing clinicians when they treat patients infected with Ebola virus and how their approaches differ depending on the resource environment.

Part II of the book focuses on filovirus pathogenesis and protection. Chapter "Ebola Virus Disease in Humans: Pathophysiology and Immunity" provides a detailed review of the human disease, including fascinating new studies of the human immune response that resulted from the West African Ebola virus outbreak. Chapters "Nonhuman Primate Models of Ebola Virus Disease" and "Small Animal Models for Studying Filovirus Pathogenesis" summarize the vast body of work using animal models, big and small, to study filovirus disease and develop experimental treatments and vaccines. Part II concludes with the chapters "Accelerating Vaccine Development During the 2013–2016 West African Ebola Virus Disease Outbreak" and "Therapeutics Against Filovirus Infection" that each provide state-of-the-art summaries of current experimental countermeasures used to combat filovirus infections, including those deployed during the Ebola virus outbreak in West Africa.

The first three chapters of Part III take us deep into the cellular level of filovirus infection. Chapter "Filovirus Strategies to Escape Antiviral Responses" provides a comprehensive account of mechanisms used by filoviruses to counteract antiviral responses. In the following two chapters, "Mechanisms of Filovirus Entry" and "Inside the Cell: Assembly of Filoviruses", we learn how filoviruses make their way into cells and which strategies they use to replicate their genomes and assemble to new particles. Finally, we reach the atomic level in "Filovirus Structural Biology: The Molecules in the Machine" which focuses on the structural analysis of filovirus proteins through the use of stunning images of these structures.

The book ends with a description of research tools used to study filoviruses. Chapter "Reverse Genetics of Filoviruses" summarizes the use of reverse genetics as a powerful tool to investigate virus replication and pathogenesis. The last chapter, "Guide to the Correct Use of Filoviral Nomenclature", is meant as a useful tool to help guide virologists through the sometimes confusing, and recently evolved, world of filovirus taxonomy.

Last but not least, we wish to thank all the authors who have contributed their work to this book. We are grateful to Jens Kuhn, who volunteered to critically read and edit almost all of the chapters, and to Jiro Wada for designing the preface figure. We also wish to state up front that any views or opinions expressed in the book do not necessarily reflect those of the editors, authors, or their respective institutions.

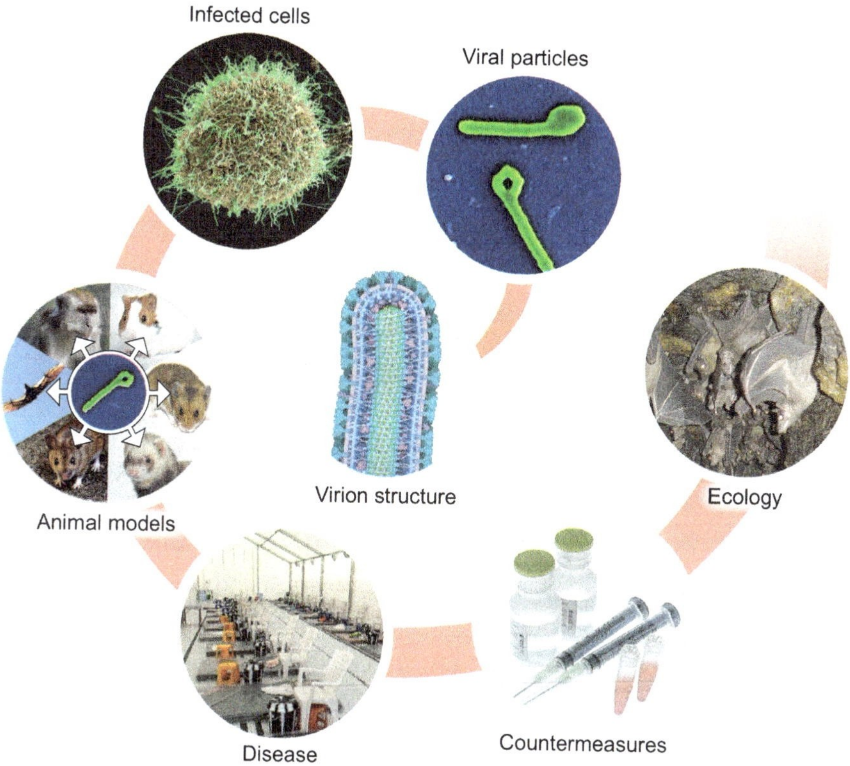

Preface Figure Marburg- and Ebolaviruses: From Ecosystems to Molecules. Figure designed by Jiro Wada, NIH/NIAID, Integrated Research Facilities. The watercolor of the virion structure was kindly provided by David S. Goodsell, RCSB Protein Data Bank. Bat photo provided by Chris Black, WHO.

Boston, USA
Frederick, USA
Atlanta, USA

Elke Mühlberger
Lisa L. Hensley
Jonathan S. Towner

Contents

Part I
Maintenance in Nature and Spillover: Ecology, Outbreaks and Clinical Management

Filovirus Research: How it Began

Werner Slenczka

Abstract The first reported filovirus outbreak occurred in August 1967, when laboratory workers in Marburg and Frankfurt, Germany, and Belgrade, Yugoslavia (now Serbia) became infected with an unknown highly pathogenic agent. The disease was characterized by high fever, malaise, rash, hemorrhagic and tetanic manifestations, and high lethality, amounting to 25%. The disease was introduced to Europe by grivets (*Chlorocebus aethiops*), which were used for biomedical research and vaccine production. The causative agent, Marburg virus, was isolated and identified by scientists of the University of Marburg, Germany in cooperation with specialists for viral electron microscopy at the Bernhard Nocht Institute in Hamburg, Germany. In this chapter, Dr. Slenczka, who was involved in the first isolation of Marburg virus in 1967, describes the desperate hunt of the causative agent of this first filovirus disease outbreak in the center of Europe, its successful isolation, the likely route of transmission from a monkey trading station to vaccine production facilities in Germany and Yugoslavia, and the consequences of this outbreak, including a shortage in the production of poliomyelitis vaccine In addition, this chapter provides insight into some of the peculiarities of filovirus infection, such as sexual virus transmission several months after recovery and the role of Ca^{2+}-loss in Marburg virus pathogenesis, which were already observed during this first well-documented Marburg virus disease outbreak.

Contents

W. Slenczka (✉)
Institute of Virology, Philipps University of Marburg, Am Weinberg 19, 35037 Marburg, Germany
e-mail: slenczka-marburg@t-online.de

Current Topics in Microbiology and Immunology (2017) 411:3–22
DOI 10.1007/82_2017_8

Published Online: 02 August 2017

1 Introduction

> One can think of the middle of the twentieth century as the end of one of the most important social revolutions in history, the virtual elimination of the infectious disease as a significant factor in social life (Sir Macfarlane Burnet) (Burnet and White 1962).

Burnet's optimistic view on the future of infectious diseases is perfectly in accordance with parts of the published opinion, which during the 60s of the past century was prevalent. Representatives of scientific organizations for oncology and for psychiatric diseases postulated that research funds for infectious disease should be cut and the money devoted to their own scientific interests. In a decade during which many insect-borne diseases had vanished due to large-scale distribution of DDT, during which bacterial infections were controlled with antibiotics, and during which the eradication of smallpox, poliomyelitis, measles, and other viral diseases had progressed from an utopian dream to a realistic prophecy, nobody was prepared to face the advent of a completely unknown infectious disease with an extremely high lethality in the center of Europe.

The story of the 1967 Marburg virus disease (MVD) outbreak has often been told and it is not the aim of this article to repeat all the known details. Instead, the intention is to restore some forgotten or neglected details and to give a personal view on some events.

2 An Unknown Disease

The summer of 1967 was very hot in Marburg, Germany, and whoever was able to leave the town went to the seaside or the mountains. In August 1967, 20 people, who lived in small villages surrounding Marburg, fell ill with fever, malaise, headache, vomiting, rash (Fig. 1), and conjunctivitis, the tentative diagnosis was "summer diarrhea" (often caused by enteroviruses or by coliform bacteria) or dysentery. Initially, despite the gravity of the symptoms, these patients were treated in their homes for up to a week. The patients were admitted, on average, on day 5 after the onset of symptoms to the hospital of the University of Marburg which had an isolation ward for the treatment of infectious diseases (Fig. 2). In total, fourteen men and six women were admitted between August 15 and the end of the month.

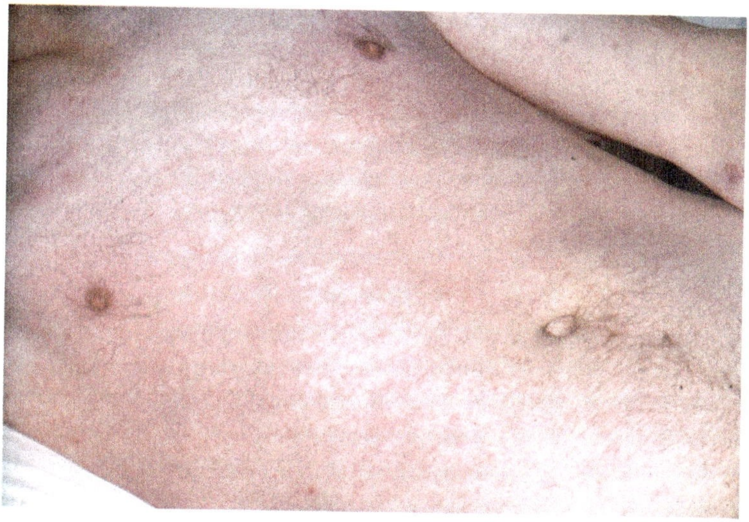

Fig. 1 Characteristic rash which was seen with all Marburg virus disease patients during the 1967 epidemic. Kindly provided by Gerhard Baltzer, Marburg, 1967

Fig. 2 Isolation ward of the hospital of the University of Marburg in 1967

It soon became clear that they all were employees of Behringwerke AG, a pharmaceutical company founded in 1904 by Emil von Behring. All of these patients had been involved in the production of poliomyelitis vaccine which relied on the use of primary simian cell cultures for propagation of the attenuated vaccine strains (Sabin 1–3). It soon became evident that concurrently four patients with similar clinical signs were being treated at the university hospital in Frankfurt, Germany (Siegert et al. 1967, 1968; Martini et al. 1968a, b; Stille et al. 1968). These patients were employees of the Paul Ehrlich Institute, a governmental institution responsible for the approval of sera and vaccines. The Frankfurt patients had also been working with monkeys and with simian cell cultures that were needed for safety control procedures and standardization of poliomyelitis vaccines. In addition to these primary cases, four members of the hospital staff, two each in Frankfurt and in Marburg, acquired nosocomial infections. In another case of secondary infection, the spouse of one of the primary cases fell ill 67 days after her husband's disease resolved and, intriguingly, sexual transmission was supposed to be the route of infection (Slenczka et al. 1968; Siegert et al. 1968; Martini and Schmidt 1968). This is the first case described in the literature of a virus causing a systemic infection that was transmitted by the sexual route even several months after convalescence. The seminal fluids of nine other convalescents were tested for the presence of Marburg virus at the same time, but infectious virus or viral antigen was not found (Siegert and Slenczka 1971). Two female convalescents gave birth to healthy children about 18 months after they had survived the disease. No virus or virus antigen was found in the placentae or umbilical cords. Antiviral IgG, but not IgM, was found in the blood (Siegert and Slenczka 1971).

During the course of the disease, most patients developed hemorrhages varying in intensity from discrete petechiae to bleeding from needle puncture sites and massive bleedings from the gastrointestinal, respiratory, and urogenital tracts. A 39-year-old patient died of massive intraventricular hemorrhage. Eventually, five of the primary cases in Marburg and two of the cases in Frankfurt succumbed to the disease. The onset of massive hemorrhagic disease proved to predict an unfavorable outcome; none of the survivors developed severe hemorrhages (Martini et al. 1968a).

In addition to the cases in Marburg and Frankfurt, two cases occurred in Belgrade, Yugoslavia (now Serbia). The two patients in Belgrade, a veterinarian and his wife, worked at the Torlak Institute, an institution devoted to the production and safety control of poliomyelitis vaccines. The female patient was infected while caring for her husband. Both patients survived the infection (Todorovic et al. 1969; Stojkovic et al. 1971).

In 1982, we tested the serum of a man who claimed that he had the disease in 1967. Although he had been severely ill, he had been treated at home. He, too, worked in the cell culture lab at Behringwerke AG and had been exposed to infected simian cell cultures. Out of more than 120 contact persons who were retrospectively tested for Marburg virus antibodies, he was the only one who was seropositive, indicating that he had indeed been infected with Marburg virus. In summary, there were 26 primary and 6 secondary cases during the MVD outbreak in 1967. None of the secondary cases succumbed to the disease, but seven of the primary cases were fatal, amounting to an overall case fatality rate of 21.9% (26.9% for the primary cases only) (Slenczka and Klenk 2007).

In all three institutions, the common epidemiological denominator of the primary cases was contact with grivets (*Chlorocebus aethiops*), or handling monkey organs and cell cultures derived from these. The monkeys had been imported from Uganda and were used for the production of kidney cell cultures, to be used for the production of vaccines against measles and poliomyelitis (Hennessen et al. 1968).

3 Role of Ca^{++}-Loss in the Pathogenesis of Marburg Virus Disease

In the treatment of patients, the hemorrhagic disease was neither mitigated by blood transfusions, nor by treatment with coagulation factors (frozen plasma) or by applying thrombocyte concentrates. However, Martini noted that the drop in the concentration of plasmatic coagulation factors was not so remarkable as to explain the severe hemorrhagic diathesis (Martini 1971).

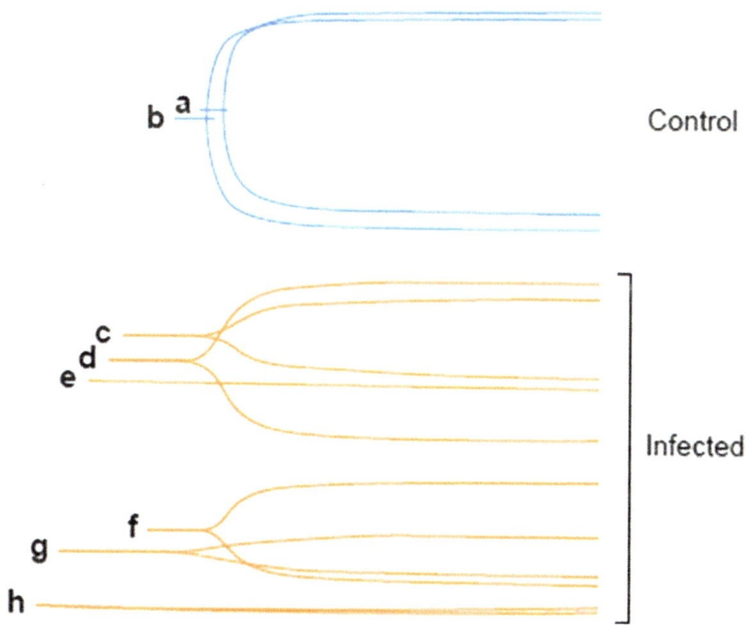

Fig. 3 Thromboelastograms (TEGs) performed with blood from one noninfected (*top*) and two Marburg virus-infected guinea pigs. *a* + *b* citrate plasma TEGs from a noninfected guinea pig. *c* − *h* TEGs from Marburg virus-infected guinea pigs on day 4 after onset of fever. *e* + *h* unmodified plasma TEGs without additional substances. *c* + *d* unmodified plasma TEGs with 0.1 ml (*c*) or 0.2 ml (*d*) of 0.1 mM solution of $CaCl_2$ per ml plasma; *f* + *g* citrate plasma TEGs with 0.1 ml (*f*) and 0.2 ml (*g*) 0.1 of mM solution of $CaCl_2$ per ml plasma. *Source* Egbring et al. (1971)

In case descriptions of the 1967 MVD outbreak, the reported symptoms include generalized paraesthesia, restless legs, sleeplessness in spite of fatigue, hyperesthesia of the skin and the feeling to "lie on crumbs". These symptoms are typical for tetany, a diagnosis, which was not verified and not even suspected at that time (Martini et al. 1968b). Tetany results when the concentration of Ca^{2+} in plasma is reduced to less than 50% of the normal value. Since ionized calcium is an important clotting factor (factor IV), tetany can be associated with spontaneous bleeding. But, where did the Ca^{++} go? Using von Kossa stain, Zlotnik (1969) and Korb et al. (1971) observed extravascular deposits of calcium in necrotic tissues of Marburg virus-infected guinea pigs and MVD patients.

In an early study on the coagulopathy in Marburg virus-infected guinea pigs, Egbring, Slenczka, and Baltzer found that on days 4 and 5 after onset of fever, coagulation was no longer detectable using thromboelastography. However, 50% of the coagulation capacity could be restored when ionized calcium ($CaCl_2$) at a final concentration of 10 mM was added directly into the reaction vessel (Fig. 3) (Egbring et al. 1971). It should be emphasized that restitution of clotting capacity by substituting calcium was performed in vitro using plasma from infected animals that contained less than 20% of clotting factors and platelets. Based on these results, MVD is the first disease in the literature for which it was shown that a consumption coagulopathy can be caused by loss of ionized calcium (Egbring et al. 1971).

4 Search for the Etiologic Agent

Diagnostic laboratory tests in search of the etiologic agent of the unknown disease were conducted in the clinical microbiology laboratories at the university hospitals in Frankfurt and Marburg. In Marburg, all this work was carried out by technicians using classical microbiology techniques and—at least initially—without awareness of the high risk. Masks and gloves were only used by laboratory scientists/technicians after the first patients had died. Laboratory infections did not occur and no seroconversions were found in these technicians.

The initial diagnostic arsenal included tests for salmonellosis, shigellosis, rickettsiosis, chlamydiosis, yellow fever, and many other infectious diseases. The negative results of these tests eventually convinced the medical staff that this was not a domestic but rather an exotic disease. Therefore, serum specimens were sent to 12 international laboratories known to be experienced in the diagnosis of zoonotic or tropical diseases. The focus of these investigations was predominantly on arboviruses and on bacterial and viral agents known to cause hemorrhagic fever. Serologic tests for antibodies against more than 80 arboviruses and other tropical viruses were negative (Siegert et al. 1968).

The serological tests lead to a single trace for a tentative diagnosis: Dr. L. Popp at the *Staatliches Medizinaluntersuchungsamt* (Governmental Medicinal Office of Investigations) in Braunschweig, Germany, an expert in leptospirosis, detected anti-leptospiral antibodies in some of the patient sera. Although this finding was not unexpected in serum specimens from laboratory staff working with wild-caught animals, it served as an incentive to start a search for leptospira. Since the guinea pig model was routinely used for isolating leptospira, Dr. W. Mannheim, who was in charge of the bacteriology laboratory at the University of Marburg, inoculated guinea pigs with blood from one of the patients on August 22, 1967. The animals did not develop overt clinical disease but showed a moderate increase in temperature for 2–3 days. However, neither leptospira nor any other microorganisms were detected in the blood of these animals (Siegert et al. 1967; Slenczka et al. 1968).

Meanwhile, on August 24, four of the hospitalized patients, two in Frankfurt and two in Marburg, succumbed to the disease after showing signs of severe hemorrhagic shock. None of the patients had responded to any therapeutic measures. At this point, it had become clear that the patients suffered from an unknown agent causing severe hemorrhagic fever and that Dr. Mannheim had most probably transmitted this agent to guinea pigs. However, it also had become clear that none of the agents which were, at that time, known to cause hemorrhagic fever were involved (Siegert et al. 1967). In 1967, various bacteria, including members of the *Borrelia*, *Rickettsia*, and *Leptospira* genera, were known to cause hemorrhagic disease. The concept of viral hemorrhagic fevers was first introduced in 1948 by the virologist Čumakov (1948). Later, Daniel Carleton Gajdusek, who worked on the etiology of Korean hemorrhagic fever, postulated that Omsk hemorrhagic fever, Crimean Congo hemorrhagic fever, and hemorrhagic fever with renal syndrome were caused by at least three different viruses (Gajdusek 1962).

The death of four patients within 2–3 weeks after onset of the disease was a shock and raised concerns that an unknown agent with high pathogenicity might be distributed in the general population. Considering the high pathogenicity of the agent, the inappropriate biosafety conditions of the diagnostic laboratories in Frankfurt and in Marburg, and their location in the centers of these towns, it was decided that further diagnostic work in search of the etiologic agent should not be conducted in these laboratories but rather in institutions which had more experience and were better equipped for work with agents of extreme pathogenicity. Diagnostic specimens were sent to the Institute Pasteur in Dakar, Senegal, the Microbiological Research Establishment in Porton Down, Salisbury, UK, the CDC in Atlanta, GA, USA, the Middle America Research Unit, Balboa Heights, Canal Zone, Panama, the Poliomyelitis Research Foundation in Johannesburg, South Africa, and the Poliomyelitis Institute Moscow, Soviet Union (now Russia) (Siegert et al. 1968).

All these institutions had offered their help in search of the causal agent and were informed on the details of the clinical presentation and results of diagnostic tests, which up to that time had been carried out. Specifically, they were informed that guinea pigs developed a fever in response to infection. The material destined for the Microbiological Research Establishment was taken to London by Dr. Dick and then shipped to Porton Down.

Inclusion of a Soviet institute among those to receive specimens raised some concerns. At this time, there was a deep distrust between western and eastern countries due to the Cold War. Newspapers in the Eastern Bloc had already claimed that the events in Marburg and in Frankfurt were due to an accident in western research facilities working on biological weapons. Shipping patient specimens to Moscow was meant as a confidence-building measure to show that there was nothing to conceal. Some years later, the strain isolated from this material, the Popp strain—the name refers to a Frankfurt patient—was used by Russian scientists for developing biological weapons (Leitenberg et al. 2012).

Three weeks later, in mid-September, the intensity and spread of the disease had dissipated. Sadly, two patients in Frankfurt and five in Marburg had succumbed to the disease, but the majority of the hospitalized patients had survived. Most of them had recovered and had been dismissed from the infection wards in a rather good physical condition. Transmission in the general population had not occurred and was no longer to be feared.

At this point, the guinea pig experiments were resumed to identify the unknown agent. In Marburg, Prof. Dr. R. Siegert, (head of the Institute of Hygiene and Medical Microbiology at Marburg University) and a research assistant in his laboratory, Dr. H.L. Shu started to inoculate guinea pigs with diagnostic material. Body temperature was monitored daily, and when the animals developed a fever, they were bled by cardiac puncture. Guinea pigs that were infected with the original material developed a fever on day 3 after inoculation without showing any other symptoms and survived the disease. However, guinea pigs, which were treated with passaged material (passage 3–5), became severely ill and had clinical signs similar to those observed in patients, fulfilling one of the Henle–Koch postulates (Koch 1890), later modified by Rivers for viral diseases (Rivers 1937).

Animals that had survived the disease were resistant to exposure with high passage material. Moreover, guinea pigs were protected from the disease by administration of convalescent sera from patients (Siegert et al. 1968).

These experiments were a step in the right direction, but the agent remained unidentified. To facilitate identification, blood from infected animals, taken at the climax of the disease, was mixed with glutaraldehyde and formaldehyde to inactivate and preserve the unknown agent and was sent to the electron microscopy (EM) laboratories at the University of Marburg and the Bernhard Nocht Institute for Tropical Medicine in Hamburg, Germany for analysis: Dr. D. Peters, head of the Virology Department at the Bernhard Nocht Institute, was a renowned virological electron microscopist and highly experienced in analyzing viral structures.

Of course, EM-based search for an unknown pathogen in biological materials can be extremely fatiguing and—in case of a negative result—very frustrating.

In the case of the guinea pig material, an additional obstacle became evident soon. A serious complication, often encountered in the search for unknown pathogens, is contamination by organisms unrelated to the disease. These "pick-up" contaminants may interfere with the etiological agent or may even cause disease themselves. Contamination may occur as a result of a preexisting infection. During passage to new animals, contaminants might be transferred with a higher efficiency than the unknown etiological agent. The risk of cultivating a contaminating agent may be reduced by using animals from an SPF (specific pathogen free) breed. In 1967, SPF guinea pigs could not be afforded in Marburg. Instead, the animals were purchased from local breeding stations which did not control for infections. Therefore, in the experiments carried out by Drs. Siegert and Shu, it happened that microbial contaminations, including pseudomonads, pasteurellae, and in some cases paramyxoviruses were present in the guinea pigs. Although it was quite clear that these well-known organisms were not the etiological pathogen, the presence of these contaminants complicated data interpretation.

When Drs. Siegert and Shu were unable to identify the unknown agent, I came into play. Up to that point in time I had not been involved in the search for the causative agent for two reasons. First, I was a research assistant receiving my salary from the German Research Foundation for conducting research and not for performing diagnostic work. Second, I had a family at home with three little children and it was an accepted policy not to expose parents to the dangerous agent. During my time as a postdoctoral fellow in Dr. F. Lehmann-Grube's laboratory, I had developed a diagnostic assay for lymphocytic choriomeningitis virus based on immunofluorescence. Today, the reagents needed for this technique can be easily obtained from commercial suppliers in excellent quality. However, in 1967, although it was possible to purchase secondary antibodies, these reagents generally led to unsatisfactory results due to insufficient quality. Therefore, researchers usually prepared their own antibody reagents to fractionate immunoglobulins and to conjugate them with fluorescein isothiocyanate (FITC). This was not a simple task, as it was not easy to find a supplier offering coupling agent of satisfactory quality. I had spent several months adapting these techniques and was thus prepared to make use of immunofluorescence analysis as a tool to identify the unknown agent.

Taking convalescent sera from humans and from infected guinea pigs, as well as the corresponding negative controls, I prepared IgG-fractions, coupled the antibodies to FITC, and purified them from unbound FITC and from nonspecific binding substances to improve serologic specificity.

Organs from infected and noninfected guinea pigs, especially livers and spleens, were used to make imprint preparations on microscopic slides, which were then air-dried and fixed with ice-cold acetone. Since we did not know if the unknown agent would be killed by acetone, we handled these slides with extreme caution. It took 3 weeks of hard work before we had the first results. I found brilliantly fluorescent cytoplasmic inclusions in liver cells from an infected guinea pig. Since all the controls were negative, I was sure I had found antigenic structures of the unknown pathogen (Fig. 4). At this time, it was not yet possible to tell whether these inclusions, which resembled the Negri bodies found in rabies virus-infected cells (Goldwasser et al. 1959), were indicative of a viral or bacterial infection. However, it was clear that I had detected something that nobody had seen before; structures of an unknown agent causing a deadly disease (Slenczka et al. 1968).

Using this assay, it was now possible to identify those animals which were infected with this agent to select material for EM investigations. Once again, guinea pig blood treated with glutaraldehyde and formaldehyde was sent to the Bernhard Nocht Institute in Hamburg for EM analysis. Dr. D. Peters, together with a technician, analyzed negative stained material for more than a day but did not observe anything reminiscent of a viral structure. On the second day of his search, Dr. Peters

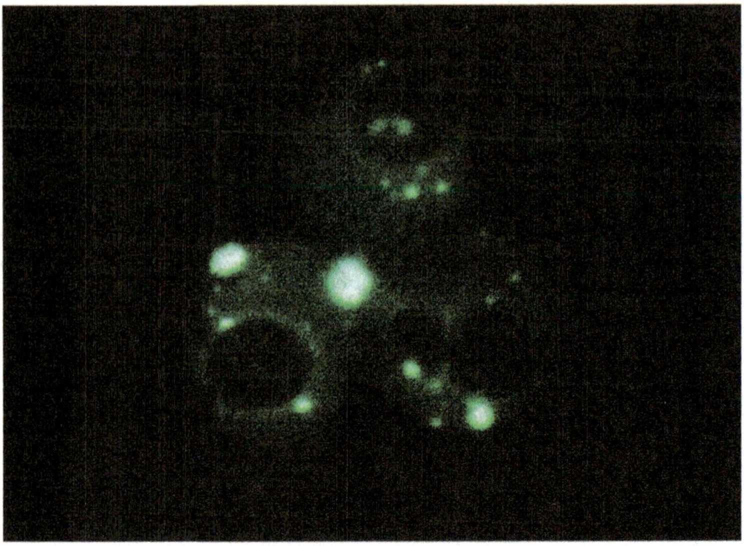

Fig. 4 Immunofluorescence analysis showing liver cells from a guinea pig infected with Marburg virus. Immunofluorescence analysis was performed using FITC-conjugated immunoglobulins derived from a Marburg virus disease survivor. Marburg virus forms large inclusions in the cytoplasm of the infected cells. W. Slenczka, October 1967 (Siegert et al. 1967, 1968)

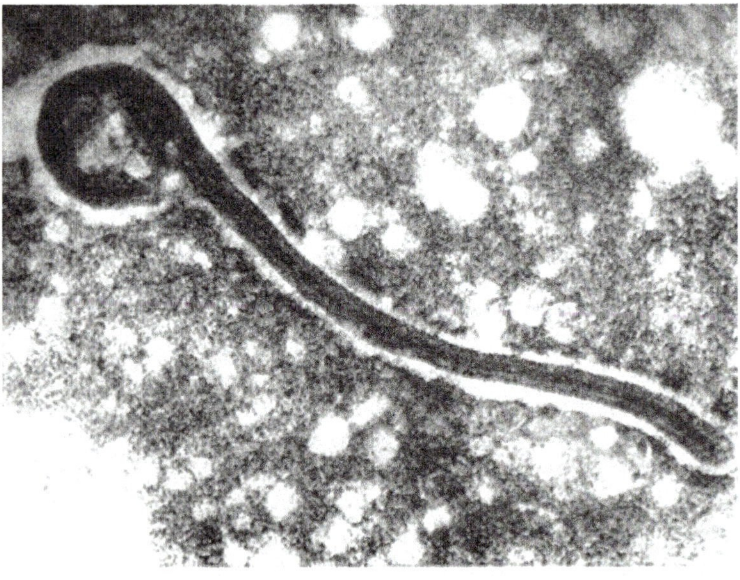

Fig. 5 First electron micrograph of a Marburg virus particle. Picture taken on November 20, 1967 by G. Müller, Bernhard Nocht Institute, Hamburg, Germany at 60,000 magnification (Siegert et al. 1967, 1968; Brauburger et al. 2012)

left the laboratory for a lunch break and handed the specimen to his coworker, Dr. G. Müller, asking him to continue the search. In less than an hour, Dr. Müller had succeeded in finding viral particles that, due to their sizes and unique morphologies, were identified as the products of an unknown virus (Fig. 5). When Dr. Peters returned from his lunch break, Müller showed him the new virus. It is not clear why Dr. Peters had not found the viral particles when he examined the samples. The most probable explanation seems to be that the particles had spontaneously sedimented to the bottom of the tube and Dr. Peters took material from the top only.

5 History of Publication

On November 27, 1967, press conferences were held in Marburg and Hamburg to announce the identification of the etiological agent, which had caused the "monkey disease". At this time, the virus was named "Marburg virus" as a reference to the town, where the greatest number of cases had occurred and to the place where the causative agent had been isolated and identified.

The first scientific communication on the isolation and identification of the Marburg virus was made at the IV Congreso Latinamericano de Microbiologia in Lima, Peru held from November 26 to December 2, 1967. This paper was published in the proceedings of this congress (Siegert, R., Shu H.L., Slenczka, W., Peters, D., and Müller, G. Detection of the so-called green monkey agent.). Three weeks later, on December 21, 1967, the groups in Marburg and Hamburg published a detailed report on the discovery of the Marburg virus in a German medical journal (Siegert et al. 1967) with the English translation following on January 1, 1968 (Siegert et al. 1968). Reprints of these papers containing the first electron micrographs of the new virus were sent to all the groups and institutions that had taken part in the efforts to isolate this agent and also to WHO (1968).

On November 29, 1967 the group in Porton Down published their results in a Lancet paper (Smith et al. 1967) in which they suggested that a member of the *Rickettsia* or *Chlamydia* genera was the etiologic agents of the "vervet monkey disease". In addition, it was stated in this paper that "our virological findings were negative" (Smith et al. 1967). Although Smith and colleagues did not identify the causative agent of MVD, their paper unfortunately has been frequently cited as the first report describing the isolation of Marburg virus.

During the first 6 months of 1968, six groups published confirmations of our findings (Kissling et al. 1968; Kunz et al. 1968; Strickland-Cholmley and Malherbe 1970; May et al. 1968). While some acknowledged the first description of Marburg virus by Siegert and colleagues, others did not.

The researchers who deserve credit for isolating and identifying Marburg virus are Walter Mannheim, University of Marburg for successful transmission to guinea pigs, Werner Slenczka, University of Marburg for detecting and identifying the Marburg virus antigen by immunofluorescence analysis, and Gerhard Müller,

Bernhard Nocht Institute for identifying the virus by EM. Walter Mannheim, a bacteriologist, was uninterested in co-authoring publications despite his involvement in the virus isolation.

6 The Monkeys

Originally, during the first decades of the nineteenth century, Indian rhesus monkeys (*Macaca mulatta*) were the preferred nonhuman primates (NHPs) used in biomedical research. However, when monkey kidney cells were needed for vaccine production, it was clear that these monkeys were not convenient because the members of the *Macaca* genus are natural hosts of herpes B virus (Sabin and Wright 1934). Therefore grivets, which were believed to be free from viruses pathogenic to humans, were used as a source to prepare kidney cell cultures for vaccine production. The attenuated vaccine strains of poliovirus (Sabin 1–3) could only be propagated in primary monkey kidney cells. Therefore, grivets were imported in great numbers from Eastern Africa to be used for the production and also for the safety testing of poliovirus vaccines.

Grivets are part of a group colloquially often referred to as “African green monkeys”. African green monkeys were previously classified as a single species (*Cercopithecus aethiops*). This classification has since then been revised repeatedly and the species was split into at least six species (Groves 2001). Grivets (*C. aethiops*) are endemic in Ethiopia, Eritrea, and the Sudans. In 1962, about 25,000 grivets were exported from Uganda to a number of countries for biomedical purposes, mainly for vaccine production. Due to loss of habitat and being hunted as a source of meat, they are now endangered. In 1967, there were no restrictions or regulations for the export or import of these monkeys.

The grivets which played a role in the 1967 MVD outbreak were exported from Entebbe, Uganda by F.R. Mann, a relative of the famous German novelist Thomas Mann. They were imported to Germany by the animal transportation company Samen-Eckers, Viersen, Germany. Normally, these animals were transported by Lufthansa on a direct flight from Entebbe to Frankfurt or to Düsseldorf, Germany. Before leaving Entebbe, a certificate of good health from the Ugandan veterinary office was needed. In 1967, however, due to the Six Days War (June 5–10, 1967), the direct Lufthansa flight was discontinued and therefore the monkeys had to be transported via London Heathrow to their German destinations. Usually, the shipments to Germany included approximately 100 monkeys. Due to technical problems, the animals stayed in London for 9–46 h before they were shipped to Germany. During this time the monkeys were transported to an animal house of the Royal Society for Prevention of Cruelty to the Animals (RSPCA) outside the airport area. This was allowed under condition that they would not have contact with other animals and especially not with rhesus monkeys. However, on one occasion the grivets were kept overnight in a room together with two Hanuman langurs (*Semnopithecus* sp.) from Ceylon (now Sri Lanka). One of the langurs became sick and died soon after, but the cause of its sickness and death was not disclosed. During another transport, grivets were caged in a room together with finches of non-European origin and with other animals. The circumstances of these stays in London and the contacts with animals from other continents caused considerable difficulties in the search for the geographic provenance of the agent of disease. On one occasion, two monkeys escaped during the transport to Germany on June 28, 1967. Luckily, they monkeys were captured quickly and arrived at Düsseldorf airport on June 29. Several journalistic conspiracy theories arose from this event.

Mr. Mann from Entebbe insisted on the validity of the Ugandan health certificate: "presumably a document of doubtful quality" (Hennessen). He also stated that a similar disease had never occurred among the 500 Ugandans he employed as monkey trappers. Later, however, it became known that there had been an epizootic event with many dead monkeys in the trapping areas around Lake Kyoga. The monkeys were captured in three places: Namasale, Kidera, and Ndolwa, prior to transport to Entebbe. With rare exceptions, the monkeys were held in individual cages during their stay in Uganda. As a rule, animals that would not thrive or showed other signs of sickness were shipped to a "Monkey island" in Lake Victoria where they were set free to live or to die. Whenever there were not enough monkeys to complete an order, the trappers would go to the "Monkey island" and would capture some healthy looking monkeys and bring them to Entebbe.

The final proof that Africa was the origin of Marburg virus was not obtained until 1975, when a 20-year-old Australian man was admitted to a hospital in Johannesburg, South Africa, where he was diagnosed with MVD and later died. His 19-year-old female travel companion and a 20-year-old nurse caring for both patients acquired secondary infections. Both women survived the infection and recovered (Gear et al. 1975). The Australian travelers had been hitchhiking in Rhodesia (now Zimbabwe). The source of their infection was never determined.

The number of monkeys imported to Germany was negligible before 1960. Beginning that year, air transportation facilitated the import of up to 5,000 animals per year for biomedical research and for vaccine production, mainly rhesus monkeys and grivets. At the Paul-Ehrlich Institute in Frankfurt monkeys were euthanized on two days per week, at Behringwerke AG in Marburg, on five days. On August 21, 1967, 18 monkeys were euthanized at on two days per week Behringwerke AG for diagnostic purposes. Postmortem specimens were collected and sent to the Institute of Hygiene in Freiburg, Germany, to the Institute of Hygiene in Vienna, Austria, to the Poliomyelitis Research Foundation in Johannesburg, South Africa, and to the Institute Pasteur in Dakar, Senegal. During this activity, a 39-year-old veterinarian was infected in Marburg and succumbed to the infection on September 3.

On August 25, 1967, per governmental order all the remaining monkeys at Behringwerke AG were killed: 235 rhesus monkeys and 266 grivets. Out of the latter group, 110 were being kept in a quarantine station outside the facility and 156 animals were held in a house in the area of the facility. None of the individuals who cared for the monkeys and cleaned the cages got the disease. The animal care takers used gloves and face masks resembling those which are worn for welding work and these are not aerosol-tight. Only animal care takers and laboratory workers who had contact with blood or organs of the monkeys were at risk of getting the disease. Three men who opened the skulls of killed monkeys to remove the brains were infected. This work was done without personal protective equipment (PPE) since the monkeys were believed to be healthy. Animal handling included to capture the monkeys with a net and to anesthetize them by electric shock, to exsanguinate the animals by opening the carotid artery, to fix the monkeys on the table, to perform ventral nephrectomies, to perform necropsies, to remove the cadavers from the table, and to transport them to the incinerator. PPE was mandatory for all these activities. For capturing and anesthetizing the monkeys as well as for the nephrectomies protective gowns and sterile masks were used in addition. Other employees were most likely exposed to the virus when they transported the kidneys to the cell culture lab or during the necropsies. On August 29, it was suspected that kidney cell cultures derived from these monkeys were infectious and therefore all remaining cell cultures, including 1000 glass vessels and 9000 glass tubes were inactivated by autoclaving. A 19-year-old student who worked at Behringwerke AG during the summer break accidentally broke one of the tubes, was infected with Marburg virus, and died on September 10. Three employees, one male, two females, were infected by contact with cell cultures. Most of the female primary cases were infected by cleaning the glassware used in the lab.

Summarizing the use of protective measures, it can be concluded that PPE was exclusively directed toward protection of monkeys, monkey organs, and cell cultures from contamination. Since the monkeys were regarded to be healthy, PPE was not used when skulls were opened, when cadavers were discarded, or when glass vessels were cleaned (Hennessen et al. 1968). Hennessen and coworkers published a thorough analysis of the sources of infection for the employees in Frankfurt and in Marburg and came to the conclusion that a small number of animals from transports that had arrived on July 21 and July 28 must have carried the virus (Hennessen et al. 1968) (Table 1). The last case that occurred in Frankfurt, however, must have been infected by contact with a monkey that had arrived on August 10.

Table 1 Human cases resulting from direct or indirect contact with monkeys from two shipments

Shipment arrival	7/28	Day 11	Day 19	Day 20	Day 21	Day 22	Day 24	Day 25	Day 31
First day of illness		8/8	8/16	8/17	8/18	8/19	8/21	8/22	8/28
Direct contact		1	3	1	2	1	2	2	1

Shipment arrival	7/21	Day 32	Day 38	Day 41	Day 49
First day of illness		8/22	8/25	8/28	9/5
Direct contact			1	2	
Indirect contact		1		2	1

The time between arrival of the shipments and the first day of reported illness of patients is indicated. Direct contact indicates contact with monkey blood or organs. Indirect contact indicates exposure by handling cell cultures or cleaning contaminated glassware. Table 1 was compiled based on data published in Hennessen et al. (1968)

The shortest period of time between arrival of the monkeys (July 28) and the first signs of human disease (August 8) was 11 days (Table 1). An interval of 32 days between the arrival of the grivets and the onset of symptoms from direct contact with the animals implies that there must have been horizontal spread of the virus among the grivets after their arrival in Marburg. From these data, it might appear that quarantine was not appropriately applied and that 4 weeks of quarantine might not have been sufficient to prevent the outbreak, especially since it cannot be excluded that latent infections and sexual transmission might have played a role in virus spread (see below).

Monkeys from the same transports were also delivered to the Torlak Institute in Belgrade, where a veterinarian and his wife were infected and survived. At the Torlak Institute, these monkeys had health issues—the lethality among them was 33%. A delegation comprising representatives of the animal handlers, from Behringwerke AG and the Paul-Ehrlich Institute visited Belgrade and stated that these monkeys had not died from a viral infection but from insufficient air conditioning due to a defective AC unit. Later, Stojkovic and his coworkers took blood from 48 surviving monkeys and tested it in a complement fixation test (CFT) using antigen prepared from infected guinea pigs (Stojkovic et al. 1971). They detected Marburg virus-specific antibodies in 88% of the samples and concluded that the infection had spread in the cages during the quarantine. We received two of these sera to test them with indirect immunofluorescence and in a CFT using our Vero cell-based antigen; our tests indicated that they were both negative for Marburg virus antibody (Slenczka et al. 1971).

7 Speculations on Persistently Infected Monkeys as Reservoirs of Marburg Virus

The most intriguing question regarding these monkeys is their state of health. Where and at what time did they acquire the virus? Why did they not show signs of disease at any location; not when they were in Entebbe, not upon their arrival in London, Frankfurt, Marburg, or Belgrade, and not when they were finally euthanized? The lethality of imported NHPs was about 5% at that time. An increase in lethality of imported NHPs should certainly have raised suspicion. Exact lethality data of the incriminated monkeys were never communicated. There can be no doubt that the animals were inspected carefully before they were used.

It was shown for humans, who survived MVD, that the virus was able to persist in the testes and probably other organs (anterior eye chamber and liver) for weeks or even months after recovery (Slenczka et al. 1968; Siegert et al. 1968; Martini and Schmidt 1968). Per analogy, it is reasonable to assume that some of the monkeys, which arrived on July 21 and 28, might have been recovering from disease caused by Marburg virus infection and were able to infect their cage companions during transport to or after their arrival in Germany. Beginning at their arrival at the collecting stations, the animals were held in individual cages. During the flight and after arrival at their final destinations, the monkeys were kept two per cage. It seems that they were not separated by sex. If a convalescent monkey was caged together with a naïve monkey, sexual transmission could have happened at any time during transport or after the animals had arrived at their destination. If sexual transmission happened early, the acutely infected companion would have fallen ill and would most probably have died during quarantine. If, however, transmission occurred about a week before the monkeys were euthanized, they could have been viremic but appeared clinically unremarkable, although in this case they would most likely have developed a fever at the time when they were euthanized. Thus, simply using a thermometer might have prevented the MVD outbreak. Admittedly, the use of a conventional thermometer in these monkeys would have only been possible if they were anesthetized, since infrared thermometers were not available at that time.

Assuming that some persistently infected monkeys from "Monkey island" could have been the source for importing Marburg virus to Europe might help to explain some peculiarities of this outbreak. It is known that a large number of monkeys from the same source in Uganda were transported to Sweden, Japan, Czechoslovakia, Italy, Switzerland, and England at the same time and for the same purpose: to prepare cell cultures. But no outbreaks were reported at any of these locations. It is possible that when the shipments to Germany and to Yugoslavia were assembled, there were not enough monkeys left at the collecting station and therefore, that animals from "Monkey island" were used to supplement the shipment. Among the animals captured from "Monkey island" were possibly some which had survived an infection with Marburg virus but appeared to be healthy. This might explain why Marburg virus was exclusively transported with shipments to locations in Germany and in Yugoslavia.

Admittedly, the above-formulated hypothesis is based on assumptions. But it offers an intriguing explanation addressing many of the open questions regarding this outbreak that, until now, have remained unanswered.

Retrospectively, it is astonishing and hard to believe how firmly scientists trusted in grivets as a safe NHP that did not pose a risk to human health. One possible explanation for this lapse in judgment might be that the scientists were blind to the risks because of their passion to achieve the overarching goal to eliminate poliomyelitis. This goal must have been hypnotizing. As Christian Morgenstern put it aptly in his famous poem *Die unmögliche Tatsache* (The impossible fact): "*Weil, so schließt er messerscharf, nicht sein kann, was nicht sein darf.*" (For, he reasons pointedly, that which must not, cannot be).

8 Shortage of Poliomyelitis Vaccine

A special and very serious problem emanating from the Marburg virus outbreak was the shortage and questionable safety of monkey kidney-based vaccines against measles and poliomyelitis, since the release of vaccines had been stopped due to the outbreak. In 1967, 60 cases of poliomyelitis were registered in West Germany, a number that indicates that more than 6000 persons were infected and could spread the virus. The immunization rate of the population was far from sufficient to prevent the spread of polioviruses. The vaccination campaign was to be continued in fall and winter after the end of the epidemic season. As a consequence of the MVD outbreak, the import of grivets into Germany was no longer permitted by governmental order starting in August 1967. After intensive discussions and consultations, this problem was solved in a very pragmatic way. It was decided that vaccine batches, which had been licensed prior to the MVD outbreak, could be used. The decisive argument for the use of these vaccine stocks was that injection of the vaccine into guinea pigs, a test designed to rule out contamination with tuberculosis, was a sufficient safety precaution. This test procedure, however, would not have been adequate for detecting Marburg virus. Only in 1968, control procedures were changed so that Marburg virus might have been detected.

Attempts to propagate the vaccine strains in human fibroblasts were not successful. To address this issue, Behringwerke AG obtained permission to import 200 grivets from Uganda on December 8, 1967. In the aftermath of the outbreak, Behringwerke AG chose to establish their own breed of SPF monkeys, which were delivered by Caesarean section. Other vaccine producers decided to use crab-eating (cynomolgus) macaques (*Macaca fascicularis*) from Southeastern Asia for vaccine production. These animals were regarded to be safe until in 1989 Reston virus, an Ebola virus and therefore relative of Marburg virus, was introduced by crab-eating macaques into the USA (Jaax and Jaax 2016). A small number of animal care takers may have been infected as shown by the presence of anti Reston virus antibodies, but fortunately it did not cause disease in the infected humans (Fisher-Hoch et al. 1992).

9 Conclusions

Marburg virus, the type virus of the family *Filoviridae*, was introduced in summer 1967 to Germany and to Yugoslavia by grivets (*C. aethiops*). The monkeys were imported from Uganda for biomedical purposes and were delivered to three institutions, Behringwerke AG in Marburg, Germany, Paul Ehrlich-Institute in Frankfurt, Germany and Institute Torlak in Belgrade, Yugoslavia. Employees of these three institutions, at least 26 persons, were infected directly by contact with blood, organs and cell cultures of these monkeys and developed a severe form of a viral hemorrhagic fever. The case fatality rate of the primary cases was 27%. In addition, five cases of (secondary) nosocomial infections and one case of sexual transmission were observed. There were no fatalities among the secondary cases. The etiologic virus was isolated and identified in a combined effort by scientists at the University of Marburg, Germany and the Institute for Tropical Medicine and Hygiene in Hamburg, Germany.

Since that time, marburgviruses and a second group of filoviruses, ebolaviruses, have caused outbreaks with a total number of more than 31,311 cases of viral hemorrhagic fever in several African countries and elsewhere with case fatality rates of around 42%. In some of the outbreaks, more than 90% of the patients succumbed to the disease. The largest outbreak of Ebola virus disease, which occurred in Guinea, Liberia and Sierra Leone 2013–2016, has caused at least 28,599 cases with a case fatality rate of 40%.

A seroepidemiological survey carried out in Mobai, Eastern province of Sierra Leone, already indicated the presence of Ebola virus in this region in 1984. This area is in the vicinity of Gueckedou/Guinea, where the 2014 outbreak originated. Out of 556 sera collected in Mobai in 1983/4, 1.8% were found to be seropositive for Ebola virus (Slenczka et al. 1984). The reemergence of Ebola virus in 2014 shows seroepidemiologic surveys might help predict future outbreaks.

References

Brauburger K et al (2012) Forty-five years of Marburg virus research. Viruses 4(10):1878–1927

Burnet FM, White DO (1962) Natural history of infectious disease, 3rd edn. Cambridge University Press, Cambridge

Čumakov MP (1948) Etiology and epidemiology of hemorrhagic fevers (Russian). Ter Arh 20:85–86

Egbring R, Slenczka W, Baltzer G (1971) Clinical manifestations and mechanism of the haemorrhagic diathesis in Marburg virus disease. In: Martini GA, Siegert R (eds) Marburg virus disease. Springer, Berlin, Heidelberg, New York, pp 41–49

Fisher-Hoch SP et al (1992) Pathogenic potential of filoviruses: role of geographic origin of primate host and virus strain. J Infect Dis 166(4):753–763

Gajdusek DC (1962) Virus hemorrhagic fevers. Special reference to hemorrhagic fever with renal syndrome (epidemic hemorrhagic fever). J Pediatr 60:841–857

Gear JS et al (1975) Outbreake of Marburg virus disease in Johannesburg. Br Med J 4(5995): 489–493

Goldwasser RA et al (1959) Fluorescent antibody staining of rabies virus antigens in the salivary glands of rabid animals. Bull World Health Organ 20:579–588

Groves CP (2001) Primate taxonomy. Smithsonian series in comparative evolutionary biology. Smithsonian Institution Press, Washington, DC, 350 S

Hennessen W, Bonin O, Mauler R (1968) On the epidemiology of the monkey transmitted disease in humans. Deutsche Medizinische Wochenschrift 93(12):582–589

Jaax J, Jaax NK (2016) An Ebola filovirus is discovered in the USA: Reston, Virginia, USA, 1989. Vet Herit 39(1):16–19

Kissling RE et al (1968) Green monkey agent of disease. Science 161(3848):1364

Koch R (1890) Über bakteriologische Forschung Vortrag in der 1. allgemeinen Sitzung des X. internationalen medicinischen Congresses am 4. August 1890. Berlin: Hirschwald. 15 S

Korb G et al (1971) Animal experiments on the "Marburg-virus"-hepatitis. Studies on guinea pigs. Virchows Arch A Pathol Pathol Anat 353(2):169–184

Kunz C et al (1968) Biological and morphological characteristics of the "hemorrhagic fever" virus occurring in Germany. Wien Klin Wochenschr 80(9):161–2 passim

Leitenberg M, Zilinskas RA, Kuhn JH (2012) The soviet biological weapons program a history. Harvard University Press, Cambridge, Massachusetts, p xvi, 921 pages

Martini GA (1971) Marburg virus disease. Clinical syndrome. In: Martini GA, Siegert R (eds) Marburg virus disease. Springer, Berlin, Heidelberg, New York, pp 1–9
Martini GA, Schmidt HA (1968) Spermatogenic transmission of the "Marburg virus" (Causes of "Marburg simian disease"). Klinische Wochenschrift 46(7):398–400
Martini GA et al (1968a) Apropos of an infectious disease, unknown until now, possibly imported by monkeys (called "the Marburg monkey disease"). Rev Med Chir Mal Foie 43(4):165–182
Martini GA et al (1968b) A hitherto unknown infectious disease contracted from monkeys "Marburg-virus" disease. Ger Med Mon 13(10):457–470
May G et al (1968) Electron microscopical findings in a communicable disease in monkeys (*Cercopithecus aethiops*). Zentralbl Bakteriol Orig 207(2):145–151
Rivers TM (1937) Viruses and Koch's postulates. J Bacteriol 33(1):1–12
Sabin AB, Wright AM (1934) Acute ascending myelitis following a monkey bite, with the isolation of a virus capable of reproducing the disease. J Exp Med 59(2):115–136
Siegert R, Slenczka W (1971) Laboratory diagnosis and pathogenesis. In: Martini GA, Siegert R (eds) Marburg virus disease. Springer, Berlin, Heidelberg, New York, pp 157–160
Siegert R et al (1967) On the etiology of an unknown human infection originating from monkeys. Deutsche Medizinische Wochenschrift 92(51):2341–2343
Siegert R et al (1968a) The aetiology of an unknown human infection transmitted by monkeys (preliminary communication). Ger Med Mon 13(1):1–2
Siegert R, Shu HL, Slenczka W (1968b) Detection of the "Marburg Virus" in patients. Ger Med Mon 13(11):521–524
Siegert R, Shu HL, Slenczka W (1968c) Isolation and identification of the "Marbury virus". Ger Med Mon 13(11):514–518
Slenczka W, Klenk HD (2007) Forty years of Marburg virus. J Infect Dis 196(Suppl 2): S131–S135
Slenczka W et al (1968) "Marburg virus" antigen demonstration by immunofluorescence in the organs of infected guinea pigs. Deutsche Medizinische Wochenschrift 93(12):612–616
Slenczka W, Wolff G, Siegert R (1971) A critical study of monkey sera for the presence of antibody against the Marburg virus. Am J Epidemiol 93(6):496–505
Slenczka W et al (1984) Seroepidemiologic studies on the presence of antibodies against Marburg and Ebola viruses in Africa (German). Mitt Österr Ges Tropenmed Parasitol 6:53–60
Smith CE et al (1967) Fatal human disease from vervet monkeys. Lancet 2(7526):1119–1121
Stille W et al (1968) An infectious disease transmitted by *Cercopithecus aethiops* ("Green monkey disease"). Ger Med Mon 13(10):470–478
Stojkovic L et al (1971) Two cases of Cercopithecus-monkeys-associated haemorrhagic fever. In: Martini GA, Siegert R (eds) Marburg virus disease. Springer, Berlin, Heidelberg, New York, pp 24–33
Strickland-Cholmley M, Malherbe H (1970) Marburg virus. Lancet 1(7644):476
Todorovic K et al (1969) Unknown viral disease transmitted from infected monkeys to man. Glas. Srpska Akademija Nauka i Umetnosti. Odeljenje Medicinskih Nauka 22:91–101
WHO (1968) Outbreak in laboratory personnel working with Cercopithecus monkeys from East Africa. Wkly Epidemiol Rec 43(4):61–62
Zlotnik I (1969) Marburg agent disease: pathology. Trans R Soc Trop Med Hyg 63(3):310–327

Ecology of Filoviruses

Brian R. Amman, Robert Swanepoel, Stuart T. Nichol and Jonathan S. Towner

Abstract Filoviruses can cause severe and often fatal disease in humans. To date, there have been 47 outbreaks resulting in more than 31,500 cases of human illness and over 13,200 reported deaths. Since their discovery, researchers from many scientific disciplines have worked to better understand the natural history of these deadly viruses. Citing original research wherever possible, this chapter reviews laboratory and field-based studies on filovirus ecology and summarizes efforts to identify where filoviruses persist in nature, how virus is transmitted to other animals and ultimately, what drivers cause spillover to human beings. Furthermore, this chapter discusses concepts on what constitutes a reservoir host and highlights challenges encountered while conducting research on filovirus ecology, particularly field-based investigations.

B.R. Amman (✉) · S.T. Nichol · J.S. Towner (✉)
Viral Special Pathogens Branch, Division of High Consequence Pathogens and Pathology, National Center for Emerging and Zoonotic Infectious Diseases, Centers for Disease Control and Prevention, 1600 Clifton Rd. Ne, Atlanta, GA, USA
e-mail: cxx1@cdc.gov

J.S. Towner
e-mail: jit8@cdc.gov

S.T. Nichol
e-mail: stn1@cdc.gov

R. Swanepoel
Department of Medical Virology, Faculty of Health Sciences, University of Pretoria, PO Box X323, Arcadia, Pretoria 0007, RSA
e-mail: bob.swanepoel@up.ac.za

J.S. Towner
Department of Pathology, College of Veterinary Medicine, University of Georgia, 501 D.W. Brooks, Athens, GA, USA

Current Topics in Microbiology and Immunology (2017) 411:23–62
DOI 10.1007/82_2017_10

Published Online: 15 July 2017

Contents

1 Introduction

Ecology can be defined broadly as the scientific study of the interactions between organisms and their environment, while infectious disease ecology has a more refined emphasis on the study of the interactions between pathogens, their host organisms, and the environment. There have been many reviews of filovirus ecology covering all aspects affecting circulation of the viruses in nature, including the search for natural reservoirs, virus and potential host dynamics, known or suspected transmission dynamics, and environmental considerations with respect to outbreak potential (Monath 1999; McCormick 2004; Feldmann et al. 2004; Wolfe et al. 2005; Pourrut et al. 2005; Groseth et al. 2007; Miranda and Miranda 2011; Brauburger et al. 2012; Olival et al. 2012, 2015; Smith and Wang 2013; Olival and Hayman 2014; Han et al. 2016; Leendertz et al. 2016). Presented in this chapter is an update on filovirus ecology research.

Filoviruses (order *Mononegavirales*; family *Filoviridae*) are negative sense RNA viruses included in the genera *Ebolavirus*, *Marburgvirus*, and *Cuevavirus*. Collectively, the zoonotic viruses of the genus *Ebolavirus* (referred to as ebolaviruses) and the genus *Marburgvirus* (referred to as marburgviruses) have caused over 40 disease outbreaks in humans, with case fatality ratios as high as 90% (Table 1). The majority of cases result from human to human transmission. However, the initial spillover events occur when humans come into contact with either the natural reservoirs of the filovirus or secondary amplifying hosts, typically through hunting or utilizing bushmeat (Leroy et al. 2009).

There has been lack of consensus on criteria that define animals as natural reservoir hosts of infectious agents. A commonly accepted assumption is that infection is likely to be unapparent or benign in reservoir hosts, presumably as a result of coevolution with the pathogen concerned. The reservoir hosts develop transient or chronic infection that facilitates transmission to further reservoir or susceptible hosts. However, since the majority of pathogens infect hosts of multiple species, it is argued that reservoir hosts can only be defined with respect to specific target populations (Haydon et al. 2002). Moreover, the reservoir may comprise one or more epidemiologically connected populations of the same or different species,

Table 1 Historical chronology of filovirus outbreaks (CDC 2014, 2016)

Outbreak year (s)	Country	Virus	Number of cases	Deaths fatality ratio
1967	West Germany and Yugoslavia	Marburg virus	32[a]	7 (23%)
1975	Johannesburg, South Africa	Marburg virus	3	1 (33%)
1976	England	Sudan virus	1	0
1976	Sudan (South Sudan)	Sudan virus	284	151 (53%)
1976	Zaire (Democratic Republic of the Congo)	Ebola virus	318	280 (88%)
1977	Zaire	Ebola virus	1	1 (100%)
1979	Sudan (South Sudan)	Sudan virus	34	22 (65%)
1980	Kenya	Marburg virus	2	1 (50%)
1987	Kenya	Marburg virus	1	1 (100%)
1989–1990	Philippines	Reston virus	3 (asymptomatic)	0
1989	USA	Reston virus	0	0
1990	USA	Reston virus	4 (asymptomatic)	0
1990	Russia	Marburg virus	1	1 (100%)
1992	Italy	Reston virus	0	0
1994	Côte d'Ivoire	Taï Forest virus	1	0
1994	Gabon	Ebola virus	52	31 (60%)
1995	Democratic Republic of the Congo	Ebola virus	315	250 (81%)
1996	Russia	Ebola virus	1	1 (100%)
1996	Philippines	Reston virus	0	0
1996	USA	Reston virus	0	0
1996	South Africa	Ebola virus	2	1 (50%)
1996–1997 (July–January)	Gabon	Ebola virus	60	45 (74%)
1996 (January–April)	Gabon	Ebola virus	37	21 (57%)
1998–2000	Democratic Republic of Congo	Marburg virus	154	128 (83%)
2000–2001	Uganda	Sudan virus	425	224 (53%)
October 2001–March 2002	Republic of the Congo	Ebola virus	57	43 (75%)
October 2001–March 2002	Gabon	Ebola virus	65	53 (82%)

(continued)

Table 1 (continued)

Outbreak year (s)	Country	Virus	Number of cases	Deaths fatality ratio
December 2002–April 2003	Republic of the Congo	Ebola virus	143	128 (89%)
November–December 2003	Republic of the Congo	Ebola virus	35	29 (83%)
2004	Russia	Ebola virus	1	1 (100%)
2004	Sudan	Sudan virus	17	7 (41%)
2004–2005	Angola	Marburg virus	252	227 (90%)
December 2007–January 2008	Uganda	Bundibugyo virus	149	37 (25%)
2007	Democratic Republic of the Congo	Ebola virus	264	187 (71%)
2007	Uganda	Marburg virus	4	1 (25%)
December 2008–February 2009	Democratic Republic of the Congo	Ebola virus	32	15 (47%)
November 2008	Philippines	Reston virus	6 (asymptomatic)	0
2008	USA ex Uganda	Marburg virus	1	0
2008	Netherlands ex Uganda	Marburg virus	1	1 (100%)
May 2011	Uganda	Sudan virus	1	1 (100%)
November 2012–January 2013	Uganda	Sudan virus	6[b]	3[a] (50%)
June–November 2012	Democratic Republic of the Congo	Bundibugyo virus	36[b]	13[a] (36.1%)
June–October 2012	Uganda	Sudan virus	11[b]	4[a] (36.4%)
2012	Uganda	Marburg virus	15	4 (27%)
August–November 2014	Democratic Republic of the Congo	Ebola virus	66	49 (74%)
(Dec) 2013–2016	Guinea, Sierra Leone, Liberia	Ebola virus	28,652	11,325 (39%)
2014	Uganda	Marburg virus	1[b]	1 (100%)

[a]The original 1967 Marburg virus disease outbreak case number of 31 was amended to 32 after another Marburg positive case was identified retrospectively (Slenczka et al. 2007)
[b]Laboratory confirmed cases only

potentially including vectors, in which a pathogen can be maintained in perpetuity and transmitted to a defined target population. The pathogenicity of the infectious agent for the reservoir host may be irrelevant provided the population exceeds the critical community size. Animals that transmit infection to the target population are termed the source population, and may form part of the maintenance community, or merely represent a transmission link.

More succinctly, therefore, the natural reservoir of infection can be defined as an ecologic system in which an infectious agent survives indefinitely (Ashford 2003). It follows that there has to be circulation and transmission of the infectious agent in the reservoir population, so that during appropriate field studies the agent should be detected regularly over time. Moreover, there should be a higher cumulative seroprevalence than active infection prevalence, as demonstrated for marburgviruses in bats (Swanepoel et al. 2007; Towner et al. 2009; Amman et al. 2012). There may be local extirpation of infection, for example through 'immunological exhaustion' (cumulative acquisition of immunity) of the population, but separate maintenance populations in close proximity, part of the maintenance community, may act as refugia from which re-introduction of infection is possible (Glass et al. 2007). Moreover, susceptible individuals can be recruited to the maintenance population through reproduction or migration.

With specific reference to Ebola virus (EBOV), it has been hypothesized that there is long-term local persistence of the virus in a cryptic and infrequently contacted reservoir host, but that following recent introductions into susceptible hosts there has been directional spread of virus through ape and human populations (Groseth et al. 2007). There has been mounting evidence to support the role of bats as cryptic hosts of filoviruses, and analogies have been made with rodents as hosts of arenaviruses and hantaviruses: each filovirus and its natural reservoir will have its own set of special considerations with respect to transmission cycles, ecological dynamics and natural histories of reservoir hosts, as discussed below (Olival et al. 2012, 2015; Luis et al. 2013; Olival and Hayman 2014).

Ostensibly there have been numerous ecological investigations, with field studies aimed at identifying the natural reservoirs of the viruses attempted in association with most filovirus outbreaks. Typically, the investigations incorporated the collection of samples from fauna and even flora in and around the outbreak area, ideally as close as possible to the locations where primary infections putatively occurred. These samples were then tested for evidence of filovirus infection. In reality, many of the studies were delayed, inappropriate, deficient, or never completed and published.

Filovirus outbreaks have tended to occur erratically, in widely separated geographic locations at unpredictable intervals, rendering systematic investigation difficult. Almost invariably the outbreaks have occurred in remote and poorly resourced locations, and months have elapsed before information reached the outside world to prompt international responses. The 2013–2016 outbreak of Ebola virus disease (EVD) in West Africa was no exception in this respect. Consequently, the investigations are often conducted in a season when the ecological circumstances that triggered the outbreak no longer remain. Ample time has elapsed for infected animals to die, clear infections, or simply leave the outbreak area (Swanepoel et al. 1996). Often outbreaks have occurred in regions of civil or even international strife, where the securing of safe working conditions, communications, transport, fuel, sanitation, plus medical, food, water and electricity supplies are matters of primary concern. Under these circumstances resources are preferentially devoted to controlling the outbreak: isolation and treatment of patients, tracing and monitoring of contacts of infection, and providing safe burial of the dead. Ecological investigations are considered to be of secondary importance and must await the decline or cessation of the outbreak.

Moreover, it is frequently unsafe to conduct ecological investigations in or near communities during filovirus outbreaks as the local population may be suspicious of any new or unusual activities, such as the collection and dissection of wildlife by researchers wearing a variety of otherworldly personal protective equipment (PPE), often at night. Rumors and misinformation can inflame fears and superstitions, rendering the investigations difficult and hazardous for researchers, and leading to long delays between the onset of the outbreak and active investigations to find the animal reservoir. Sometimes the investigations have even occurred years later in locations remote from the outbreak. In contrast, where attempts are made to conduct studies at regular seasonal intervals over years, research institutions and funding agencies are reluctant to commit resources, and host countries are reluctant to permit investigations, when there are no current outbreaks.

Apart from considerations of timing and logistics, one of the most obvious confounders has simply been the biological diversity in the geographic areas where outbreaks occur, and the numerous possibilities that exist for transmission of infection from animals to humans. Mounting evidence that bats play a key role in the maintenance of filoviruses has helped narrow and focus the scope of field investigations. Nevertheless, capturing and processing individuals from this diverse mammalian order (Chiroptera) presents many challenges depending on the target species and region of investigation. Trapping and sampling small terrestrial mammals with restricted habitats and home ranges can usually be managed with facility, depending on population sizes. However, collecting volant (flying) small mammals presents new challenges with greater and less predictable variables. Bats are not confined by geographic barriers that limit movements of terrestrial mammals, such as bodies of water, patches of deforestation, or minor geologic formations. Nor are they constrained by political boundaries; they simply fly from point to point and beyond. The challenge is to determine where target points are located to encounter the largest number of target species. Moreover, instead of luring animals into baited traps in areas inhabited by suspected terrestrial reservoirs, optimal capture equipment varies with type of bat, and must be placed taking into account 3-dimensional space and flight paths. An exception to this scenario occurs where target species form colonies at specific roosting sites, which provide singular locations to focus collection efforts, and typically these provide capture numbers sufficient to provide definitive results.

2 Reservoir Studies of Marburgviruses

Between the first outbreak of Marburg virus disease (MVD) in 1967 and the Durba outbreak in 1998, there had been only three known, naturally occurring MVD outbreaks, and all involved three cases or less. Each of the primary cases were tourists with known travel histories, an epidemiologic feature that narrowed the scope of the ensuing ecological investigations. Analyses of tissues collected from arthropods and bats, and serum samples from wild and domestic animals, failed to produce any positive results by virus isolation or antibody testing that could implicate any taxon as the source of the infections (Conrad et al. 1978; Smith et al. 1982). However, highly sensitive analyses such as RT-PCR and qRT-PCR were not available at that time. In light of the fact that the outbreaks in 1980 and 1987 were associated with visits to Kitum cave at Mt. Elgon, researchers attempted to identify fauna in the cave as the source of infection by placing caged animals, 9 nonhuman primates (five Sykes monkeys, two baboons, two vervet monkeys) and 20 guinea pigs inside the cave for 22 days, with some placed directly under roosting Egyptian rousette (*Rousettus aegyptiacus*) bats (J.C. Morrill, pers com). These attempts were unsuccessful, but the results were important by showing that marburgviruses were not likely transmitted via aerosols.

The initial breakthrough towards identifying the Marburg virus (MARV) and Ravn virus (RAVV) natural reservoir occurred following a prolonged MVD outbreak in the gold mining village of Durba, Democratic Republic of the Congo (DRC) (October 1998–September 2000) where at least nine marburgvirus lineages were found circulating among miners and their contacts (Bausch et al. 2006). A rigorous ecological investigation ensued, and researchers collected and analyzed fauna associated within and around the Goroumbwa mine where the outbreak was centered. The collection effort included bats, rodents, amphibians and arthropods (Swanepoel et al. 2007). Using antibody ELISA and nested VP35 RT-PCR, bats from two insectivorous species and one frugivorous species tested positive for MARV. Eloquent horseshoe bats (*Rhinolophus eloquens*) tested positive by serology (20/206) and RT-PCR (7/197), while greater long-fingered bats (*Miniopterus inflatus*) bats were positive by RT-PCR only (1/33). Egyptian rousettes also tested positive by serology (32/156) and RT-PCR (4/127). Sequence analysis of PCR fragments revealed the presence of genetically diverse MARV, but not RAVV, in the bats tested. No other animal samples tested positive and several bat-derived virus sequence fragments closely matched those from human MARV isolates obtained during the outbreak.

Further evidence linking MARV to cave-dwelling, but not arboreal, bats was obtained when over 1100 bats representing 10 species were collected in Gabon and the northwest Republic of the Congo and tested for virus-specific RNA and antibody. Evidence of infection was found only in Egyptian rousettes, in which 4/238 samples tested positive for MARV RNA and 29/242 sera samples showed reactivity to MARV antigen (Towner et al. 2007). These data extended the geographic range of potential MARV spillover to the western half of Africa where a MVD outbreak of uncertain origin occurred in Angola two years prior (Towner et al. 2006). Interestingly, the two MARV sequence fragments found in the Gabonese bats phylogenetically aligned most closely to virus sequences obtained from miners in Durba, DRC (Swanepoel et al. 2007) and tourists infected thousands of miles away in 1975 in Rhodesia (now Zimbabwe), suggesting the virus common ancestor was carried by a highly mobile animal reservoir. Follow-up surveys between 2003 and 2010 in Gabon and Republic of the Congo again found Egyptian rousettes positive for MARV by antibody ELISA (7%) and qRT-PCR (4.1%) (Pourrut et al. 2009; Maganga et al. 2011). Of note, the study by Pourrut et al. (2009) reported 1% of hammer-headed fruit bats (*Hypsignathus monstrosus*) bats to also have antibody reactivity to MARV antigen, but none were positive by RT-PCR. These data, along with the findings of Towner et al. (2007), represent the westernmost distribution of MARV-positive bats to date.

Perhaps the most convincing set of data implicating *R. aegyptiacus* bats as a bona-fide natural reservoir for marburgviruses came after a series of discrete spillover events occurred in miners and tourists in Uganda. The first incidents happened in July and September of 2007, when miners working in Kitaka mine in southwest Uganda became separately infected with MARV and RAVV, respectively (Adjemian et al. 2011). This lead and gold mine was occupied by Egyptian rousettes and a smaller insectivore, Sundevall's leaf-nosed bats (*Hipposideros caffer*). The Egyptian rousette population was statistically approximated to be more than 100,000 animals (Towner et al. 2009), and over two collection periods spanning 8 months, samples from 611 Egyptian rousettes and 609 Sundevall's leaf-nosed bats were tested for marburgviruses. Of these, 31 Egyptian rousettes and 1 Sundevall's leaf-nosed bat tested positive by qRT-PCR. Moreover, live MARV and RAVV were isolated directly from bat tissues (n = 5 bats) providing critical evidence that infectious virus could be maintained within the bat population over time (Towner et al. 2009). The single MARV-positive Sundevall's leaf-nosed bat, along with the MARV-positive insectivorous bats from DRC and Gabon, are the likely result of spillover events to members of non-reservoir species and not considered part of the marburgvirus maintenance cycle per se. However, further field studies examining marburgvirus spillover to cohabitating bats of other species is warranted.

The second series of spillover events occurred at Python Cave in Uganda, a popular tourist destination and home to approximately 40,000 Egyptian rousettes. The cave is located in the Maramagambo Forest in Queen Elizabeth National Park, approximately 50 km away from Kitaka mine. There, a Dutch tourist became infected with MARV after visiting the cave in July 2008 (Timen et al. 2009). At about the same time, a tourist from the United States was also identified, albeit retrospectively, as having been infected with MARV following a visit to the same cave six months prior in December 2007 (CDC 2009). In response to these events, over 1600 Egyptian rousettes, the only bat present in Python cave, were captured and sampled four times between August 2008 and November 2009. Collectively, 2.5% (40/1622) of the bats tested positive by qRT-PCR for marburgviruses (Amman et al. 2012). Virus isolates, of both MARV and RAVV, were again obtained from liver and spleen samples (n = 12 bats), supporting the findings of

Towner et al. (2009) that marburgvirus circulation and transmission is maintained in Egyptian rousette populations in southwest Uganda.

Interestingly, when the investigations at Kitaka mine and Python cave were analyzed together, it became apparent that the prevalence of active infection spiked on a seasonal basis, and these pulses occurred in juvenile bats during the biannual birthing seasons when the bats are roughly 4–6 months in age. Active infections among this age cohort spiked to an average high of 12.4% during peak birthing months (mid-June through mid-September and mid-December through mid-March). Moreover, when human marburgvirus spillover data were analyzed, 83.1% of known spillover dates occurred within the months surrounding the biannual birthing seasons (Amman et al. 2012), suggesting older juvenile Egyptian rousettes represent a major driver for virus spillover to humans.

3 Reservoir Studies of Ebolaviruses

Early investigations into the natural sources of ebolaviruses consisted of sample collections and testing over many years from numerous vertebrate and invertebrate taxa, both sylvan and domestic (Arata et al. 1978; Germain 1978; WHO 1978; Breman et al. 1999; Leirs et al. 1999; Reiter et al. 1999). All of these investigations were unable to detect ebolavirus positive samples. Serum from one Lord Derby's scaly-tailed squirrel (*Anomalurus derbianus*) reportedly showed reactivity to EBOV antigen via immunofluorescent assay (IFA), but the positive antibody result could not be confirmed by radioimmunoassay (Breman et al. 1999). Fragments of EBOV RNA were reportedly detected in Peters's mice (*Mus setulosus*), two *Prayomys sp*, and a greater forest shrew (*Sylvisorex ollula*) captured in the Central African Republic. Sequences were generated that matched (at that time) the Zaire and Gabon lineages and examination of spleen samples with electron microscopy detected tubular structures similar to filovirus nucleocapsids in one of the *Prayomys sp*. However, these findings lacked corroboration by serology, virus isolation or antigen detection (Morvan et al. 1999).

Bats are becoming increasingly recognized as reservoirs for a variety of zoonotic viruses (Dobson 2005, Calisher et al. 2006; Smith and Wang 2013; Drexler et al. 2014) and have been reported to host a significantly higher number of viruses compared to rodents (Luis et al. 2013). Of late, bats have been tentatively recognized as ancestral hosts of mammalian paramyxoviruses in a study characterizing many new paramyxoviruses in bats from all over the world (Drexler et al. 2012). Several pteropid fruit bats are reservoirs for Nipah virus (Johara et al. 2001; Shirai et al. 2007) and Hendra virus (Halpin et al. 2000; Drexler et al. 2009). Recently, a new pathogenic paramyxovirus, Sosuga virus, was detected in Egyptian rousettes, a natural reservoir for marburgviruses (Amman et al. 2015a). Given the close phylogenetic relationships between the family *Paramyxoviridea* and *Filoviridae* in the order *Mononegavirales* (Cleveland et al. 2011; Kuhn et al. 2013; Li et al. 2015), it is conceivable that bats are also ancestral reservoirs for filoviruses.

The increasing suspicion that bats are natural reservoirs for filoviruses has been evident in several collection efforts over the last few decades. Over 1000 animals, including 679 bats, 222 birds, and 129 small terrestrial vertebrates were tested during an EVD outbreak investigation in Gabon from 2001 to 2003 (Table 2) and resulted in the detection of antibodies reactive to EBOV antigen and EBOV-specific RNA in pteropodids of three different species (little collared fruit bats [*Myonycteris torquata*], 4/141; Franquet's epauletted fruit bats [*Epomops franqueti*], 5/117; and hammer-headed fruit bats, 5/111; Leroy et al. 2005). This nucleic acid evidence, coupled with positive serological results for bats of the same three species, provided the first solid evidence of chiropteran involvement in the EBOV transmission cycle. A few years later, the possible epidemiological evidence linking direct human contact with bats to an outbreak of EVD in DRC was reported by Leroy et al. (2009), citing that the presumed primary case may have purchased freshly killed fruit bats from a local hunter. Since that time, the majority of ebolavirus ecology investigations have focused on bats with varying results.

Serological reactivity to EBOV antigen using whole cell lysate (EBOV antigens diluted 1:1000) was also reported for Peters's lesser epauletted fruit bat (*Micropteropus pusillus*), Egyptian rousettes, and unidentified microchiropteran bats in Gabon (Pourrut et al. 2007, 2009). African straw-colored fruit bats (*Eidolon helvum*), Gambian epauletted fruit bats (*Epomophorus gambianus*), Franquet's epauletted fruit bats, hammer-headed fruit bats, and Veldkamp's dwarf epauletted fruit bats (*Nanonycteris veldkampii*) bats from Ghana were reported to have antibodies reactive to EBOV using indirect fluorescent assays (IFA) and recombinant nucleoproteins (Hayman et al. 2010, 2012). Ogawa et al. (2015) reported finding *E. helvum* bats in Zambia with antibody reactive to EBOV (19/48), Sudan virus (SUDV; 19/748), Taï Forest virus (TAFV; 9/748), Bundibugyo virus (BDBV; 8/748), Reston virus (RESTV; 9/748), and MARV (7/748) using recombinant viral glycoproteins. None of the bats were positive by PCR (Table 2). An important consideration when interpreting bat serology results is that in humans, there is a high degree of cross reactivity between ebolaviruses (MacNeil et al. 2011). In this light, caution should be exercised when using serology to determine past infection by a specific filovirus.

The most recent EVD outbreak was the largest filovirus-associated outbreak on record, resulting in 15,261 laboratory confirmed cases (28,652 suspected, probable, and confirmed) and 11,325 deaths (CDC; http://www.cdc.gov/vhf/ebola/outbreaks/2014-west-africa/index.html accessed on 23 Aug. 2016). During the initial stages of the outbreak, a team of researchers set out to determine the source of the virus. During this process, 169 bats from 17 species were captured and tested, none of which tested positive for EBOV RNA (Saez et al. 2015). This collection included a single Angolan free-tailed bat (*Mops condylurus*), the species the presumed primary case was reportedly in contact with. Identified through deep sequencing of bat mitochondrial DNA found in ash from a burned out tree, Angolan free-tailed bats were shown to be roosting at the site where the alleged 2-year-old primary case had been playing and handling bats (Saez et al. 2015). With regard to potential reservoir status, it must be borne in mind that the geographic range of Angolan free-tailed

Table 2 Bats tested for filoviruses

PCR							IgG									
Ref	Species tested	EBOV	RESTV	LLOV	MARV	Fn	EBOV	SUDV	TAFV	BDBV	RESTV	MARV	IFA	Iso	IHC	CC
WHO (1978)	Unidentified bats													0/7		COD
	Total													0/7		
Arata et al. (1978)	*Epomophevus anupus*[a]													0/2		SSD
	Micropteropus pusillus													0/2		
	Taphozous mauritianus													0/12		
	Scotoecus hirundo													0/12		
	Scotoecus hindei													0/2		
	Tadarida (Mops) nanulus													0/8		
	Tadarida (Mops) trevori													0/140		
	Total													0/185		
Conrad et al. (1978)	Unidentified bats													0/6		ZWE
	Total													0/6		
Formenty et al. (1999)[b]	Unidentified bats															CIV
	Total						0/652									

(continued)

Table 2 (continued)

PCR							IgG									
Ref	Species tested	EBOV	RESTV	LLOV	MARV	Fn	EBOV	SUDV	TAFV	BDBV	RESTV	MARV	IFA	Iso	IHC	CC
Morvan et al. (1999)	*Epomophorus gambianus*	0/3														CAF
	Epomops franqueti	0/1														
	Hipposideros sp.	0/1														
	Hypsignathus monstrosus	0/1														
	Micropteropus pusillus	0/2														
	Myonycteris torquata	0/13														
	Rousettus aegyptiacus	0/3														
	Vespertilionidae	0/4														
	Total	0/23														
Breman et al. (1999)	*Chaerephon major*													0/26		COD
	Chaerephon chapini													0/23*		
	Chalinolobus sp.													0/15		
	Eidolon helvum													0/6		
	Epomops franqueti													0/21		
	Eptesicus[c] *tenuipinnis/rendalli*													0/22		COD
	Hipposideros commersoni													0/23*		
	Hipposideros cyclops													0/52		

(continued)

Table 2 (continued)

PCR							IgG									
Ref	Species tested	EBOV	RESTV	LLOV	MARV	Fn	EBOV	SUDV	TAFV	BDBV	RESTV	MARV	IFA	Iso	IHC	CC
	Hipposideros ruber													0/17		
	Hipposideros sp.													0/23*		
	Hypsignathus monstrosus													0/23*		
	Kerivoula lanosa													0/23*		
	Megaloglossus woermanni													0/23*		
	Mops condylurus													0/54		
	Mops congicus													0/20		
	Mops nanulus													0/15		
	Mops sp.													0/23*		
	Mops thersites													0/69		
	Myonycteris torquata													0/23*		
	Myotis bocagii													0/17		
	Nycteris sp.													0/14		
	Pipistrellus[c] *nanus*													0/73		
	Scotophilus sp.													0/10		
	Taphozous mauritianus													0/23*		
	Taphozous[d] *peli*													0/9		
	Total													0/463		

(continued)

Table 2 (continued)

PCR							IgG									
Ref	Species tested	EBOV	RESTV	LLOV	MARV	Fn	EBOV	SUDV	TAFV	BDBV	RESTV	MARV	IFA	Iso	IHC	CC
Leirs et al. (1999)	*Casinycteris argynnis*						0/1							0/2		COD
	Chaerephon ansorgei						0/120							0/121		
	Chaerephon pumilus						0/210							0/211		
	Epomops franqueti						0/2							0/2		
	Eptesicus[c] *somalicus*						0/0							0/1		
	Eptesicus[c] *tenuipinnis*						0/1							0/1		
	Megaloglossus woermanni						0/38							0/43		
	Micropteropus pusillus						0/78							0/78		
	Miniopterus minor						0/2							0/2		COD
	Mops condylurus						0/10							0/10		
	Mops nanulus						0/14							0/14		
	Mops niveiventer						0/3							0/3		
	Mops thersites						0/1							0/1		
	Myopterus whitleyi						0/1							0/2		
	Myotis bocagii						0/22							0/22		
	Nycteris hispida						0/1							0/2		
	Pipistrellus nanus						0/2							0/2		
	Scotophilus dinganii						0/19							0/20		
	Total						0/525							0/537		

(continued)

Table 2 (continued)

PCR							IgG									
Ref	Species tested	EBOV	RESTV	LLOV	MARV	Fn	EBOV	SUDV	TAFV	BDBV	RESTV	MARV	IFA	Iso	IHC	CC
Leroy et al. (2005)	*Epomops franqueti*	5/117					8/117							ND		COD
	Hypsignathus monstrosus	4/21					4/17							ND		
	Myonycteris torquata	4/141					4/58							ND		
	Unidentified bats	0/131					0/25							0/679		
	Total	13/410					16/217							0/679		
Swanepoel et al. (2007)	*Hipposideros caffer*				0/7							0/10		0/13		COD
	Hipposideros commersoni				0/13							0/16		0/17		
	Lissonycteris angolensis				0/3							0/3		0/3		
	Miniopterus inflatus				1/33							0/34		0/38		
	Nycteris hispida				0/1							0/1		0/1		
	Rhinolophus eloquens				7/197							20/206		0/222		
	Rhinolophus landeri				ND							ND		0/1		
	Rousettus aegyptiacus				4/127							32/156		0/230		
	Total				12/382							52/426		0/524		

(continued)

Table 2 (continued)

PCR							IgG									
Ref	Species tested	EBOV	RESTV	LLOV	MARV	Fn	EBOV	SUDV	TAFV	BDBV	RESTV	MARV	IFA	Iso	IHC	CC
Pourrut et al. (2007)	*Epomops franqueti*						25/725									GAB
	Hypsignathus monstrosus						5/111									
	Myonycteris torquata						34/554									
	Total						64/1390									
Towner et al. (2007)	*Casinycteris argynnis*				0/2							0/0				GAB
	Eidolon helvum				0/35							0/33				
	Epomops franqueti				0/296							0/47				
	Hipposideros gigas				0/1							0/1				
	Hypsignathus monstrosus				0/56							0/12				
	Megaloglossus woermanni				0/37							0/20				
	Micropteropus pusillus				0/149							0/19				
	Microchiroptera				0/15							0/9				
	Myonycteris torquata				0/264							0/55				
	Rousettus aegyptiacus				4/283							29/242				
	Total				4/1138							29/438				

(continued)

Table 2 (continued)

PCR							IgG									
Ref	Species tested	EBOV	RESTV	LLOV	MARV	Fn	EBOV	SUDV	TAFV	BDBV	RESTV	MARV	IFA	Iso	IHC	CC
Pourrut et al. (2009)	*Casinycteris argynnis*						0/18					0/18				GAB
	Eidolon helvum						0/49					0/47				
	Epomops franqueti						36/805					2/679				
	Hypsignathus monstrosus						9/125					1/103				
	Megaloglossus woermanni						0/49					0/39				
	Microchiroptera						3/24					0/20				
	Micropteropus pusillus						4/197					1/177				
	Myonycteris torquata						19/573					0/493				
	Rousettus aegyptiacus						24/307					21/299				
	Total						95/2147					25/1876				
Towner et al. (2009)	*Rousettus aegyptiacus*				31/611							13/546		5/31	2/31	UGA
	Hipposideros sp.				1/609							0/609				
	Total				32/1220							13/1155		5/31	2/31	
Hayman et al. (2010)	*Eidolon helvum*												1/262			GHA
	Hypsignathus monstrosus												0/3			
	Total												1/265			

(continued)

Table 2 (continued)

PCR							IgG									
Ref	Species tested	EBOV	RESTV	LLOV	MARV	Fn	EBOV	SUDV	TAFV	BDBV	RESTV	MARV	IFA	Iso	IHC	CC
Taniguchi et al. (2011)	*Cynopterus brachyotis*										0/35					PHL
	Eonycteris spelaea										0/5					
	Ptenochirus jagori										0/38		0/38			PHL
	Haplonycteris fischeri										0/6		0/6			
	Macroglossus minimus										0/2		0/2			
	Rhinolophus rufus										0/2		0/2			
	Rhinolophus arcuatus										0/1		0/1			
	Emballonura alecto										0/9		0/9			
	Rousettus amplexicaudatus										10/16		3/16			
	Pipistrellus javanicus										0/2		0/2			
	Scotophilus kuhlii										0/5		0/5			
	Miniopterus australis										0/8		0/8			
	Miniopterus schreibersii										0/8		0/8			
	Miniopterus tristis tristis										0/1		0/1			
	Hipposideros diadema										0/1		0/1			

(continued)

Table 2 (continued)

PCR							IgG									
Ref	Species tested	EBOV	RESTV	LLOV	MARV	Fn	EBOV	SUDV	TAFV	BDBV	RESTV	MARV	IFA	Iso	IHC	CC
	Myotis macrotarsus										0/1		0/1			
	Unidentified bat										0/1		0/1			
	Total										10/141		3/141			
Negredo et al. (2011)	*Miniopterus schreibersii*			24/70												ESP
	Myotis myotis			0/9												
	Unidentified bats			0/1250												
	Total			24/1329												
Hayman et al. (2012)	*Epomophorus gambianus*						5/37				3/37					GHA
	Epomops buettikoferi						0/1				0/1					
	Epomops franqueti						3/27				10/27					
	Hypsignathus monstrosus						2/16				1/16					
	Nanonycteris veldkampii						0/4				0/4†					
	Total						9/13				3/13					

(continued)

Table 2 (continued)

PCR							IgG									
Ref	Species tested	EBOV	RESTV	LLOV	MARV	Fn	EBOV	SUDV	TAFV	BDBV	RESTV	MARV	IFA	Iso	IHC	CC
Saez et al. (2015)	*Casinycteris*[e] *ophiodon*	0/1														GIN
	Chaerephon pumilus	0/1														
	Eidolon helvum	0/6														
	Epomops buettikoferi	0/17														GIN
	Hipposideros caffer/ruber	0/18														
	Hipposideros cyclops	0/23														
	Hipposideros jonesi	0/1														
	Hipposideros sp.	0/2														
	Hypsignathus monstrosus	0/1														
	Kerivoula sp.	0/1														
	Megaloglossus azagnyi	0/3														
	Mops condylurus	0/1														
	Myonycteris angolensis	0/45														
	Myonycteris leptodon	0/21														
	Nanonycteris veldkampii	0/17														
	Nycteris sp.	0/6														
	Rhinolophus sp.	0/5														
	Total	0/169														

(continued)

Table 2 (continued)

PCR							IgG									
Ref	Species tested	EBOV	RESTV	LLOV	MARV	Fn	EBOV	SUDV	TAFV	BDBV	RESTV	MARV	IFA	Iso	IHC	CC
Ogawa et al. (2015)	*Eidolon helvum*	0/367			0/367		19/748	19/748	9/748	8/748	9/748	7/748				ZMB
	Total	0/367			0/367		19/748	19/748	9/748	8/748	9/748	7/748				
Jayme et al. (2015)	*Acerodon jubatus*		0/56								3/56					PHL
	Chaerephon plicatus		1/82‡								0/82					
	Cynopterus brachyotis		1/37‡								0/37					
	Eonycteris robusta		0/11								0/11					
	Eonycteris spelaea		0/1								0/1					
	Hipposideros ater		0/1								0/1					
	Hipposideros diadema		0/6								0/6					
	Hipposideros pygmaeus		0/8								0/8					
	Hipposideros sp.		0/1								0/1					
	Miniopterus australis		1/70‡								0/70					
	Miniopterus schreibersii		3/44								0/44					
	Murina cyclotis		0/1								0/1					
	Myotis horsfieldii		0/5								0/5					PHL
	Ptenochirus jagori		0/52								0/52					
	Pteropus vampyrus		0/5								1/5					
	Rhinolophus arcuatus		0/31								0/31					
	Rhinolophus philippinensis		0/1								0/1					

(continued)

Table 2 (continued)

PCR							IgG									
Ref	Species tested	EBOV	RESTV	LLOV	MARV	Fn	EBOV	SUDV	TAFV	BDBV	RESTV	MARV	IFA	Iso	IHC	CC
	Rhinolophus rufus		0/6								0/6					
	Rhinolophus virgo		0/1								0/1					
	Rousettus amplexicaudatus		0/42								0/42					
	Tylonycteris robustula		0/3								0/3					
	Total		3/464								4/464					
Amman et al. (2012)	*Rousettus aegyptiacus*				40/1622							250/1622		7/40	4/80	UGA
	Total				40/1622							250/1622		7/40	4/80	
Yuan et al. (2012)	*Cynopterus sphinx*										2/2					CHN
	Hipposideros armiger										0/41					
	Hipposideros cineraceus										0/111					
	Hipposideros larvatus										0/21					
	Hipposideros pomona										3/39					
	Hipposideros sp.										1/15					
	Miniopterus schreibersii										1/23					
	Myotis chinensis										0/6					
	Myotis daubentonii										0/24					
	Myotis davidii										0/5					

(continued)

Table 2 (continued)

PCR							IgG									
Ref	Species tested	EBOV	RESTV	LLOV	MARV	Fn	EBOV	SUDV	TAFV	BDBV	RESTV	MARV	IFA	Iso	IHC	CC
	Myotis fimbriatus										0/2					
	Myotis ricketti										4/83					
	Myotis sp.										3/118					
	Pipistrellus pipistrellus										4/35					
	Rhinolophus affinis										1/69					
	Rhinolophus ferrumequinum										0/15					
	Rhinolophus pearsonii										0/3					
	Rhinolophus pusillus										0/14					CHN
	Rhinolophus sinicus										0/6					
	Rhinolophus sp.										0/15					
	Rousettus leschenaultii										11/126					
	Scotophilus kuhlii										1/25					
	Unidentified bats										0/2					
	Total										32/843					
Olival et al. (2013)	*Cynopterus* sp.		0/75				0/75				0/75					BGD
	Macroglossus sobrinus		0/1				0/1				0/1					
	Megaderma lyra		0/56				0/56				0/56					
	Rousettus leschenaultii		0/141				4/141				1/141					
	Total		0/276				4/276				1/276					

(continued)

Table 2 (continued)

PCR							IgG									
Ref	Species tested	EBOV	RESTV	LLOV	MARV	Fn	EBOV	SUDV	TAFV	BDBV	RESTV	MARV	IFA	Iso	IHC	CC
Amman et al. (2014)	*Rousettus aegyptiacus*				53/400									9/53		UGA
	Total				53/400									9/53		
He et al. (2015)	*Rousettus leschenaultii*					1/29										CHN
	Total					1/29										

[a]*Epomophezus anupus* is a misspelling in Arata et al. (1978). This most likely refers to *Epomophorus anurus*, which was once thought to be a distinct species in the genus *Epomophorus* (Koopman 1975), but is now a subsumed under *E. labiatus* as a synonym (Wilson and Reeder 2005). Other minor misspellings in this and other sources have been corrected in this table

[b]P. Formenty, P. Jahrling, C. Rossi, H. Artsob, R. Swanepoel, B. LeGuenno, K. Steele, N. Woolen, M. Gartshore, C. Boesch, C. Noé, P. Barriere, O. Perpete, N. Jaax, A. Kuhene, C. Akoua-Koffi, M. Colyn, Search for the Ebola virus reservoir in Taï Forest, Côte d'Ivoire: 1996–1997, preliminary results, in: XIth International Congress for Virology, & International Union of Microbiological Societies. (1999). Abstract book: XIth International Congress of Virology, 9–13 August 1999, Sydney. Sydney: publisher not identified. As cited in Pourrut et al. (2005)

[c]Genus *Neoromicia*

[d]*Saccolaimus peli*

[e]*Casinycteris ophiodon* is misidentified and should be *Scotonycteris ophiodon*

* All species part of an undifferentiated testing pool of 23 total bats

† One *N. veldkampii* tested positive for the EBOV/RESTV mix, but negative by each individually

‡ Uncorroborated PCR and RT-PCR positive bats

ND No Data reported; *Fn* Filovirus *novo* (Bt-DH04); *IFA* Indirect Fluorescence Assay; *Iso* Virus isolation; *IHC* Immunohistochemical analysis (staining); *CC* Country code

COD Yambuku, Zaire (DRC); *SSD* Nzara, Sudan; *ZWE* (Rhodesia) Zimbabwe; *CIV* Côte d'Ivoire; *CAF* Bangui, Central African Republic; *COD* DRC, Cameroon; *COD* Kikwit, DRC; *COG* Republic of the Congo; *COD* DRC; *GAB* Gabon and north COG; *UGA* Kitaka Mine, Uganda; *GHA* Ghana; *PHL* Philippines; *ESP* Spain; *GHA* Ghana; *GIN* Guinea; *ZMB* Zambia; *PHL* Philippines; *UGA* QENP, Uganda; *CHN* China; *BGD* Bangladesh; *UGA* Kitaka Mine, Uganda bats is reported to not include a large portion of the Congo basin where many past outbreaks of EVD have occurred. Further, Angolan free-tailed bats are reported to have peridomestic tendencies and can often be found roosting in groups of several hundred in buildings and houses (Kingdon et al. 2013), which would predict higher occurrences of EBOV spillover events in human populations than have actually been recorded. Consistent with this natural history, Kingdon et al. (2013) reports that these bats have not been recorded in undisturbed rainforests or montane habitats, where numerous outbreaks in populations of gorillas and chimpanzees have been documented. Regardless, the study by Saez et al. (2015) demonstrated that humans regularly interact with bats in Western Africa through hunting or other means, and that there was no clear linkage to a dead or diseased nonhuman primate.

RESTV was the first of three non-African endemic filoviruses to be identified. It was discovered after an outbreak occurred in crab-eating macaques (*Macaca fascicularis*) imported from the Philippines to a primate facility in Reston, Virginia (Jahrling et al. 1990). This newly discovered filovirus, while lethal in nonhuman primates, remains the only filovirus that appears to be non-pathogenic in humans (Hayes et al. 1992; WHO 2009; Miranda and Miranda 2011). Ecological studies focused on finding the RESTV natural reservoir began in earnest almost 20 years after its discovery, when RESTV was detected in domestic swine in the Philippines (Barrette et al. 2009). Over 140 bats of multiple species were captured and tested for antibodies against RESTV. Only sera from Geoffroy's rousettes (*Rousettus amplexicaudatus*) (10/16) tested positive (Taniguchi et al. 2011). Results from another ecologic study reported RESTV RNA in bats of multiple species from the Philippines, including Schreibers's long-fingered bats (*Miniopterus schreibersii*), little long-fingered bats (*Miniopterus australis*), lesser short-nosed fruit bats (*Cynopterus brachyotis*), and wrinkle-lipped free-tailed bats (*Chaerephon plicatus*), although only the Schreibers's long-fingered bats results were corroborated (Jayme et al. 2015). Antibodies reactive to RESTV recombinant nucleoproteins were also reported in golden-capped fruit bats (*Acerodon jubatus*) and large flying foxes (*Pteropus vampyrus*) (Jayme et al. 2015). Surprisingly, these data, if correct, seem to imply that RESTV infections are widespread among taxonomically diverse bats in the Philippines.

The second non-African filovirus, Lloviu virus (LLOV), was discovered in Spain after an investigation into the cause of massive bat mortalities in Schreibers's long-fingered bat populations inhabiting the local caves (Negredo et al. 2011). Tissues from both live and dead Schreibers's long-fingered bats and mouse-eared myotis (*Myotis myotis*) collected at Cueva del Lloviu were analyzed using qRT-PCR. Only tissues from carcasses of deceased Schreibers's long-fingered bats (but not live bats) tested positive for LLOV, perhaps indicating a spillover event into these bats from another source.

More geographically widespread evidence of filovirus infection in bats has been reported, including in Bangladesh and China. Olival et al. (2013) reported antibodies reactive with a mixture of purified RESTV and EBOV nucleoproteins in Leschenault's rousettes (*Rousettus leschenaultii*) bats in Bangladesh. Interestingly, all but one of the confirmed antibody-positive samples reacted more strongly to purified recombinant EBOV nucleoprotein than to that of RESTV. However, no bats were found positive by PCR. Yuan et al. (2012) sampled 843 bats from several provinces in southern China between 2006 and 2009 and found antibodies reactive against RESTV nucleoprotein in 32 bats representing several species, with highest prevalence in Leschenault's rousettes (8.73%), common pipistrelles (*Pipistrellus pipistrellus*) (11.43%), and various species of *Myotis* (2.54%) bats. Of those 32 positives, 25 also had antibodies reactive with EBOV nucleocapsid protein. Later, He et al. (2015) also reported finding filovirus RNA in Leschenault's rousettes. Based on partial sequences from NP and VP35, the authors suggested this sequence represents a new filovirus, Bt-DH04, which is phylogenetically placed between marburgviruses and a clade containing LLOV, SUDV, and RESTV. Leschenault's rousettes are common across southern China and Southeast Asia, ranging south into the Malaysian Archipelago and west across most of the Indian subcontinent. These findings and those of Jayme et al. (2015) and others identify a common and abundant chiropteran genus, *Rousettus*, as a potential taxon for filovirus hosts, which may one day fill the current geographic and phylogenetic gap between the African filoviruses and their Asian counterparts.

4 Public Health Risk Mitigation Efforts, Lessons Learned

The MVD outbreaks at Kitaka mine and Python cave were linked to human activity in and around large populations of Egyptian rousette bats. With that in mind, the most sensible and straight forward public health message is to avoid contact with these and other large colonies of Egyptian rousettes. In certain instances, like Python cave, this goal can be achieved by not allowing tourist to enter caves where these bats roost. Prior to the outbreaks, guided tours were routinely taken to the entrance and interior of the cave to allow tourists to experience the bats and the large African rock pythons up close (Timen et al. 2009; CDC 2009). After ecological investigations determined the bats' potential to harbor marburgviruses (Amman et al. 2012), an enclosed viewing platform was constructed to allow viewing of the cave entrance, the bats, and occasionally the pythons, at a safe distance from the opening of the cave. This, in turn, allowed tourism at this site to continue while reducing the risk of exposure to infected bats. In the approximate 10 years since, no known incidences of MVD have been linked to Python Cave.

Subterranean mining presents an entirely different challenge. After finding marburgviruses in the large Egyptian rousette population in Kitaka mine (Towner et al. 2009), the site was closed to all mining activity, eliminating the only source of revenue for the already low-income miners. The mine owner believed that depopulating the mine of all bats would allow operations to resume in a safe environment. Sometime between May and August of 2008, the mine was sealed, destroying most, if not all, of the 100,000+ Egyptian rousettes (Amman et al. 2014). Unfortunately this type of response to a human threat is typical in many resource-poor countries. In some instances, the culling of wildlife to control disease in populations is known to be effective (Lachish et al. 2010), but other studies have shown that culling can increase disease prevalence (Laddomada 2000; Donnelly et al. 2003; Swanepoel 2004). Regardless, the Kitaka mine is clearly an example of a failed attempt of wildlife culling to reduce zoonotic disease. The mine remained sealed for several years, but time and the elements deteriorated the papyrus reed and plastic tarp barriers that were used to seal the mine. Bats from surrounding populations began to repopulate the roosting site. Amman et al. (2012) described marked Egyptian rousettes from Kitaka mine being captured at Python Cave over 50 km to the southwest, suggesting routine travel between the two sites, and probably others within range. Mining operations resumed in 2010, which further opened the mine to repopulation. Not long after, in October 2012, Uganda experienced the largest MVD outbreak in its history with 15 laboratory confirmed cases in the nearby town that supported the miners and general mining operations (Albarino et al. 2013). The subsequent ecological investigation found the mine repopulated with Egyptian rousettes, albeit at greatly reduced numbers (Amman et al. 2014). Importantly, the prevalence of actively infected Egyptian rousettes, as determined by qRT-PCR, was more than double that reported previously (Towner et al. 2009). Of the 400 Egyptian rousettes captured in November 2012, 13.7% were positive compared to the 5.1% from 2007 to 2008 outbreak. Nine marburgvirus isolates were obtained from bat tissue and some sequences were nearly identical to those obtained from human cases (Amman et al. 2014). The apparent spike in the prevalence of active infection was not limited to only marburgviruses. An investigation later that year of Sosuga virus in Egyptian rousettes showed the prevalence of active infection in samples from Kibaale, Uganda to be 2.5% (3/122) and nearby Python cave to be 2.1% (18/809), whereas the prevalence of Sosuga infection at the newly repopulated Kitaka mine was 10.2% (41/400) (Amman et al. 2014). This bat culling and subsequent repopulation appears to have increased the risk for human infection with marburgviruses or other zoonotic agents. While a direct epidemiological link to the mine and the bat population was never definitively established, there is no doubt that increased levels of active infection in the bat population pose serious health threats to the miners and other humans in surrounding villages.

5 Filoviruses in Nonhuman Secondary Hosts

While bats are considered by many to be natural reservoirs for filoviruses, the majority of EVD outbreaks that can be traced back to an animal source other than bats, typically through handling animals recently hunted or found dead (Leroy et al. 2004, 2009). Most often, these animals are acutely infected nonhuman primates and duikers. To date, only two outbreaks of EVD are reported to have epidemiologic linkage, albeit tenuous, to exposure to bats (Leroy et al. 2009; Saez et al. 2015). Large declines in ape populations due to infection with ebolaviruses have been reported and are at the forefront of conservation and public health discussions (Formenty et al. 1999; Huijbregts et al. 2003; Walsh et al. 2003; Bermejo et al.

2006). Some outbreaks of human EVD have been preceded by wildlife deaths in the forest (Leroy et al. 2004). These wildlife mortalities present a risk to human health because infectious dead animals are often found and butchered for consumption or sale at markets. Further, these ebolavirus-related wildlife mortalities have dire consequences for the animal populations from a conservation perspective. The animals with slower reproductive rates such as chimpanzees and gorillas can see large and long lasting effects. Duiker populations, having faster reproductive rates, may recover more quickly from ebolavirus-related mortalities (Leroy et al. 2004).

Risk of filovirus transmission may also come from domestic animals, including livestock. In a review of animal sampling during EVD outbreaks in humans, Olson et al. (2012) reports that antibodies reactive with EBOV were never identified in domestic stock animals (goats, cows, pigs, and sheep). However, dogs have been reportedly exposed to EBOV. Allela et al. (2005) tested over 330 dogs from outbreak localities in Gabon and found that over 37% (127/337) had antibodies reactive with EBOV. Witnesses describe seeing dogs eating discarded viscera from EBOV positive bushmeat and licking vomit from human EBOV patients. However, the study also used samples from 102 Parisian (French) dogs, as a control group and found 2% (2/102) reactive to EBOV antigen. These results were considered false positives attributed to the low positivity cut-off value and the 1:400 serum dilutions utilized (Allela et al. 2005). Of note, there have been no known instances of dog to human ebolavirus transmission, including during the large EBOV outbreak in Western Africa, where thousands of dogs could have been exposed to infectious human material.

Although there is a lack of evidence associating domestic stock animals with EBOV, RESTV has been detected in domestic pigs (*Sus scrofa domesticus*) in the Philippines (Barrette et al. 2009) and in China (Pan et al. 2014). After an outbreak of porcine reproductive and respiratory syndrome virus (PRRSV) in pigs on rural farms in the Philippines, samples were sent for analysis to the Foreign Animal Disease Diagnostic Laboratory (FADDL) of the United States Department of Agriculture (USDA) Animal and Plant health Inspection Service (APHIS). In addition to identifying PRRSV in these samples, RESTV was isolated and identified from lymph tissue. Moreover, 6 of 141 workers from the infected pig farms had antibodies reactive with RESTV, indicating transmission of RESTV from pigs to humans (Barrette et al. 2009). Similar findings were reported following a 2011 survey for PRRSV in farms near Shanghai, China, when RESTV RNA was detected in 4/137 piglets (Pan et al. 2014).

To further investigate the potential for pigs to act as secondary hosts of ebolaviruses capable of virus transmission to other animals including humans, experimental infections of pigs with RESTV and EBOV demonstrated that both viruses were able to replicate to high levels. RESTV injected into 5-week-old piglets replicated in the viscera and nasopharynx and live virus was shed via oral and nasal secretions without overt signs of morbidity (Marsh et al. 2011). Pigs inoculated with EBOV also developed high levels of infection in the respiratory tract and demonstrated viral shedding from the oronasal mucosa. Unlike RESTV, inoculations with EBOV produced severe morbidity in the pigs, primarily in the lungs (Kobinger et al. 2011).

For marburgviruses, the frequency of known spillover to nonhuman secondary hosts appears much less. That said, the first known outbreak of MVD occurred in laboratory workers in Germany and Yugoslavia (now Serbia) following exposure to infected grivets (*Chlorocebus aethiops*) (Siegert et al. 1968) imported from Uganda. In an attempt to gain more understanding of the etiology of this new virus, sera from a variety of nonhuman primates and humans were tested for antibodies reactive with MARV. Kalter et al. (1969) reported finding MARV antibodies in African born nonhuman primates (chimpanzees, baboons, grivets, patas, and talapoins). Chimpanzees and baboons born in captivity in the United States did not have antibodies reactive to MARV. Antibodies were also detected in Asian primates (crab-eating macaques and rhesus monkeys [*Macaca mulatta*]), but they all had a history of contact with African-borne primates (Kalter et al. 1969).

6 Experimental Filovirus Infections in Captive Bats

Experimental infections of animals with zoonotic viruses are nominally performed to model human disease. However, experimental infection studies may also prove vital for measuring the replication dynamics in animals of a specific species to determine if those dynamics meet expectations for a reservoir host, namely the production of a clinically asymptomatic disease state with sufficient viral replication to produce shedding, leading to long-term maintenance of the virus within the animal population. Such data can, in turn, confirm findings from field research that identified a particular species as a potential reservoir. Additionally, this approach can be used to narrow the scope of potential target species and thereby redirect field research towards (or away from) specific taxon sampling.

The first such reservoir experiments involving the injection of filoviruses into living organisms was conducted by Swanepoel et al. (1996) in which various plants and animals were inoculated with EBOV. This study was the first laboratory-derived evidence evaluating the capacity of Angola free-tailed bats, little free-tailed bats (*Chaerephon pumilus*), and Wahlberg's epauletted fruit bats (*Epomophorus wahlbergi*) to tolerate, replicate, develop antibodies, and ultimately shed EBOV. For up to three weeks post inoculation, virus was isolated from pooled viscera and blood of all infected bats, and an isolate was also obtained from the feces of a Wahlberg's epauletted fruit bat after 21 days post inoculation (DPI), demonstrating potential for viral shedding. Pulmonary endothelial cells of one insectivorous bat exhibited limited immunohistochemical staining for EBOV antigen, without the presence of lesions, suggesting a possible respiratory or oral dissemination of virus within a colony setting (Swanepoel et al. 1996).

Almost two decades after the original plant and animal inoculation studies by Swanepoel et al. (1996), multiple laboratories independently began a series of MARV infection studies in captive Egyptian rousettes to determine if MARV replicated in a manner consistent with Egyptian rousettes being a reservoir host. The first of these experiments involved the inoculation of 29 adult and newborn bats with 2000 $TCID_{50}$ of high-passage (P38) Vero cell-adapted virus (Hogan isolate) derived from a human case (Paweska et al. 2012). Bats were inoculated by a combination of subcutaneous and intraperitoneal injection or by oronasal drip. Three bats from the group inoculated by the oronasal route showed no evidence of infection after 21 DPI and were later inoculated subcutaneously. The combination of subcutaneous and intraperitoneal inoculation produced systemic infection in the bats without overt morbidity and mortality, meeting certain expectations for a reservoir host. From 2 to 9 DPI, MARV was detected in blood and multiple tissues. IgG antibody reactive to MARV was detected from 9 to 21 DPI. However, viral shedding was not observed in any bodily secretions tested (Paweska et al. 2012).

The second Egyptian rousette experimental infection study differed in approach and produced evidence of MARV shedding (Amman et al. 2015b). This study utilized first generation captive bred juvenile Egyptian rousettes roughly 5 months old; the same age cohort as those most frequently found infected in wild populations in Uganda (Amman et al. 2012). These bats were infected with a low passage (P2) MARV originally isolated from an infected Egyptian rousette (371 bat virus) from Kitaka mine, thereby matching the virus and host as precisely as possible (Towner et al. 2009). Twenty seven Egyptian rousettes were infected with 10^4 $TCID_{50}$ MARV via a subcutaneous only route. Similar to the results of Paweska et al. (2012), infection occurred in all 27 bats with no signs of disease related morbidity or mortality. Viremia was detected as early as 1 DPI and ranged to 9 DPI with an average viremic period of 3 days. Multiple tissues showed signs of disseminated infection starting at 3 DPI (liver and spleen) and peaking at 5 DPI. Despite one spleen and kidney sample testing weakly positive via RT-PCR at 28 DPI, the mean viral loads in the majority of liver/spleen samples decreased dramatically and were cleared by 12 DPI. Organs compatible with viral shedding (salivary gland, kidney, bladder, and large intestine) also showed signs of infection as early as 3 DPI, albeit with lower viral loads, but peaked later than liver and spleen, typically around 7–9 DPI. Kidney tissue tested weakly positive as late as 28 DPI, while the remainder of the aforementioned tissues associated with shedding had cleared the virus by 12 DPI. All bats seroconverted after day 10. Importantly, MARV was detected in oral (n = 17) and rectal (n = 6) secretions. Further, virus isolates were obtained from 3 of the 17 oral swabs, all were samples that had peak levels greater than 10^4 $TCID_{50}$ (Amman et al. 2015b). These data provided the first evidence for an oral mechanism of transmission between bats in a colonizing population, perhaps through grooming or fighting. These data also provide some evidence for a fecal route of transmission, which would fit with the theory that while human MVD primary cases are often associated with bat habitats, evidence is lacking for direct contact (bites, collisions, scratches, etc.) with the bats, based on survivor interviews (Adjemian et al. 2011).

Jones et al. (2015) experimentally infected Egyptian rousettes with all five ebolaviruses (EBOV, SUDV, RESTV, TAFV and BDBV) in a comparative study to test the susceptibility of the MARV natural reservoir to infection with ebolaviruses. The rationale was based in part on the report by Pourrut et al. (2009) that 8% of Egyptian rousettes from Gabon and the Republic of Congo had antibody reactive with EBOV antigen. As expected, the control infection with MARV produced similar results to Amman et al. (2015b), with detectable MARV RNA in both oral and rectal swabs. In contrast, virus-specific RNA was not detected in oral or rectal swabs from any bats experimentally infected with any one of the five ebolaviruses. No ebolavirus-infected bat exhibited viremia and the skin at inoculation site was the only tissue that showed significant levels of virus RNA. Viral RNA was weakly detected in the axillary lymph node tissue for all bats except those infected with TAFV. Only SUDV showed any signs of infection in visceral organs with RT-PCR positive liver and spleen detected in two bats at 5 DPI, but at much lower levels than their MARV-infected counterparts. The SUDV results prompted an additional 15 day infection study in which samples were taken at closer intervals (days 3, 6, 9, 12, and 15) than that during the pilot study (days 5 and 10). With the increased sampling intervals, the results of this second infection showed SUDV had disseminated to several organs (liver, spleen, bladder, small intestines, large intestine, and gonad), but had mostly been cleared by 6 DPI. Only the axillary lymph node remained positive out to 12 DPI. Jones et al. (2015) concluded that Egyptian rousettes are unlikely sources of SUDV (or other ebolaviruses) in nature based on the fact that the SUDV-infected bats never became viremic and viral RNA was never detected in oral or rectal swabs. These findings also support the notion of a single virus–single reservoir host relationship for filoviruses (Jones et al. 2015).

The next logical step was to determine if MARV could be transmitted experimentally between infected and naïve bats housed together in direct contact with each other. Paweska et al. (2015) inoculated Egyptian rousettes with MARV at a ratio of 2 infected to 1 naïve bat and monitored for 42 days. Some bats were serially euthanized to monitor infection, and while transmission of MARV to naïve bats was not demonstrated, the study showed through post inoculation challenges that protective immunity to reinfection was present in all of the infected bats for up to 42 days, consistent with expectations for a natural reservoir. The study also confirmed the waning of maternal antibody in pups 3–5 months of age, as previously documented in naturally infected bats in Uganda (Amman et al. 2012). Moreover, conclusions presented in this report suggested that transmission might occur through mechanisms other than bat to bat contact, perhaps through hematophagous ectoparasites (Paweska et al. 2015).

A similar MARV transmission experiment was conducted on captive bred Egyptian rousettes by Schuh et al. (2017) that continued for a much longer period, 9 months, and demonstrated bat to bat transmission. Captive-borne experimentally infected donor bats (n = 12) were either cohoused 1:1 directly with naïve contact bats or placed over cages of additional naïve bats (n = 12) such that potentially infectious fruit spats, urine and feces could pass to the cage below. No bats were subjected to serial sacrifice and all were allowed to remain in their defined groups for 56 days, after which time they were gang housed for an additional 7 months. Infection was achieved in all inoculated donor bats and MARV shedding was detected in 11/12 bats by qRT-PCR in oral and/or rectal or urine samples between 5 and 19 DPI. Simultaneous to the virus shedding from the inoculated donor bats, eight of the naïve contact bats had detectable MARV RNA in oral swab samples indicating exposure to MARV, perhaps through contaminated food or water, or via social grooming. Late in the experiment, at 7 months post infection (MPI), MARV viremia was detected in three naïve contact bats, one of which also tested positive by oral swab. Two of these three bats seroconverted by 8 MPI, and by 9 MPI, nine of the 24 naïve contacts bats seroconverted, demonstrating horizontal transmission had occurred between infected donor bats and naïve contact bats. These data indicate that the incubation period in naturally infected bats may be longer than 21 days (Schuh et al. 2017).

7 Potential Role for Arthropods

Kunz and Hofmann (1971) reported the potential for natural marburgvirus transmission in culicine mosquitoes after successful propagation of MARV in experimentally inoculated yellow fever mosquitoes (*Aedes aegypti*). The mosquitoes were injected with a guinea pig serum containing 10^7 LD_{50} of MARV per ml and 11 days post inoculation they were ground and suspended in PBS. The suspension was injected into guinea pigs as well as additional mosquitoes for second passage. Virus isolates were obtained from both guinea pig and mosquito passages consistent with stability and perhaps replication of infectious virus in the mosquitoes. The same procedures were tried unsuccessfully on a mosquito of another species (*Anopheles maculipennis*) and castor bean ticks (*Ixodes ricinus*).

The majority of evidence suggests bats are the primary natural reservoirs for filoviruses. However, the potential for arthropod involvement as an intermediate host has never been definitively disproven and still requires consideration. That said, MARV can be transmitted from bat to bat under experimental conditions in the absence of arthropods (Schuh et al. 2017) and arthropods have been collected and tested as part of ecological investigations in past filovirus outbreaks, without any positive results (Conrad et al. 1978; Smith et al. 1982; Reiter et al. 1999; Breman et al. 1999). After the discovery of marburgviruses in Egyptian rousettes, bat flies (family Nycteribiidae; $n = 25$) and argasid ticks (family Argasidae, *Ornithodoros faini*; $n \approx 300$) were collected directly from bats and off the walls at the roost sites in Python Cave where active marburgvirus infection was occurring. These arthropods tested negative for marburviruses by qRT-PCR (Towner et al. 2009; Amman et al. 2012) although the number of individuals studied was deemed insufficient to reasonably conclude these hematophagous ectoparasites were not involved in the transmission cycle. More evidence was presented by Schuh et al.

(2016) after testing 3125 *O. faini* ticks collected from the walls of Python Cave where Egyptian rousettes were roosting. All pooled samples were negative for marburgviruses, indicating *O. faini* ticks do not appear to act as replication or mechanical vectors for marburgviruses in Egyptian rousette bat populations. Moreover, these results coupled with data that peak viremias in both sylvan and experimentally infected bats tend to be low and of short duration (Amman et al. 2012, 2015b), suggest it is unlikely that feeding *O. faini* ticks are able to acquire the virus and act as a mechanical vector (Schuh et al. 2016). Nevertheless, more testing of other chiropteran ectoparasites is warranted.

8 Current Endeavors and Future Directions

Any advancement in understanding filovirus ecology starts with identifying the natural reservoir. A significant amount of information has already been uncovered following the discovery of a chiropteran marburgvirus reservoir, leading to further experimentation with captive bats (Swanepoel et al. 1996; Towner et al. 2009; Amman et al. 2012; Paweska et al. 2012; Amman et al. 2015b; Jones et al. 2015; Paweska et al. 2015; Schuh et al. 2017). Unfortunately, identifying the natural reservoirs for ebolaviruses has proven difficult. The overwhelming consensus is that bats are involved in the natural transmission cycle, but beyond that, there appear to be two contrasting philosophies for how ebolaviruses are maintained in nature. One philosophy supports a multiple host transmission dynamic, where each host may play a role in long-term maintenance and the eventual spillover event leading to an outbreak in the human population (Haydon et al. 2002; Han et al. 2016; Leendertz et al. 2016). The second philosophy proposes a distinct primary bat reservoir for each filovirus, similar to other host specific zoonotic disease reservoir systems such as hantaviruses and arenaviruses (Childs and Peters 1993; Klingstrom et al. 2002; Ramsden et al. 2009; Mills et al. 2010). Recent evidence by Jones et al. (2015) supports host specificity of filoviruses by showing that, compared to MARV, Egyptian rousette bats do not support viral replication or shedding following experimental inoculation with any of the five known ebolaviruses. Similar conclusions were reached by Paweska et al. (2016) when they infected captive Egyptian rousettes with EBOV. In their study, one bat became viremic but no evidence of EBOV shedding was detected in any bat. In contrast, Swanepoel et al. (1996) reported bats of three different species were able to support EBOV replication, including one shedding virus in feces up to 21 DPI.

Another popular trend in filovirus ecology research utilizes mathematical modeling to identify the geographic range of filoviruses in Africa, and thereby predict areas at risk for outbreaks (Peterson et al. 2004, 2006; Pigott et al. 2014, 2015). Peterson et al. (2004) reports that these models increased known distributions of filoviruses to include most of sub-Saharan Africa, with EVD outbreaks more probable in humid rain forests of Middle and Western Africa and MVD outbreaks more likely in the drier portions of Middle and Eastern Africa, including portions of eastern Angola. The minimal overlap between EBOV and MARV distributions seen with their ecological niche modeling (ENM) suggest that the hosts for these viruses have pronounced ecological requirements leading to a specific set of criteria that could be used to narrow the scope of field investigations. Pigott et al. (2014) reported an EVD transmission niche spanning 22 Middle and Western African countries using species distribution models. Similar modeling results were described by Pigott et al. (2015) identifying 27 African countries at risk for MVD outbreaks using data from previous human outbreaks and reported infections in animals, including those of the natural reservoir, Egyptian rousettes. Modeling has also been used to examine the potential of certain groups of animals to act as natural reservoirs for zoonotic viruses (Luis et al. 2013). The utility of ENM towards the actual identification of reservoirs is unclear. Ultimately, these determinations will likely need to be made on the ground where individual animals can be caught, sampled and tested.

Hayman (2015) used modeling to predict that bats living in communities in excess of 20,000 individuals and exhibiting a biannual breeding/birthing reproduction cycle would need a 21+ day incubation period for virus to persist. Indeed, colonies in excess of 100,000 bats have been found in DRC and Uganda (Swanepoel et al. 2007; Towner et al. 2009). These findings support results from data collected in the field and reported by Amman et al. (2012).

More recently, Han et al. (2016) used modeling to identify potential filovirus reservoirs using life history variables of bats. They report a biological profile of potential filovirus reservoirs including such traits as larger pups, more than one litter per year, bimodal sexual maturity age, and a tendency to live in larger colonies. They also identify these potential reservoirs as having geographic distributions that overlap with regions of high mammalian diversity, similar to the findings reported by Luis et al. (2013). Different species of bats are identified as potentially hosting filoviruses based on traits similar to known PCR and antibody-positive animals. This report predicts an expanded range of potential filovirus hosts outside of sub-Saharan Africa with the majority of species overlap occurring in Southeastern Asia (Han et al. 2016).

Understanding filovirus infection dynamics in host systems can help further define mechanisms of transmission, which can then be developed into better risk reduction strategies and public health messaging. Experimental infections that result in lethal symptomatic disease, absence of viral shedding, or result in an outright failure to achieve viral replication during infection, can serve to narrow the focus of field studies by eliminating non-reservoirs from the myriad of possibilities that exist in nature. This serves not only to reduce the amount of time and money spent searching for potential reservoir hosts, it promotes the conservation of an ecologically important, and in many instances, vulnerable, group of small mammals by identifying non-target species.

Filovirus ecology is a growing and evolving concept, constantly shifting and changing with each new discovery. The future of this burgeoning field of study lies not within one single philosophy or scientific method, but within a combination of multiple supporting approaches that challenge paradigms and incorporate the possibility that filoviruses may present a heretofore unseen disease model that may only come to light with further patient and persistent effort.

Disclaimer The findings and conclusions in this review are those of the authors and not necessarily those of the Centers for Disease Control and Prevention.

References

Adjemian J, Farnon EC, Tschioko F et al (2011) Outbreak of Marburg hemorrhagic fever among miners in Kamwenge and Ibanda Districts, Uganda, 2007. J Infect Dis 204(Suppl 3):S796–S799

Albarino CG, Shoemaker T, Khristova ML et al (2013) Genomic analysis of filoviruses associated with four viral hemorrhagic fever outbreaks in Uganda and the Democratic Republic of the Congo in 2012. Virology 442(2):97–100

Allela L, Boury O, Pouillot R et al (2005) Ebola virus antibody prevalence in dogs and human risk. Emerg Infect Dis 11(3):385–390

Amman BR, Albarino CG, Bird BH et al (2015a) A recently discovered pathogenic Paramyxovirus, Sosuga virus, is present in *Rousettus aegyptiacus* fruit bats at multiple locations in Uganda. J Wildl Dis 51(3):774–779

Amman BR, Jones ME, Sealy TK et al (2015b) Oral shedding of Marburg virus in experimentally infected Egyptian fruit bats (*Rousettus aegyptiacus*). J Wildl Dis 51(1):113–124

Amman BR, Nyakarahuka L, McElroy AK et al (2014) Marburgvirus resurgence in Kitaka Mine Bat population after extermination attempts. Uganda. Emerg Infect Dis 20(10):1761–1762

Amman BR, Carroll SA, Reed ZD et al (2012) Seasonal pulses of Marburg virus circulation in juvenile *Rousettus aegyptiacus* bats coincide with periods of increased risk of human infection. PLoS Pathog 8(10):e1002877

Arata AA, Johnson B, Pattyn SR (1978) Approaches towards studies on potential reservoirs of viral haemorrhagic fever in southern Sudan. In: Pattyn SR (ed) Ebola virus haemorrhagic fever, Amsterdam 1977. Elsevier, North Holland, pp 191–200

Ashford RW (2003) When is a reservoir not a reservoir? Emerg Infect Dis 9(11):1495–1496

Barrette RW, Metwally SA, Rowland JM et al (2009) Discovery of swine as a host for the Reston Ebolavirus. Science 325(5937):204–206

Bausch DG, Nichol ST, Muyembe-Tamfum J. et al (2006) Marburg hemorrhagic fever associated with multiple genetic lineages of virus. New Engle J Med 355:909–919

Bermejo M, Rodriguez-Teijeiro JD, Illera G, Barroso A, Vila C, Walsh PD (2006) Ebola outbreak killed 5000 gorillas. Science 314(5805):1564

Brauburger K, Hume AJ, Muhlberger E, Olejnik J (2012) Forty-five years of Marburg virus research. Viruses 4(10):1878–1927

Breman JG, Johnson KM, van der Groen G et al (1999) A search for Ebola virus in animals in the Democratic Republic of the Congo and Cameroon: ecologic, virologic, and serologic surveys, 1979–1980. J Infect Dis 179(Supplement 1):S139–S147

Calisher CH, Childs JE, Field HE, Holmes KV, Schountz T (2006) Bats: important reservoir hosts of emerging viruses. Clin Microbiol Rev 19(3):531–545

CDC (2009) Imported case of Marburg hemorrhagic fever, Colorado, 2008. MMWR Morb Mortal Wkly Rep 58(49):1377–1381

CDC (2014) Outbreaks chronology: Marburg fever. https://www.cdc.gov/vhf/marburg/outbreaks/chronology.html. Accessed 2 Dec 2016

CDC (2016) Outbreaks chronology: Ebola virus disease. https://www.cdc.gov/vhf/ebola/outbreaks/history/chronology.html. Accessed 2 Dec 2016

Childs JE, Peters CJ (1993) Ecology and epidemiology of arenaviruses and their hosts. In: Salvato MS (ed) The Arenaviridae. Springer, Plenum, NY, pp 331–384

Cleveland SB, Davies J, McClure MA (2011) A bioinformatics approach to the structure, function, and evolution of the nucleoprotein of the order mononegavirales. PLoS ONE 6(5):e19275

Conrad JL, Isaacson M, Smith EB, Wulff H, Crees M, Geldenhuys P, Johnston J (1978) Epidemiologic investigation of Marburg virus disease, Southern Africa, 1975. Am J Trop Med Hyg 27 (6):1210–1215

Dobson AP (2005) What links bats to emerging infectious diseases? Science 310(5748):628–629

Donnelly CA, Woodroffe R, Cox DR et al (2003) Impact of localized badger culling on tuberculosis incidence in British cattle. Nature 426(6968):834–837

Drexler JF, Corman VM, Gloza-Rausch F et al (2009) Henipavirus RNA in African bats. PLoS ONE 4(7):e6367

Drexler JF, Corman VM, Drosten C (2014) Ecology, evolution and classification of bat coronaviruses in the aftermath of SARS. Antiviral Res 101:45–56

Drexler JF, Corman VM, Muller MA et al (2012) Bats host major mammalian paramyxoviruses. Nat Commun 3:1–12

Feldmann H, Wahl-Jensen V, Jones SM, Stroher U (2004) Ebola virus ecology: a continuing mystery. Trends Microbiol 12(10):433–437

Formenty P, Boesch C, Wyers M et al (1999) Ebola virus outbreak among wild chimpanzees living in a rain forest of Cote d'Ivoire. J Infect Dis 179(Suppl 1):S120–S126

Germain M (1978) Collection of mammals and arthropods during the epidemic of haemorrhagic fever in Zaire. In: Pattyn SR (ed) Ebola virus haemorrhagic fever, Amsterdam 1977. Elsevier, North Holland, pp 185–189

Glass GE, Shields T, Cai B, Yates TL, Parmenter R (2007) Persistently highest risk areas for hantavirus pulmonary syndrome: potential sites for refugia. Ecol Appl 17(1):129–139

Groseth A, Feldmann H, Strong JE (2007) The ecology of Ebola virus. Trends Microbiol 15 (9):408–416

Halpin K, Young PL, Field HE, Mackenzie JS (2000) Isolation of Hendra virus from pteropid bats: a natural reservoir of Hendra virus. J Gen Virol 81(Pt 8):1927–1932

Han BA, Schmidt JP, Alexander LW, Bowden SE, Hayman DTS, Drake JM (2016) Undiscovered bat hosts of filoviruses. PLoS Neglect Trop D 10(7):e0004815

Haydon DT, Cleaveland S, Taylor LH, Laurenson MK (2002) Identifying reservoirs of infection: a conceptual and practical challenge. Emerg Infect Dis 8(12):1468–1473

Hayes CG, Burans JP, Ksiazek TG et al (1992) Outbreak of fatal illness among captive macaques in the Philippines caused by an Ebola-related filovirus. Am J Trop Med Hyg 46(6):664–671

Hayman DT (2015) Biannual birth pulses allow filoviruses to persist in bat populations. Proc Biol Sci 282(1803):20142591

Hayman DT, Yu M, Crameri G, Wang LF, Suu-Ire R, Wood JL, Cunningham AA (2012) Ebola virus antibodies in fruit bats, Ghana. West Afr Emerg Infect Dis 18(7):1207–1209

Hayman DTS, Emmerich P, Yu M et al (2010) Long-term survival of an urban fruit bat seropositive for Ebola and Lagos bat viruses. Plos One 5(8)

He B, Feng Y, Zhang H et al (2015) Filovirus RNA in fruit bats. China Emerg Infect Dis 21 (9):1675

Huijbregts B, De Wachter P, Obiang LSN, Akou ME (2003) Ebola and the decline of gorilla *Gorilla gorilla* and chimpanzee *Pan troglodytes* populations in Minkebe forest, north-eastern Gabon. Oryx 37(04):437–443

Jahrling PB, Geisbert TW, Johnson ED, Peters CJ, Dalgard DW, Hall WC (1990) Preliminary report: isolation of Ebola virus from monkeys imported to USA. The Lancet 335(8688):502–505

Jayme SI, Field HE, de Jong C et al (2015) Molecular evidence of Ebola Reston virus infection in Philippine bats. Virol J 12:107

Johara MY, Field H, Rashdi AM et al (2001) Nipah virus infection in bats (order Chiroptera) in Peninsular Malaysia. Emerg Infect Dis 7(3):439–441

Jones ME, Schuh AJ, Amman BR, Sealy TK, Zaki SR, Nichol ST, Towner JS (2015) Experimental inoculation of Egyptian Rousette bats (*Rousettus aegyptiacus*) with viruses of the *Ebolavirus* and *Marburgvirus* Genera. Viruses 7(7):3420–3442

Kalter SS, Ratner JJ, Heberling RL (1969) Antibodies in primates to the Marburg virus. Exp Biol Med 130(1):10–12

Kingdon J, Happold D, Butynski T, Hoffmann M, Happold M, Kalina J (2013) Hedgehogs, shrews, and bats. In: Happold M, Happold D (eds) Mammals of Africa, vol IV. Bloomsbury, London

Klingstrom J, Heyman P, Escutenaire S, Sjolander KB, Jaegere FD, Henttonen H, Lundkvist A (2002) Rodent host specificity of European hantaviruses: evidence of Puumala virus interspecific spillover. J Med Virol 68(4):581–588

Kobinger GP, Leung A, Neufeld J et al (2011) Replication, pathogenicity, shedding, and transmission of Zaire ebolavirus in pigs. J Infect Dis 204:200–208

Koopman KF (1975) Bats of the Sudan. American Museum of Natural History, Vol 154, New York

Kuhn JH, Bekal S, Yingyun C et al (2013) *Nyamiviridae*: proposal for a new family in the order *Mononegavirales*. Arch Virol 158(10):2209–2226

Kunz C, Hofmann H (1971) Some characteristics of the Marburg virus. In: Martini GA, Siegert R (eds). Springer, Berlin, p 109–111

Lachish S, McCallum H, Mann D, Pukk CE, Jones ME (2010) Evaluation of selective culling of infected individuals to control Tasmanian devil facial tumor disease. Conserv Biol 24(3):841–851

Laddomada A (2000) Incidence and control of CSF in wild boar in Europe. Vet Microbiol 73 (2):121–130

Leendertz SA, Gogarten JF, Dux A, Calvignac-Spencer S, Leendertz FH (2016) Assessing the evidence supporting fruit bats as the primary reservoirs for Ebola viruses. EcoHealth 13:1–8

Leirs H, Mills JN, Krebs JW et al (1999) Search for the Ebola virus reservoir in Kikwit, Democratic Republic of the Congo: reflections on a vertebrate collection. J Infect Dis 179 (Supplement 1):S155–S163

Leroy EM, Epelboin A, Mondonge V, Pourrut X, Gonzalez JP, Muyembe-Tamfum JJ, Formenty P (2009) Human Ebola outbreak resulting from direct exposure to fruit bats in Luebo, Democratic Republic of Congo, 2007. Vector-Borne Zoonot 9(6):723–728

Leroy EM, Kumulungui B, Pourrut X et al (2005) Fruit bats as reservoirs of Ebola. Nature 438:575–576

Leroy EM, Rouquet P, Formenty P et al (2004) Multiple Ebola virus transmission events and rapid decline of central African wildlife. Science 303(5656):387–390

Li CX, Shi M, Tian JH et al (2015) Unprecedented genomic diversity of RNA viruses in arthropods reveals the ancestry of negative-sense RNA viruses. Elife 4:e05378

Luis AD, Hayman DT, O'Shea TJ et al (2013) A comparison of bats and rodents as reservoirs of zoonotic viruses: are bats special? Proc Biol Sci 280(1756):20122753

MacNeil A, Reed Z, Rollin PE (2011) Serologic cross-reactivity of human IgM and IgG antibodies to five species of Ebola virus. PLoS Negl Trop Dis 5(6):e1175

Maganga GD, Bourgarel M, Ella GE, Drexler JF, JP Gonzalez, Drosten C, Leroy EM (2011) Is Marburg virus enzootic in Gabon? J Infect Diss 204(Suppl 3):S800–S803

Marsh GA, Haining J, Robinson R et al (2011) Ebola Reston virus infection of pigs: clinical significance and transmission potential. J Infect Dis 204(Suppl 3):S804–S809

McCormick JB (2004) Ebola virus ecology. J Infect Dis 190(11):1895–1899

Mills JN, Amman BR, Glass GE (2010) Ecology of Hantaviruses and their hosts in North America. Vector-Borne Zoonotic Dis 10(6):563–574

Miranda ME, Miranda NL (2011) Reston ebolaviruses in humans and animals in the Philippines: a review. J Infect Dis 204(Suppl 3):S757–S760

Monath TP (1999) Ecology of Marburg and Ebola viruses: speculations and directions for future research. J Infect Dis 179(Suppl 1):S127–S138

Morvan JM, Deubel V, Gounon P et al (1999) Identification of Ebola virus sequences present as RNA or DNA in organs of terrestrial small mammals of the Central African Republic. Microbes Infect 1:1193–1201

Negredo A, Palacios G, Vazquez-Moron S et al (2011) Discovery of an ebolaviruses-like filovirus in Europe. PLoS Pathog 7(10):e1002304

Ogawa H, Miyamoto H, Nakayama E et al (2015) Seroepidemiological prevalence of multiple species of filoviruses in fruit bats (*Eidolon helvum*) migrating in Africa. J Infect Dis 212(Suppl 2):S101–S108

Olival KJ, Weekley CC, Daszak P (2015) Are bats really 'special' as viral reservoirs? What we know and need to know. In: Wang LF, Cowled C (eds) Bats and viruses: a new frontier of emerging infectious diseases. Wiley, Hoboken, NJ, p 281–294

Olival KJ, Hayman DT (2014) Filoviruses in bats: current knowledge and future directions. Viruses 6(4):1759–1788

Olival KJ, Epstein JH, Wang LF, Field HE, Daszak P (2012) Are bats exceptional viral reservoirs. In: Aguirre AA, Ostfeld RS, Daszak P (eds) New directions in conservation medicine: applied cases of ecological health, 2nd edn. Oxford University Press, Oxford, UK, pp 195–212

Olival KJ, Islam A, Yu M et al (2013) Ebola virus antibodies in fruit bats, Bangladesh. Emerg Infect Dis 19(2):270–273

Olson SH, Reed P, Cameron KN et al (2012) Dead or alive: animal sampling during Ebola hemorrhagic fever outbreaks in humans. Emerg Health Threats J 5:9134. doi:10.3402/ehtj.v5i0.9134

Pan Y, Zhang W, Cui L, Hua X, Wang M, Zeng Q (2014) Reston virus in domestic pigs in China. Arch Virol 159(5):1129–1132

Paweska JT, Jansen van Vuren P, Fenton KA et al (2015) Lack of Marburg virus transmission from experimentally infected to susceptible in-contact Egyptian Fruit bats. J Infect Dis 212 (Suppl 2):S109–S118

Paweska JT, Jansen van Vuren P, Masumu J et al (2012) Virological and serological findings in *Rousettus aegyptiacus* experimentally inoculated with vero cells-adapted Hogan strain of Marburg virus. PLoS ONE 7(9):e45479

Paweska JT, Storm N, Grobbelaar AA, Markotter W, Kemp A, Jansen van Vuren P (2016) Experimental inoculation of Egyptian Fruit bats (*Rousettus aegyptiacus*) with Ebola virus. Viruses 8(2) (2016 Jan 22)

Peterson AT, Bauer JT, Mills JN (2004) Ecologic and geographic distribution of filovirus disease. Emerg Infect Dis 10:40–47

Peterson AT, Lash RR, Carroll DS, Johnson KM (2006) Geographic potential for outbreaks of Marburg hemorrhagic fever. Am J Trop Med Hyg 75(1):9–15

Pigott DM, Golding N, Mylne A et al (2014) Mapping the zoonotic niche of Ebola virus disease in Africa. Elife 3:e04395

Pigott DM, Golding N, Mylne A et al (2015) Mapping the zoonotic niche of Marburg virus disease in Africa. Trans R Soc Trop Med Hyg:trv024

Pourrut X, Kumulungui B, Wittmann T et al (2005) The natural history of Ebola virus in Africa. Microbes Infect 7(7–8):1005–1014

Pourrut X, Delicat A, Rollin PE, Ksiazek TG, Gonzalez JP, Leroy EM (2007) Spatial and temporal patterns of Zaire ebolavirus antibody prevalence in the possible reservoir bat species. J Infect Dis 196(Suppl 2):S176–S183

Pourrut X, Souris M, Towner JS, Rollin PE, Nichol ST, Gonzalez JP, Leroy E (2009) Large serological survey showing cocirculation of Ebola and Marburg viruses in Gabonese bat populations, and a high seroprevalence of both viruses in *Rousettus aegyptiacus*. BMC Infect Dis 9(1):159

Ramsden C, Holmes EC, Charleston MA (2009) Hantavirus evolution in relation to its rodent and insectivore hosts: no evidence for codivergence. Mol Biol Evol 26(1):143–153

Reiter P, Turell M, Coleman R et al (1999) Field investigations of an outbreak of Ebola hemorrhagic fever, Kikwit, Democratic Republic of the Congo, 1995: arthropod studies. J Infect Dis 179(Suppl 1):S148–S154

Saez AM, Weiss S, Nowak K et al (2015) Investigating the zoonotic origin of the West African Ebola epidemic. EMBO Mol Med 7(1):17–23

Schuh AJ, Amman BR, Apanaskevich DA, Sealy TK, Nichol ST, Towner JS (2016) No evidence for the involvement of the argasid tick *Ornithodoros faini* in the enzootic maintenance of marburgvirus within Egyptian rousette bats *Rousettus aegyptiacus*. Parasit Vectors 9:128

Schuh AJ, Amman BR, Jones MEB et al. (2017) Modeling filovirus maintenance in nature by experimental transmission of Marburg virus between Egyptian Rousette bats. Nat Commun 8:14446

Shirai J, Sohayati AL, Mohamed Ali AL, Suriani MN, Taniguchi T, Sharifah SH (2007) Nipah virus survey of flying foxes in Malaysia. Jarq-Jpn Agric Res Q 41(1):69–78

Siegert R, Shu HL, Slenczka HL, Peters D, Muller G (1968) The aetiology of an unknown human infection transmitted by monkeys (preliminary communication). Ger Med Mon 13(1):1–2

Slenczka W, Klenk, HD (2007) Forty years of marburg virus. J Infect Dis 196 Suppl 2:S131–S135

Smith DH, Johnson BK, Isaacson M et al (1982) Marburg-virus disease in Kenya. Lancet 1 (8276):816–820

Smith I, Wang LF (2013) Bats and their virome: an important source of emerging viruses capable of infecting humans. Curr Opin Virol 3(1):84–91

Swanepoel R (2004) Rabies. In: Coetzer JAW, Tustin RC (eds) Infectious diseases of livestock, 2nd edn. Oxford University Press Southern Africa, Cape Town, pp 1123–1182

Swanepoel R, Leman PA, Burt FJ et al (1996) Experimental inoculation of plants and animals with Ebola virus. Emerg Infect Dis 2(4):321–325

Swanepoel R, Smit SB, Rollin PE et al (2007) Studies of reservoir hosts for Marburg virus. Emerg Infect Dis 13(12):1847–1851

Taniguchi S, Watanabe S, Masangkay JS et al (2011) Reston ebolavirus antibodies in bats, the Philippines. Emerg Infect Dis 17(8):1559–1560

Timen A, Koopmans MP, Vossen AC et al (2009) Response to imported case of Marburg hemorrhagic fever, the Netherlands. Emerg Infect Dis 15(8):1171–1175

Towner JS, Amman BR, Sealy TK et al (2009) Isolation of genetically diverse Marburg viruses from Egyptian fruit bats. PLoS Pathog 5(7):e1000536

Towner JS, Khristova ML, Sealy TK et al (2006) Marburgvirus genomics and association with a large hemorrhagic fever outbreak in Angola. J Virol 80(13):6497–6516

Towner JS, Pourrut X, Albarino CG et al (2007) Marburg virus infection detected in a common African bat. PLoS ONE 2(1):e764

Walsh PD, Abernethy KA, Bermejo M et al (2003) Catastrophic ape decline in western equatorial Africa. Nature 422(6932):611–614

WHO (2009) WHO experts consultation on Ebola Reston pathogenicity in humans. p 1–4

WHO/International Study Team (1978) Ebola haemorrhagic fever in Zaire, 1976. Bull World Health Organ 56(2):271–293

Wilson DE, Reeder DM (2005) Mammal species of the world (3 edn). Johns Hopkins University Press, Baltimore

Wolfe ND, Daszak P, Kilpatrick AM, Burke DS (2005) Bushmeat hunting, deforestation, and prediction of zoonotic disease. Emerg Infect Dis 11(12):1822–1827

Yuan J, Zhang Y, Li J, Zhang Y, Wang LF, Shi Z (2012) Serological evidence of ebolavirus infection in bats. China Virol J 9(1):1

West Africa 2013 Ebola: From Virus Outbreak to Humanitarian Crisis

Daniel G. Bausch

This article is dedicated to the many health workers who sacrificed their time, energy and, all too often their lives, to combat the Ebola virus disease outbreak in West Africa (Bausch et al. 2014).

Abstract The 2013 outbreak of Ebola virus disease (EVD) in West Africa constituted a major humanitarian crisis. The outbreak numbered over 28,500 cases, more than 10 times the number cumulatively registered from all previous EVD outbreaks combined, with at least 11,000 deaths, and resulted in billions of dollars of lost economic growth to an already impoverished region. The unprecedented scale of West Africa 2013 took the world by surprise and laid bare deficiencies in our response capacity to complex humanitarian disasters of highly infectious and lethal pathogens. However, the magnitude of West Africa 2013 also provided a unique opportunity and obligation to better understand not only the biology and epidemiology of EVD, but also the many scientific, economic, social, political, ethical, and logistical challenges in confronting emerging infectious diseases in the modern era.

Contents

D. G. Bausch (✉)
Department of Tropical Medicine, Tulane School of Public Health and Tropical Medicine, 1440 Canal St, Suite 2309, MC#8317, New Orleans, LA 70112, USA
e-mail: Daniel.Bausch@lshtm.ac.uk

D. G. Bausch
UK Public Health Rapid Support Team-UK PHRST, Public Health England, Wellington House, 133-155 Waterloo Road, London SE1 8UG, UK

Current Topics in Microbiology and Immunology (2017) 411:63–92
DOI 10.1007/82_2017_69

Published Online: 26 October 2017

1 First Case and Early Spread

In early December 2013, a 2-year-old boy in the village of Méliandou in the remote, largely deforested, and resource-poor Prefecture of Guéckédou, Republic of Guinea, fell sick with fever, vomiting, and black stool, dying a few days later (WHO 2014a; Baize et al. 2014; Mari Saez et al. 2014) (Figs. 1 and 2). The boy was reported to have previously played in a hollow tree housing a colony of the insectivorous bat *Mops condylurus*, a possible Ebola virus (EBOV) reservoir (Mari Saez et al. 2014). Over the following weeks and months, multiple waves of transmission and disease occurred in family members, healthcare workers (HCWs) who cared for them, and persons with contact with corpses during funeral rituals. On March 21, 2014, Guinean health authorities notified the World Health Organization (WHO) of a "rapidly evolving outbreak." On March 23, EBOV infection was confirmed on patient samples sent to Biosafety Level 4 laboratories in Lyon, France, and Hamburg, Germany, and an outbreak of Ebola virus disease (EVD) was declared. Unbeknownst to all at the time, at least 49 cases with multiple but often poorly defined chains of transmission had already occurred in Guinea and the virus had already slipped across the border, smoldering quietly at first, into neighboring Liberia and Sierra Leone (WHO 2014b, 2015c).

2 Virus Introduction

Five members of the *Filoviridae* family are known to cause disease in humans: EBOV (formerly designated Zaire Ebola virus), Bundibugyo, Sudan, and Taï Forest viruses of the genus *Ebola virus* and Marburg and Ravn viruses of the genus *Marburg virus*, with characteristic geographic distributions (Table 1 and Fig. 3). Prior to 2013, only a single case of EVD had been reported in West Africa, due to Taï Forest virus (Formenty et al. 1999a, b). When EVD appeared in neighboring Guinea in 2013, most experts predicted that Taï Forest virus would again be the culprit, and thus were surprised when the causative virus turned out to be a new variant of EBOV, subsequently named Makona after the Makona River in Guinea, close to the border with Liberia and Sierra Leone (Baize et al. 2014). How EBOV, which had never been noted outside of the Congo Basin of Equatorial Africa, found its way to West Africa remains unknown (Bausch and Schwarz 2014). Migration and transmission from infected bats, putative EBOV reservoirs, is considered the most likely modality (Mari Saez et al. 2014; Bausch and Schwarz 2014). Comparative phylogeographic analysis suggests that fruit bats of three species are

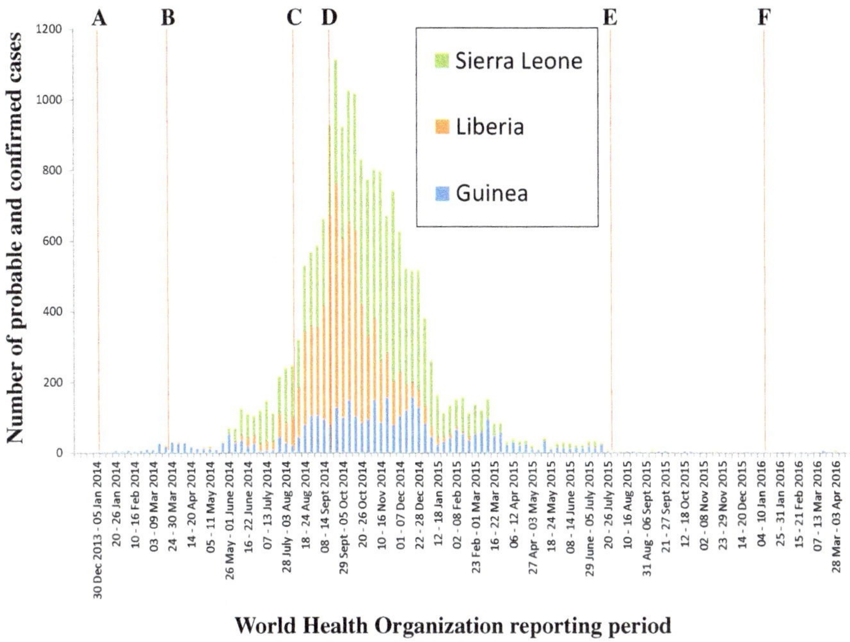

Fig. 1 Epidemiologic curve of the West Africa 2013 Ebola virus disease (EVD) outbreak. The dashed-vertical lines indicate key events during the outbreak: **a** First suspected case in Méliandou, *Guinea*, **b** Laboratory confirmation of EVD and disease reported by Guinean Health Authorities, **c** WHO declares Public Health Emergency of International Concern, **d** U.S. President Obama announces major initiative to help control EVD in *Liberia*; Creation of the United Nations Mission for Ebola Emergency, **e** Publication of preliminary results from first EVD Phase III vaccine efficacy study (rVSIV-EBOV), **f** Publication of preliminary results of first EVD Phase III therapeutic efficacy trial (convalescent plasma). *Adapted* from WHO Ebola Response Roadmap Situation Reports with publically available data. World Health Organization

theoretically capable of dispersing EBOV directly from the Congo Basin to Guinea: African straw-colored fruit bats (*Eidolon helvum*), hammer-headed fruit bats (*Hypsignathus monstrosus*), and Egyptian fruit bats (*Rousettus aegyptiacus*), although definitive evidence of EBOV infection in any type of bat anywhere has yet to be found (Hassanin et al. 2016).

Epidemiological and phylogenetic analyses during and after the West Africa EVD outbreak are consistent with all cases stemming from a single introduction of the Makona variant in Guinea in 2013 (Kuhn et al. 2014). This finding is consistent with most other EVD outbreaks, which generally result from a single introduction from the wild followed by amplification exclusively through human-to-human transmission. Molecular clock dating analyses suggest that the Makona variant diverged from other EBOVs only about a decade ago (Holmes et al. 2016). Also due to EBOV, sequence analysis makes clear that the much smaller outbreak of EVD that occurred in the Democratic of the Republic of the Congo in 2014 was a separate event unrelated to the outbreak in West Africa.

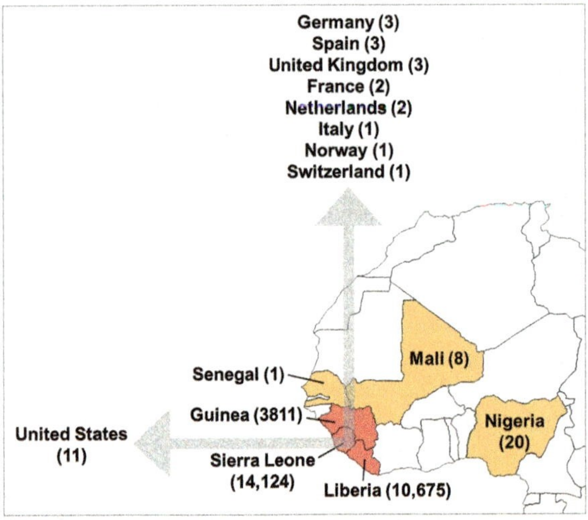

Fig. 2 Map of West Africa showing the epicenter of the 2013–2016 outbreak of Ebola virus disease (*red*) and imported cases (*orange* and *arrows*). The village of Meliandou, *Guinea*, where the first case is thought to have occurred, is indicated by a *blue star*. The total number of cases seen in each country is shown in parentheses. Reprinted with permission from Bausch and Rojek (2016)

3 Underlying Geopolitical, Social, and Cultural Factors

There has been much speculation and extensive discussion in the literature regarding the causes for the unprecedented size of West Africa 2013 (Moon et al. 2015; Bausch and Rojek 2016). The outbreak's roots are undoubtedly multifactorial, entailing a complex web of interrelated social, cultural, ecologic, and economic determinants viewed in the context of the overall geopolitical history of the region (Table 2). Many of these factors and challenges have been encountered in previous EVD outbreaks, but certainly not on the scale and with the intensity noted in West Africa.

4 A Failed Response

In the absence of effective therapeutics and vaccines (a work in progress—see below), control of EVD outbreaks is almost completely based on the classic control measures of thorough surveillance for case identification, isolation, and care in the setting of sound infection prevention and control (IPC) practices; contact tracing; and safe burials, all enhanced by effective social mobilization and public education campaigns, and coordinated through a partnership between the national government, WHO, and other international stakeholders. The international community and the governments they supported were accustomed to success in this approach, generally ending outbreaks after a few hundred cases within a few months, and indeed certainly anticipated the same result in West Africa (Table 3). However, a tragically different scenario played out.

Table 1 Laboratory-confirmed outbreaks of Ebola virus disease since discovery of the virus in 1976 through September 2017. Cases related to laboratory infections are not shown

Year of onset	Virus	Country	Epicenter (s)	# Cases (CFR [%])	Source of primary infection	Factors contributing to secondary spread	# Cases in healthcare workers[a]
1976	Ebola (formerly designated Zaire Ebola virus)	Zaire (current DRC)	Yambuku	318 (88)	Unknown	Nosocomial transmission	≥ 13
1976	Sudan	Sudan (current South Sudan)	Maridi and Nzara	284 (53)	Unknown	Nosocomial transmission	70
1977	Ebola	Zaire (current DRC)	Tandala	1 (100)	Unknown	None	0
1979	Sudan	Sudan (current South Sudan)	Maridi and Nzara	34 (65)	Unknown	Nosocomial transmission	≥ 2
1994	Ebola	Gabon	Mékouka, Ogooué-Ivindo Province	52 (60)	Infection in gold mining camps	Traditional healing practices, Nosocomial and community-based transmission	None reported
1994	Taï Forest	Côte d'Ivoire	Taï Forest	1 (0)	Scientist conducting autopsy on wild chimpanzee	None	0
1995	Ebola	Zaire (current DRC)	Kikwit	315 (81)	Unknown	Nosocomial and community-based transmission	None reported

(continued)

Table 1 (continued)

Year of onset	Virus	Country	Epicenter (s)	# Cases (CFR [%])	Source of primary infection	Factors contributing to secondary spread	# Cases in healthcare workers[a]
1996	Ebola	Gabon	Mayibout, Ogooué-Ivindo Province	21 (57)	Consumption of dead chimp	Community-based transmission	None reported
1996	Ebola	Gabon, South Africa[b]	Booué, Ogooué-Ivindo Province	62 (75)	Consumption of chimp?	Nosocomial and community-based transmission	1 (South Africa)
2000	Sudan	Uganda	Gulu	425 (53)	Unknown	Nosocomial and community-based transmission, traditional burial practices	≥ 3
2001	Ebola	Gabon and ROC	Ogooué-Ivindo Province (Gabon)	65 (82)	Hunting and consumption of nonhuman primates	Nosocomial transmission and community-based transmission, Traditional healing practices	2
2001	Ebola	ROC and Gabon	Cuvette Ouest Region (ROC)	57 (75)	Unknown	Community-based transmission	0
2002	Ebola	ROC	Mbomo and Kéllé, Cuvette Ouest Region	143 (89)	Hunting and consumption of nonhuman primates	Nosocomial and community-based transmission, Traditional healing practices	None reported

(continued)

Table 1 (continued)

Year of onset	Virus	Country	Epicenter (s)	# Cases (CFR [%])	Source of primary infection	Factors contributing to secondary spread	# Cases in healthcare workers[a]
2003	Ebola	ROC	Mbomo and Mbandza, Cuvette Ouest Region	35 (83)	Hunting and consumption of nonhuman primates	Traditional healing practices	None reported
2004	Sudan	Sudan (present South Sudan)	Yambio	17 (41)	Exposure to baboon meat?	Nosocomial transmission and community-based transmission	None reported
2007	Ebola	DRC	Kasai Occidental Province	264 (71)	Exposure to local wildlife, including bats	Nosocomial and community-based transmission	None reported
2007	Bundibugyo	Uganda	Bundibugyo	149 (25)	Unknown	Nocosomial transmission and community-based transmission	None reported
2008	Ebola	DRC	Mweka and Luebo	32 (47)	Exposure to fruit bats through hunting?	Unknown	None reported
2011	Sudan	Uganda	Luwero	1 (100)	Unknown	None	0
2012	Sudan	Uganda	Kibaale	11 (36)	Unknown	Community-based transmission	None reported
2012	Bundibugyo	DRC	Province Orientale	36 (36)	Hunted bushmeat?	Community-based transmission	≥ 13
2012	Sudan	Uganda	Luwero	6 (50)	Unknown	Unknown	None reported

(continued)

Table 1 (continued)

Year of onset	Virus	Country	Epicenter (s)	# Cases (CFR [%])	Source of primary infection	Factors contributing to secondary spread	# Cases in healthcare workers[a]
2013	Ebola	Multiple, mostly Republic of Guinea, Liberia, and Sierra Leone	Guéckédou, Guinea	> 28,616 (40)	Unknown, suspected exposure to bats	Nosocomial and community-based transmission, unsafe burial practices	≥ 874
2014	Ebola	DRC	Province Equateur	66 (74)	Hunted bushmeat?	Community-based transmission	≥ 8
2017	Ebola	DRC	Province Bas-Uele	8 (50)	Unknown	Community-based transmission	0

Abbreviations CFR—case fatality rate, DRC—Democratic Republic of the Congo, ROC—Republic of the Congo

[a]May include cleaners and other ancillary staff working in Ebola Treatment Units

[b]In addition to the 60 cases with 45 deaths in Gabon, two cases occurred in South Africa resulting from an infected traveler from Gabon, with subsequent fatal nosocomial infection in a nurse in Johannesburg

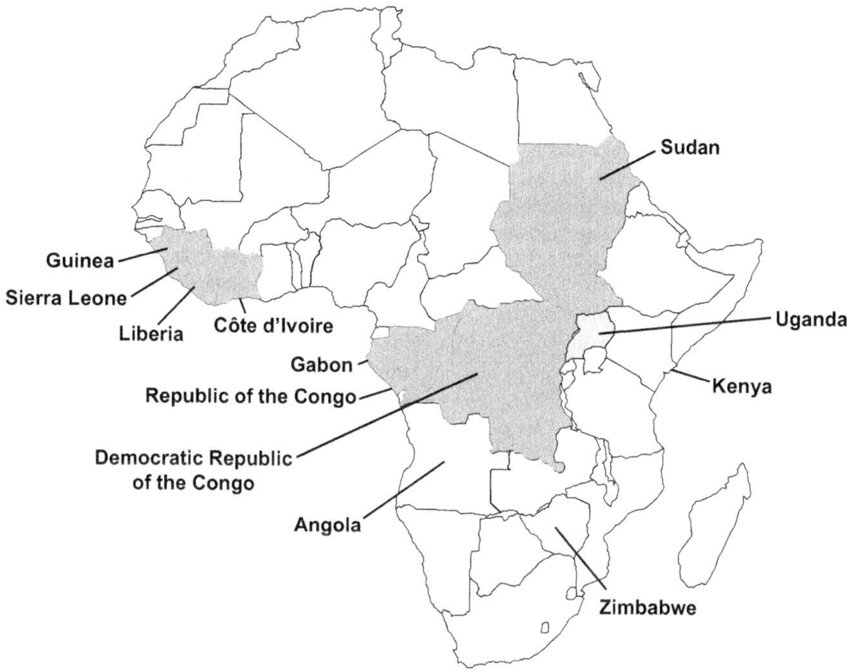

Fig. 3 Endemic areas for Filo viruses. Only Filo viruses known to cause hemorrhagic fever are shown. *Gray* indicates countries where Ebola virus disease has been seen and *diagonal lines* countries with Marburg virus disease. Incidence and risk of disease may vary significantly within each country

Due to a preexisting health project, the nongovernmental organization Médecins Sans Frontières (MSF) was present at the epicenter of the EVD outbreak in Guinea even before its onset. On March 31, seven days after the EVD outbreak was declared, MSF warned of an EVD outbreak unprecedented in magnitude and distribution. Although many international organizations, including WHO, the International Federation of Red Cross and Red Crescent Societies, the U.S. Centers for Disease Control and Prevention, the European Union, and UNICEF, quickly joined MSF and national partners in the three West African countries to mount outbreak response operations, this early effort would prove to be too slow, piecemeal, and disorganized to prevent what would ultimately become a major humanitarian crisis. The shortcomings in the international response have been extensively discussed in the literature, especially with regard to WHO who, although engaged in outbreak control activities since the beginning, did not formally declare the outbreak to be a Public Health Emergency of International Concern, as outlined under International Health Regulations, until August 8, 2014 (Bausch and Rojek 2016) (Fig. 1).

Table 2 Underlying geopolitical, social, and cultural factors implicated in the 2013–2016 West Africa Ebola virus outbreak and components of the failed international response (subject extensively reviewed in Bausch and Rojek (2016))

Factor	Antecedents and comments
Guinea, Liberia, and Sierra Leone	
Resource-poor countries with fragile healthcare, disease surveillance and response systems, physical infrastructure, and supply chains	Fledgling and fragile governments and economies emerging from civil war and unrest. Three countries rank near the bottom of the 187 nations on the United Nations Development Programme Human Development Index, with majority of population living below national poverty line
High population density	Population density more than twice and surface area much smaller and more navigable compared to areas of Equatorial Africa where EVD typically noted (Fig. 6). Longer distance between the outbreak epicenter and surrounding populations may impede virus transmission
Highly mobile population	Common ethnic groups spanning contiguous borders of the three countries who cross regularly for both social and economic motives. Weak border control and little history of cross-border government communication and surveillance made more challenging by highly polyglot population and different national languages (French in Guinea, English in Liberia and Sierra Leone). Also frequent travel between remote rural and major urban areas due to relatively short distances and low cost of transport (Fig. 7)
Insufficient healthcare workforce prior to outbreak	Doctors per 10,000 population: Guinea-940, Sierra Leone-95, Liberia-51. Amongst the lowest health workforce coverage in the world
Loss of healthcare workers to EVD or fear of EVD	Estimated over 900 cases of EVD with 500 deaths in healthcare workers. Undeveloped medical culture and infrastructure for infection prevention and control, even for simple necessities such as soap, clean water, and sterile needles
Cultural beliefs and behavioral practices leading to risk of infection	Traditional healing and unsafe burial practices that may hold extreme cultural significance but may often involve high-risk behaviors, such as touching the corpse
Historical distrust of authority	History of slave trade and colonialism followed by failed government and civil war/ unrest after independence, leading to resistance to control measures, sometimes culminating in violence

(continued)

Table 2 (continued)

Factor	Antecedents and comments
International community	
Shallow pool of experts in the management of EVD	Small number of experienced personnel in organizations typically responding to EVD outbreaks, further complicated by frequent turnover within organizations and consequent loss of institutional memory. Long outbreak duration exhausted availability of experienced personnel
Slow and disorganized response	International community generally slows to appreciate gravity of situation and differences from previous EVD outbreaks. World Health Organization delay in declaration of Public Health Emergency of International Concern (see text for details)
Insufficient supply and confusing directions for use of personal protective equipment	No organized supply chain or surge capacity. Varied and sometimes conflicting approaches and guidelines for use from international partners (see text for details) (Fig. 4)
Insufficient funding for global health preparedness and outbreak response	Frequency of outbreaks increasing but minimal growth in funding for global health preparedness and response since 2009. Funding for World Health Organization plateaued or decreased since 2010
Poorly crafted public health messages and failure to engage community as partner and stakeholder	EVD often described as "incurable, with no treatment or vaccine," enhancing fear and potentially discouraging early presentation. Failure to understand extreme cultural importance of funeral rituals, sometimes compounded by efforts to drastically change centuries-old customs (such as mandating cremation), often setting community opposed to outbreak control efforts. Non-evidence-based policies, sometimes politically motivated, such as mandatory quarantine of persons returning from West Africa regardless of possible exposures or symptoms, posed challenges to recruiting international volunteers
Ease of international air travel	Despite exit and entry screening, air travel resulted in exportation of cases to Nigeria, the United States, and various countries in Europe (Fig. 2)

Abbreviations EVD, Ebola virus disease

The dramatic rise in cases in West Africa, projections of millions of cases if an aggressive response was not mounted (Meltzer et al. 2014), and increasing numbers of imported cases into surrounding countries in West Africa, the United States, and Europe, finally stirred the international community to more concerted action.

Table 3 Comparison of various parameters and possible determinants of previous outbreaks of Ebola virus disease and the 2013–2016 outbreak in West Africa

Aspect	Previous outbreaks	West Africa 2013–2016
Location	Remote areas of Equatorial Africa/Congo Basin	More connected regions of West Africa
Number of cases	Typically 100–200. Largest on record prior to West Africa 2013–2016 was 425	28,616 officially recorded, widely thought to be a underestimate
Time from first clinical case until laboratory confirmation and outbreak declaration	Mean of ~2 months	~3½ months
Number of countries involved at a time	Usually 1, sometimes extending across a contiguous border	Widely distributed across 3 countries in West Africa, with imported cases to 3 other countries in Africa, 8 countries in Europe, and the United States
Duration	2–4 months	2½ years
Exported cases outside area	Extremely rare	27 exported or medically evacuated cases to countries outside West Africa (Fig. 2)
Community cooperation	Occasional resistance	Frequent resistance
Organizations responding	5–10	>100
Urban areas	Spared	Heavily involved
Number of Ebola treatment units	Typically 1–2	>50
Number of diagnostic labs	Typically 1–2	>50
Cost	<$5 million	>$3.6 billion in response, with $2.2 lost in gross domestic product in West Africa

Responses generally aligned with historical connections between the United States and European countries and their colonial-era African counterparts. In September 2014, United States President Obama committed to the construction of seventeen 100-bed Ebola Treatment Units (ETUs) in Liberia, deployment of up to 3000 medical military and support personnel, and support to train 500 HCWs a week. The United Kingdom and France soon followed with commitments to combat EVD in their ex-colonies of Sierra Leone and Guinea, respectively.

Ultimately, a vast array of at least 100 government and nongovernmental organizations, including over 5000 military personnel, contributed to the outbreak response, establishing over 70 ETUs, 800 community care centers, a vast network of over 50 laboratories (generally providing reliable diagnostic results within 24 h after receipt of a specimen), and an extensive surveillance and contact tracing operation across the three implicated countries. However, the response remained agonizingly slow, hampered by the logistical challenges of operationalizing work in the poorest countries in the world with fledging governments and poor infrastructure. The dispersal of cases across three West Africa countries, and in both remote rural and densely populated urban areas, ultimately presented too many "battle fronts," outstripping both local and international capacities. Beds for patients with EVD (Table 4), HCWs to care for them, and field workers to undertake surveillance and contact tracing were woefully insufficient. Consequently, highly infectious patients remained untreated in the community and patients who were admitted to the drastically understaffed ETUs could expect little more than palliative care. Even after laboratories began being rapidly established, the steep increase in the number of samples exceeded local diagnostic capacities in many areas until well into the outbreak. With cases of EVD in HCWs mounting, some ETUs opted to enhance safety by proscribing close contact with patients, including the very controversial measure of not placing IVs for fluid repletion. This move, while perhaps unavoidable, likely further undermined the local population's already shaky faith in the response operation.

Coordination of the outbreak response was a major challenge, with each organization largely acting independently or in bilateral concert with the government. In August 2014, the United Nations appointed a Special Envoy on Ebola, followed by the creation in September 2014 of a coordination body, the United Nations Mission for Ebola Emergency, headquartered in Ghana (Fig. 1). Opinions vary on the efficacy of these measures. Without doubt, the enormous scale and complexity of the outbreak, and the sheer number of organizations involved (far more than had ever been involved in an EVD outbreak before and at times compounded by historical frictions between them) made coordination a substantial challenge.

Although HCW infections have occurred in virtually every EVD outbreak to date (Table 1), prior to West Africa 2013, they were relatively uncommon once international support and resources arrived to assist with establishing ETUs and appropriate IPC measures. In contrast, one of the tragic consequences of the poor baseline infrastructure and failed response West Africa 2013 was EBOV infection in over 900 HCWs during the outbreak, including two cases contracted in the United States and one in Spain, with over 500 deaths (WHO 2015a). In the vast majority of HCW infections, no clear exposure risk, such as a needle stick or blood splash to mucous membranes, could be identified. The high number of HCW

Table 4 Bed capacity and bed requirements for patients with Ebola virus disease in West Africa in October, 2014. Bed capacity in each district was planned on the basis of a need assessment carried out by the relevant Ministry of Health. *Source* WHO: Ebola Response Roadmap Situation Report, October 8, 2014, World Health Organization

Country	Current number of beds	Estimated number of beds required	Current capacity/estimated demand (%)
Guinea	160	210	76
Liberia	620	2930	21
Sierra Leone	304	1148	26

infections engendered speculation that EBOV Makona is more transmissible than other variants, although no supporting data for this theory are available.

Although IPC entails many diverse measures, many of which were inadequate during the outbreak, most of the focus has been on the issue of personal protective equipment (PPE). In addition to shortages of PPE, especially early in the outbreak, the situation was complicated by the diversity of PPE types advocated by different groups, causing significant confusion with training and safe use (Bausch and Rojek 2016; WHO 2008; Franklin 2016) (Fig. 4). Although PPE guidelines were eventually produced, which for the first time included technical specifications for PPE, the lack of evidence precluded a consensus on efficacy (Hersi et al. 2015). The procedure for doffing contaminated PPE, often considered a confusing and a vulnerable point for infection, is logically a focus of attention but, again, no data are available. Furthermore, it is not always clear the HCW infection resulted from exposure in the ETU. Local HCWs are members of the same communities where EBOV may be circulating during an outbreak, and thus may share many of the same risks. There are also many anecdotal reports of HCWs seeing patients in their homes, where the uses of full PPE and other IPC measures are unlikely to be adequate (Faye et al. 2015; Brainard et al. 2016). More in-depth investigations are needed, and indeed are ongoing, to reveal vulnerable points for HCW infection in the care of EVD, and to develop evidence-based uniform PPE standards to protect them.

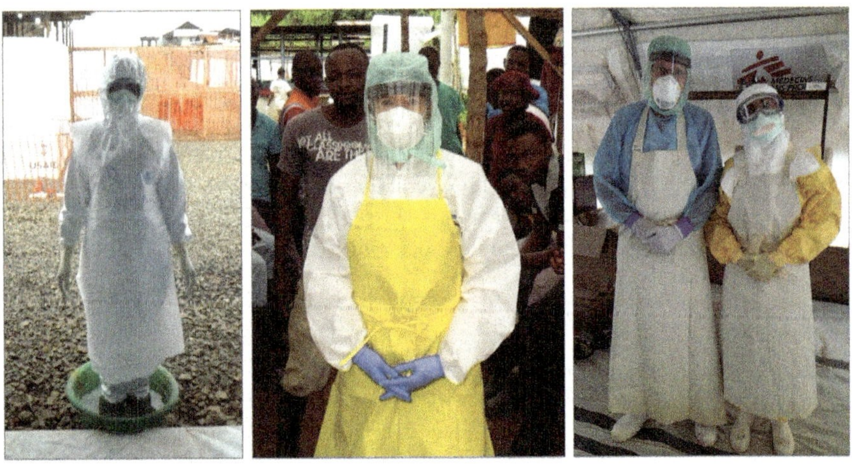

Fig. 4 Examples of various types of personal protective equipment used during the care of patients with Ebola virus disease during the 2013–2016 outbreak in West Africa. The equipment shown are for demonstration only, and should not be construed as implying as advocating or confirming efficacy for any specific equipment. Photos by Thomas Fletcher and Frederique Jacquerioz

5 Virus Evolution During the Outbreak

Whole genome sequencing was performed on hundreds of samples from West Africa 2013, a far greater number than had been sequenced from all previous EVD outbreaks combined (Gire et al. 2014; Park et al. 2015; Matranga et al. 2014). There are conflicting reports and considerable controversy over whether EBOV Makona evolved genetically more rapidly during West Africa 2013 relative to the causative viruses of other EVD outbreaks (Holmes et al. 2016) and whether virus adaptation to humans occurred (Li et al. 2016; Urbanowicz et al. 2016). What is clear, however, is that the prolonged course of the outbreak provided sufficient time for the emergence of at least three distinct viral lineages (Gire et al. 2014; Urbanowicz et al. 2016; Carroll et al. 2015; Quick et al. 2016; Simon-Loriere et al. 2015). Most of these major lineages circulated locally, with only sporadic cross-border transmission.

The highest level of EBOV genetic amino acid diversity generated during West Africa 2013 occurred in the EBOV glycoprotein (GP) (Holmes et al. 2016; Ning et al. 2017). Because of its key role in virus–host interactions, with the potential for altered interaction with the EBOV host receptor Niemann-Pick C1, GP sequence variation is of particular interest. In the laboratory, minor changes in the GP have been shown to impact viral entry into cells from different mammalian species (Ning et al. 2017). Pseudo-typed virion particles incorporating synthetically generated amino acid substitutions observed during the outbreak more efficiently entered human cells, with possible implications for viral fitness, host specificity, and transmissibility (Urbanowicz et al. 2016). Early in the West Africa outbreak, a variant in lineage SL2 emerged with sequence changes in the GP receptor-binding site (Holmes et al. 2016).

Despite evidence for a degree of EBOV evolution during West Africa 2013, no clear phenotypic significance (i.e., changes in transmissibility, virulence, antigenicity, or influence on the efficacy of diagnostic assays, vaccines, or therapeutics) has been noted between the lineages of EBOV Makona or between Makona and other variants of EBOV; the range of reported case fatality rates in West Africa (31–76%), calculated basic reproduction number R_0 (1.5–2.5), and duration of virus shedding are comparable to those noted in previous outbreaks, with no evidence for heritable changes during the course of the outbreak (Holmes et al. 2016; Bausch and Rojek 2016). An EBOV vaccine developed against the 1995 Kikwit variant of EBOV was 100% protective against the Makona variant in both animal models (Marzi et al. 2015a) and a Phase III trial in humans (Henao-Restrepo et al. 2016), suggesting that the genetic differences between the two virus variants did not result in significant differences in immunogenicity. Nor have laboratory studies shown EBOV Makona to be appreciably different from other variants of EBOV with regard to virulence in nonhuman primates (Marzi et al. 2015b), entry into cells (Dunham et al. 2015; Hofmann-Winkler et al. 2015), or detection via polymerase chain reaction assays (Sozhamannan et al. 2015). One exception, however, is the results from experiments in humanized laboratory mice, in which a longer mean-time-to-death was noted with EBOV Makona compared to the 1976 Yambuku variant (Bird et al. 2016), leading to speculation that the seemingly high attack rates in West Africa 2013 could be the result of prolonged virus shedding and thus opportunity for transmission. Despite high-profile speculation to the contrary, there is no evidence or reason to believe that EBOV Makona or any other EBOV variant or even any other Ebola virus has or will evolve naturally to be capable of aerosol or airborne transmission (Osterholm et al. 2015).

6 A Heavy Toll Before Final Outbreak Control

The EVD outbreak in West Africa would ultimately last 3 years and officially result in over 28,616 cases and 11,310 deaths (numbers widely considered underestimates), eclipsing by far and by every measure all previous EVD outbreaks combined (Table 1). In addition to the toll in terms of cases counted and lives lost, the outbreak resulted in billions of dollars in lost economic growth in West Africa (Bank TW 2016), upward of 3500 orphaned children (UNICEF 2016), delayed or impaired child development since school was canceled for a year, widespread job loss resulting in economic and food insecurity, and deep but less easily measurable mental health and sociocultural impacts. Furthermore, as the region's resources were funneled to EVD, an estimated 10,000+ excess deaths occurred due to untreated malaria, HIV/AIDS, and tuberculosis (Parpia et al. 2016). Reductions in vaccination coverage and a rise in teenage pregnancy were also noted (Elston et al. 2015). In addition to the cases in Africa, there were 27 cases, 5 of which were fatal, imported, or medically evacuated to the United States and Europe (Fig. 2) (System Ebola Epidemiology Team IM 2014; WHO 2015b)—a pittance compared to the massive humanitarian disaster in West Africa, but a situation that nevertheless fomented considerable panic and expenditure of resources in those industrialized areas of the world.

The success of EVD outbreak control measures is highly dependent on community engagement to arrive at a common understanding of the nature of the disease threat and cooperation with the plan for control. A combination of historical distrust of authority and slow and poor messaging conspired to impede such engagement, understanding, and cooperation in many communities during West Africa 2013, providing a major impediment. Whether the major factors in ultimate control of the outbreak were the classic response measures implemented by national and international public health agencies or a more grassroots autonomous behavioral adaptation of the indigenous population is a matter of debate, and certainly varies by country and community. In some cases, additional measures were implemented, such as quarantine of affected villages and exit and entrance screening for fever and other clinical manifestations of EVD of travelers between affected regions and in airports, often controversially and with uncertain effect. One novel control method that may have had a significant impact on transmission in the later stages of the outbreak was vaccination (see below and chap. 9 by Higgs).

7 Sequelae, Virus Persistence, and Recrudescence

One of the unforeseen issues in West Africa 2013 was the high frequency of sequelae, virus persistence, and, more rarely, recrudescence in EVD survivors. Although a host of both short- and long-term post-EVD sequelae has been noted dating back to the first recognized outbreak in Zaire (current Democratic Republic of the Congo) in 1976, little attention was typically afforded to survivors, in part due to the limited infrastructure for study in the outbreak areas (Vetter et al. 2016b). However, early anecdotal observations and subsequent more formal study on the estimated over 15,000 EVD survivors from West Africa 2013 reveal a wide range of medical and psychosocial challenges, including persistent arthralgia, ocular complications (including potentially sight-threatening uveitis that may result in early cataract formation) abdominal pain, extreme fatigue, and anorexia, sleep and memory disturbances, anxiety disorders, depression, post-traumatic stress disorder, and survivors' guilt in not only survivors, but also other family and community members (Vetter et al. 2016b). The challenges facing survivors created an urgent moral imperative to provide clinical care, only partially met by national programs with international support, including clinical care guidelines for EVD survivors developed by (WHO 2016a).

The underlying pathogenesis of EVD sequelae is not well understood, but anecdotal observations increasingly suggest that at least some relate to persistent virus in selected immunologically protected tissue compartments and fluids, including the testes/semen, chambers of the eye, cerebrospinal fluid (CSF), and the fetus, placenta, and amniotic sac/fluid of women infected during pregnancy (Vetter et al. 2016a, b). EBOV RNA has been found by RT-PCR in a host of body fluids for weeks or even months after resolution of acute disease and clearance of virus from the blood (Fig. 5), although the significance of these findings is often unknown since, in most cases, infectious virus could not be isolated by cell culture after a few weeks after disease onset.

Virus persistence in the semen is of most concern since it has occasionally resulted in sexual transmission, sometimes initiating small case clusters in the wake of the acute outbreak (Eggo et al. 2015; Christie et al. 2015; Mate et al. 2015). Albeit in low copy numbers and in a small minority of EVD survivors, EBOV RNA has been detected in the semen up to a year or more and infectious virus by cell culture up to 82 days after acute disease (Deen et al. 2015). A few cases of recrudescence associated with prolonged virus persistence have been noted, including uveitis with EBOV isolated from the aqueous humor of the eye at 14 weeks after disease onset (Varkey et al. 2015) and severe meningitis with seizures with isolation of virus from the CSF 9 months after resolution of acute disease (Jacobs et al. 2016). Anecdotal reports exist of recrudescent disease and viremia in West Africa, in some cases thought to be related to underlying HIV-1 infection, although this association remains to be validated (Howlett et al. 2016). A few cases have been noted in which women infected with EBOV during pregnancy, possibly with no or atypically mild disease, have recovered and remained pregnant, only to

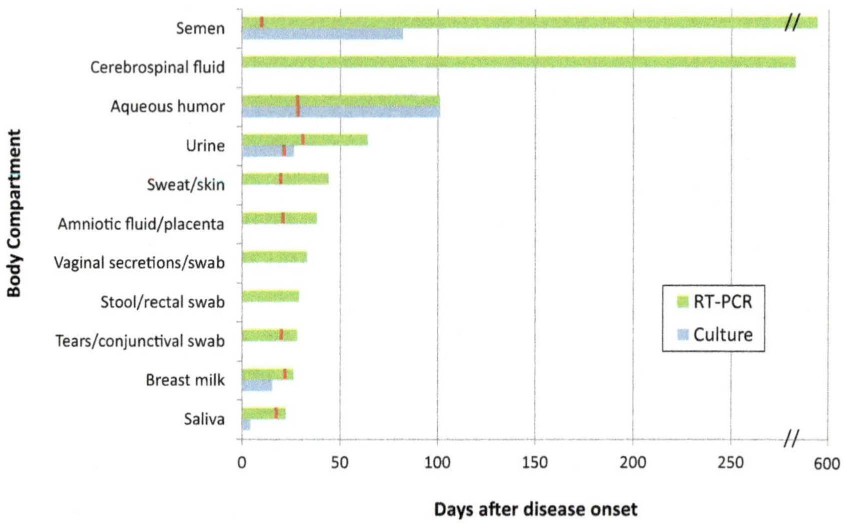

Fig. 5 Virus persistence after the day of disease onset in various body compartments in survivors of Ebola virus disease as detected by reverse-transcription polymerase chain reaction (*RT-PCR, green*) and cell culture (*blue*). *Red bars* represent the day of the first negative *RT-PCR* detection in the patient's blood, when available. Updated and reprinted with permission from Vetter et al. (2016b)

spontaneously abort a macerated and nonviable fetus in subsequent weeks or months (Bower et al. 2015; Caluwaerts et al. 2015). EBOV RNA was found in the products of conception, although cell culture results confirming the presence of infectious virus were generally not reported (Bower et al. 2015; Caluwaerts et al. 2015; Baggi et al. 2014; Oduyebo et al. 2015). With the exception of sexual transmission, no cases of secondary transmission resulting from EVD survivors have been suspected. Nevertheless, the possibility of virus persistence and renewed transmission from EVD survivors illustrate the need for continued non-stigmatizing but heighted surveillance even after the immediate threat of EVD from more common modes of transmission has been extinguished.

8 Research During the Outbreak

As the gravity of the situation in West Africa rose, the global community felt increasingly compelled to consider use of various experimental therapeutics and vaccines. Albeit unwelcome, the magnitude of the outbreak provided an important opportunity and obligation for prospective clinical research that had never before been possible. In August 2014, WHO convened a meeting in Geneva, Switzerland, of the diverse stakeholders, including representatives from the ministries of health, pharmaceutical companies, drug regulatory agencies, nongovernmental organizations

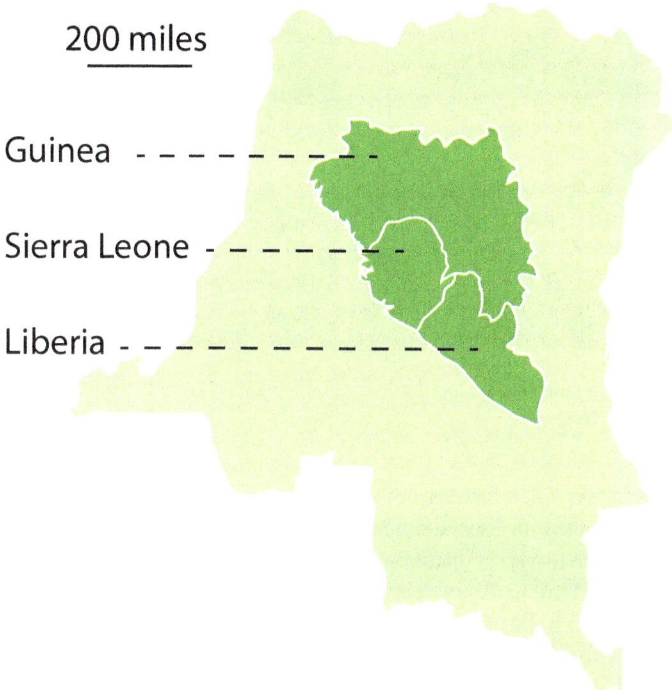

	Republic of the Congo	Guinea-Liberia-Sierra Leone
Population	68 million	22 million
Surface Area (Km^2)	2.3 million	430,000
Population Density (persons/ Km^2)	**30**	**51**

Fig. 6 Sizes and *population densities* of *Guinea*, *Liberia*, and *Sierra Leone* combined compared with the Democratic *Republic of the Congo*. To illustrate the difference in size, the three West African countries are shown superimposed on the Democratic. Republic of the Congo. Reprinted with permission from Bausch and Rojek (2016)

providing clinical care, and experts in virology and medical ethics. WHO then quickly created a Scientific and Technical Advisory Committee for Ebola Experimental Interventions to guide the process, which required consideration not only of the evidence for safety and efficacy, but also the anticipated feasibility and utility of conducting clinical trials in the setting of limited production capacities or intermittent drug availability. While various therapeutic trials were undertaken (Table 5), the many complex scientific, logistical, and sociocultural challenges could not be met

Fig. 7 "Bush taxis" in Guinea traveling back-and-forth between remote areas and major cities. Photos by Frederique Jacquerioz

quickly enough to take full advantage of the large case numbers potentially affording statistical power. By the time most therapeutic trials were implemented, case counts in West Africa had fallen to a level insufficient to meet clinical endpoints. There was also an opportunity missed to enroll more patients in clinical trials in resource-rich settings.

While there is disappointment that therapeutic trials during West Africa 2013 did not produce definitive evidence of an efficacious drug for EVD, the experience cannot be considered completely futile. Many difficult but valuable lessons were learned regarding the challenges of inconsistent reproducibility of in vitro experiments, poorly predictive animal models, and the operational demands of conducting trials overseas in an ETU during an outbreak without any preexisting research infrastructure. Rigorous debate continues regarding the scientific and ethical merit of the various clinical trial designs used in this outbreak. Nevertheless, numerous drug candidates progressed through Phase I, II, and III clinical trials at an unprecedented pace and the recognition that some agents are ineffective, along with promising interim results for a few, provide a starting point for prioritization in future outbreaks. However, much work remains to be done to capitalize on the lessons learned from West Africa 2013 and make the accelerated pace of therapeutic trials during outbreaks the norm, including prioritizing drug candidates, working out trial designs, prepositioning protocols and ethics committee reviews, and setting logistical frameworks for rapid operationalization.

As with therapeutics, the urgency of West Africa 2013 thrust vaccines for EVD from a conventional protracted research and development timeline into high gear. After rapid Phase I and II clinical trials were undertaken at various sites in the United States, Europe, and Africa (outside the EVD epidemic zone), a large Phase III trial of an experimental vaccine composed of a recombinant vesicular stomatitis Indiana virus expressing the EBOV GP was implemented in Guinea with a ring vaccination approach. The trial showed 100% vaccine efficacy (Henao-Restrepo et al. 2016) and was employed in the later stages of the outbreak to help stem spread from reintroduced virus from sexual transmission. Adverse

Table 5 Registered clinical trials reporting results of experimental therapeutics for Ebola virus disease during the 2013–2016 West Africa outbreak. Adapted and reprinted with permission from (Hayden et al. 2017)

	Trial characteristics			Dose characteristics			
Agent	Name	Sponsor and/or Funder	Design	Route	Regimen	Outcomes/Primary endpoint	Comment
ZMapp	PREVAIL II	National Institute of Allergy and Infectious Diseases, USA	Open-label RCT with adaptive design, comparison to optimized SOC alone (including favipiravir in Guinea)	Intravenous	50 mg/kg within 24 h of enrolment, followed by 2 more doses, every third day	Enrolment not met (72 of 200 targeted). Overall mortality by day 28 after EVD onset: 13/35 (37%) in SOC group vs 8/36 (22%) SOC + ZMapp group. Mortality among those with high virus levels (Ct $\leq$ 22) at entry, 9/15 (60%) in SOC group vs 7/15 (47%) in SOC + ZMapp group	No statistically significant survival benefit in patients with EVD but underpowered. Infusions require 2–12 h and may be associated sometimes with systemic reactions; these can be ameliorated by pretreatment with antihistamines and antipyretics. Specific for Zaire EBOV including Makona variant
TKM 130803	RAPIDE-TKM	University of Oxford, UK	Open-label, single arm with historical and concurrent controls, as part of a multistage approach	Intravenous	0.3 mg/kg once daily for up to 7 days	Halted after meeting prespecified futility endpoint (survival to day 14 of $\leq$55%). Survival at day 14 after EVD onset in 3/12 (25%), after excluding 2 who	No apparent survival benefit compared to historical controls, but potential confounding by enrollment of patients with high viral loads and late-stage disease. Dose-limiting

(continued)

Table 5 (continued)

	Trial characteristics			Dose characteristics			
						died <48 h after enrollment. Infusions generally well tolerated, except for one possible reaction	systemic infusion reactions (acute cytokine release syndrome) in healthy volunteers. Infused over minimum 2 h. EBOV Makona specific
Favipiravir	JIKI	Institut National de la Santé et de la Recherche Medicale, France	Open-label, single arm with historical controls	Oral	6 g on day 1, followed by 2.4 g/day on days 2–10 in divided doses	Completed. Among 99 evaluable adults and adolescents, mortality was 20% (95% CI, 11.6–32.4%) in those with a Ct $\geq$ 20 and 91% (95% CI, 78.8–91.1%) in those with a Ct < 20. Viral RNA loads and mortality were not significantly different between 31 adults starting favipiravir within <72 h of symptom onset and 68 who started later. No grade 3 or 4 clinical AEs	Mortality did not significantly differ from the predefined target values of 30% for high patients with high Ct values and 85% for patients with low Ct values. Much less active against EBOV than influenza virus in preclinical models. Dose regimen was approximately 50% higher than that tested in phase 3 trials of uncomplicated influenza but appeared to be generally well tolerated

(continued)

Table 5 (continued)

	Trial characteristics			Dose characteristics			
Convalescent plasma	Ebola-Tx	Institute of Tropical Medicine, Belgium	Open-label, single arm with historical controls	Intravenous	Two transfusions of 200–250 mL of ABO-compatible convalescent plasma, with each plasma unit obtained from a separate donor, given within 2 days of EVD diagnosis; those weighing <45 kg received two transfusions of 10 mL/kg body weight	Completed. Among 84 evaluable subjects, the mortality rate from day 3 to day 16 after diagnosis was 31% in the convalescent plasma group and 38% in the control group (odds ratio 0.88 [95% CI, 0.51–1.51], adjusted for Ct values and age). No serious AEs related to the infusions	No overall survival benefit as compared to historical controls, but levels of anti-EBOV antibodies were not determined in the plasma units. No survival or antiviral effects seen in EBOV-infected NHPs given convalescent blood with high titers of neutralizing antibodies, but hyperimmune globulin effective in NHPs. Transfusion-associated acute lung injury reported in a separate patient with EVD given convalescent plasma
Brincidofovir	RAPIDE-BCV	University of Oxford, UK	Open-label, single arm with historical controls, as part of a multistage approach	Oral	200 mg as a loading dose on day 1, followed by 100 mg on days 4, 8, 11, and 15; further adjusted for patients weighing <50 kg	Recruitment halted by manufacturer after 4 patients enrolled. Survival at day 14 after EVD onset in 0/4 patients. No serious or unexpected AEs	Variable, assay-dependent antiviral activity and selectivity for EBOV in cell culture. Antiviral action linked to brincidofovir's lipid moiety. No survival

(continued)

Table 5 (continued)

	Trial characteristics			Dose characteristics			
							benefit in murine model studies at nontoxic doses. Unable to be studied in NHPs owing to pharmacokinetic profile
rIFN-β1a[a]	–	Canadian Institutes of Health Research	Open-label, single arm, single center with historical controls	Subcutaneous	30 μg [6 × 10^6 IU] rIFN ß-1a daily for up to 10 days	Enrollment of 9 patients. Primary outcome of blood viral load reduction based on Ct values appeared faster than controls. Mortality rates of 84% among 38 controls and in 33% of rIFN-β1a recipients	Patients enrolled within 6 days of symptom onset. Analysis to address differences in baseline CT values revealed that the probability of dying in the untreated was 1.8 times that in the treated group

Abbreviations: AE, adverse event; CI, confidence interval, Ct, cycle threshold; EBOV, Ebola virus; EVD, Ebola virus disease; NHP, nonhuman primate, RCT, randomized controlled trial, rIFN-β1a, recombinant interferon β1a; SOC, standard of care. EVD—Ebola virus disease

[a]Data are from E. Fish, personal communication, September 16, 2016

effects were frequent but mostly minor, although vaccine-induced arthritis, dermatitis, and vasculitis were reported (Henao-Restrepo et al. 2016; Huttner et al. 2015).

In addition to the aforementioned prospective clinical investigations on therapeutics and vaccines, a vast amount of information was generated from informal observations and empiric experience with the large number of cases. This included noting relatively rare clinical presentations—at times challenging accepted case definitions—sequelae, and modes of transmission (Bausch and Rojek 2016). In particular, the spectrum of disease and transmission modes in pregnant women and their offspring brought many new and often unforeseen challenges (Akerlund et al. 2015). The 20 medically evacuated cases to the United States and Europe were generally cared for in advanced medical settings that allowed for more detailed clinical observation and laboratory analysis of both acute disease and sequelae than was typically possible in West Africa (Uyeki et al. 2016).

Although, as discussed above, clinical trials generally evolved too slowly to provide firm conclusions on efficacy, it is interesting to note that the CFR in the 27 cases who received care in the United States and Europe was only 18.5% (Uyeki et al. 2016), compared to 31–76% reported from West Africa 2013, depending upon the specific ETU and time during the outbreak. It is unknown whether this discrepancy in outcome relates to use of experimental therapies (of which 85% of patients in high-resource setting received one or more), better fluid and electrolyte monitoring and organ support (including mechanical ventilation and renal replacement therapy), genetic predisposition, and/or diminished comorbidities relative to the West African population. Observational studies on sequelae and persistence in EVD survivors, including the Partnership for Research on EBOV in Liberia (PREVAIL) III, a large multiyear controlled cohort study of EVD being undertaken in Liberia, promise to eventually yield a wealth of information (NIH 2015).

Significant progress was made on development and validation of new laboratory diagnostic platforms for nucleic acid detection of EBOV as well as progress on rapid tests (Dhillon et al. 2015). West Africa 2013 also set a new benchmark in providing real-time large-scale molecular epidemiologic data to guide response efforts *during* an outbreak (Holmes et al. 2016; Matranga et al. 2014). Such high-resolution genetic analysis is generally available only retrospectively. However, despite often difficult conditions, toward the end of the West Africa outbreak, novel field-applicable genome sequencing platforms were developed and deployed that were capable of generating results in less than 24 h (Holmes et al. 2016; Quick et al. 2016). These genomic data may allow accurate ongoing estimates of various important outbreak parameters, including of reproduction numbers R_0 and R_1, to determine the impact of specific interventions such as border closures and quarantine; elucidation of transmission chains and virus provenance when classically collected field epidemiologic data are unclear, including identification of super-spreaders and cases where sexual or other transmission from survivors is suspected; identification of signatures of host adaptation; identification and monitoring of diagnostic targets; and characterization of responses and resistance to treatments and vaccines (Matranga et al. 2014; Quick et al. 2016; Mate et al. 2015; Whitmer et al. 2016; Lau et al. 2017; Keita et al. 2016; Diallo et al. 2016).

9 Conclusions and Future Challenges

The unprecedented scale of West Africa 2013 took the world by surprise and sadly added another tragic event to a region already struggling to escape decades of poverty and war. The outbreak also shook the international response community, laying bare deficiencies in our response capacity to complex humanitarian crises involving highly infectious and lethal pathogens. Although much remains to be learned, West Africa 2013 also represents a tragic but nevertheless watershed moment in our understanding not only of the biology and epidemiology of EVD, but equally important, the many economic, social, political, ethical, and logistical challenges in confronting emerging diseases in the modern era. As the global population surges and becomes more interconnected, the risk of such outbreaks is destined to increase. In the absence of redoubled efforts to build capacity for surveillance and response, outbreaks such as West Africa 2013 threaten to become the "new norm." One need not look much further for the proof than to the Zika virus disease outbreak that swept through the Caribbean and Latin America starting in 2015.

West Africa 2013 has challenged the world to respond better. The pressure is on to capitalize on the lessons learned during the outbreak both from failures and the glimpses of innovation and research progress to create a new norm of comprehensive surveillance and organized response. In response to numerous internal as well as external evaluations (Moon et al. 2015), WHO has created a new Health Emergencies Programme designed to streamline response operations to such crises under one clear line of authority (WHO 2016b). However, the funding to fully implement the new program is still in question. Furthermore, reform and improved performance must extend far beyond only WHO. Many national governments and independent stakeholders have also created rapid response teams. Whether this revamped structure and national capacity will result in a more effective response to the next outbreak remains to be seen. Coordination of the many different partners and programs during the next outbreak will undoubtedly remain one of the key challenges.

Lastly, let us remember that, while important, science and technological advancement alone will never be sufficient; poverty and lack of the fundamental human right to health consistently underlie outbreaks of emerging pathogens (Nations 1948). EVD is but the proverbial "canary in the coal mine," indicative of the world's most vulnerable populations. We must advocate for and work toward restitution of the right to health in low- and middle-income countries (LMICs). This will entail much more than simply building a laboratory or conducting a research project. Local educational institutions must be strengthened and career opportunities created to stop the "brain drain" of HCWs to high-income countries and produce future "home grown" leaders in the health sciences. Novel and technology-appropriate approaches to local problems must be sought, as well as the funding mechanisms that enable their execution. Furthermore, after the major struggle to implement the *quantity* of medical care necessary in West Africa 2013, the outbreak rightly brought up the issue of *quality* of care. Implicit in this is a just rejection of a perhaps long-held but implicit acceptance of disparate qualities of care between patients in LMICs and resource-rich countries, an archaic notion whose time must now be passed. Regardless of country of origin or personal wealth, patients should have the right to HCWs with the right training for their condition and who implement evidence-based standards of care. Of course, this gap between rich and poor cannot be closed overnight. There is much work to be done with regard to both scientific research to generate the best evidence and advocacy and organization to ensure thorough and equitable implementation. Responsibility falls also on LMICs to create strong and transparent governmental and public health administrative frameworks capable of capitalizing on international collaboration and support. Long after West Africa 2013 is over, these will be our true measures of success.

Acknowledgements The author thanks Ian Crozier and Jens Kuhn for critical review of the manuscript and Lara Schwarz for administrative assistance.

References

Akerlund E, Prescott J, Tampellini L (2015) Shedding of Ebola virus in an asymptomatic pregnant woman. N Engl J Med 372(25):2467–2469

Baggi FM, Taybi A, Kurth A, Van Herp M, Di Caro A, Wolfel R et al (2014) Management of pregnant women infected with Ebola virus in a treatment centre in Guinea, June 2014. Euro Surveill: Bull Eur Sur Les Mal Transm = European Commun Dis Bull 19(49)

Baize S, Pannetier D, Oestereich L, Rieger T, Koivogui L, Magassouba N et al (2014) Emergence of Zaire Ebola virus disease in Guinea. N Engl J Med 371(15):1418–1425

Bank TW (2016) The world bank fact sheet. The world bank

Bausch DG, Rojek A (2016) West Africa 2013: re-examining Ebola. Microbiol Spectr 4(3)

Bausch DG, Schwarz L (2014) Outbreak of Ebola virus disease in Guinea: where ecology meets economy. PLoS Negl Trop Dis 8(7):e3056

Bausch DG, Bangura J, Garry RF, Goba A, Grant DS, Jacquerioz FA et al (2014) A tribute to Sheik Humarr Khan and all the healthcare workers in West Africa who have sacrificed in the fight against Ebola virus disease: Mae we hush. Antiviral Res 111C:33–35

Bird BH, Spengler JR, Chakrabarti AK, Khristova ML, Sealy TK, Coleman-McCray JD et al (2016) Humanized mouse model of Ebola virus disease mimics the immune responses in human disease. J Infect Dis 213(5):703–711

Bower H, Grass JE, Veltus E, Brault A, Campbell S, Basile AJ et al (2015) Delivery of an Ebola virus-positive stillborn infant in a rural community health center, Sierra Leone, January 2015. Am J Trop Med Hyg

Brainard J, Pond K, Hooper L, Edmunds K, Hunter P (2016) Presence and persistence of Ebola or Marburg virus in patients and survivors: a rapid systematic review. PLoS Negl Trop Dis 10(2): e0004475

Caluwaerts S, Fautsch T, Lagrou D, Moreau M, Modet Camara A, Gunther S et al (2015) Dilemmas in managing pregnant women with Ebola: 2 case reports. Clin Infect Dis: Off publ Infect Dis Soc Am

Carroll MW, Matthews DA, Hiscox JA, Elmore MJ, Pollakis G, Rambaut A et al (2015) Temporal and spatial analysis of the 2014–2015 Ebola virus outbreak in West Africa. Nature 524 (7563):97–101

Christie A, Davies-Wayne GJ, Cordier-Lasalle T, Blackley DJ, Laney AS, Williams DE et al (2015) Possible sexual transmission of Ebola virus—Liberia, 2015. MMWR Morb Mortal Wkly Rep 64(17):479–481

Deen GF, Knust B, Broutet N, Sesay FR, Formenty P, Ross C et al (2015) Ebola RNA persistence in semen of Ebola virus disease survivors—preliminary report. N Engl J Med

Dhillon RS, Srikrishna D, Garry RF, Chowell G (2015) Ebola control: rapid diagnostic testing. Lancet Infect Dis 15(2):147–148

Diallo B, Sissoko D, Loman NJ, Bah HA, Bah H, Worrell MC et al (2016) Resurgence of Ebola virus disease in Guinea linked to a survivor with virus persistence in seminal fluid for more than 500 days. Clin Infect Dis: Off Publ Infect Dis Soc Am 63(10):1353–1356

Dunham EC, Banadyga L, Groseth A, Chiramel AI, Best SM, Ebihara H et al (2015) Assessing the contribution of interferon antagonism to the virulence of West African Ebola viruses. Nat Commun 6:8000

Eggo RM, Watson CH, Camacho A, Kucharski AJ, Funk S, Edmunds WJ (2015) Duration of Ebola virus RNA persistence in semen of survivors: population-level estimates and projections. Euro surveillance: Bull Eur Sur Les Mal Transm = Eur Commun Dis Bull 20(48)

Elston JW, Moosa AJ, Moses F, Walker G, Dotta N, Waldman RJ et al (2015) Impact of the Ebola outbreak on health systems and population health in Sierra Leone. J Public Health

Faye O, Boelle PY, Heleze E, Faye O, Loucoubar C, Magassouba N et al (2015) Chains of transmission and control of Ebola virus disease in Conakry, Guinea, in 2014: an observational study. Lancet Infect Dis 15(3):320–326

Formenty P, Hatz C, Le Guenno B, Stoll A, Rogenmoser P, Widmer A (1999a) Human infection due to Ebola virus, subtype Cote d'Ivoire: clinical and biologic presentation. J Infect Dis 179 (Suppl 1):S48–S53

Formenty P, Boesch C, Wyers M, Steiner C, Donati F, Dind F et al (1999b) Ebola virus outbreak among wild chimpanzees living in a rain forest of Cote d'Ivoire. J Infect Dis 179(Suppl 1): S120–S126

Franklin SM (2016) A comparison of personal protective standards: caring for patients with Ebola virus. Clin Nurse Spec 30(2):E1–E8

Gire SK, Goba A, Andersen KG, Sealfon RS, Park DJ, Kanneh L et al (2014) Genomic surveillance elucidates Ebola virus origin and transmission during the 2014 outbreak. Science (New York, NY) 345(6202):1369–1372

Hassanin A, Nesi N, Marin J, Kadjo B, Pourrut X, Leroy E et al (2016) Comparative phylogeography of African fruit bats (Chiroptera, Pteropodidae) provide new insights into the outbreak of Ebola virus disease in West Africa, 2014–2016. CR Biol 339(11–12):517–528

Hayden FG, Friede M, Bausch DG (2017) Experimental therapies for Ebola virus disease: what have we learned? J Infect Dis

Henao-Restrepo AM, Camacho A, Longini IM, Watson CH, Edmunds WJ, Egger M et al (2016) Efficacy and effectiveness of an rVSV-vectored vaccine in preventing Ebola virus disease: final results from the Guinea ring vaccination, open-label, cluster-randomised trial (Ebola Ca Suffit!). Lancet

Hersi M, Stevens A, Quach P, Hamel C, Thavorn K, Garritty C et al (2015) Effectiveness of personal protective equipment for healthcare workers caring for patients with filovirus disease: a rapid review. PLoS ONE 10(10):e0140290

Hofmann-Winkler H, Gnirss K, Wrensch F, Pohlmann S (2015) Comparative analysis of host cell entry of Ebola virus from Sierra Leone, 2014, and Zaire, 1976. J Infect Dis 212(Suppl 2): S172–S180

Holmes EC, Dudas G, Rambaut A, Andersen KG (2016) The evolution of Ebola virus: insights from the 2013–2016 epidemic. Nature 538(7624):193–200

Howlett P, Brown C, Helderman T, Brooks T, Lisk D, Deen G et al (2016) Ebola virus disease complicated by late-onset encephalitis and polyarthritis Sierra Leone. Emerg Infect Dis 22 (1):150–152

Huttner A, Dayer JA, Yerly S, Combescure C, Auderset F, Desmeules J et al (2015) The effect of dose on the safety and immunogenicity of the VSV Ebola candidate vaccine: a randomised double-blind, placebo-controlled phase 1/2 trial. Lancet Infect Dis 15(10):1156–1166

Jacobs M, Rodger A, Bell DJ, Bhagani S, Cropley I, Filipe A et al (2016) Late Ebola virus relapse causing meningoencephalitis: a case report. Lancet

Keita M, Duraffour S, Loman NJ, Rambaut A, Diallo B, Magassouba N et al (2016) Unusual Ebola virus chain of transmission, Conakry, Guinea, 2014–2015. Emerg Infect Dis 22 (12):2149–2152

Kuhn JH, Andersen KG, Baize S, Bao Y, Bavari S, Berthet N et al (2014) Nomenclature- and database-compatible names for the two Ebola virus variants that emerged in Guinea and the Democratic Republic of the Congo in 2014. Viruses 6(11):4760–4799

Lau MS, Dalziel BD, Funk S, McClelland A, Tiffany A, Riley S et al (2017) Spatial and temporal dynamics of superspreading events in the 2014–2015 West Africa Ebola epidemic. Proc Natl Acad Sci USA 114(9):2337–2342

Li X, Zai J, Liu H, Feng Y, Li F, Wei J et al (2016) The 2014 Ebola virus outbreak in West Africa highlights no evidence of rapid evolution or adaptation to humans. Sci Rep 6:35822

Mari Saez A, Weiss S, Nowak K, Lapeyre V, Zimmermann F, Dux A et al (2014) Investigating the zoonotic origin of the West African Ebola epidemic. EMBO Mol Med 7(1):17–23

Marzi A, Robertson SJ, Haddock E, Feldmann F, Hanley PW, Scott DP et al (2015a) EBOLA VACCINE. VSV-EBOV rapidly protects macaques against infection with the 2014/15 Ebola virus outbreak strain. Science (New York, NY) 349(6249):739–742

Marzi A, Feldmann F, Hanley PW, Scott DP, Gunther S, Feldmann H (2015b) Delayed disease progression in cynomolgus macaques infected with Ebola virus makona strain. Emerg Infect Dis 21(10):1777–1783

Mate SE, Kugelman JR, Nyenswah TG, Ladner JT, Wiley MR, Cordier-Lassalle T et al (2015) Molecular evidence of sexual transmission of Ebola virus. N Engl J Med

Matranga CB, Andersen KG, Winnicki S, Busby M, Gladden AD, Tewhey R et al (2014) Enhanced methods for unbiased deep sequencing of Lassa and Ebola RNA viruses from clinical and biological samples. Genome Biol 15(11):519

Meltzer MI, Atkins CY, Santibanez S, Knust B, Petersen BW, Ervin ED et al (2014) Estimating the future number of cases in the Ebola epidemic–Liberia and Sierra Leone, 2014–2015. MMWR Suppl 63(3):1–14

Moon S, Sridhar D, Pate MA, Jha AK, Clinton C, Delaunay S et al (2015) Will Ebola change the game? Ten essential reforms before the next pandemic. The report of the Harvard-LSHTM Independent Panel on the Global Response to Ebola. Lancet 386(10009):2204–2221

Nations U (ed) (1948) The universal declaration of human rights. UN Gen Assem, New York, United Nations

NIH (2015) Study of Ebola survivors opens in Liberia [cited 2016 April 21]. http://www.nih.gov/news-events/news-releases/study-ebola-survivors-opens-liberia

Ning YJ, Deng F, Hu Z, Wang H (2017) The roles of Ebolavirus glycoproteins in viral pathogenesis. Virol Sin. 32(1):3–15

Oduyebo T, Pineda D, Lamin M, Leung A, Corbett C, Jamieson DJ (2015) A pregnant patient with Ebola virus disease. Obstet Gynecol

Osterholm MT, Moore KA, Kelley NS, Brosseau LM, Wong G, Murphy FA et al (2015) Transmission of Ebola viruses: what we know and what we do not know. mBio 6(2):e00137

Park DJ, Dudas G, Wohl S, Goba A, Whitmer SL, Andersen KG et al (2015) Ebola virus epidemiology, transmission, and evolution during seven months in Sierra Leone. Cell 161 (7):1516–1526

Parpia AS, Ndeffo-Mbah ML, Wenzel NS, Galvani AP (2016) Effects of response to 2014–2015 Ebola outbreak on deaths from Malaria, HIV/AIDS, and tuberculosis, West Africa. Emerg Infect Dis 22(3):433–441

Quick J, Loman NJ, Duraffour S, Simpson JT, Severi E, Cowley L et al (2016) Real-time, portable genome sequencing for Ebola surveillance. Nature 530(7589):228–232

Simon-Loriere E, Faye O, Faye O, Koivogui L, Magassouba N, Keita S et al (2015) Distinct lineages of Ebola virus in Guinea during the 2014 West African epidemic. Nature 524 (7563):102–104

Sozhamannan S, Holland MY, Hall AT, Negron DA, Ivancich M, Koehler JW et al (2015) Evaluation of signature erosion in Ebola virus due to genomic drift and its impact on the performance of diagnostic assays. Viruses 7(6):3130–3154

System Ebola Epidemiology Team IM (2014) Update: Ebola virus disease epidemic—west Africa, November 2014. MMWR Morbidity and mortality weekly report 63(46):1064–1066

UNICEF (2016) Children hardest hit New York, USA [cited 2017 April 23]. https://www.unicef.org/emergencies/ebola/75941_76202.html

Urbanowicz RA, McClure CP, Sakuntabhai A, Sall AA, Kobinger G, Muller MA, et al (2016) Human adaptation of Ebola virus during the West African outbreak. Cell 167(4):1079–1087 e5

Uyeki TM, Mehta AK, Davey RT Jr, Liddell AM, Wolf T, Vetter P et al (2016) Clinical management of Ebola virus disease in the United States and Europe. N Engl J Med 374 (7):636–646

Varkey JB, Shantha JG, Crozier I, Kraft CS, Lyon GM, Mehta AK et al (2015) Persistence of Ebola virus in ocular fluid during convalescence. N Engl J Med

Vetter P, Fischer WA, 2nd, Schibler M, Jacobs M, Bausch DG, Kaiser L (2016a) Ebola virus shedding and transmission: review of current evidence. J Infect Dis

Vetter P, Kaiser L, Schibler M, Ciglenecki I, Bausch DG (2016b) Sequelae of Ebola virus disease: the emergency within the emergency. Lancet Infect Dis

Whitmer SL, Albarino C, Shepard SS, Dudas G, Sheth M, Brown SC et al (2016) Preliminary evaluation of the effect of investigational Ebola virus disease treatments on viral genome sequences. J Infect Dis 214(suppl 3):S333–S341

WHO (2008) Interim infection control recommendations for care of patients with suspected or confirmed Filovirus (Ebola, Marburg) Haemorrhagic Fever

WHO (2014a) Ground zero in Guinea: the Ebola outbreak smoulders—undetected—for more than 3 months. http://www.who.int/csr/disease/ebola/ebola-6-months/guinea/en/

WHO (2014b) Ebola virus disease in Guinea. http://www.afro.who.int/en/clusters-a-programmes/dpc/epidemic-a-pandemic-alert-and-response/outbreak-news/4063-ebola-hemorrhagic-fever-in-guinea.html

WHO (2015a) Ebola Situation Report—4 November 2015. http://apps.who.int/ebola/current-situation/ebola-situation-report-4-November-2015

WHO (2015b) Ebola situation report—2 December 2015. http://apps.who.int/ebola/current-situation/ebola-situation-report-2-December-2015

WHO (2015c) Ebola virus disease (EVD) in West Africa: an extraordinary epidemic. Releve epidemiologique hebdomadaire/section d'hygiene du secretariat de la societe des nations = weekly epidemiological record/ health section of the secretariat of the league of nations. 90 (10):89–96

WHO (2016a) Clinical care for survivors of Ebola virus disease. Interim guidance Geneva, Switzerland: WHO [cited 2016 January 24, 2016]. http://apps.who.int/csr/resources/publications/ebola/guidance-survivors/en/index.html

WHO (2016b) WHO health emergencies program geneva, Switzerland: WHO [cited 2017 April 23]. http://www.who.int/about/who_reform/emergency-capacities/emergency-programme/en/

Clinical Management of Ebola Virus Disease Patients in Low-Resource Settings

Armand Sprecher, Michel Van Herp and Pierre E. Rollin

Abstract The low-resource environment deprives healthcare providers caring for patients with Ebola virus disease (EVD) of many of the means employed for the critically ill that are available in better resourced settings, such as advanced therapeutic interventions and abundant staff. In addition to these limitations may be added those imposed by the remote tropical locations, where EVD outbreaks occur. In this setting, a safe environment is created where healthcare workers may care for their patients over the evolving course of their acute illness into their convalescent period. Clinical management of EVD combines supportive and symptomatic care while also addressing the patient's emotional and mental health needs. A variety of specific therapies directly targeting the virus has become available, but none of these has, as of yet, conclusively demonstrated an impact. Healthcare workers caring for EVD patients must be constantly aware that they are part of a larger epidemic control operation, and their actions have consequences that go beyond their patients to their families and the community affected by the outbreak.

A. Sprecher (✉) · M. Van Herp
Médecins Sans Frontières, Rue Arbre Benit 46, 1050 Brussels, Belgium
e-mail: armand.sprecher@brussels.msf.org

M. Van Herp
e-mail: michel.van.herp@brussels.msf.org

P.E. Rollin
Viral Special Pathogens Branch, Division of High Consequence Pathogens and Pathology, National Center for Emerging Zoonotic Infectious Diseases, Centers for Disease Control and Prevention, 1600 Clifton Road, Atlanta, GA 30333, USA
e-mail: pyr3@cdc.gov

Current Topics in Microbiology and Immunology (2017) 411:93–114
DOI 10.1007/82_2017_18

Published Online: 24 June 2017

Contents

1 Introduction

Since the discovery of Ebola virus (EBOV) in 1976, nearly all patients with Ebola virus disease (EVD) have been cared for in low-resource settings. It may be assumed that the overwhelming majority of future patients will also be cared for in these settings. EVD is a very severe illness caused by four distinct ebolaviruses (Bundibugyo virus, Ebola virus, Sudan virus, and Taï Forest virus) that poses a significant threat to healthcare providers in any setting. However, the location where this care usually occurs imposes some additional challenges.

First, as the name implies, low-resource settings provide medical professionals with less or different material to work with than would be the case in better resourced locations. Many of the typical tools of the critical care trade are unlikely to be readily available: mechanical ventilation, invasive monitoring, and renal replacement therapy to name but a few. Given the remote locations where EVD outbreaks often occur, therapeutic options that may be available within the country may be difficult to bring to bear, e.g., supplemental oxygen. These difficulties in applying advanced therapeutics are not insurmountable, and many elements of modern critical care were made available for EVD patients in the West African outbreak (Wong et al. 2015). However, these modalities are not easily deployed, and their availability during outbreaks that are not of very long duration may be difficult to arrange.

Similar limitations are to be expected in the human resources needed to provide advanced levels of care. The countries where ebolaviruses are likely to reemerge in the future often have insufficient doctors and nurses to care for their ordinary health needs, let alone the additional burden imposed by EVD outbreaks. Doctors and nurses trained in critical care are exceedingly rare in Central Africa. As such, the most valuable asset that critical care units in well-resourced settings have, sufficient well-trained staff to allow for intensive attention to their patients' needs, is not often available.

Compounding the limitation imposed by the scant medical human resources available is another inconvenient feature of the locations, where EVD outbreaks typically occur: tropical heat. Healthcare workers providing direct care for EVD patients wear protective equipment designed to keep infectious fluids from coming into contact with their skin and mucous membranes. This protective gear is, regrettably, equally effective in preventing the wearer's sweat from providing evaporative cooling. In settings where daytime temperatures are usually in the upper thirties, the need for cooling is not insignificant. The inability to cool through sweat evaporation limits the amount of time that protective equipment can be worn, and so the amount of time the wearer can engage in patient care. The human resources that can be assembled to care for EVD patients are thus further limited in the person-hours of patient contact time they can provide each shift.

For those organizing care for EVD patients in resource-limited settings, the inadequacies in tools and manpower are accompanied by another inevitable complication—EVD cases are encountered during EVD outbreaks. This has at least four consequences. One, the number of patients requiring attention can be large relative to the capacity of the medical service providing care. Second, there is an urgent need to put safe and effective patient care services into place rapidly. Three, normal healthcare services are frequently disrupted, and many people with health problems endemic to the region will need care that is unavailable through the usual means. Many of these conditions are not easily distinguished from EVD, and these patients will often present to the ETU. Finally, the treatment of EVD patients is not an isolated instance of medical care, but rather an element in the larger outbreak control response, usually in a setting where the epidemic has provoked a high level of fear and suspicion in the population. How care is delivered will be under a certain amount of scrutiny by the community affected by the outbreak, and their willingness to participate in outbreak control activities may depend on how they perceive that medical care is being provided.

The medical practitioner who may be coming from a well-resourced environment to the settings where EVD outbreaks naturally occur to treat EVD patients must be ready to quickly do more with less in a way that is transparently compassionate while producing results that will encourage those in the community to come to them for care when they are ill. This is the complex challenge of care for EVD in resource-limited settings.

2 Objectives of Clinical Management in Low-Resource Settings

The actions undertaken in the clinical management of EVD are in pursuit of one or more of three objectives. First, the spread of the virus must be controlled. In the clinical care setting, this is done by protecting care providers and the patients in the care unit who may not have EVD. As part of the outbreak control framework, the Ebola [virus disease] treatment unit (ETU) is itself a means of limiting viral spread by preventing uninfected persons from coming into unprotected contact with virus through limiting and controlling access to acute EVD patients, the environment into which they are shedding virus, and the infectious waste generated in their care.

Second, and perhaps obviously, the care provided to the EVD patients is intended to maximize their chance of survival, by providing supportive care, addressing potentially complicating coinfections, and eventually employing therapeutics that directly address the viral infection.

Third, and certainly not least, the ETU care providers play an important role in reducing suffering. This is not limited to addressing the physical and psychological suffering of their patients, but also, by providing compassionate care, care providers help the suffering of their patients' families and loved ones, and, by providing hope and showing solidarity, the suffering of the larger community.

This chapter provides an overview of EVD clinical management principles in resource-limited settings. It is not a substitute for guidelines that go into the appropriate level of detail needed to set up and run a medical service for EVD patients. Such guidelines are available from the World Health Organization (WHO 2016) or from non-governmental agencies experienced in the management of EVD patients.

3 Infection Control

An unfortunate hallmark of EVD outbreaks is the death of healthcare staff caring for patients. Ebolaviruses are the archetypal biohazards, and management of this hazard and the protection of workers is the defining feature of response to EVD outbreaks. Worker protection agencies have long recognized a hierarchy of measures to limit worker exposures to hazards (NIOSH 2015), and a range of exposure controls is employed in providing EVD care. Although a full treatment of these measures is beyond the scope of this chapter, it is important to cover them briefly, as they are critical to the safe care of the EVD patient and to the safe management of an ETU.

The protection of the personnel (and patients), the environment and the public from exposure to ebolaviruses is based on risk assessment and has resulted in an appropriate selection of facility safeguards, safety equipment, and practices and procedures used by the personnel.

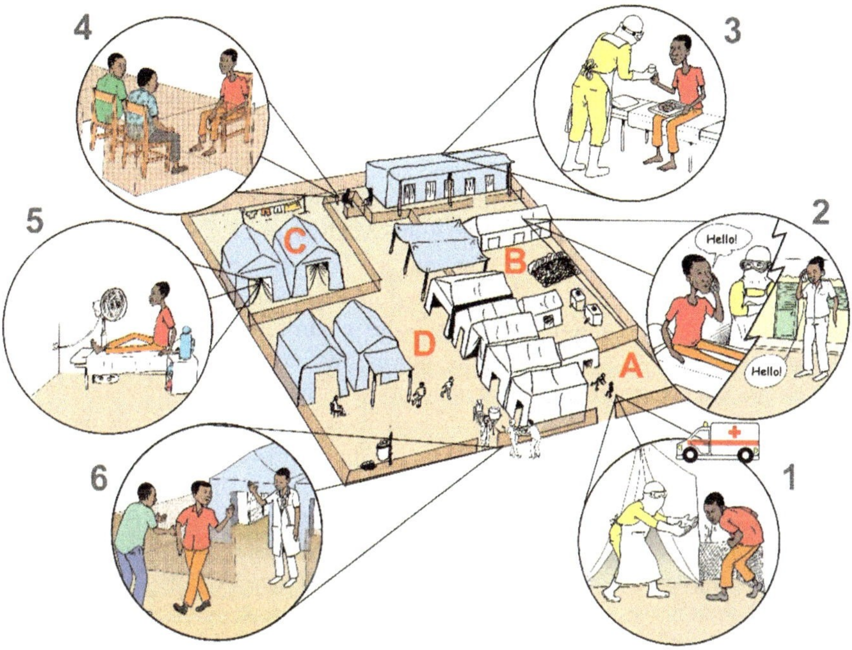

Fig. 1 Schematic drawing of an Ebola [virus disease] treatment unit by "Charlie" Guinean artist and cartoonist. Some of the important features are detailed. *A* Screening and triage area, *B* High-risk zone: admission areas and hospitalization for suspect or probable patients waiting for their laboratory results, *C* High-risk zone for confirmed patients, *D* Low-risk zone (no patient contact). *1* Admission and examination of suspect patients, *2* Transfer zone to the high-risk confirmed area. Throughout the hospitalization the patients can communicate with outside family with provided phones, *3* During the hospitalization, patients receive treatment and nutritional support, *4* Zone of interaction with family and friends, *5* During the hospitalization supportive care and psychological support is assured, *6* Discharged patients are followed up to ensure their continued health and well-being

The most effective means of preventing exposure to hazards like ebolaviruses is appropriate engineering of the healthcare environment (Fig. 1). The ETU is a workplace designed to minimize the risk of exposure of uninfected individuals to ebolaviruses. The ETU provides a means to limit access to EVD patients though restricted points of entry. The unit is divided into two zones: a high-risk zone, where the EVD patients are cared for and infectious material rendered safe, and a low-risk zone, where all of the other elements of care that do not require patient contact are found, e.g., nursing stations, laundry, medical and material supplies, etc. Entry to and exit from the high-risk zone is arranged to create unidirectional flow through the high-risk area, moving from areas of lowest infection risk to higher infection risk. Patients who are suspected of having EVD awaiting laboratory confirmation of their status are separated from those who have been confirmed to have it. Patients in the suspect area are given sufficient space from each other, ideally with individual rooms, to limit cross-infection risk. The entire high-risk zone is well lit with enough space to allow unimpeded movement. Patient care areas are arranged to be visible to healthcare providers in the low-risk area, so that healthcare workers do not need to be in protective equipment to interact with their patients. Exit from the high-risk area is designed to allow supervised decontamination and safe doffing of personal protective equipment (PPE).

Management of an ETU creates a large amount of liquid and solid waste. Just as patient care pathways follow a nonoverlapping pathway from lower to higher risk, so do waste removal pathways, leading to waste disposal areas within the high-risk zone. Large quantities of chlorine solutions are used for decontamination, and the resulting inactivated waste is usually burned in large, supervised incineration pits.

The most visible component of exposure control is the PPE worn by the healthcare workers. The coverall, boots, gloves, and covering for the head and face prevent the wearer's skin and, especially, mucous membranes, from coming into contact with infectious materials. This garb is a mixed blessing; its protection comes not only at the cost of impaired cooling that carries a risk of heat illness, but also with reduced vision, hearing, and sense of touch that limit the wearer's situational awareness, also a hazard in the high-risk environment (Sprecher et al. 2015).

Administrative controls in place in the ETU prevent entry to the high-risk area by anyone who has not undergone training on the use of PPE and safe work practices within the high-risk area. Entry into the high-risk area is only by pairs of workers who observe each other throughout their stay to ensure adherence to safe work practices and watch for dangerous unintentional actions. Worker shifts and breaks are arranged to minimize worker fatigue and maximize alertness. Safe work procedures include such practices as the minimization of interventions requiring sharp objects, carrying out hazardous procedures with observation and assistance from a partner and with good ergonomics, rapid cleanup of infectious material, and limiting time spent in PPE. In addition to the involvement of each worker, relentless supervision by qualified and respected individuals is a critical and mandatory requirement of infection control of an ETU. These behavioral controls complement the engineering controls.

The advent of a vaccine that looks to be effective in protecting against EVD may offer further protection for healthcare workers (Henao-Restrepo et al. 2015). However, it will not be until such efficacy is well demonstrated, and can be confirmed or safely assumed in individual workers that onerous exposure control measures, such as stifling PPE, may be scaled back to configurations that allow the doctors and nurses a greater liberty to interact longer and more closely with their patients.

4 Management of the Patient

Patients with EVD have an imprecise clinical presentation, and no truly distinguishing feature allows clinicians great certainty as to whether or not someone has the disease. After an incubation period of 6–12 days on average (range 2–21), patients will first experience fever, malaise, weakness, and body aches (Bray and Chertow 2016, Kortepeter et al. 2011). Gastrointestinal signs and symptoms may ensue in the next few days with anorexia, nausea, epigastric and abdominal pain, vomiting, and diarrhea. Some patients may have conjunctival injection and, briefly, a rash. Some patients will in the following days, around day 7 of illness, have delirium and shock with multi-organ failure. Although long classified as a viral hemorrhagic fever, usually only a minority of EVD patients will have hemorrhagic signs (McElroy 2015), chiefly gastrointestinal bleeding.

Given the vague nature of this illness, anyone who is ill in or near to an EVD epidemic zone must be evaluated cautiously. Similarly, no one can be simply assumed to have the disease, such that any exposure to ebolaviruses would be of no further consequence. Management of patients who may have EVD involves both treating them as if they had the disease for the purposes of their treatment and protecting healthy persons and healthcare workers, while also treating them as if they did not have the disease for the purpose of protecting them from other infected patients and their environment.

Faced with an ill patient during an EVD outbreak, a clinician must decide if the level of suspicion that the patient has EVD is sufficient to warrant admission to the ETU, which places the patient at some risk of becoming infected by another patient if they do not in fact have EVD. For this reason, until rule-out and discharge occur, the ETU must be a safe place for such patients with adequate alternative therapeutic options. The clinician must also decide if the level of suspicion is low enough to allow the patient to be safely managed outside the ETU by those who are not as well protected from ebolaviruses.

Determination of whether someone should be suspected of having EVD uses both an assessment of the clinical presentation for consistency with the disease and an evaluation of the epidemiologic risk in the patient's history. The suspect case definition for EVD during an epidemic is either: (a) a patient having contact with a known case AND a fever, (b) a patient with fever AND 3 or more of: headache, anorexia, weakness, difficulty swallowing, difficulty breathing, myalgia/arthralgia, nausea, vomiting, diarrhea, or abdominal pain, (c) a patient with unexplained bleeding, or (d) sudden unexplained death (WHO 2014). Often, miscarriage or abortion is also a condition that is sufficient for being considered a suspect case. Case definitions are tools for epidemiologists and outbreak managers to have a uniform way of counting cases. They may be a guide to clinicians in patient management, but they should not be considered as strict admission criteria for an ETU or a substitute for clinical judgment. For example, if someone who is a close contact of a known EVD case happens to be afebrile but otherwise has a clinical picture compatible with EVD (e.g., weakness and gastrointestinal signs and symptoms), they would not meet the case definition because of the absence of fever, but nevertheless they should be admitted for testing because of the risk for EVD.

An important aid to clinical decision making is good communication with the epidemiologists engaged in outbreak control. These epidemiologists may be able to inform the clinician if their patient is a contact of a known case, resides in a neighborhood where ongoing transmission is known to occur, or fits a profile known to be high risk in that epidemic, e.g., traditional healer.

If a medical team has been dispatched to investigate a concerning patient in the community, sources of information beyond the patient may be available to aid determination of appropriate disposition. Can the patient or their friends, family, or neighbors tell the medical team if the patient has recently been exposed to someone who may have been shedding virus—e.g., cared for anyone sick or attended a funeral? If the patient spontaneously presents or is brought to the ETU, sources of information may be more limited.

Prior to admission to the ETU, patients should receive an intake briefing to familiarize them with the hygiene practices of the ETU that will keep them at lowest risk for infection with ebolaviruses if they are not yet infected. This should cover appropriate use of showers and latrines, safe interactions with other patients, limitations on their movements about the high-risk area, areas designed for them to safely receive visits from friends and family, and some words about the need for the caregivers' PPE and its occasionally frightening aspect. Family members should also be aware of these procedures and offered a possibility to contact the patient and remotely provide psychological support.

Admitted patients should have an initial blood specimen taken for diagnostic testing and be placed into the part of the ETU reserved for patients whose ebolavirus infection status is uncertain. Patients whose test for ebolaviruses is positive should be moved to the area reserved for confirmed patients. Patients whose test is definitively negative, i.e., a negative result on a specimen drawn >72 h after onset of symptoms, should be discharged. Those with an equivocal test, i.e., a negative test <72 h after onset of symptoms, should be kept in the suspect patient area and retested in 24–48 h.

In an ETU that has not been set up to provide each suspected patient with an individual room, cohorting of patients based on the clinician's assessment of their probability of having EVD or the degree to which their clinical presentation make it likely that they are shedding significant amounts of virus, e.g., patients with vomiting, diarrhea, or advanced stages of illness, is advisable. Patients deemed to be at lower risk should be placed earlier on the care provision pathway through the ETU high-risk area, and patients felt to be at higher risk should be the last ones seen by medical personnel before moving on to the confirmed patients.

4.1 End-Stage and Deceased Patients

The death of a large number of patients has been an invariant feature of every EVD outbreak. Palliative care of the dying is one aspect of EVD patient care that can be equally well carried out in the resource-limited as well-resourced settings (see Sect. 5.2).

EVD survivors have reported that one of their most significant sources of stress in the ETU is their exposure to deceased patients (Evlampidou et al. 2016). To reduce this, clinicians who have determined a patient to be end-stage or at significant risk for death in the near term may wish to move them to an area within the ETU that affords greater privacy. Deceased patients should be moved to the morgue as quickly as possible.

4.2 Discharge

Criteria for discharge have been a matter of debate and have changed over the years. One of two matters of concern, if the patient is well enough to no longer need the care provided in the ETU, is a matter of clinical judgment and not a subject of contention. The second matter, whether the patient is no longer an infection hazard to those around them, is a thornier matter. Very little is known about when patients stop shedding active virus in a way that would allow for casual social contact with little risk of transmitting disease. However, no cases of EVD have been linked to exposure only to convalescent patients, aside from rare cases of sexual transmission.

A proxy measure of infection risk has been the PCR testing of blood specimens in recovering patients. Patients who have recovered enough to care for themselves and who have a negative blood PCR test, have been considered safe to discharge. PCR testing criteria have in some settings been set to a higher standard of two negative tests at least 48 h apart when an added measure of caution has been felt appropriate by health authorities. In other settings when there has been pressure to open beds in the ETU for newly arriving patients, the criteria for discharge have been loosened, and having a blood specimen PCR result above a cycle threshold (Ct) value of 36 has been used. This latter criterion has been based on laboratory studies that have found that materials with a Ct value of 36 or higher have not produced infection in experimental animals (Spengler et al. 2015).

PCR testing of other specimens, e.g., sweat, urine, saliva, have been used to inform discharge decision-making in highly resourced settings (Kreuels et al. 2014), but this has not been standard practice in low-resource settings. The infection risk of convalescing patients with detectable viral RNA in these body fluids is currently unknown, and there have not been reported EVD cases linked to exposure to these fluids.

Late transmission of filoviruses through sexual contact was documented as early as the 1967 Marburg virus disease outbreak (Martini and Schmidt 1968). However, it was felt to be rare enough that this risk never impeded discharge of convalescent patients. In fact, no case of sexual transmission of ebolaviruses was documented until 2015 (Mate et al. 2015). Nevertheless, male patients discharged from the ETU have always received discrete counseling about this risk, been provided with condoms, and given instructions on how to safely dispose of potentially infectious material. Several cases of sexual transmission have been documented or suspected towards the end of the West African outbreak, but as of yet, the risk of sexual transmission has still not precluded discharge from the ETU. That being said, male survivors should be followed up after their discharge to evaluate their ongoing potential for sexual transmission of ebolaviruses through semen testing. A future role of EVD antiviral therapy may be in the reduction of this potential.

4.3 Survivor Care

Care of the EVD patient does not end upon discharge from the ETU. EVD survivors can have disease sequelae that cause ongoing suffering and disability. Arranging post-discharge care for survivors is important, as counseling and symptomatic treatment can be a great benefit to the vast majority of patients suffering from persistent symptoms of their disease. Common physical post-EVD signs and symptoms are arthralgia, ocular problems (eye pain, conjunctivitis, uveitis, loss of acuity), auditory problems (tinnitus, hearing loss), and neurological sequelae (cerebrovascular accident, encephalopathy, meningitis) (Vetter et al. 2016). The resource-limited setting presents difficulty in caring for some of these issues due to a scarcity of relevant specialist physicians, such as ophthalmologists. Many patients will experience a full recovery from these physical ailments.

Psychological consequences are also common among EVD survivors. Two-thirds of EVD survivors will suffer from depression and/or posttraumatic stress disorder (Rabelo et al. 2016). Many patients will have lost friends and family to EVD, been deprived of an opportunity to grieve with their communities, and may not have graves to visit. The stigma these survivors face is an additional stressor. Mental health services for these patients are an important part of their care, and mental health professionals may also be in short supply in resource-limited settings.

The social consequences are another significant burden for survivors. Many EVD survivors return to their homes only to be shunned by their neighbors or threatened with violence, as they are seen as a continuing threat to the community. The loss of family members and being ostracized from their communities leaves many EVD survivors in a precarious situation, and so EVD survivor care programs should also include social support.

4.4 Children

About 7% of confirmed patients treated by MSF in the West African outbreak were less than 5-years old (MSF in press), which is much less than the 17% of the general West African population that is under 5-years old (United Nations 2015). This may be because most infections occur in the setting of provision of care for the ill and funeral attendance, which are not activities typical for young children. In contrast to their relatively lower numbers in the ETU, young children have a higher case fatality ratio than adults (MSF, in press).

Young children can present a challenge in the ETU, as they require more assistance with care and constant supervision. In the absence of adult supervision, uninfected children are at risk for becoming exposed if the ETU does not have controls in place to limit their movement. Some ETU's have had dedicated pediatric areas to allow closer supervision of children. In other units adult convalescent patients have remained in the ETU past the time when they were ready for discharge to serve as childcare providers.

It is not unusual for a female EVD patient to come to the ETU with her asymptomatic child if no one else is available to care for them. It is common practice to separate the child from the mother on the assumption that the child may not yet be infected. This poses another management problem, as these children will need someone to care for them. Because they are at high risk of becoming ill with little warning, this creates an infection risk for the caregiver. One successful solution to this problem has been the creation of limited access daycare units staffed by female EVD survivors, who are assumed to have protective immunity.

4.5 *Pregnant Women*

Pregnant EVD patients present another complicated problem for the ETU. In past outbreaks they appear to have had a higher risk of death than nonpregnant patients (Mupapa et al. 1999a), but an analysis of 77 pregnant EVD patients during the West African outbreak suggests their risk for death may be similar to that of the general population (Caluwaerts et al. 2016). As demise of the fetus in utero is very common, and survival of the newborn beyond the very early neonatal period has only been recorded once (Dörnemann et al. 2017) the focus of management is on maternal survival. Ideally delivery is managed after the mother has survived her disease, as women delivering while acutely ill are at increased risk of complications from hemorrhage and DIC.

Every ETU should be prepared for pregnant EVD patients and be ready to manage a delivery. Testing of amniotic fluid from EVD patients has found high levels of viral RNA (Caluwaerts et al. 2016), and exposure to amniotic fluid of a pregnant EVD patient has been a source of healthcare worker infections in past outbreaks (Wamala et al. 2010; Maganga et al. 2014). ETU staff should be prepared to handle the risk posed by preforming the delivery and be ready to carry out post-delivery decontamination procedures.

EVD patients who survive their disease with their pregnancy intact will not, as a rule, carry their baby to term to deliver a healthy newborn. As such, and given the danger posed to attendants by the eventual delivery, it is recommended that this delivery take place in the ETU. The option of termination of pregnancy should be discussed with the patient.

5 Medical Care of the EVD Patient

5.1 *Supportive Care*

Supportive care of EVD patients is based upon the similarity of its pathogenesis to septic shock (Mahanty and Bray 2004), and clinical illness is consistent with this similarity. Given this, a model for supportive care is that of management of septic shock (Dellinger et al. 2013). There are occasional presentations that deviate somewhat from an uncomplicated septic shock syndrome. Some patients will have a greater degree of gastrointestinal fluid losses, which will require volume replacement to compensate for these. Other patients may have an encephalopathic component to their illness, and so may need control of neurologic signs. With all but a handful of EVD patients having gone through their full course of illnesses in resource-limited settings without access to a clinical laboratory, EVD patients have most often had their care guided by clinical signs and symptoms, though more recently the use of biochemistry testing has provided additional guidance.

5.1.1 Hemodynamic Support

Intravenous crystalloid has long been the mainstay of management of hemodynamic instability in the critically ill, and the current guidelines for management of septic shock continue to support this. That being said, the liberty to aggressively pursue crystalloid therapy is somewhat constrained in the resource-limited setting by the relative unavailability of the modalities often used to correct overhydration, e.g., mechanical ventilation and renal replacement therapy. Added to this are recent concerns that moderation is called for in the use of crystalloid therapy (Marik and Bellomo 2016; Maitland et al. 2011). Clinicians caring for EVD patients should endeavor to be up to date on the most recent understanding of the resuscitation of patients with septic shock when deciding on the volumes of intravenous fluid therapy to employ.

5.1.2 Laboratory Support

In previous outbreaks, a diagnostic laboratory capable of safely handling specimens from suspected EVD patients was often either missing completely or located well away from the site of EVD treatment. In addition, beside a few point of care tests (malaria, pregnancy), laboratory assays were limited to the confirmation of the EVD, usually by molecular techniques. Having laboratory support close to the ETU has become a more routine feature of recent outbreaks. As well, the use of point of care biochemistry testing for EVD patients in resource-limited settings became commonplace toward the end of West African EVD outbreak. These machines may be safely employed if they are placed within the safety enclosure employed by the field diagnostic labs or inside the high-risk area of the ETU.

5.1.3 Correction of Biochemical Derangements

Biochemistry testing has revealed that EVD patients may have correctable conditions such as hypokalemia and hypoglycemia. Hyperkalemia is also not uncommon, presumably due to cell lysis and renal dysfunction.

Kidney function testing often reveals a degree of renal impairment suggestive of a mixture of pre-renal and intrinsic renal failure. Some biochemistry derangements, such as elevations of aspartate transferase and creatine phosphokinase, may reflect the degree of cytotoxicity being caused by the virus, and so may have prognostic value (Rollin et al. 2007; Sissoko et al. 2016).

5.1.4 Nutritional Support

Another important difference of the resource poor setting is the baseline nutritional status of the average central African prior to disease onset. The countries vulnerable to EVD outbreaks are also among those with the worst population nutritional status indicators (IFPRI 2016). Guidance for nutritional support of the EVD patient in resource-limited settings is based upon standard care for the severely ill and malnourished in these same settings (WHO/UNICEF/WFP 2014).

For patients who are very weak and have little appetite, therapeutic milk-based feeding, such as with F75/F100 is appropriate. For patients with some appetite who are able to eat solid or semi-solid foods, cereal or peanut-based ready to use therapeutic foods are a good option. Familiar foods that are nutrient dense and appealing to the patient can be substituted if tolerated by the patient. These can also be enriched with therapeutic supplements (e.g., Plumpy-doz®, Nuttributter®). Allowing the family to assist with patient care by preparing some meals at home to bring to the ETU can be a comfort to them and to the patient. Meals should be prepared with medical guidance, and the ETU staff should still maintain primary responsibility for ensuring proper patient nutrition.

5.2 *Symptomatic Care*

Whatever the impact of supportive care on patient outcomes, their suffering can be alleviated. EVD patients are in variable degrees of distress which may change over the course of their illness. As arthralgia, myalgia, headache, and sore throat are common clinical signs, pain management is an area open to medical intervention. The use of aspirin and nonsteroidal anti-inflammatory drugs is generally discouraged in EVD because of the bleeding complications and renal compromise that may be worsened. Paracetamol/acetaminophen is often sufficient for milder symptoms, and opiates employed for more severe pain. Opiate treatment protocols should be discussed with local medical staff, as palliative care may not be well accepted in some settings where EVD outbreaks occur. Opiates are also useful in relieving the discomfort of patients with terminal dyspnea.

Nausea and vomiting have been treated in the past with metoclopramide, however, benzodiazepines and haloperidol are commonly available and may be more effective. 5-HT3 receptor antagonists such as ondasetron are still more effective, and were employed in the West African EVD outbreak to good effect. Adequate control of nausea and vomiting provides not only relief to the patient, but also an opportunity to resume eating and drinking.

A not uncommon symptom in end-stage patients is terminal delirium. Whether caused by reduced cerebral perfusion, electrolyte disturbances, metabolic acidosis, or medications that may be employed to control other symptoms (e.g., opiates, benzodiazepines), this can be profoundly disturbing. Mistaking this delirium for anxiety and treating the patient with benzodiazepines may sedate the patient, but can also exacerbate the delirium. Treatment with antipsychotics such as haloperidol can provide relief and calm the patient during their final hours.

5.3 Systematic Care

A number of illnesses that are commonplace in the areas of central Africa, where EVD outbreaks occur, are clinically very similar to EVD. A patient who presents with a vague febrile illness without an obvious localizing feature or severe illness with or without gastrointestinal signs may have malaria, typhoid fever, shigellosis, or similar illness. Patients who present to the ETU with these diseases instead of EVD may get admitted to rule out EVD. As an epidemic wanes, the positive predictive value of the admitting criteria often drops, and an increasing number of patients will have non-EVD pathologies. Even patients who do have EVD may be coinfected with *Plasmodium* or a bacterial pathogen.

Given the diagnostic limitations common in the EVD care setting, it is common practice to treat patients upon admission with a clinical syndrome compatible with EVD presumptively with antimalarials and antibiotics covering locally relevant pathogens to reduce morbidity from treatable conditions. These may be present on admission, but treatable conditions may arise during the course of hospitalization, e.g., from transluminal migration of bowel flora. Clinicians should be alert for sudden changes in their patients' clinical evolution that may herald the onset of a secondary infection, which may also arise from common causes, such as indwelling medical devices. Where malaria testing can be done safely, e.g., in field lab bio-containment, targeted treatment for malaria for patients testing positive is an option.

5.4 Specific Care

Until the West African EVD outbreak, care for patients had been almost entirely supportive. Very few patients received therapeutic agents that directly addressed the causative ebolavirus or its known pathology (Mupapa et al. 1999b; Emond et al. 1977). This changed during the West African outbreak, and a number of patients received viral polymerase inhibitors, antisense therapy, convalescent plasma, or monoclonal antibodies (Sissoko et al. 2016; Dunning et al. 2016; van Griensven et al. 2016; Davey et al. 2016). The majority of these patients were treated in resource-limited contexts, many in the setting of a clinical trial of the agent employed. None of these therapies demonstrated significant efficacy at the time, but it may be hoped that the field of targeted ebolavirus therapy will advance.

A review of current ebolavirus-specific therapeutics is beyond the scope of this chapter, and would quickly become outdated. Any healthcare provider working with EVD patients should be aware of the current leading therapeutic options and planned clinical trials.

Important factors to consider when deciding whether or not to employ a novel therapeutic, in addition to the ethical concerns, are the route and duration of therapy and the patient monitoring requirements, as this will determine much of the associated workload. During an outbreak in a resource poor setting, this may be a non-trivial concern that requires a good deal of planning. As well, the ETU must be ready to manage adverse reactions and meet cold chain requirements.

5.5 Psychological Support

EVD patients suffer not only from the physical effects of the disease but also from the fear and anxiety that come with being an EVD patient. Patients are usually admitted without knowing if they have EVD or not, and this uncertainty while awaiting test results is a stressor. Knowing little about the disease they may or may not have adds to this stress. The patients' greatest fear is of dying, but this is also accompanied by fear of how one will be treated as a survivor in the community upon release. Patients are also stressed by isolation from friends and family during their illness. Helping patients cope with their fear and anxiety can reduce their suffering.

Addressing the EVD patients' psychological issues begins with the design of the ETU. Providing a space where patients can talk safely and comfortably with their friends and family outside will help reduce their sense of isolation. Having separate space with some degree of privacy for counselors to talk with patients and families facilitates their interactions.

Upon arrival, giving each patient and their family a briefing on the nature of the disease, the infection control measures in place, and the sequence of events that will occur (placement in the suspect patient's area, testing, and movement to confirmed or discharged based upon results, etc.) can provide patients with a sense of safety and lessen their uncertainty about what will happen to them during their stay in the ETU.

During the patient's stay in the ETU, the patient's emotional distress, can be minimized by providing their family with regular updates on the patient's condition and being attentive to changes in the patient's condition and any anxiety this may provoke. Critical moments to ensure psychological support to the patient are during the initial entry into the ETU and during the disclosure and explanation of test results. The patient's family will similarly need support at such moments, as well as when they are informed of the patient's death if this occurs. Presence among the health providers of former EVD patients certainly helps with answering patient questions and transferring relevant information to patients and families. Where direct contact between patients and family is not practical, the use of remote means, e.g., mobile telephone, video link via tablet device, etc., can be very helpful. The need for psychological support does not end at discharge, as many patients will need help with reintegration into their home community and adjustment to life as an EVD survivor.

5.6 *Impact of Disease on Care Provision*

The infection control procedures, infrastructure, and equipment required to care for EVD limit the proximity of healthcare staff to their patients. The design of the ETU will determine how well the staff in the low-risk area will be able to see and communicate with their patients. The PPE will limit how much time the healthcare staff can spend at the patients' bedsides. These in turn limit how well the staff can respond to their patients' needs. As with many illnesses, these needs evolve over the course of the disease, and understanding these changing needs may help healthcare workers plan the delivery of their care. Patients at different stages of EVD progression will have different care requirements. If patients in similar phases of their illness are grouped together, the delivery of their care may be made more efficient.

5.6.1 Early

Patients in the early stage of their illness do not yet suffer from significant disability. They are able to walk, drink, feed themselves, and engage in many aspects of self-care. They do not usually suffer yet from nausea and vomiting, and so they can still benefit from oral medications, oral fluid therapy, and meet their nutritional needs.

5.6.2 Patients with GI Compromise and Weakness

Many patients will proceed to a phase of illness where weakness and apathy will impair their ability to eat, drink, and take care of themselves. Eating and drinking are also made difficult by anorexia, sore throat, and difficulty swallowing common in EVD. Patients are often still able to eat and drink if they are given encouragement by the staff, and not infrequently by fellow patients. The onset of nausea and vomiting in many patients further complicates oral intake. Antiemetic therapy may provide some relief to the patient and allow further feeding and drinking, but weakness and apathy may require that they are assisted.

Diarrhea is another gastrointestinal symptom that often appears at this stage. This not only adds to fluid losses, but it also generates a volume of infectious waste that needs removal, as patients are often too weak to walk and may end up lying in their own feces until they can be cared for.

5.6.3 Patients in Shock

Patients who go into shock will need closer monitoring and hemodynamic support. Close monitoring of the patients' condition may be facilitated by ETU design that allows observation of the patient and monitoring equipment from the low-risk area. Noninvasive hemodynamic monitoring, although not routinely available in low-resource settings, was employed during the West African outbreak (Wong et al. 2015). Moving critically ill patients together to a part of the ETU where more labor-intensive care is delivered may allow for more efficient use of staff limited by the time that may be spent in PPE.

5.6.4 Recovering Patients

Patients who survive the acute phase of their illness and begin to recover can resume some degree of self-care. They will need more frequent contact with psychological and social support staff who can prepare them for their eventual return to their communities. Clinicians should be aware of the possibility of rare complications, such as late onset neurological disease (Jacobs et al. 2016; Howlett et al. 2016).

6 Management Issues Beyond the Patient

Treatment of EVD patients in the setting of an EVD outbreak places the provision of clinical care in the delicate position of being the most visible aspect of EVD outbreak control. How the care is perceived to be provided will play a role in public acceptance of outbreak control measures. As failure to successfully engage the community has had a negative effect on past attempts at outbreak control, clinicians caring for EVD patients should be aware of how their work is being perceived (Marais et al. 2016). Interactions with families, the community, religious leaders, and sometimes traditional healers, should include visits (taking safety concerns into account), and active communication with the help of survivors, if available.

Mistrust of outbreak control agencies, and especially those providing clinical care, is an all too common feature of EVD epidemics. Medical care providers have been accused of stealing patients' blood for pharmaceutical companies and killing their patients to sell their organs on the black market. Transparent provision of compassionate care is thus not only of benefit to the patient, but also good public relations.

Care providers should assume that word of their actions will reach the community in one form or another. Ensuring that all staff interact respectfully and compassionately with patients, responding to their needs in a timely manner, and reducing stressors in their environment, e.g., exposure to dying or deceased patients, will reflect positively on the ETU.

Excessive isolation is also an important source of patient stress, and an important link to the community is through the patients' friends and family. Infection control measures will limit their access to the patient, but good ETU design will allow for areas where a patient in the high-risk zone can be accessible, usually over a pair of low fences spaced two meters apart, to those who wish to visit them. This is not only a great comfort to the patient and their visitors, but it also allows members of the community to have an understanding of what goes on beyond their sight inside the ETU.

Just as the patients often have a great deal of stress and anxiety as a consequence of their illness, so do their friends and families. The same mental health services available at the ETU for the care of patients can be of great benefit to these people as well. This also provides an opportunity to ensure that there are no misunderstandings as to what is going on within the ETU.

6.1 Clinical Care as Part of Outbreak Response

Clinical care is an important part of outbreak control, not only because it brings the sick to a place where they can have better chance of survival and no longer infect others in the community, but also because it is a necessary element of winning over the community as a partner in outbreak control. Community engagement with control measures depends on their trust in outbreak response agencies, and this trust is earned, to an extent, by how these agencies are perceived to care for the sick. However, the ETU is not the only place clinicians are needed in EVD outbreak response.

Medical outreach, where a multidisciplinary team goes into the community to evaluate potential EVD cases identified by surveillance systems, profits greatly from the presence of clinicians experienced in the care of EVD patients. The determination in the community of whether someone needs to be admitted to the ETU benefits from a wealth of information available from family, friends, and neighbors who may not otherwise be available. As well, clinicians can respond to the questions and concerns of these same people.

For patients who refuse to be taken to the ETU, care of the EVD patient by the family in the home can be initiated as a second, if less desirable, option. Overly coercive attempts to bring the patient to the ETU can be counterproductive and break surveillance contact with the patient and their family. By allowing for the option of home-based care, the outbreak control team can maintain contact with the patient. This requires the explanation of infection control measure adapted for the home environment, designation of a single care provider, regular resupply of consumable materials, and ongoing medical follow up of the patient, as well as continued reassessment of willingness to come to the ETU.

The employment of EVD survivors on the outreach teams is invaluable, as they are able to explain to members of their community, in their own language and in familiar terms, what is to be expected in the ETU and how care there is delivered.

6.2 Care of Healthcare Providers

No EVD outbreak can be controlled, and no medical care delivered, without the able assistance of local staff. These people work under very adverse circumstances. They are often made unwelcome in their home community because of their work. At the same time, they are not infrequently called upon by their neighbors to carry out informal medical work in this same setting, where they do not have the protections in place that they would in the ETU. Their work exposes them on a regular basis to the suffering and death of people from their own community. The mental health of the staff suffers from all of this. They will, as well, occasionally become sick, both with ordinary illness, which they may mistake for EVD, and, occasionally with EVD itself. International staff also suffer from the harsh and traumatic conditions surrounding an EVD outbreak. A service for the care of everyone working in and around the ETU, both for their mental health and for any physical illness they may have, is essential. In addition to this, daily monitoring of the working and off-duty personnel allows an early notification of potential incident and should be used to reinforce proper behavior and procedures. Preventive vaccination of the staff, when available, will add additional relief.

7 Conclusion

It is tempting to entertain the idea that care of EVD patients in a resource-limited setting would somehow be a less complex undertaking than in settings where resources for advanced care are available. However, care of EVD patients in low-resource settings is invariably entangled with the demands of the outbreak. Many patients at different phases of their illness will need care that requires proper procedures and training, and the rapid construction of a structure fit for purpose which is situated in the midst of a community experiencing a traumatic event that looks to outside agencies for relief. Tropical heat and personal protective equipment will require the efficient organization of healthcare delivery to make the best use of available human resources. Caring for suffering patients while protecting medical staff and providing hope for a traumatized community is an undertaking that goes well beyond simply deciding on appropriate treatment for an often deadly disease.

Disclaimer The findings and conclusions in this chapter are those of the author(s) and do not necessarily represent the official position of the Centers for Disease Control and Prevention

References

Bray M, Chertow D (2016) Filoviruses. In: Richman D, Whitley R, Hayden F (eds) Clinical virology, 4th edn. ASM Press, Washington, DC, pp 1–28

Caluwaerts S, Fautsch T, Lagrou D (2016) Dilemmas in managing pregnant women with Ebola: 2 case reports. Clin Infect Dis 62(7):903–905

Davey RT, The Multi-National PREVAIL II Study Team (2016) PREVAIL II: A randomized controlled trial of ZMapp™ in Acute Ebola virus infection. Presented at the conference on retroviruses and opportunistic infections, Boston, 22–25 Feb 2016

Dellinger RP, Levy MM, Rhodes A (2013) Surviving sepsis campaign: international guidelines for management of severe sepsis and septic shock: 2012. Crit Care Med 41(2):580–637

Dörnemann J, Burzio C, Ronsse A et al (2017) First newborn baby to receive experimental therapies survives Ebola virus disease. J infect Dis 215(2):171–174

Dunning J, Sahr F, Rojek A et al (2016) Experimental treatment of Ebola virus disease with TKM-130803: a single-arm phase 2 clinical trial. PLoS Med 13(4):e1001997

Emond RT, Evans B, Bowen ET et al (1977) A case of Ebola virus infection. Br Med J 2:541–544

Evlampidou I, Rabelo I, Lee V (2016) Psychological suffering among Ebola virus disease survivors in Monrovia, Liberia: a mixed methods study. Front Public Health 4:142

Henao-Restrepo AM, Longini IM, Egger M et al (2015) Efficacy and effectiveness of an rVSV-vectored vaccine expressing Ebola surface glycoprotein: interim results from the Guinea ring vaccination cluster-randomised trial. Lancet 386(9996):857–866

Howlett P, Brown C, Helderman T (2016) Ebola virus disease complicated by late-onset encephalitis and polyarthritis, Sierra Leone. Emerg Infect Dis 22(1):150–152

International Food Policy Research Institute (IFPRI) (2016) Global nutrition report 2016: from promise to impact: ending malnutrition by 2030. Washington, DC

Jacobs M, Rodger A, Bell DJ (2016) Late Ebola virus relapse causing meningoencephalitis: a case report. Lancet 388(10043):498–503

Kortepeter MG, Bausch DG, Bray M (2011) Basic clinical and laboratory features of filoviral hemorrhagic fever. J Infect Dis 204(Suppl 3):S810–S816

Kreuels B, Wichmann D, Emmerich P et al (2014) A case of severe Ebola virus infection complicated by gram-negative septicemia. N Engl J Med 371(25):2394–2401

Maganga GD, Kapetshi J, Berthet N et al (2014) Ebola virus disease in the Democratic Republic of Congo. N Engl J Med 371(22):2083–2091

Mahanty S, Bray M (2004) Pathogenesis of filoviral haemorrhagic fevers. Lancet Infect Dis 4(8):487–498

Maitland K, Kiguli S, Opoka RO et al (2011) Mortality after fluid bolus in African Children with severe infection. N Engl J Med 364(26):2483–2495

Marais F, Minkler M, Gibson N (2016) A community-engaged infection prevention and control approach to Ebola. Health Promot Int 31(2):440–449

Marik P, Bellomo R (2016) A rational approach to fluid therapy in sepsis. Br J Anaesth 116(3):339–349

Martini GA, Schmidt HA (1968) Spermatogenic transmission of the "Marburg virus": causes of "Marburg simian disease". Klin Wochenschr 46:398–400 (In German)

Mate SE, Kugelman JR, Nyenswah TG et al (2015) Molecular evidence of sexual transmission of Ebola virus. N Engl J Med 373(25):2448–2454

McElroy A (2015) Understanding bleeding in Ebola virus disease. Clin Adv Hematol Oncol 13(1):29–31

Mupapa K, Mukundu W, Bwaka MA et al (1999a) Ebola hemorrhagic fever and pregnancy. J Infect Dis 179(Suppl 1):S11–S12

Mupapa K, Massamba M, Kibadi K et al (1999b) Treatment of Ebola hemorrhagic fever with blood transfusions from convalescent patients. J Infect Dis 179(Suppl 1):S18–S23

The National Institute for Occupational Safety and Health (NIOSH) (2015) Hierarchy of controls. https://www.cdc.gov/niosh/topics/hierarchy/. Accessed 10 Aug 2016

Rabelo I, Lee V, Fallah MP et al (2016) Psychological distress among Ebola survivors discharged from an Ebola treatment unit in Monrovia, Liberia—a qualitative study. Front Public Health 4(142)

Rollin PE, Bausch DG, Sanchez A (2007) Blood chemistry measurements and D-Dimer levels associated with fatal and nonfatal outcomes in humans infected with Sudan Ebola virus. J Infect Dis 196(Suppl 2):S364–S371

Sissoko D, Laouenan C, Folkesson E et al. (2016) Experimental treatment with Favipiravir for Ebola virus disease (the JIKI Trial): a historically controlled, single-arm proof-of-concept trial in Guinea. PLoS Med 13(3). doi:10.1371/journal.pmed.1001967

Spengler JR, McElroy AK, Harmon JR et al (2015) Relationship between Ebola virus real-time quantitative polymerase chain reaction-based threshold cycle value and virus isolation from human plasma. J Infect Dis 212(Suppl 2):S98–S100

Sprecher AG, Caluwaerts A, Draper M et al (2015) Personal protective equipment for filovirus epidemics: a call for better evidence. J Infect Dis 212(Suppl 2):S346–S349

United Nations, Department of Economic and Social Affairs, Population Division (2015) World population prospects: the 2015 revision. https://esa.un.org/unpd/wpp/. Accessed 10 Aug 2016

van Griensven J, Edwards T, de Lamballerie X et al (2016) Evaluation of convalescent plasma for Ebola virus disease in Guinea. N Engl J Med 374(1):33–42

Vetter P, Kaiser L, Schibler M et al (2016) Sequelae of Ebola virus disease: the emergency within the emergency. Lancet Infect Dis 16(6):e82–e91

Wamala JF, Lukwago L, Malimbo M et al (2010) Ebola hemorrhagic fever associated with novel virus strain Uganda 2007-2008. Emerg Infect Dis 16(7):1087–1092

WHO/UNICEF/WFP (2014) Interim guideline: nutritional care of children and adults with Ebola virus disease in treatment centres. Geneva, Switzerland

Wong KK, Perdue CL, Malia J et al (2015) Monrovia medical unit. Supportive care of the first 2 Ebola virus disease patients at the Monrovia medical unit. Clin Infect Dis 61(7):e47–e51

World Health Organization (2014) Case definition recommendations for Ebola or Marburg virus diseases. http://www.who.int/csr/resources/publications/ebola/case-definition/en/. Accessed 10 Aug 2016

World Health Organization (WHO) (2016) Clinical management of patients with viral haemorrhagic fever: a pocket guide for the front-line health worker. Geneva, Switzerland

Clinical Management of Patients with Ebola Virus Disease in High-Resource Settings

G. Marshall Lyon, Aneesh K. Mehta and Bruce S. Ribner

Abstract Like most viral illnesses in humans, supportive care of the patient is the mainstay of clinical care for patients with Ebola virus disease (EVD). The goal is to maintain and sustain the patient until a specific immune response develops and clears the viral infection. Clearly, antiviral therapy may eventually help speed recovery, but supportive care will likely always be the centerpiece of care of the patient with EVD. While terrible in terms of human suffering and loss, the EVD outbreak of 2014–2016 provided an unheralded opportunity to advance our understanding in the care of patients (WHO 2016). Regardless of the care setting, resource-rich or resource-constrained, it is beneficial to have an established team of care providers. This team should consist of nurses and physicians who are familiar with clinical care of patients with EVD and have demonstrated competency using necessary personal protective equipment (PPE). Consideration should be given to having several physician specialties on the team, including critical care, infectious diseases, and anesthesiology. Additional individuals in other medical specialties should be identified in case needed during the course of caring for a patient. The National Ebola Training and Education Center (NETEC) has detailed guidance on preparations for developing a high-containment unit and care team (NETEC 2016).

G. Marshall Lyon
Division of Infectious Diseases, Emory University School of Medicine, 101 Woodruff Circle, WMRB 2101, Atlanta, GA 30322, USA

A.K. Mehta
Department of Medicine, Division of Infectious Diseases, Emory University School of Medicine, 101 Woodruff Circle, WMRB 2101, Atlanta, GA 30322, USA
e-mail: aneesh.mehta@emory.edu

B.S. Ribner (✉)
Department of Medicine, Division of Infectious Diseases, Emory University School of Medicine, Emory University Hospital, 1364 Clifton Road NE, Suite B705, Atlanta, GA 30322, USA
e-mail: bribner@emory.edu

Current Topics in Microbiology and Immunology (2017) 411:115–138
DOI 10.1007/82_2017_19

Published Online: 11 June 2017

Contents

1 Symptoms and Clinical Manifestations

After an incubation period of two to 21 days (mode 8–10), patients will have the onset of symptoms (Fig. 1). Initial symptoms are not specific and can have the appearance of a myriad of tropical infections, including those which are most common, i.e., malaria, typhoid, influenza. These initial symptoms include fever, headache, myalgias, malaise, and sore throat (Del Rio et al. 2014). Toward the end of the first week, fever, headache, and malaise continue and the onset of the gastrointestinal (GI) phase begins (Chertow et al. 2014; Lyon et al. 2014). After the first few days of illness, a rash may also appear. The rash can be either a classic viral exanthem or petechial. The rash may begin as an exanthem but progress to petechial as the platelets drop and vascular leakage becomes more pronounced.

The GI phase is critical for patient care as this is when many patients become critically ill with volume and electrolyte derangements. The first days of the GI phase are characterized by both vomiting and diarrhea. The vomiting makes oral rehydration difficult as the patient is losing 2–3 L per 24-h period orally (Chertow

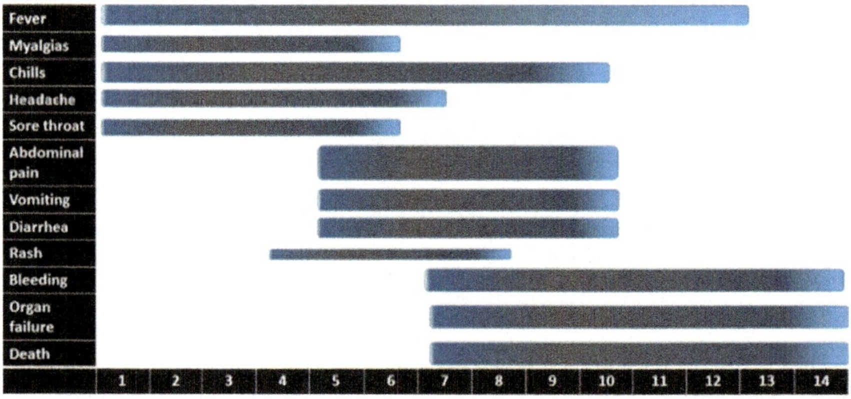

Fig. 1 Relative timing of symptoms with Ebola virus disease (EVD) as well as the relative intensity of symptoms

et al. 2014). After 2–3 days, the vomiting begins to improve, but the diarrhea continues to worsen. It is common for many patients to continue losing between 3 and 5 L of fluid per day and in the severest cases can be up to 10 L of output per day per rectum (Kraft et al. 2015). Massive fluid loss along with vascular leakage which occurs in Ebola virus disease (EVD) can lead to intravascular hypovolemic dehydration. Uncorrected, these derangements may lead to cardiac arrhythmias which can be fatal (Liddell et al. 2015; Lyon et al. 2014).

It is during the GI phase when laboratory abnormalities start to become evident. One of the hallmarks of EVD is a rise in aspartate aminotransferase (AST) concentrations, which is markedly higher than the rise of alanine aminotransferase (ALT) concentrations. Thrombocytopenia is also an early abnormal laboratory finding which appears to precede clinical bleeding (Kraft et al. 2015; Lyon et al. 2014). With fluid loss also comes abnormal low electrolyte concentrations, most notably potassium and calcium (Lyon et al. 2014). Albumin concentrations are also low early in the illness. Hypoalbuminemia may be a result of loss through the GI tract, decreased nutritional intake, or due to vascular leakage. Low albumin concentrations complicate the physiology and management of patients with EVD as it leads to low oncotic pressure and contributes to hypovolemia and hypotension.

The GI phase lasts into the first part of the second week. As a consequence of the GI phase and viral sepsis, intravascular hypovolemia and vascular leakage, organ failure can develop. Coagulopathy, renal failure, and respiratory failure are the most commonly encountered organ systems which fail. Coagulopathy typically manifests as clinical bleeding. GI bleeding is the most common manifestation. Patients may also experience oozing from venipuncture sites and intravenous catheter sites (Uyeki et al. 2016b; Respiratory vaccine shows promise 2014; Qin et al. 2015; Bah et al. 2015). In previous EVD outbreaks, bleeding from gums and the urinary tract have also been noted. Bleeding complications has also been linked to poor outcomes (Ndambi et al. 1999). Renal failure is an uncommon complication of EVD occurring in about 5% of cases (Bah et al. 2015). Although uncommon, renal failure has historically been associated with increased mortality (Schieffelin et al. 2014). Renal failure is felt to be either pre-renal from shock (Bah et al. 2015) or from acute tubular necrosis (ATN) from direct infection of the renal tubules by Ebola virus (Martines et al. 2015).

Death from EVD has been associated with several clinical features or demographic characteristics. Older age is associated with a higher risk of dying from EVD. For patients who have or might get EVD, forty years is old. This has been noted across multiple outbreaks of EVD with different variants of Ebola viruses, from Kikwit in the 1990s to West Africa in 2014–2016. In addition to age, severity of illness is another predictor of mortality. For EVD, this is borne out as higher viral load, higher temperature, higher blood urea nitrogen (BUN) and creatinine (Cr) concentrations, and higher AST concentrations. Higher BUN and Cr concentrations represent renal failure, and higher AST concentration represents increased liver injury. Clinically evident hemorrhagic manifestations such as melena, petechiae, ecchymoses, and oozing puncture sites are all associated with increased mortality (Roddy et al. 2012; Geisbert 2015). The 2014–2016 outbreak in West Africa may have indicated that the level of care is also associated with mortality. The overall mortality from EVD in West Africa was 40% (WHO 2016). However, in persons who were evacuated from West Africa to either Europe or the USA, the mortality rate was 18.5% (Uyeki et al. 2016b). This suggests that the higher level of care provided outside of West Africa decreased the mortality rate by half. Of the 27 patients who were evacuated out of Africa, five died and three of those had respiratory and renal failure (Uyeki et al. 2016b). However, there were nine total patients who had respiratory failure and nine patients who had oliguria indicating possible renal failure. Seven of the nine with respiratory failure received mechanical ventilation, and five of the nine with renal failure received dialysis (Uyeki et al. 2016b). These patients had severe illness yet the mortality rate was half of what was seen in West Africa. What is not clear is whether these patients would have lived or died had they remained in West Africa. But, it seems plausible that improving the level of supportive care in West Africa may reduce the mortality associated with any future outbreaks of EVD.

2 Management of Complications

Given the severity of illness many patients with EVD have, it is important to have an almost "expectant management" philosophy. Expectant management means anticipating potential problems and pitfalls and taking corrective action before they actually arise. In this way, problems can be avoided or mitigated. Knowing that patients with EVD can develop hypokalemia, providers may choose to supplement potassium as the gastrointestinal phase develops. Similarly, expecting hemorrhage to potentially be an issue, transfusing platelets when the platelet counts reach 30–50 K/μL (as opposed to <20 K/μL), or providing fresh frozen plasma when the international normalized ratio (INR) is 1.5 should be strongly considered. Anticipating problems and intervening at an earlier time point than in usual clinical practice may help prevent a very sick patient from becoming a critically ill patient, hopefully saving the patient's life.

Volume and electrolytes

Volume loss can be significant in EVD during the gastrointestinal phase with up to 10 L of fluid loss in a 24-h period. Matching this volume loss with oral rehydration or intravenous fluids is challenging. However, it should be the goal of fluid replacement to maintain intravascular volume to avoid shock. If the patient has a lot of vomiting, it may be very difficult to do this with oral rehydration, and intravenous access should be obtained before significant dehydration develops. Intravenous replacement fluids should ideally be done with balanced crystalloid solutions such as lactated ringers, though the use of normal saline is certainly appropriate if balanced crystalloid solutions are not readily available. There is concern that relying solely on normal saline or similar fluids can result in hyperchloremia and potentially induce or exacerbate renal failure (Guidet et al. 2010). However, in EVD patients where colloidal products may also be needed, using normal saline as a carrier or to supplement colloids is likely very appropriate. Normal saline should not be avoided at the expense of maintaining intravascular volume.

Electrolyte loss can be significant during the GI phase especially if diarrhea is voluminous on several consecutive days. Of patients medically evacuated to Europe or the USA, hyponatremia, hypokalemia, hypocalcemia, hypomagnesemia, and hypoalbuminemia were seen in the majority of patients (Uyeki et al. 2016b; Lyon et al. 2014; Kraft et al. 2015; Wolf et al. 2015). Whether a consequence of the electrolyte imbalance or as a part of the EVD, cardiac arrhythmias were seen in almost half of the patients medically evacuated out of West Africa (Uyeki et al. 2016b).

One word of caution on aggressive rehydration—in resource-constrained areas overly aggressive rehydration with intravenous fluids can lead to additional problems. As the septic phase begins, patients develop vascular leakage. While this is not an issue in subcutaneous tissues, it can be problematic in the lungs. Pulmonary edema can lead to hypoxia and the need for supplemental oxygen or mechanical ventilation. Clinicians need to be aware of the potential to overhydrate and cause additional problems. In areas where renal replacement therapy is not readily available, correcting overhydration may not be easily accomplished.

Gastrointestinal

The GI phase of the disease is where significant morbidity and mortality happens with EVD (Dallatomasina et al. 2015). In one center with a mortality rate of about 50%, nearly 90% of the deaths occurred during the GI phase (Dallatomasina et al. 2015). It should again be noted that the later part of the GI phase overlaps with the beginning of sepsis and multi-organ failure. Therefore, aggressive care during the GI phase may help ameliorate mortality in patients with EVD.

Antiemetic medications can help reduce the amount of vomiting experienced by patients. Clinicians should be proactive when using antiemetic medications especially if they have to rely heavily on oral rehydration. Similarly, antidiarrheal medications may help diminish volume loss and electrolyte loss (Chertow et al. 2015). Although antidiarrheal medications like loperamide may not abate diarrhea altogether, one or two fewer liters of diarrheal loss may be the difference between life and death. However, the use of loperamide and similar agents has not been definitively shown beneficial in EVD.

Hemorrhage

Hemorrhage is historically a hallmark of EVD. In studies which have examined the correlation, there is an association between hemorrhage and mortality (Ndambi et al. 1999). Therefore, correcting coagulopathy is a goal of supportive care in EVD. Again, early transfusion of platelets or infusion of clotting factors via fresh frozen plasma should be a goal when feasible. This may be more easily achieved in resource-rich environments where platelet counts and INR can be more closely monitored. As a surrogate, oozing from intravenous catheter sites or phlebotomy sites should prompt transfusion of platelets and fresh frozen plasma. If frank hemorrhage occurs as manifested by visible bleeding, hematemesis or melena, transfusion of platelets, clotting factors, and red blood cells is likely warranted. Here, the goal is to maintain acceptable hemoglobin levels and correct coagulopathy. Transfusion will likely also help maintain intravascular hydration and blood pressure.

Organ failure

Organ failure is a rare but serious, life-threatening complication in EVD.

Renal failure develops in about 5% of patients with EVD historically (Bah et al. 2015), but was seen in 20% of medical evacuees in the 2014–2016 West Africa outbreak (Uyeki et al. 2016b). In patients evacuated to Europe or the USA, 20% had renal failure and two were described as having renal failure with anuria (Wolf et al. 2015; Connor et al. 2015). Both of these patients received renal replacement therapy with dialysis. Both patients also had recovery of renal function such that dialysis was no longer needed by 2 months past onset of symptoms. In both of these cases, dialysate fluid was tested for the presence of Ebola virus (EBOV) RNA and was found in one instance at one of the institutions (Connor et al. 2015; Wolf et al. 2015). It should be noted that dialysis membranes are designed to allow the passage of electrolytes but not larger molecules such as proteins or viruses. Therefore, it would be expected that dialysate fluid should be without EBOV. But both institutions autoclaved dialysate fluid so as to eliminate the potential for environmental contamination (Wolf et al. 2015; Connor et al. 2015). While renal failure is associated with a higher risk of dying from EVD, aggressively supporting the patient with dialysis or continuous renal replacement therapy can have positive, life-saving effects. Bah et al. (2015) reported severe pre-renal kidney failure in one of their patients. This patient responded well to intravascular rehydration with crystalloid intravenous solutions.

Respiratory failure is also an uncommon but ominous development in patients with EVD. In patients who were medically evacuated to either Europe or the USA, respiratory failure was described as hypoxic or mixed hypercapneic/hypoxic respiratory failure (Sueblinvong et al. 2015; Kraft et al. 2015; Uyeki et al. 2016b; Wolf et al. 2015). Patients who had mild isolated hypoxia often responded to supplemental oxygen (Uyeki et al. 2016b). Some of these evacuated patients progressed to respiratory failure, requiring either noninvasive positive pressure ventilation or endotracheal intubation with mechanical ventilation (Uyeki et al. 2016b; Sueblinvong et al. 2015; Wolf et al. 2015). Despite going on to need mechanical ventilation, the majority of patients eventually recovered from their respiratory failure.

Cardiac abnormalities were frequently noted (41%) in patients medically evacuated from West Africa to Europe or the USA (Uyeki et al. 2016b). Most of these abnormalities were arrhythmias seen on cardiovascular monitoring or electrocardiogram (Uyeki et al. 2016b). However, life-threatening arrhythmias, such as atrial fibrillation, are also possible and require intervention in order to correct the abnormal rhythm (Kraft et al. 2015; Sueblinvong et al. 2015). Fatal arrhythmias developed in patients who die from EVD.

Vascular leakage is a common phenomenon in EVD. In some patients, this can lead to third spacing of fluid and resultant intravascular hypovolemia. This vascular leakage typically develops about the same time as the GI phase and quickly reverses once recovery begins (Lyon et al. 2014). Two patients evacuated to Atlanta in the USA were noted to quickly mobilize and auto-diurese about 10 L of fluid in less than 48 h (Lyon et al. 2014). In patients who develop sepsis cardiovascular collapse may occur (Wolf et al. 2015; Sueblinvong et al. 2015; Kraft et al. 2015). Aggressively supporting the patient with the use of inotropes and volume resuscitation can result in survival and recovery (Kraft et al. 2015; Sueblinvong et al. 2015; Wolf et al. 2015).

Coincidental infections in patients with EVD are frequently noted. The WHO Handbook for EVD care recommends giving antimalarial therapy and antibacterials empirically because of the high rates of malaria, typhoid, and other infections in Africa. At least one patient who was medically evacuated from West Africa to Germany developed sepsis and was treated with several sequential and concomitant antibiotics (Wolf et al. 2015). While only one blood culture was positive for a *Staphylococcus* species (Wolf et al. 2015), processing of blood cultures is difficult in high-containment units due to a lack of availability of automated blood culture incubation systems. The sensitivity of detecting bacterial sepsis is likely lower than in normal care settings, and routine blood culturing of patients with EVD in resource-constrained areas is unlikely to be feasible.

Neurologic complications are commonly noted in patients with EVD. Most patients display some degree of cognitive impairment, likely from dehydration and severe illness. Seizures have been seen as well (Bah et al. 2015). At least one patient medically evacuated to the USA was felt to have hemorrhagic encephalitis based on magnetic resonance imaging obtained after recovery from EVD (personal knowledge of authors). One survivor of EVD had a late-onset meningoencephalitis due to EBOV (Jacobs et al. 2016). This resulted in prolonged illness with neurologic complications and deficits which slowly resolved over time (Jacobs et al. 2016).

Post-Ebola syndrome

The most common symptoms experienced by survivors of EVD are fatigue, musculoskeletal pain (70%), headache (48%), and eye abnormalities (14%) (Scott et al. 2016). The musculoskeletal pain is often joint pain primarily affecting large joints (Scott et al. 2016). Pain in tendon insertions, or enthesitis, is also commonly noted. Eye disease has been described as either an anterior, posterior, or pan uveitis and can be associated with virus detected by PCR in conjunctival swabs, or detection of virus by PCR in the aqueous fluid of the eye (Varkey et al. 2015; Kibadi et al. 1999). Post-Ebola syndrome is poorly understood, and at the time of this writing, there are ongoing efforts to better characterize the symptoms of post-Ebola syndrome as well as attempts to gain an understanding of the pathophysiology.

3 Therapeutics for Ebola Virus Disease

EVD can be caused by any of four ebolaviruses: Ebola virus (EBOV, previously known as Zaire ebolavirus), Sudan virus (SUDV, previous known as Sudan ebolavirus), Bundibugyo virus (BDBV), and Taï Forest virus (TAFV). In addition, there exists an additional ebolavirus, Reston virus (RESTV) that is only known to cause disease in non-human primates. While aggressive supportive care remains the only therapeutic modality proven to be efficacious in treating patients with EVD, there have been ongoing efforts to develop agents specifically for EVD. The magnitude and longevity of the 2014–2016 West Africa EVD outbreak highlighted the need for efficacious therapeutics against Ebola viruses to augment the supportive care measures in the treatment of individual patients and also to assist in diminishing the spread of the virus in communities. Thus, the 2014–2016 EVD outbreak allowed for greater attention and support for research into the development of these therapeutics, many of which are now in clinical studies (WHO Web site http://www.who.int/medicines/emp_ebola_q_as/en/). These therapeutics fall into three main categories: immunologic therapies, antiviral molecules, and vaccines.

3.1 Passive Immune Therapy

Convalescent blood and plasma

The idea of transferring protective immunity from survivors of an infectious disease to patients with an active infection has been studied and used since the beginning of the twentieth century, including such pathogens as polioviruses, influenza viruses, hepatitis B virus, cytomegalovirus, and Ebola viruses (Winkler and Koepsell 2015). The first known use of this modality in EVD occurred during the 1976 outbreak when a laboratory worker had an occupational exposure and was infected with an Ebola virus. This patient received transfusions of two units of convalescent plasma and had a full recovery (Emond et al. 1977; Winkler and Koepsell 2015). Following this experience, eight patients received whole blood transfusions obtained from five different EVD survivors during the 1995 EVD outbreak in Kikwit, Democratic Republic of Congo (Mupapa et al. 1999). While only 1 patient who received these transfusions succumbed to EVD, this cohort received significantly higher levels of care than the majority of patients in the Kikwit outbreak (Mupapa et al. 1999). Since this time, multiple animal studies have shown the potential efficacy of transferring a high-quality, species-specific convalescent serum/plasma for EBOV infections (Winkler and Koepsell 2015; Zeitlin et al. 2016).

Once the 2014–2016 EVD outbreak caught international attention, there were calls from WHO and other agencies to explore the utilization of convalescent blood products for the multitude of patients in the EVD treatment units (ETUs) in affected nations (World Health Organization (WHO) 2014). The ability to safely separate and store vast amounts of whole blood or plasma from EVD survivors proved to be difficult in these resources-constrained environments undergoing a devastating epidemic. As a few nationals of European nations and the USA responding to the outbreak became infected with Ebola and were repatriated for treatment of EVD, the receiving centers began to explore collecting convalescent plasma from their recovered patients. The first use of a convalescent blood product for EVD outside of Africa occurred at the University of Nebraska Medical Center (UNMC), when plasma collected from a previous EVD survivor was transfused after informed consent to another patient in that ETU (Kraft et al. 2015). Since that time, a national repository of Ebola Convalescent Plasma (ECP) has been created at Emory University in Atlanta, GA, USA, from survivors living in the USA, and multiple units have been distributed for the care of EVD patients in the USA (Kraft et al. 2015; Florescu et al. 2015; Liddell et al. 2015; Winkler and Koepsell 2015). There were also several patients in Europe who received convalescent blood products (Uyeki et al. 2016b). However, these few cases, in the absence of controlled clinical trials, do not provide adequate data to assess efficacy or safety of these products. Three clinical trials of ECP were initiated in West Africa, though enrollment was slow as the outbreak began to wane and data are not yet available. Until clinical trials of standardize convalescent plasma products are performed, it will remain unclear what is the role of these products in the treatment of EVD.

Ebola virus-specific antibodies

Another approach to passive immunity is the use of monoclonal antibodies (mAb) with known activity against EBOV. The challenges to isolating and producing EBOV-specific antibodies with activity in humans have been recently reviewed by Zeitlin et al. (2016). A significant advance in this area came as the developers of two different mAb cocktails, ZMab and MB-003, which had shown good antiviral activity in NHP models, developed a collaboration to identify the best mABs to include into a cocktail (Zeitlin et al. 2016). This novel combination, later termed ZMapp™ (Mapp Biopharma), was tested in a pivotal study by Qiu et al. (2014), showing 100% protection in EBOV-infected rhesus macaques (*Macaca mulatta*) when treatment with this cocktail was initiated within 5 days of infection. Furthermore, the study demonstrated rescue of animals even with advanced EVD (transaminitis and hemorrhages) with ZMapp™ (Qiu et al. 2014).

This study of ZMapp™ in non-human primates (NHP) had just completed, when in August of 2014 two humanitarian aid workers from the USA developed EVD. The medical team caring for these two patients was offered the use of ZMapp™, and after consenting the patients for the compassionate use of this experimental agent, they became the first human recipients of ZMapp™ (Lyon et al. 2014). Since this experience, EBOV-specific monoclonal cocktails, including ZMapp™, ZMab, and MIL77, have been used in 11 other patients evacuated from West Africa (Uyeki et al. 2016b). However, given that all of this experience was outside of controlled clinical trials, it is impossible to ascertain the impact of any of these agents on the outcomes of these patients.

While sources (i.e., recovered patients) of convalescent blood products may or may not be readily available during a progressive outbreak, these products will always remain difficult to use given the measures needed to collect, store, and administer them safely. In addition, there is heterogeneity between batches. The advent of recombinant mAb, with known specificities and activities and more ease of storage and delivery, will likely prove to be a promising therapeutic strategy for EVD and other similar outbreaks in the near future. Hopefully, manufacturing techniques to scale up production when needed and the studies needed to assess the efficacy and safety of these products will be ready for next outbreak.

3.2 Antiviral Agents

Pharmaceutical antiviral agents for EBOV generally fall into two categories: compounds that inhibit viral replication and small-molecule inhibitors of virus entry and endosomal escape.

Inhibitors of Ebola virus replication

Favipiravir (T-705; Toyama Chemical Co Ltd) is a broad-spectrum antiviral purine analog, which becomes phosphorylated into its active form intracellularly and inhibits RNA-dependent RNA polymerase (Furuta et al. 2013). Favipiravir has many appealing characteristics to support its use in EVD: a good safety profile in humans, large amounts of the drug are readily available (having been produced for pandemic influenza), and good preliminary data in animal models of EVD (Madelain et al. 2016). The drug was used in 10 of the 27 EVD patients evacuated to the USA and Europe, and was overall well tolerated (Uyeki et al. 2016b). In addition, a non-comparative proof-of-concept trial (JIKI trial) was conducted in Guinea, in which all patients received favipiravir along with standardized care. While the study design did not allow for assessing the efficacy of the drug, overall it was well tolerated by the study participants (Sissoko et al. 2016).

BCX4430 (BioCryst Pharmaceuticals), an adenosine nucleoside analogue, inhibits viral RNA-dependent RNA polymerase activity indirectly through non-obligate RNA chain termination and has shown a broad spectrum of activity against RNA viruses, including filoviruses, flaviviruses, bunyaviruses, arenaviruses, paramyxoviruses, picornaviruses, and coronaviruses (Taylor et al. 2016). The efficacy of BCX4430 has been studied in two different NHP models of EVD (Madelain et al.

2016). In a cynomolgus macaque (*Macaca fascicularis*) model, animals which were given BCX4430 48 h after an otherwise lethal EBOV exposure had prolonged survival time but a similar survival rate (Madelain et al. 2016). In a rhesus macaque model, animals were given high doses of BCX4430 intramuscularly after an otherwise lethal EBOV exposure and 67% of treated animals survived versus none of the untreated animals. A phase I study to evaluate the safety, tolerability, and pharmacokinetics of BCX4430 is ongoing (Madelain et al. 2016).

Small-molecule inhibitors of Ebola virus

TKM-Ebola (Tekmira Pharmaceuticals, British Columbia, Canada) is a lipid nanoparticle containing a combination of small interfering RNA (siRNA) molecules that block the translation of three Ebola viral proteins: RNA-dependent RNA polymerase, VP24, and VP35 (Geisbert et al. 2010; Mendoza et al. 2016). The original formulation, TKM-100802, was designed to target these three genes of the 1995 Kikwit EBOV variant. It was shown to prevent disease in 66 and 100% of rhesus macaques given four or seven, respectively, treatments of the agent after an otherwise lethal EBOV infection (Geisbert et al. 2010). The first human use of TKM-100802 occurred in two patients evacuated to the USA from West Africa in late 2014 (Kraft et al. 2015) and subsequently in three additional evacuated patients (Uyeki et al. 2016b). However, these patients received advanced supportive and other experimental agents, making it difficult to ascertain any treatment effect of TKM-100802. Investigations in the matching of TKM-100802 to the new West African EBOV variant (Makona) revealed mismatches for the polymerase gene and the VP35 gene (Thi et al. 2015). A new formulation, TKM-130803, was designed to better match the circulating EBOV strain and demonstrated good protection in a small study of rhesus macaques given an otherwise lethal inoculum of EBOV (Thi et al. 2015). Based on these results, the Rapid Assessment of Potential Interventions and Drugs for Ebola (RAPIDE-TKM) trial was designed and launched in early 2015. In this single-arm phase 2 trial, patients with EVD were administered a 0.3 mg/kg intravenous infusion of TKM-130803 daily for up to seven days (Dunning et al. 2016). The study enrolled 14 patients, of whom 11 died and 3 survived, and subsequently, the trial was halted due to meeting its pre-specified futility threshold (Dunning et al. 2016).

AVI-7537 (Sarepta Therapeutics Inc, Cambridge, Massachusetts) is an antisense small nucleic acid oligomer, known as a phosphorodiamidate morpholino oligomer (PMO), which targets the VP24 gene of EBOV and interferes with its translation (Madelain et al. 2016). AVI-7537 has been evaluated in NHP models of EBOV infection, though usually in combination with AVI-7539, a PMO targeting the VP35 gene of EBOV (Madelain et al. 2016). However, in one study, rhesus monkeys received either AVI-7537+AVI-7539, AVI-7537, AVI-7539, or placebo and found to have survival rates of 62.5, 75, 0, and 0%, respectively (Warren et al. 2015). In addition, viral loads were similar in the AVI-7537 and AVI-7537 +AVI-7539 groups, but lower than the AVI-7539 and control groups (Warren et al. 2015). These data indicated that AVI-7537 was potentially sufficient to prevent disease and thus moved forward for clinical development (Madelain et al. 2016). AVI-7537 (in combination with AVI-7539) was evaluated in a phase I human study, showing it was well tolerated and without severe adverse reactions (Heald et al. 2014). However, plans for further clinical trials are unknown at this time.

4 Vaccines

Developing an effective vaccine against EBOV is a paramount aim for preventing outbreaks of EVD and may play an important role tempering an ongoing outbreak and caring for those who have been exposed to EBOV. While research into EBOV vaccines was ongoing, the West African EVD outbreak brought much needed attention and resources to these endeavors. Currently, almost 20 different vaccines are in preclinical or clinical development, almost 10 times the number in 2013 (Ohimain 2016). Of these, EBOV vaccines utilizing recombinant vesicular stomatitis virus (rVSV), chimpanzee adenovirus (ChAd), or adenovirus type 26 vectors have made the most progress to date.

Recombinant vesicular stomatitis virus-vectored vaccine

VSV belongs to the *Rhabdoviridae* virus family and causes vesicles and ulcerations of the mouth, feet, and teats of livestock (Roberts et al. 1999). A replication-competent recombinant VSV (rVSV)-vectored EBOV vaccine (rVSV-ZEBOV) was developed by the Public Health Agency of Canada (PHAC). In August of 2014, PHAC donated 800 vaccine doses to the WHO to conduct rapid human studies. The VSV Ebola Consortium (VEBCON) designed a group of studies of these vaccines to assess the safety and immunogenicity of various doses of rVSV-ZEBOV in Gabon, Kenya, Germany, and Switzerland (Agnandji et al. 2016). A total of 158 healthy adults received either rVSV-ZEBOV in doses of ranging from 300,000 to 50 million plaque-forming units (PFU) or placebo. EBOV-specific antibody responses were detected in all the study participants with higher neutralizing antibody titers seen at higher doses of the vaccine (Agnandji et al. 2016). The vaccine was also found to have significant rates of reactogenicity, with fever in 30% of vaccine recipients and vaccine related arthritis in 22% of recipients at one center (Agnandji et al. 2016). rVSV was detected in the blood of 95% of vaccinees, skin vesicles of two recipients and in the synovial fluid aspirate of an additional subject (Agnandji et al. 2016). Contemporaneous to these studies, the rVSVΔG-ZEBOV-GP group conducted two phase 1, placebo-controlled, trials of the same vaccine in the USA, at Walter Reed Medical Center and at the National Institutes of Health (Regules et al. 2017). A total of 40 adults received the rVSV-ZEBOV vaccine at a dose of 3 million or 20 million PFU. All vaccine recipients developed EBOV-specific antibodies, with recipients of the higher doses having statistically higher antibody titers and higher levels of neutralization (Regules et al. 2017). Again, reactogenicity was very common with 55% of recipients reporting fevers and >75% reporting significant injection site pain (Regules et al. 2017). rVSV viremia was found in all vaccinees, though no significant arthritis nor skin disease was reported in this study (Regules et al. 2017). Given the concerns of arthritis from the previous studies, the VEBCON group developed a dose comparative study, evaluating 300,000 versus 10 million versus 50 million PFU dosages of the vaccine (Huttner et al. 2015). As expected, EBOV-specific antibody titers and neutralizing titers were lower with the low-dose vaccine (Huttner et al. 2015). Reactogenicity remained common even at the lower dose (88 vs. 98% in the higher doses), though fevers, chills, and myalgias were all significantly lower in the low-dose group. Importantly, the investigators found no significant reduction in the rates of vaccine-related arthritis or dermatitis (Huttner et al. 2015). Taking these data together, the developers of this vaccine (NewLink Genetics [Ames, IA, USA] and Merck Sharp and Dohme [Kenilworth, NJ, USA]) moved forward trials of 20 million PFU dosage of the rVSV-ZEBOV, renamed V920.

The pivotal study of this vaccine was conducted in Guinea with an innovative cluster-ring strategy trial. In this open-label study, 90 clusters of contacts of known cases of EVD, 7651 subjects in total, were randomized to immediate vaccination or 21-day delay of vaccination with V920 at 20 million PFU (Henao-Restrepo et al. 2015). In the immediate vaccination group, there were no cases of EVD, but in the delayed vaccination group, 16 cases of EVD from seven clusters were diagnosed. These data indicated 100% efficacy of the vaccine (Henao-Restrepo et al.). From these results, V920 was granted "Breakthrough Therapy Designation" from the FDA and PRIME (Priority Medicines) Status from European Medicines Agency (EMA) (New Link Genetics Corporation 2016).

Adenovirus-vectored vaccines

Vaccines utilizing adenoviruses as vectors have been in development for many different pathogens; however, these efforts have been hampered by preexisting immunity to several adenoviruses in the majority of the humans (Martins et al. 2016). For protection against EBOV, two different strategies have been developed to utilize these vectors, a chimpanzee adenovirus and human adenoviruses with little preexisting immunity.

A vaccine candidate employing chimpanzee adenovirus 5 (ChAd5) has been developed with the NIH's National Institute of Allergy and Infectious Diseases Vaccine Research Center (Martins et al. 2016). In a phase 1 study of this non-replicating VRC-EBOADC076-00-VP (ChAd3-EBOZ; GlaxoSmithKline) vaccine, three different doses (10 billion, 25 billion, and 50 billion viral particles) were given to 20 participants each at one center in the UK (Ewer et al. 2016). Then, a booster of modified vaccinia Ankara (MVA) vector (MVA-BN Filo; Bavarian Nordic), which encodes the same EBOV isolate glycoprotein antigen as that encoded by the ChAd3-EBOZ vaccine, in addition to glycoproteins of Sudan virus and Marburg virus and the nucleoprotein of TAFV, was given to 10 participants in each dose group. EBOV-specific antibody titers and neutralization activity were similar to those reported in rVSV-ZEBOV studies. The boosting with the MVA-BN Filo significantly increased EBOV-specific antibody and CD8+ T cell responses, although the antibody responses were more long lived (Ewer et al. 2016). Overall, the vaccine was well tolerated (Ewer et al. 2016). In the combined data from two phase 1 dose-finding studies, one in the USA (n = 20) and the other in Mali (n = 91), with 52 of the Malians receiving MVA-BN-Filo boosters, the 1 billion viral particle dose was found to be the optimal dose in phase 3 efficacy trials (Tapia et al. 2016). Currently, the ChAd3-EBOZ vaccine is being tested in several clinical trials and has been included as a comparator arm in the PREVAIL study (Partnership for Research on Ebola Virus in Liberia) evaluating rVSV-ZEBOV.

Crucell Holland B.V. has developed another potential vaccine candidate, Ad26. ZEBOV, utilizing their human adenovirus 26 vector (Martins et al. 2016). In the first phase 1 study, 87 subjects at a single center received a priming vaccination of either Ad26.ZEBOV or MVA-BN-Filo, followed by a booster of the other vaccine or placebo (Milligan et al. 2016). One month after primary immunization, 97% of the Ad26.ZEBOV and 23% of MVA-BN-Filo recipients had detectable EBOV-specific antibodies, and all recipients had detectable antibodies 21 days after the booster vaccination. The majority of subjects demonstrated EBOV-specific T-cell responses as well (Milligan et al. 2016). Fevers were reported in 9% of Ad26. ZEBOV and none of MVA-BN-Filo recipients (Milligan et al. 2016). This heterologous prime-boost model is being further assessed in phases 2 and 3 studies.

5 Infection Control

Although patients with EVD have been cared for in resource-constrained environments for nearly 40 years, it was not until 2014 that two patients were transported to the USA to be cared for in a resource-rich environment at Emory University Hospital in Atlanta, GA (Lyon et al. 2014). Over the course of the 2014–2016 EVD outbreak that devastated West Africa, leading to 28,616 cases and 11,310 deaths, 27 individuals received their medical care in resource-rich environments (Uyeki et al. 2016b). While the care of these 27 individuals taught us much about the aggressive care required to improve patient survival, it also helped to clarify the key elements of infection control in caring for these patients.

Healthcare workers have historically been one of the highest risk groups for acquiring EVD from infected patients. In one outbreak in 1995, 25% of the infected were healthcare workers (Khan et al. 1999). The role of appropriate infection control precautions was demonstrated during this outbreak, as there were 80 infected healthcare workers prior to the implementation of appropriate barrier precautions, but only one additional case after barrier precautions were instituted. In the most recent 2014–2016 outbreak, there were 876 confirmed and probable cases of EBOV infection in healthcare workers of whom 509 died (WHO 2016). Healthcare workers were between 21 and 32 times more likely to be infected with EBOV than people in the general adult population, and nearly 60% of those infected healthcare workers died. The importance of infection control in this setting was, as in the 1995 outbreak, demonstrated by the fact that healthcare workers represented 12% of all EVD patients in July 2014, but decreased to 1% by February 2015 as more rigorous infection control and occupational health and safety strategies were implemented. While the source of healthcare worker infection may be difficult to trace in locations where there is an ongoing community outbreak, such transmission is easier to identify when there is no ongoing community disease. The number of infected healthcare workers in resource-rich settings with no ongoing community disease was markedly less than the number in resource-constrained settings. However, one nurse in Spain and two nurses in the USA acquired EBOV infection from their patients who had acquired EBOV infection in West Africa. For these reasons, infection control has received a great deal of attention in the care of patients infected with Ebola viruses in both resource-rich and resource-constrained settings.

Ebolaviruses, with rare exceptions, are transmitted by contact with blood, body fluids, and contact with skin. In addition to blood, infectious virus has been documented in patients' saliva, stool, rectal swabs, vaginal secretions, breast milk, tears (conjunctival swab), urine, and seminal fluid, but not vomit or sweat (Bausch et al. 2007; Dowell et al. 1999). Although aerosol transmission has been suspected in outbreaks involving RESTV in non-human primates, such transmission of other Ebola viruses must be rare if it occurs at all in humans. There is no support for the transmission of the EBOV via the airborne route (Peters et al. 1996).

The role of fomites in Ebola virus transmission is unclear. A recent study found that the Ebola virus strain associated with the 2014–2016 West Africa outbreak was able to persist on stainless steel, plastic, and Tyvek® under environmental conditions reflective of the high temperatures and relative levels of humidity found in the outbreak regions (Fischer et al. 2015). However, in a study evaluating 31 environmental specimens from an EBOV isolation ward that were not visibly bloody, all specimens were negative by RT-PCR, suggesting that fomites in a clinical setting where cleaning and decontamination would be frequent are unlikely to be capable of EBOV transmission (Bausch et al. 2007). Similarly, multiple environment samples obtained in a patient care room of a patient treated for severe EVD were negative by RT-PCR (Varkey 2016).

Persistence of infectious virus in immune-privileged sites has also been documented. EBOV has been detected in seminal fluid for months following patient recovery (Uyeki et al. 2016a). A secondary case has occurred following sexual exposure to semen 155 days after the source patient had cleared virus from his blood (Mate et al. 2015). One case of uveitis was diagnosed with infectious virus in the aqueous humor 9 weeks after the patient had cleared his viremia (Varkey et al. 2015). A nurse was diagnosed with meningitis, with recovery of EBOV from her cerebrospinal fluid, 9 months after clearing her viremia (Jacobs et al. 2016). While such events are concerning, the role of such persistence in person-to-person transmission is still being evaluated. However, patients who have recovered from

EBOV infection who require an invasive procedure of an immune-privileged site need to have special attention paid to the infection control measures followed during such a procedure.

EBOV blood levels in infected patients increase rapidly during the first week after the onset of symptoms, going from undetectable to many log orders of virus (Towner et al. 2004). During the most symptomatic period, viral levels in the blood may reach 10^8 virions per milliliter of blood (Iwen et al. 2015; CDC 2016a), which is many logs higher than is seen with other blood borne pathogens. This helps to explain why individuals exposed to blood and body fluids of patients with Ebola infection during the early days of illness frequently do not become infected, while healthcare workers caring for the patient during the later stages of illness are at greatest risk. When this high viral load is added to the observation that symptomatic patients may excrete 8–10 L of infected fluid daily, one can readily understand why healthcare workers who do not use optimal PPE are at increased risk of occupationally acquired infection.

The importance of appropriate PPE and adequate training in using this equipment has been recognized for many years. In resource-constrained units, such equipment frequently consists of double gloving, impermeable suits, aprons, and surgical masks. In the absence of air conditioning, healthcare workers wearing such equipment are frequently limited to short periods of time in the patient care unit due to heat exhaustion, representing a challenge in the delivery of the aggressive supportive care required for optimal patient outcomes. Recommendations for PPE in resource-rich environments evolved rapidly in 2014. As the USA prepared for its first EBOV-infected patients in 2014, the official recommendation from the US Public Health Service was that healthcare workers should use standard contact precautions, with goggles or face shields, fluid resistant gowns, face masks, and gloves generally in use in healthcare facilities in the USA. Unfortunately, there was little emphasis on covering all skin surfaces and proper training in the use of PPE. The occurrence of EBOV infection in 2 nurses caring for a critically ill patient in October 2014 lead to a reconsideration of these recommendations. What evolved was a new appreciation for effective training and complete covering of all skin and mucous membranes when caring for a patient with Ebola virus infection. New guidelines stress: Healthcare workers caring for patients with EVD must have received comprehensive training and demonstrated competency in performing EVD-related infection control practices and procedures; PPE that covers the clothing and skin and completely protects mucous membranes is required when caring for patients with EVD; personnel providing care to patients with EVD must be supervised by an onsite manager at all times, and a trained observer must supervise each step of every PPE donning/doffing procedure to ensure established PPE protocols are completed correctly; individuals unable or unwilling to adhere to infection control and PPE use procedures should not provide care for patients with EVD (CDC 2016a). In addition, there is a renewed emphasis on frequent cleaning of the floors and surfaces in the patient room and doffing area. Policies should be in place to limit room entry to only those healthcare workers essential to the patient's care and restrict non-essential personnel and visitors from the patient care area.

Such policies will need to balance any conflicts between healthcare worker safety and local work and union regulations.

Cleaning of floors and surfaces on a frequent basis requires special considerations. No environmental disinfectant has been labeled as effective in inactivating the Ebola virus. Although Ebola viruses produce enveloped virions, EPA has elected to take a cautious approach and require that all environmental disinfectants used in ETUs should meet the following criteria: (1) the use of an EPA-registered hospital disinfectant with a label claim for use against a non-enveloped virus (e.g., norovirus, rotavirus, adenovirus, poliovirus) and (2) the product label use directions for the non-enveloped viruses should be followed when disinfecting against EBOV (EPA 2016).

Delivery of aggressive supportive care requires access to laboratory testing. Institutions need to evaluate whether such testing will occur in a core hospital laboratory or in a point of care laboratory. Such decisions will need to be based on a risk assessment to determine the potential for exposure from sprays, splashes, or aerosols generated during all laboratory processes, procedures, and activities (Iwen et al. 2015). Many instruments require that the samples be in open vials or centrifuged, which must occur in an enclosed environment. Consideration will also need to be given to whether the use of the equipment for the testing of blood in patients who may have serious communicable pathogens requires that equipment to be removed from service if the patient tests positive. Laboratory personnel need to be trained in the donning and doffing of appropriate PPE, such training to be similar to that required of patient care personnel. There must also be an appropriate location for those activities to occur.

The disposal of regulated medical waste has been a major issue for units caring for patients with EVD and other serious communicable diseases. When the first patients with EBOV infection were brought to the USA in 2014, contractors initially refused to take the medical waste generated by their care. Eventually, contractors agreed to take the waste if it was first autoclaved prior to transport. However, even then such waste was transported separately from other regulated medical waste and only accepted for incineration in special facilities. For institutions that do not have the capability of autoclaving medical waste on site, contractors require that the waste be immersed in disinfectant inside 55-gallon metal drums, the drums sealed, and then transported by dedicated trucks to specially designated facilities (Stericycle). The expense of such special handling has been a major issue for patient care units. More recently, contractors have demonstrated a willingness to accept unit waste as standard regulated medical waste after it has been autoclaved. However, units without autoclave capability on site continue to be faced with logistic issues. All personnel who handle unit waste prior to autoclaving must be included in the unit PPE training. They must also be monitored using the same procedures as those used for other personnel in the unit.

Employee monitoring

Due to the concern about healthcare worker acquisition of Ebola virus infection, all personnel who have had potential exposure to the virus, either through patient care activity, laboratory processing of specimens, or exposure to potentially contaminated materials or surfaces, must be monitored for potential illness, starting with their first potential exposure and ending 21 days after their last potential exposure. While this monitoring can be performed by facility Employee Health Services, many facilities have utilized the services of local or state health departments to perform such monitoring.

Disposal of liquid and solid waste

Public health service policies have traditionally stated that patient liquid and solid waste may be discharged through municipal sanitary sewers, as such systems are designed to inactivate viruses (CDC 2016b). However, facilities must ensure that local and state regulations do not impose additional restrictions. In addition, concern has been raised due to the potential for plumbing leaks and stoppages within the patient care facility. For these reasons, most facilities have chosen to inactivate waste with an appropriate disinfectant before disposing of the waste into the sanitary sewers.

References

Agnandji ST, Huttner A, Zinser ME, Njuguna P, Dahlke C, Fernandes JF, Yerly S, Dayer J-A, Kraehling V, Kasonta R, Adegnika AA, Altfeld M, Auderset F, Bache EB, Biedenkopf N, Borregaard S, Brosnahan JS, Burrow R, Combescure C, Desmeules J, Eickmann M, Fehling SK, Finckh A, Goncalves AR, Grobusch MP, Hooper J, Jambrecina A, Kabwende AL, Kaya G, Kimani D, Lell B, Lemaître B, Lohse AW, Massinga-Loembe M, Matthey A, Mordmüller B, Nolting A, Ogwang C, Ramharter M, Schmidt-Chanasit J, Schmiedel S, Silvera P, Stahl FR, Staines HM, Strecker T, Stubbe HC, Tsofa B, Zaki S, Fast P, Moorthy V, Kaiser L, Krishna S, Becker S, Kieny M-P, Bejon P, Kremsner PG, Addo MM, Siegrist C-A (2016) Phase 1 trials of rVSV Ebola vaccine in Africa and Europe. N Engl J Med 374 (17):1647–1660. doi:10.1056/NEJMoa1502924

Bah EI, Lamah MC, Fletcher T, Jacob ST, Brett-Major DM, Sall AA, Shindo N, Fischer WA 2nd, Lamontagne F, Saliou SM, Bausch DG, Moumie B, Jagatic T, Sprecher A, Lawler JV, Mayet T, Jacquerioz FA, Mendez Baggi MF, Vallenas C, Clement C, Mardel S, Faye O, Faye O, Soropogui B, Magassouba N, Koivogui L, Pinto R, Fowler RA (2015) Clinical presentation of patients with Ebola virus disease in Conakry, Guinea. N Engl J Med 372(1):40–47. doi:10.1056/NEJMoa1411249

Bausch DG, Towner JS, Dowell SF, Kaducu F, Lukwiya M, Sanchez A, Nichol ST, Ksiazek TG, Rollin PE (2007) Assessment of the risk of Ebola virus transmission from bodily fluids and fomites. J Infect Dis 196(Suppl 2):S142–S147. doi:10.1086/520545

CDC (2016a) Guidance on Personal Protective Equipment (PPE) to be used by healthcare workers during management of patients with confirmed Ebola or Persons under Investigation (PUIs) for Ebola who are clinically unstable or have bleeding, vomiting, or diarrhea in U.S. Hospitals, Including procedures for Donning and Doffing PPE. http://www.cdc.gov/vhf/ebola/healthcare-us/ppe/guidance.html. Accessed 1 June 2016

CDC (2016b) Interim guidance for environmental infection control in hospitals for Ebola Virus. http://www.cdc.gov/vhf/ebola/healthcare-us/cleaning/hospitals.html. Accessed 1 June 2016

Chertow DS, Kleine C, Edwards JK, Scaini R, Giuliani R, Sprecher A (2014) Ebola virus disease in West Africa–clinical manifestations and management. N Engl J Med 371(22):2054–2057. doi:10.1056/NEJMp1413084

Chertow DS, Uyeki TM, DuPont HL (2015) Loperamide therapy for voluminous diarrhea in Ebola virus disease. J Infect Dis 211(7):1036–1037. doi:10.1093/infdis/jiv001

Connor MJ Jr, Kraft C, Mehta AK, Varkey JB, Lyon GM, Crozier I, Stroher U, Ribner BS, Franch HA (2015) Successful delivery of RRT in Ebola virus disease. J Am Soc Nephrology: JASN 26(1):31–37. doi:10.1681/ASN.2014111057

Dallatomasina S, Crestani R, Sylvester Squire J, Declerk H, Caleo GM, Wolz A, Stinson K, Patten G, Brechard R, Gbabai OB, Spreicher A, Van Herp M, Zachariah R (2015) Ebola outbreak in rural West Africa: epidemiology, clinical features and outcomes. Trop Med Int Health 20(4):448–454. doi:10.1111/tmi.12454

Del Rio C, Mehta AK, Lyon GM 3rd, Guarner J (2014) Ebola hemorrhagic fever in 2014: the tale of an evolving epidemic. Ann Intern Med 161(10):746–748. doi:10.7326/M14-1880

Dowell SF, Mukunu R, Ksiazek TG, Khan AS, Rollin PE, Peters CJ (1999) Transmission of Ebola hemorrhagic fever: a study of risk factors in family members, Kikwit, Democratic Republic of the Congo, 1995. Commission de Lutte contre les Epidemies a Kikwit. J Infect Dis 179 Suppl 1:S87–91. doi:10.1086/514284

Dunning J, Sahr F, Rojek A, Gannon F, Carson G, Idriss B, Massaquoi T, Gandi R, Joseph S, Osman HK, Brooks TJG, Simpson AJH, Goodfellow I, Thorne L, Arias A, Merson L, Castle L, Howell-Jones R, Pardinaz-Solis R, Hope-Gill B, Ferri M, Grove J, Kowalski M, Stepniewska K, Lang T, Whitehead J, Olliaro P, Samai M, Horby PW, for the R-TKMtt (2016) Experimental treatment of Ebola Virus disease with TKM-130803: a single-arm phase 2 clinical trial. PLoS Med 13(4):e1001997. doi:10.1371/journal.pmed.1001997

Emond RT, Evans B, Bowen ET, Lloyd G (1977) A case of Ebola virus infection. BMJ 2 (6086):541–544. doi:10.1136/bmj.2.6086.541

EPA (2016) List L: disinfectants for use against the Ebola virus. https://www.epa.gov/pesticide-registration/list-l-disinfectants-use-against-ebola-virus. Accessed 1 June 2016

Ewer K, Rampling T, Venkatraman N, Bowyer G, Wright D, Lambe T, Imoukhuede EB, Payne R, Fehling SK, Strecker T, Biedenkopf N, Krähling V, Tully CM, Edwards NJ, Bentley EM, Samuel D, Labbé G, Jin J, Gibani M, Minhinnick A, Wilkie M, Poulton I, Lella N, Roberts R, Hartnell F, Bliss C, Sierra-Davidson K, Powlson J, Berrie E, Tedder R, Roman F, De Ryck I, Nicosia A, Sullivan NJ, Stanley DA, Mbaya OT, Ledgerwood JE, Schwartz RM, Siani L, Colloca S, Folgori A, Di Marco S, Cortese R, Wright E, Becker S, Graham BS, Koup RA, Levine MM, Volkmann A, Chaplin P, Pollard AJ, Draper SJ, Ballou WR, Lawrie A, Gilbert SC, Hill AVS (2016) A monovalent chimpanzee adenovirus Ebola vaccine boosted with MVA. N Engl J Med 374(17):1635–1646. doi:10.1056/NEJMoa1411627

Fischer R, Judson S, Miazgowicz K, Bushmaker T, Prescott J, Munster VJ (2015) Ebola virus stability on surfaces and in fluids in simulated outbreak environments. Emerg Infect Dis 21 (7):1243–1246. doi:10.3201/eid2107.150253

Florescu DF, Kalil AC, Hewlett AL, Schuh AJ, Stroher U, Uyeki TM, Smith PW (2015) Administration of brincidofovir and convalescent plasma in a patient with Ebola virus disease. Clin Infect Dis 61(6):969–973. doi:10.1093/cid/civ395

Furuta Y, Gowen BB, Takahashi K, Shiraki K, Smee DF, Barnard DL (2013) Favipiravir (T-705), a novel viral RNA polymerase inhibitor. Antiviral Res 100(2):446–454. doi:10.1016/j.antiviral.2013.09.015

Geisbert TW, Lee ACH, Robbins M, Geisbert JB, Honko AN, Sood V, Johnson JC, de Jong S, Tavakoli I, Judge A, Hensley LE, MacLachlan I (2010) Postexposure protection of non-human primates against a lethal Ebola virus challenge with RNA interference: a proof-of-concept study. The Lancet 375(9729):1896–1905. doi:10.1016/S0140-6736(10)60357-1

Geisbert TW (2015) Marburg and Ebola Hemorrhagic Fevers (Marburg and Ebola Viral Diseases) (Filoviruses)-In: Bennett JE, Dolin RMD, Blaser MJ (eds) Mandell, Douglas, and Bennett's Principles and practice of infectious diseases, Updated Edition, 166, 1995–1999. iv

Guidet B, Soni N, Rocca GD, Kozek S, Vallet B, Annane D, James M (2010) A balanced view of balanced solutions. Crit Care 14(5):1–12. doi:10.1186/cc9230

Heald AE, Iversen PL, Saoud JB, Sazani P, Charleston JS, Axtelle T, Wong M, Smith WB, Vutikullird A, Kaye E (2014) Safety and pharmacokinetic profiles of phosphorodiamidate morpholino oligomers with activity against ebola virus and marburg virus: results of two single-ascending-dose studies. Antimicrob Agents Chemother 58(11):6639–6647. doi:10.1128/aac.03442-14

Henao-Restrepo AM, Longini IM, Egger M, Dean NE, Edmunds WJ, Camacho A, Carroll MW, Doumbia M, Draguez B, Duraffour S, Enwere G, Grais R, Gunther S, Hossmann S, Kondé MK, Kone S, Kuisma E, Levine MM, Mandal S, Norheim G, Riveros X, Soumah A, Trelle S, Vicari AS, Watson CH, Kéïta S, Kieny MP, Røttingen J-A (2015) Efficacy and effectiveness of an rVSV-vectored vaccine expressing Ebola surface glycoprotein: interim results from the Guinea ring vaccination cluster-randomised trial. Lancet 386 (9996):857–866. doi:10.1016/S0140-6736(15)61117-5

Huttner A, Dayer J A, Yerly S, Combescure C, Auderset F, Desmeules J, Eickmann M, Finckh A, Goncalves AR, Hooper JW, Kaya G, Krähling V, Kwilas S, Lemaître B, Matthey A, Silvera P, Becker S, Fast PE, Moorthy V, Kieny MP, Kaiser L, Siegrist C-A (2015) The effect of dose on the safety and immunogenicity of the VSV Ebola candidate vaccine: a randomised double-blind, placebo-controlled phase 1/2 trial. Lancet Infect Dis 15 (10):1156–1166. doi:10.1016/S1473-3099(15)00154-1

Iwen PC, Smith PW, Hewlett AL, Kratochvil CJ, Lisco SJ, Sullivan JN, Gibbs SG, Lowe JJ, Fey PD, Herrera VL, Sambol AR, Wisecarver JL, Hinrichs SH (2015) Safety considerations in the laboratory testing of specimens suspected or known to contain Ebola virus. Am J Clin Pathol 143(1):4–5. doi:10.1309/AJCP26MIFUIETBPL

Jacobs M, Rodger A, Bell DJ, Bhagani S, Cropley I, Filipe A, Gifford RJ, Hopkins S, Hughes J, Jabeen F, Johannessen I, Karageorgopoulos D, Lackenby A, Lester R, Liu RS, MacConnachie A, Mahungu T, Martin D, Marshall N, Mepham S, Orton R, Palmarini M, Patel M, Perry C, Peters SE, Porter D, Ritchie D, Ritchie ND, Seaton RA, Sreenu VB, Templeton K, Warren S, Wilkie GS, Zambon M, Gopal R, Thomson EC (2016) Late Ebola virus relapse causing meningoencephalitis: a case report. Lancet 388(10043):498–503. doi:10.1016/S0140-6736(16)30386-5

Khan AS, Tshioko FK, Heymann DL, Le Guenno B, Nabeth P, Kerstiens B, Fleerackers Y, Kilmarx PH, Rodier GR, Nkuku O, Rollin PE, Sanchez A, Zaki SR, Swanepoel R, Tomori O, Nichol ST, Peters CJ, Muyembe-Tamfum JJ, Ksiazek TG (1999) The reemergence of Ebola hemorrhagic fever, Democratic Republic of the Congo, 1995. Commission de Lutte contre les Epidemies a Kikwit. J Infect Dis 179 Suppl 1:S76–86. doi:10.1086/514306

Kibadi K, Mupapa K, Kuvula K, Massamba M, Ndaberey D, Muyembe-Tamfum JJ, Bwaka MA, De Roo A, Colebunders R (1999) Late ophthalmologic manifestations in survivors of the 1995 Ebola virus epidemic in Kikwit, democratic Republic of the Congo. J Infect Dis 179(Suppl 1): S13–S14. doi:10.1086/514288

Kraft CS, Hewlett AL, Koepsell S, Winkler AM, Kratochvil CJ, Larson L, Varkey JB, Mehta AK, Lyon GM, 3rd, Friedman-Moraco RJ, Marconi VC, Hill CE, Sullivan JN, Johnson DW, Lisco SJ, Mulligan MJ, Uyeki TM, McElroy AK, Sealy T, Campbell S, Spiropoulou C, Stroher U, Crozier I, Sacra R, Connor MJ, Jr., Sueblinvong V, Franch HA, Smith PW, Ribner BS, Nebraska Biocontainment U, the Emory Serious Communicable Diseases U (2015) The Use of TKM-100802 and convalescent plasma in 2 patients with Ebola Virus disease in the United States. Clin Infect Dis 61 (4):496–502. doi:10.1093/cid/civ334

Liddell AM, Davey RT Jr, Mehta AK, Varkey JB, Kraft CS, Tseggay GK, Badidi O, Faust AC, Brown KV, Suffredini AF, Barrett K, Wolcott MJ, Marconi VC, Lyon GM 3rd, Weinstein GL, Weinmeister K, Sutton S, Hazbun M, Albarino CG, Reed Z, Cannon D, Stroher U, Feldman M, Ribner BS, Lane HC, Fauci AS, Uyeki TM (2015) Characteristics and clinical management of a cluster of 3 patients with Ebola Virus disease, including the first domestically acquired cases in the United States. Ann Intern Med 163(2):81–90. doi:10.7326/M15-0530

Lyon GM, Mehta AK, Varkey JB, Brantly K, Plyler L, McElroy AK, Kraft CS, Towner JS, Spiropoulou C, Stroher U, Uyeki TM, Ribner BS (2014) Emory Serious Communicable Diseases U (2014) Clinical care of two patients with Ebola virus disease in the United States. N Engl J Med 371 (25):2402–2409. doi:10.1056/NEJMoa1409838

Madelain V, Nguyen THT, Olivo A, de Lamballerie X, Guedj J, Taburet A-M, Mentré F (2016) Ebola virus infection: review of the pharmacokinetic and pharmacodynamic properties of drugs considered for testing in human efficacy trials. Clin Pharmacokinet 55(8):907–923. doi:10.1007/s40262-015-0364-1

Martines RB, Ng DL, Greer PW, Rollin PE, Zaki SR (2015) Tissue and cellular tropism, pathology and pathogenesis of Ebola and Marburg viruses. J Pathol 235(2):153–174. doi:10.1002/path.4456

Martins KA, Jahrling PB, Bavari S, Kuhn JH (2016) Ebola virus disease candidate vaccines under evaluation in clinical trials. Expert Rev Vaccines 15(9):1101–1112. doi:10.1080/14760584.2016.1187566

Mate SE, Kugelman JR, Nyenswah TG, Ladner JT, Wiley MR, Cordier-Lassalle T, Christie A, Schroth GP, Gross SM, Davies-Wayne GJ, Shinde SA, Murugan R, Sieh SB, Badio M, Fakoli L, Taweh F, de Wit E, van Doremalen N, Munster VJ, Pettitt J, Prieto K, Humrighouse BW, Stroher U, DiClaro JW, Hensley LE, Schoepp RJ, Safronetz D, Fair J, Kuhn JH, Blackley DJ, Laney AS, Williams DE, Lo T, Gasasira A, Nichol ST, Formenty P, Kateh FN, De Cock KM, Bolay F, Sanchez-Lockhart M, Palacios G (2015) Molecular evidence of sexual transmission of Ebola Virus. N Engl J Med 373(25):2448–2454. doi:10.1056/NEJMoa1509773

Mendoza EJ, Qiu X, Kobinger GP (2016) Progression of Ebola therapeutics during the 2014–2015 outbreak. Trends Mol Med 22(2):164–173. doi:10.1016/j.molmed.2015.12.005

Milligan ID, Gibani MM, Sewell R et al (2016) Safety and immunogenicity of novel adenovirus type 26—and modified vaccinia ankara—vectored Ebola vaccines: a randomized clinical trial. JAMA 315(15):1610–1623. doi:10.1001/jama.2016.4218

Mupapa K, Massamba M, Kibadi K, Kuvula K, Bwaka A, Kipasa M, Colebunders R, Muyembe-Tamfum JJ (1999) Treatment of Ebola hemorrhagic fever with blood transfusions from convalescent patients. International Scientific and Technical Committee. J Infect Dis 179 Suppl 1:S18–23. doi:10.1086/514298

Ndambi R, Akamituna P, Bonnet MJ, Tukadila AM, Muyembe-Tamfum JJ, Colebunders R (1999) Epidemiologic and clinical aspects of the Ebola virus epidemic in Mosango, Democratic Republic of the Congo, 1995. J Infect Dis 179(Suppl 1):S8–10. doi:10.1086/514297

NETEC (2016) National Ebola Training & Education Center. http://www.netec.org. Accessed 30 Aug 2016

NewLink Genetics Corporation (2016) NewLink genetics announces merck receives breakthrough therapy designation from FDA and PRIME status from EMA for Investigational Ebola Zaire Vaccine (V920). http://investors.linkp.com/releasedetail.cfm?ReleaseID=980896. Accessed 15 Sept 2016

Ohimain EI (2016) Recent advances in the development of vaccines for Ebola virus disease. Virus Res 211:174–185. doi:10.1016/j.virusres.2015.10.021

Peters CJ, Jahrling PB, Khan AS (1996) Patients infected with high-hazard viruses: scientific basis for infection control. Arch Virol Suppl 11:141–168

Qin E, Bi J, Zhao M, Wang Y, Guo T, Yan T, Li Z, Sun J, Zhang J, Chen S, Wu Y, Li J, Zhong Y (2015) Clinical features of patients with Ebola virus disease in sierra leone. Clin Infect Dis 61 (4):491–495. doi:10.1093/cid/civ319

Qiu X, Wong G, Audet J, Bello A, Fernando L, Alimonti JB, Fausther-Bovendo H, Wei H, Aviles J, Hiatt E, Johnson A, Morton J, Swope K, Bohorov O, Bohorova N, Goodman C, Kim D, Pauly MH, Velasco J, Pettitt J, Olinger GG, Whaley K, Xu B, Strong JE, Zeitlin L, Kobinger GP (2014) Reversion of advanced Ebola virus disease in nonhuman primates with ZMapp. Nature 514(7520):47–53. doi:10.1038/nature13777

Regules JA, Beigel JH, Paolino KM, Voell J, Castellano AR, Muñoz P, Moon JE, Ruck RC, Bennett JW, Twomey PS, Gutiérrez RL, Remich SA, Hack HR, Wisniewski ML, Josleyn MD, Kwilas SA, Van Deusen N, Mbaya OT, Zhou Y, Stanley DA, Bliss RL, Cebrik D, Smith KS, Shi M, Ledgerwood JE, Graham BS, Sullivan NJ, Jagodzinski LL, Peel SA, Alimonti JB, Hooper JW, Silvera PM, Martin BK, Monath TP, Ramsey WJ, Link CJ, Lane HC, Michael NL, Davey RTJ, Thomas SJ (2017) A recombinant vesicular stomatitis virus Ebola Vaccine—Preliminary report. N Engl J Med 0 (0):null. doi:10.1056/NEJMoa1414216

Respiratory vaccine shows promise (2014). Can Nurse 110 (9):8

Roberts A, Buonocore L, Price R, Forman J, Rose JK (1999) Attenuated vesicular stomatitis viruses as vaccine vectors. J Virol 73(5):3723–3732

Roddy P, Howard N, Van Kerkhove MD, Lutwama J, Wamala J, Yoti Z, Colebunders R, Palma PP, Sterk E, Jeffs B, Van Herp M, Borchert M (2012) Clinical manifestations and case management of Ebola haemorrhagic fever caused by a newly identified virus strain, bundibugyo, uganda, 2007–2008. PLoS ONE 7(12):e52986. doi:10.1371/journal.pone.0052986

Schieffelin JS, Shaffer JG, Goba A, Gbakie M, Gire SK, Colubri A, Sealfon RS, Kanneh L, Moigboi A, Momoh M, Fullah M, Moses LM, Brown BL, Andersen KG, Winnicki S, Schaffner SF, Park DJ, Yozwiak NL, Jiang PP, Kargbo D, Jalloh S, Fonnie M, Sinnah V, French I, Kovoma A, Kamara FK, Tucker V, Konuwa E, Sellu J, Mustapha I, Foday M, Yillah M, Kanneh F, Saffa S, Massally JL, Boisen ML, Branco LM, Vandi MA, Grant DS, Happi C, Gevao SM, Fletcher TE, Fowler RA, Bausch DG, Sabeti PC, Khan SH, Garry RF, Program KGHLF, Viral Hemorrhagic Fever C, Team WHOCR (2014) Clinical illness and outcomes in patients with Ebola in Sierra Leone. N Engl J Med 371(22):2092–2100. doi:10.1056/NEJMoa1411680

Scott JT, Sesay FR, Massaquoi TA, Idriss BR, Sahr F, Semple MG (2016) Post-ebola syndrome Sierra Leone. Emerg Infect Dis 22(4):641–646. doi:10.3201/eid2204.151302

Sissoko D, Laouenan C, Folkesson E, M'Lebing A-B, Beavogui A-H, Baize S, Camara A-M, Maes P, Shepherd S, Danel C, Carazo S, Conde MN, Gala J-L, Colin G, Savini H, Bore JA, Le Marcis F, Koundouno FR, Petitjean F, Lamah M-C, Diederich S, Tounkara A, Poelart G, Berbain E, Dindart J-M, Duraffour S, Lefevre A, Leno T, Peyrouset O, Irenge L, Bangoura NF, Palich R, Hinzmann J, Kraus A, Barry TS, Berette S, Bongono A, Camara MS, Chanfreau Munoz V, Doumbouya L, Souley H, Kighoma PM, Koundouno FR, Réné L, Loua CM, Massala V, Moumouni K, Provost C, Samake N, Sekou C, Soumah A, Arnould I, Komano MS, Gustin L, Berutto C, Camara D, Camara FS, Colpaert J, Delamou L, Jansson L, Kourouma E, Loua M, Malme K, Manfrin E, Maomou A, Milinouno A, Ombelet S, Sidiboun AY, Verreckt I, Yombouno P, Bocquin A, Carbonnelle C, Carmoi T, Frange P, Mely S, Nguyen V-K, Pannetier D, Taburet A-M, Treluyer J-M, Kolie J, Moh R, Gonzalez MC, Kuisma E, Liedigk B, Ngabo D, Rudolf M, Thom R, Kerber R, Gabriel M, Di Caro A, Wölfel R, Badir J, Bentahir M, Deccache Y, Dumont C, Durant J-F, El Bakkouri K, Gasasira Uwamahoro M, Smits B, Toufik N, Van Cauwenberghe S, Ezzedine K, Dortenzio E, Pizarro L, Etienne A, Guedj J, Fizet A, Barte de Sainte Fare E, Murgue B, Tran-Minh T, Rapp C, Piguet P, Poncin M, Draguez B, Allaford Duverger T, Barbe S, Baret G, Defourny I, Carroll M, Raoul H, Augier A, Eholie SP, Yazdanpanah Y, Levy-Marchal C, Antierrens A, Van Herp M, Günther S, de Lamballerie X, Keïta S, Mentre F, Anglaret X, Malvy D, Group JS (2016) Experimental treatment with favipiravir for ebola virus disease (the jiki trial): a historically controlled, single-arm proof-of-concept trial in guinea. PLoS Med 13 (3): e1001967. doi:10.1371/journal.pmed.1001967

Sueblinvong V, Johnson DW, Weinstein GL, Connor MJ Jr, Crozier I, Liddell AM, Franch HA, Wall BR, Kalil AC, Feldman M, Lisco SJ, Sevransky JE (2015) Critical care for multiple organ failure secondary to Ebola virus disease in the united states. Crit Care Med 43(10):2066–2075. doi:10.1097/CCM.0000000000001197

Tapia MD, Sow SO, Lyke KE, Haidara FC, Diallo F, Doumbia M, Traore A, Coulibaly F, Kodio M, Onwuchekwa U, Sztein MB, Wahid R, Campbell JD, Kieny M-P, Moorthy V, Imoukhuede EB, Rampling T, Roman F, De Ryck I, Bellamy AR, Dally L, Mbaya OT,

Ploquin A, Zhou Y, Stanley DA, Bailer R, Koup RA, Roederer M, Ledgerwood J, Hill AVS, Ballou WR, Sullivan N, Graham B, Levine MM (2016) Use of ChAd3-EBO-Z Ebola virus vaccine in Malian and US adults, and boosting of Malian adults with MVA-BN-Filo: a phase 1, single-blind, randomised trial, a phase 1b, open-label and double-blind, dose-escalation trial, and a nested, randomised, double-blind, placebo-controlled trial. Lancet Infect Dis 16 (1):31-42. doi:10.1016/S1473-3099(15)00362-X

Taylor R, Kotian P, Warren T, Panchal R, Bavari S, Julander J, Dobo S, Rose A, El-Kattan Y, Taubenheim B, Babu Y, Sheridan WP (2016) BCX4430—A broad-spectrum antiviral adenosine nucleoside analog under development for the treatment of Ebola virus disease. J Infect Public Health 9(3):220–226. doi:10.1016/j.jiph.2016.04.002

Thi EP, Mire CE, Lee ACH, Geisbert JB, Zhou JZ, Agans KN, Snead NM, Deer DJ, Barnard TR, Fenton KA, MacLachlan I, Geisbert TW (2015) Lipid nanoparticle siRNA treatment of ebola-virus-makona-infected nonhuman primates. Nature 521(7552):362–365. doi:10.1038/nature14442

Towner JS, Rollin PE, Bausch DG, Sanchez A, Crary SM, Vincent M, Lee WF, Spiropoulou CF, Ksiazek TG, Lukwiya M, Kaducu F, Downing R, Nichol ST (2004) Rapid diagnosis of Ebola hemorrhagic fever by reverse transcription-PCR in an outbreak setting and assessment of patient viral load as a predictor of outcome. J Virol 78(8):4330–4341

Uyeki TM, Erickson BR, Brown S, McElroy AK, Cannon D, Gibbons A, Sealy T, Kainulainen MH, Schuh AJ, Kraft CS, Mehta AK, Lyon GM 3rd, Varkey JB, Ribner BS, Ellison RT 3rd, Carmody E, Nau GJ, Spiropoulou C, Nichol ST, Stroher U (2016) Ebola virus persistence in Semen of male survivors. Clin Infect Dis 62(12):1552–1555. doi:10.1093/cid/ciw202

Uyeki TM, Mehta AK, Davey RT, Jr., Liddell AM, Wolf T, Vetter P, Schmiedel S, Grunewald T, Jacobs M, Arribas JR, Evans L, Hewlett AL, Brantsaeter AB, Ippolito G, Rapp C, Hoepelman AI, Gutman J, Working Group of the USECNoCMoEVDPitUS, Europe (2016b) Clinical management of Ebola virus disease in the United States and Europe. N Engl J Med 374 (7):636–646. doi:10.1056/NEJMoa1504874

Varkey JB (2016)

Varkey JB, Shantha JG, Crozier I, Kraft CS, Lyon GM, Mehta AK, Kumar G, Smith JR, Kainulainen MH, Whitmer S, Stroher U, Uyeki TM, Ribner BS, Yeh S (2015) Persistence of Ebola Virus in Ocular fluid during convalescence. N Engl J Med 372(25):2423–2427. doi:10.1056/NEJMoa1500306

Warren TK, Whitehouse CA, Wells J, Welch L, Heald AE, Charleston JS, Sazani P, Reid SP, Iversen PL, Bavari S (2015) A single phosphorodiamidate morpholino oligomer targeting VP24 protects Rhesus Monkeys against Lethal Ebola Virus Infection. mBio 6 (1):e02344–02314. doi:10.1128/mBio.02344-14

WHO (2016) Ebola situation report. http://apps.who.int/ebola/ebola-situation-reports. Accessed 30 Aug 2016

Winkler AM, Koepsell SA (2015) The use of convalescent plasma to treat emerging infectious diseases: focus on Ebola virus disease. Curr Opin Hematol 22(6):521–526. doi:10.1097/moh.0000000000000191

Wolf T, Kann G, Becker S, Stephan C, Brodt HR, de Leuw P, Grunewald T, Vogl T, Kempf VA, Keppler OT, Zacharowski K (2015) Severe Ebola virus disease with vascular leakage and multiorgan failure: treatment of a patient in intensive care. Lancet 385(9976):1428–1435. doi:10.1016/S0140-6736(14)62384-9

World Health Organization (WHO) (2014) Use of convalescent whole blood or plasma collected from patients recovered from Ebola Virus disease for transfusion, as an empirical treatment during outbreaks. http://www.euro.who.int/__data/assets/pdf_file/0011/268787/Use-of-convalescent-whole-blood-or-plasma-collected-from-patients-recovered-from-ebola-virus-disease-for-transfusion,-as-an-empirical-treatment-during-outbreaks-eng.pdf. Accessed 10 July 2016

Zeitlin L, Whaley KJ, Olinger GG, Jacobs M, Gopal R, Qiu X, Kobinger GP (2016) Antibody therapeutics for Ebola virus disease. Current Opin Virol 17:45–49. doi:10.1016/j.coviro.2016.01.006

Part II
Disease: Pathogenesis and Protection

Ebola Virus Disease in Humans: Pathophysiology and Immunity

César Muñoz-Fontela and Anita K. McElroy

Abstract Viruses of the *Ebolavirus* genus cause sporadic epidemics of severe and systemic febrile disease that are fueled by human-to-human transmission. Despite the notoriety of ebolaviruses, particularly Ebola virus (EBOV), as prominent viral hemorrhagic fever agents, and the international concern regarding Ebola virus disease (EVD) outbreaks, very little is known about the pathophysiology of EVD in humans and, in particular, about the human immune correlates of survival and immune memory. This lack of basic knowledge about physiological characteristics of EVD is probably attributable to the dearth of clinical and laboratory data gathered from past outbreaks. The unprecedented magnitude of the EVD epidemic that occurred in West Africa from 2013 to 2016 has allowed, for the first time, evaluation of clinical, epidemiological, and immunological parameters in a significant number of patients using state-of-the-art laboratory equipment. This review will summarize the data from the literature regarding human pathophysiologic and immunologic responses to filoviral infection.

Contents

C. Muñoz-Fontela (✉)
Laboratory of Emerging Viruses, Heinrich Pette Institute, Leibniz Institute for Experimental Virology, Martinistraße 52, 20251 Hamburg, Germany
e-mail: cesar.munoz-fontela@hpi.uni-hamburg.de

A.K. McElroy (✉)
Infectious Diseases, Department of Pediatrics, Emory University School of Medicine, 2015 Uppergate Drive NE, Atlanta, GA 30322, USA
e-mail: akmcelr@emory.edu

Current Topics in Microbiology and Immunology (2017) 411:141–170
DOI 10.1007/82_2017_11

Published Online: 27 June 2017

1 Introduction

Ebola virus (EBOV) is the prototypic member of the *Ebolavirus* genus in the *Filoviridae* family of negative-sense, single-stranded RNA viruses. Discovered in 1976 during the first documented outbreak of Ebola virus disease (EVD) in the town of Yambuku in northern Zaire (today Democratic Republic of the Congo), EBOV has since caused sporadic human disease outbreaks of varying magnitude in Equatorial African countries (Sanchez et al. 2007a). In March 2014, an EBOV variant later named EBOV Makona was first detected in Guinea. This variant was responsible for a 3-year-long epidemic that affected tens of thousands of people in several West African countries, collapsing the healthcare systems of three of them. EBOV Makona rampaged through both rural and urban areas, and underscored previously poorly characterized features of EVD, like sexual transmission and virus persistence after recovery (Bausch et al. 2007; Rowe et al. 1999; Chughtai et al. 2016; Deen et al. 2015; Fischer et al. 2016; Rodriguez et al. 1999; Varkey et al. 2015; Uyeki et al. 2016a).

The scientific and clinical knowledge of human EVD before its appearance in West Africa was very limited. The scarcity of human cases and their occurrence in rural areas of Equatorial Africa limited research, as did confinement of filovirus research to biosafety level 4 containment laboratories. In addition, basic studies on EVD pathophysiology have been hampered by the lack of susceptible small animal models with competent immunity. For example, laboratory mice, a commonly used disease model, are completely resistant to nonadapted EBOV.

Before 2014, EVD was described as an acute hemorrhagic fever, thus earning its former name Ebola hemorrhagic fever (EHF); case fatality rates of up to 90% had been reported. The disease was characterized by lymphopenia, disseminated intravascular coagulation (DIC), immunosuppression, and a systemic inflammatory response resembling septic shock (Feldmann and Geisbert 2011). While many of these observations have been strengthened by findings from the West African EVD outbreak, some of the previous hypotheses have been revised. Perhaps one of the most surprising findings has been the low overall number of human cases presenting with bleeding (Schieffelin et al. 2014), as well as the lack of correlation between bleeding and disease severity (Schieffelin et al. 2014; McElroy et al. 2014a, b). These findings triggered the change in disease nomenclature from Ebola hemorrhagic fever to Ebola virus disease. Moreover, the finding that EVD correlates with robust immune activation rather than immunosuppression (Ruibal et al.

2016; McElroy et al. 2015a), and the ability of the virus to persist in several body fluids long after recovery (Varkey et al. 2015; Uyeki et al. 2016a; Sow et al. 2016; Green et al. 2016; Deen et al. 2015) have changed our current view of EVD and have prompted new directions in research and new public health policies. Here we will aim to integrate these novel findings within the current human EVD model, and will discuss future research directions.

Several ebolaviruses cause EVD, and while differences may exist between the diseases caused by the individual viruses, this review will focus on EVD as a disease caused by all known viruses in the *Ebolavirus* genus (ebolaviruses) that are pathogenic for humans. The reader will note that most of the available data come from infections caused by EBOV rather than the other pathogenic viruses in this genus: Sudan virus (SUDV), Bundibugyo virus (BDBV), and Taï Forest virus (TAFV). The related marburgviruses, Marburg virus (MARV) and Ravn virus (RAVV), will be mentioned where appropriate data are available, but unfortunately, information on Marburg virus disease (MVD), which is caused by both of them, is still lacking.

2 Portals of Ebolavirus Entry

Epidemiological data collected over the last 40 years indicate that human infection with EBOV occurs mainly through close contact with infected body fluids. This probably occurs during both spillover events (e.g., contact with infected blood during butchering of bushmeat) and human-to-human transmission. There is no evidence that direct contact with bats causes EBOV spillover into humans (Mari Saez et al. 2015; Leroy et al. 2009), but infection with MARV and RAVV via direct or indirect contact with Egyptian rousettes (fruit bats of the species *Rousettus aegyptiacus*) (Amman et al. 2012; Schuh et al. 2017), has been documented. Human visits to caves or mines in which these bats roost have been directly associated with the development of MVD (Bausch et al. 2003; Centers for Disease and Prevention 2009; Adjemian et al. 2011), strongly indicating that mucosal or skin contact with bat droppings is sufficient to initiate MARV infection in humans.

With the exception of the first EVD outbreaks in Zaire, which were linked to substantial percutaneous needle transmission (Ebola haemorrhagic fever in Zaire 1976/1978), most of the data since the early 1990s suggest that exposing skin and mucosae to EBOV while conducting activities like body washing during traditional funerals or caring for sick relatives in the household is sufficient for human-to-human transmission of EBOV (Bausch et al. 2007; Dowell et al. 1999; Francesconi et al. 2003). Early data collected from a laboratory exposure to SUDV even suggest that skin abrasions may not be necessary to allow ebolavirus entry through the skin (Emond et al. 1977). These findings raise questions regarding how ebolavirus infection takes place in skin and mucosae, and which cells are involved in the primary amplification of the virus.

Antigen-presenting cells are a putative initial target of EBOV infection and previous research in animal models of disease has indicated that dendritic cells (DCs) and macrophages are early and preferred targets of EBOV and support virus replication (Geisbert et al. 2003a). Both DCs and macrophages can also be productively infected by EBOV in vitro (Gupta et al. 2001, 2007; Mahanty et al. 2003; Bosio et al. 2003), and EBOV prevents activation of in vitro-derived DCs, mainly through the action of VP24 and VP35 (Yen et al. 2014; Jin et al. 2010; Ilinykh et al. 2015).

To further complicate things, a great deal of research over the last decade has been devoted to defining the ontogeny and specific function of DC subsets in mice and humans. The emerging picture is that several cell subsets exist with overlapping and nonoverlapping functions, and these subsets can be roughly classified into classical, plasmacytoid, and inflammatory DCs in humans (see (Haniffa et al. 2013) for an excellent review). Whether EBOV can equally infect different DC subsets is not known, but some of the existing evidence suggest that it cannot (Leung et al. 2011). For example, a number of cellular receptors have been involved in the attachment of EBOV virions to target cells. These receptors include several C-type lectins present on the surface of DCs, such as dendritic cell-specific ICAM-3-grabbing non-integrin (DC-SIGN) (Simmons et al. 2003) and liver/lymph node SIGN (L-SIGN) (Alvarez et al. 2002). DCs of the epidermis and mucosal epithelium do not express these molecules, but Langerhans cells in the skin and CD141$^+$ DCs in mucosal epithelium do express the C-type lectin langerin (Merad et al. 2008). In fact, studies in monkeys and pigs have indicated that DC-SIGN$^+$ cells are scarce in the dermis and the lamina propria or submucosa in the steady state (Schwartz et al. 2002; Huang et al. 2009), suggesting that other cell types may be targets for early EBOV replication. Initiating EBOV infection may depend on attachment to target cells via TIM-1 and TIM-4, which are highly expressed in mucosal epithelia (Rhein et al. 2016; Kondratowicz et al. 2011). Initial virus amplification could then lead to inflammation and infiltration of a high number of myeloid cells expressing DC-SIGN and other described EBOV attachment factors, like triggering receptor expressed on myeloid cells 1 (TREM-1) expressed by neutrophils (Mohamadzadeh et al. 2006), and human macrophage C-type lectin specific for galactose/N-acetylgalactosamine (hMGL) expressed by macrophages (Takada et al. 2004). The elucidation of the initial steps by which EBOV establishes productive infection in a host organism is highly needed to understand the mechanisms by which the virus disseminates from the initial site of entry to the body, and perhaps to design medical countermeasures aimed at preventing virus spread.

3 Virus Dissemination and Initiation of EBOV-Specific Immunity

As mentioned above, DCs and macrophages are early targets of EBOV infection. Due to the migratory potential of DCs, these immune cells may participate in disseminating EBOV from the initial points of entry to the draining lymph nodes

(Geisbert et al. 2003a). This strategy is commonly used by other viruses, including SARS coronavirus (Liu et al. 2015), Toscana virus (Cusi et al. 2016), and measles virus (Mesman et al. 2012), for dissemination in the host. However, perhaps due to the lack of suitable in vivo models for kinetic studies of EBOV, the involvement of DCs in EBOV dissemination has not been experimentally addressed. In any case, the specific subsets of cells responsible for EBOV dissemination remain to be identified. It is plausible that tissue-resident DCs or inflammatory DCs derived from infiltrating monocytes are important for EBOV dissemination. Both myeloid DC populations are migratory and can transport a variety of antigens from inflamed tissues to the draining lymph nodes (Leon et al. 2007; Ersland et al. 2010). Conversely, macrophages and neutrophils are less likely to participate in EBOV dissemination due to their low mobility and nonproductive infection, respectively (Mohamadzadeh et al. 2006).

Recent studies have demonstrated that human EVD is associated with loss of peripheral blood monocytes, in particular nonclassical $CD16^+$ monocytes (Ludtke et al. 2016) which have been proposed as the main antiviral monocyte subset (Cros et al. 2010). Even though this study did not demonstrate direct infection of $CD16^+$ by EBOV, it raised the possibility that this cell subset could be involved in virus dissemination. In fact, $CD16^+$ monocytes are also called patrolling monocytes due to their ability to attach to endothelial cells in a LFA-1-dependent manner and to extravasate into inflamed tissues where they differentiate into inflammatory DCs and macrophages (Cros et al. 2010; Auffray et al. 2007). This hypothesis is also substantiated by a previous study that demonstrated that EBOV particles can attach to monocytes and enter these cells only when the monocyte differentiation program has started, that is, during their differentiation into DCs and macrophages in inflamed tissues (Martinez et al. 2013).

The identification of the DC subsets specifically involved in filovirus dissemination is a highly relevant topic of study, because the function of DCs can be enhanced or inhibited, ex vivo or in vivo, by antigen delivery or use of molecules and antibodies. Therefore, DCs are putative immunotherapeutic targets for post-exposure EVD treatment (Klechevsky and Banchereau 2013). For example, the ligand of the DC co-stimulatory molecule CD40 (sCD40L) is commonly used to enhance DC-mediated antigen presentation (Kornbluth and Bot 2012), and previous studies have demonstrated a correlation between circulating levels of sCD40L and survival after SUDV infection (McElroy et al. 2014a, b). In addition, poor activation profiles of circulating antigen-presenting cells have been correlated with severe EVD (Ludtke et al. 2016). These findings provide a rationale for the use of DC enhancers as immunotherapy candidates in filovirus disease.

Another key driver of EBOV dissemination may be the cytokine microenvironment, since these innate immune signaling molecules play an important role in recruiting myeloid cells, which are putative EBOV targets, to sites of inflammation. A considerable body of research exists on cytokine and chemokine responses during EVD. Despite some conflicts, in general, these data correlate fatal outcomes during EVD with high concentrations of pro-inflammatory cytokines (e.g., IL-6), pro-inflammatory chemokines (e.g., IP-10), and anti-inflammatory cytokines (IL-RA and IL-10), overall suggesting a general dysregulation in the expression of these key immune signaling molecules (Hutchinson and Rollin 2007; Gupta et al. 2012; Wauquier et al. 2010; Baize et al. 2002; Villinger et al. 1999). An inability to control viral replication is likely leading to continued innate immune stimulation. Data from asymptomatic human cases have shown an even greater magnitude of cytokine and chemokine upregulation, followed by rapid downregulation of this response in association with control of viral replication (Leroy et al. 2000, 2001), suggesting that cytokine/chemokine dysregulation is a consequence of uncontrolled viral replication rather than a primary mediator of pathogenesis.

Type I interferons (IFN-I) are key antiviral cytokines, and perhaps one of the more conflicting aspects of comparing EBOV experimental and clinical human data is the role of IFN-I in EBOV immunity and pathogenesis. Importantly, IFN-I not only induces an antiviral state in infected and bystander cells during early virus infection, but also is a key modulator in the transition between innate and adaptive immunity. By enhancing natural killer cell function, antigen presentation by DCs, and expansion of effector T cells, IFN-I bridges natural and acquired antiviral immunity (see (McNab et al. 2015) for a review), so elucidating its functions during human EVD is highly relevant to understanding disease pathogenesis. IFN is critical in protecting laboratory mice from EBOV (Brannan et al. 2015; Bray et al. 2002), but data are somewhat conflicting in nonhuman primates (NHPs) and humans (McElroy et al. 2016; Smith et al. 2013; Villinger et al. 1999; Yen et al. 2011). Higher levels of IFN-alpha were associated with fatal EVD cases (Villinger et al. 1999), but higher IFN-beta was associated with less severe EVD (McElroy et al. 2016) and IFN-beta administration prolonged survival in nonhuman primates (Smith et al. 2013). However, as IFN responses are highly dynamic, drawing conclusions regarding human pathogenesis is difficult. For example, EVD survivors may mount early and robust IFN responses that keep viral replication at bay, while patients who succumb to EVD may display higher IFN levels later on due to increased viral replication and inflammation.

Interestingly, recent studies have shown that patients who survive and patients who succumb to EVD both show robust T cell activation (McElroy et al. 2015a; Ruibal et al. 2016). Since DCs are the only antigen presenting cells capable of priming naïve T cells (Banchereau and Steinman 1998), these results suggest at least two possibilities. On one hand, infected DCs may retain their capacity to initiate T cell-specific responses, as has been shown in other viral infections (Wahid et al. 2005; Rivera and McGuire 2005; Kvale et al. 2006). On the other hand, some DC subsets may be spared from infection and thus able to prime EBOV-specific T cells. The generation of this EBOV-specific adaptive immunity is the topic of the next two sections.

4 Adaptive Immunity: Human Antibody Responses

While innate immune responses may play a chief role in controlling early EBOV replication in humans, the current model identifies the character (though not necessarily the magnitude) of adaptive immunity as the main factor driving viral clearance and recovery. Both humoral and cellular immunity seem to be required for EBOV clearance in humans, a hypothesis strengthened by the finding that EVD patients mount robust adaptive immune responses (McElroy et al. 2015a; Ruibal et al. 2016) with high numbers of circulating plasmablasts and EBOV-specific T cells.

The more difficult task is assessing whether adaptive immune responses mark substantial differences between fatal and surviving patients. Initial studies supported the idea that early development of IgM and isotype switching to IgG correlated with positive outcome. Indeed, a high percentage of patients with fatal outcomes do not seem to develop IgM (Ksiazek et al. 1999). These field studies are also in agreement with findings in patients evacuated into Europe or the US for medical treatment during the recent West African EVD outbreak. Surviving EVD patients mounted early IgM responses and showed upregulation of serum IgG over the course of the disease, which was correlated with viral clearance (Kreuels et al. 2014; Wolf et al. 2015). Conversely, deficient or diminished IgM and IgG responses have been reported in fatal cases of both EVD and MVD (van Paassen et al. 2012; Baize et al. 1999).

However, limited field data also indicate survivors who did not develop IgG, as well as patients who died after developing detectable circulating anti-EBOV antibodies (Onyango et al. 2007). In addition, limited clinical data obtained from the SUDV-caused EVD outbreak in Gulu, Uganda, did not reveal significant differences between the humoral responses in fatal and nonfatal EVD, with very late expression of IgG in both groups that was unrelated with viral clearance (Towner et al. 2004). As in many other aspects of EVD immunology, the kinetics of antibody responses in a statistically relevant cohort of acute-stage patients with defined outcomes must be studied.

Antibodies play many roles during the immune response to pathogens, including neutralization and antibody-dependent killing of virus-infected cells by targeting them to Fc receptor bearing cells (ADCC) or complement (CMC). Neutralizing antibodies (N_{AB}) probably play a small role in recovery from acute EVD, since in many survivors N_{AB} are not detectable until weeks or even months after recovery (Luczkowiak et al. 2016; Sobarzo et al. 2012). This is a puzzling and as yet unexplained finding. One of the plausible hypotheses is that disrupting lymphoid architecture during acute EVD infection may compromise germinal center formation and B cell affinity maturation, a feature that has been observed during Lassa fever (Carrion et al. 2007). However, this hypothesis does not reconcile easily with the levels of circulating plasmablasts in patients in the acute stage of illness (McElroy et al. 2015a), or with the limited focal necrosis observed in human biopsy samples of lymphoid tissues (Martines et al. 2015). One interesting possibility is that the long filament shape of filovirus virions may require a highly diverse repertoire of N_{AB} for effective neutralization. In fact, studies have found greater B cell clonality in EVD survivors than in individuals with B cell memory against HIV-1 or influenza A virus (Bornholdt et al. 2016b). Still, this cannot be the whole story, as similar delays in N_{AB} production have been described in other viral hemorrhagic fevers (e.g., Lassa fever). Perhaps more importantly, delayed N_{AB} production strongly suggests long-term virus or antigen persistence, which is in agreement with duration of post-EVD sequelae (see below).

Nevertheless, long-term survivors develop effective N_{AB}, mainly directed against several epitopes of the ebolaviral $GP_{1,2}$ (Bornholdt et al. 2016a, b; Misasi et al. 2016; Corti et al. 2016). Many of the described N_{AB} isolated from survivors are directed against the $GP_{1,2}$ glycan cap as well as against the region bridging GP1 and GP2, which seem to be epitopes amenable for antibody-based therapeutics like ZMapp (Murin et al. 2014). Importantly, a number of studies in surviving patients have highlighted the presence of naturally occurring N_{AB} with cross-reactivity against other ebolaviruses and even against MARV (Olal et al. 2012; Misasi et al. 2016; Bornholdt et al. 2016a). These findings strongly suggest that antibody-mediated immune memory may provide long-term protection against secondary infection with filoviruses, and may have important implications for public health measures (e.g., recruiting survivors as caregivers in future outbreaks).

Currently, the relative importance of N_{AB} versus other antibody-mediated mechanisms, such as ADCC and CMC, is unclear, even though most protective antibodies probably act through both neutralization and ADCC/CMC activation (Schmaljohn and Lewis 2016). Antibodies with both neutralizing and ADCC capacity have been detected in EVD survivors more than a decade after recovery (Corti et al. 2016), and ADCC is probably an important mechanistic feature of ZMapp (Olinger et al. 2012). In addition to dissecting whether or not neutralization and ADCC specifically contribute to EVD immunity, it is important to determine the kinetics of antibody-mediated immunity from acute infection to long-term recovery. Lack of information on EVD antibody kinetics and EBOV-specific quantitative activity is probably largely responsible for the lack of protective effect demonstrated by convalescent plasma therapy (van Griensven et al. 2016). This therapeutic strategy has putative applicability for field outbreak conditions, but requires characterization of the virus specific activity in the product as well as optimization to ensure transfer of sufficient quantities of protective antibodies.

5 Adaptive Immunity: Human T Cell Responses

T cells, particularly CD8 T cells, are essential for clearance of acute viral infections. Naïve T cells react to stimulation with pathogen-specific peptides by massively expanding, differentiating into effector cells and migrating to peripheral infection sites for elimination of infected cells (see (Zhang and Bevan 2011) for a review). Because naïve CD8 T cells can be activated only by DCs (Banchereau and

Steinman 1998), the initial assumption was that EBOV-induced DC inactivation would in turn result in poor T cell priming and overall inability of the host to eliminate infection. This hypothesis was substantiated by early studies demonstrating that, despite being spared from infection, many T cells underwent apoptosis during human EVD (Baize et al. 1999; Wauquier et al. 2010). While this observation is still valid, data gathered mainly during the recent West African outbreak have suggested that lymphocytes in general display very dynamic kinetics during EVD, which may include early proliferation followed by lymphopenia (Kreuels et al. 2014; Wolf et al. 2015). Similar dynamics occur during other systemic viral infections and during pro-inflammatory disorders such as sepsis (Luan et al. 2015), suggesting that perhaps lymphopenia is not a differential characteristic of EVD.

A substantial difference between earlier studies and those carried out in the context of the recent outbreak has been the application of multiparametric flow cytometry, which has allowed for the first time collection of phenotypic and functional information from single cells in unprecedented detail. These studies have revealed that, in fact, EVD is characterized by massive T cell activation rather than inhibition in both surviving and fatal cases. Co-expression of activation markers such as CD38 and HLA-DR, as well as proliferation markers like Ki-67, were detected in a significant percentage of CD4 and CD8 T cells in EVD patients (McElroy et al. 2015a; Ruibal et al. 2016) and were comparable with the magnitude of activation observed in other acute infections or after vaccination (Lindgren et al. 2011; Miller et al. 2008). Of note, since co-expression of CD38 and HLA-DR is correlated with engagement of the T cell receptor (Appay et al. 2002), these findings strongly suggest that proper T cell priming by antigen-presenting cells occurs during EVD in humans. Additionally, these findings were comparable between patients receiving experimental therapy (McElroy et al. 2015a) and those who received supportive care in the field (Ruibal et al. 2016), indicating that robust T cell activation is a characteristic of EVD unrelated to treatment.

A paramount question, therefore, is why robust T cell activation does not lead to viral clearance during EVD. To some extent, this lack of T cell effectiveness may be related to defects in negative immune checkpoints, namely the molecular mechanisms that control the transition from activation to immune homeostasis and that are essential for autoimmune control (Buchbinder and Desai 2016). Two such mechanisms are triggered by the T cell co-inhibitor molecules, programmed cell death-1 (PD-1) and cytotoxic T-lymphocyte-associated protein 4 (CTLA-4). An earlier review already hypothesized that T cell dysfunction during filovirus infection could be related to high expression of PD-1 and CTLA-4 in T cells (Mohamadzadeh et al. 2007), which leads to a nonfunctional but reversible status termed T cell exhaustion (Wherry 2011). Studies from the recent West African EVD outbreak found that peripheral blood T cells from EVD patients expressed high levels of PD-1 and CTLA-4 (McElroy et al. 2015a; Ruibal et al. 2016), which were significantly higher in fatal cases (Ruibal et al. 2016). As a follow-up to these observational studies, determining the correlation between high expression of T cell inhibitory molecules and T cell function and apoptosis will be important. Determining this correlation will most likely require relevant in vivo models that can reproduce this T cell phenotype. Utilizing immunotherapeutic approaches to block PD-1 and CTLA-4 function during postexposure filovirus infection treatment may provide an interesting opportunity. Several therapeutic products are licensed extensively to block PD-1 and CTLA-4 in several types of cancer, thereby restoring T cell function (see (Sharma and Allison 2015) for a review).

Another important and related question is whether broad and polyfunctional T cell responses lead to increased disease manifestations, like in hantavirus (cardio) pulmonary syndrome (Terajima and Ennis 2011), or to decreased susceptibility, as in dengue virus 1–4 infections (Weiskopf et al. 2013). To answer this question, an exhaustive analysis of EBOV T cell immunodominance in humans must be performed, which is still not available. Previous evidence shows that the viral nucleoprotein (NP) drives most of the CD8 T cell response (McElroy et al. 2015a; Sundar et al. 2007; Wilson and Hart 2001). This finding is consistent with the observation that HLA alleles recognizing conserved filovirus NP epitopes provide protection against SUDV infection (Sanchez et al. 2007b). Additional studies of HLA association with EVD outcomes in a statistically significant cohort of patients are highly needed to strengthen these initial observations. The finding that NP drives most of the CD8 T cell response also has significant implications for vaccine design and may explain, at least to some extent, why most $GP_{1,2}$-based vaccines induce poor T cell immunity (Agnandji et al. 2016; Ewer et al. 2016; Zhu et al. 2015).

6 Electrolyte Imbalances

The degree to which electrolyte abnormalities contribute to EVD pathogenesis was not appreciated in earlier outbreak responses because real-time serum electrolyte data were not available. The ability to acquire these measurements in patients during the West African EVD outbreak, as well as the degree of profuse watery diarrhea that was reported, have brought to the forefront the severity of electrolyte imbalances in EVD and the impact electrolytes could have on patient outcome. Such data were first collected during the SUDV outbreak in Gulu, Uganda, in 2000–2001, during which elevated BUN/Cre levels and hypocalcemia were associated with severe disease and fatal outcomes (Rollin et al. 2007). Data from African cohorts (Hunt et al. 2015) combined with data from repatriated patients who were cared for in developing nations (Uyeki et al. 2016b) have revealed potassium abnormalities, hyponatremia, hypomagnesemia, and hypocalcemia. Some of these alterations may be related to acute renal injury that is also common among severely ill patients; others might be related to volume and electrolyte imbalance secondary to profuse watery diarrhea. The clinical consequences of electrolyte imbalances could include cardiac arrhythmias, seizures, or coma. Indeed, 41% of the repatriated patients exhibited arrhythmia or electrocardiographic changes, one patient had seizures, and three were in a coma (Uyeki et al. 2016b). Electrolyte levels are easily measured blood chemistry parameters that can be corrected with electrolyte and fluid administration. Such measures may have contributed to improved outcomes during the recent outbreak, as one EVD treatment center that incorporated these data into patient management had a case fatality rate of only 37%, significantly lower than the 50 and 74% rates reported from two other treatment centers in Sierra Leone (Hunt et al. 2015; Schieffelin et al. 2014; Lanini et al. 2015).

7 Endothelial Dysfunction

Multiple lines of evidence have suggested that the endothelium is dysfunctional during EVD. While endothelial cells are directly infected, they do not show significant cytopathic effect, and endothelial infection is thought to occur during the terminal phase of the illness (Martines et al. 2015; Geisbert et al. 2003c). The overall dysfunction of the endothelium is thought to be an indirect effect of pro-inflammatory cytokines like TNF-alpha (Villinger et al. 1999; Feldmann et al. 1996), or other molecules, like nitric oxide, that increase the permeability of the endothelium during inflammation (Sanchez et al. 2004). Increased levels of several pro-inflammatory cytokines and chemokines are associated with EVD-related deaths (Baize et al. 2002; Wauquier et al. 2010; Gupta et al. 2012; Hutchinson and Rollin 2007; McElroy et al. 2014a, b). Infected antigen-presenting cells, such as macrophages, DCs, or monocytes are the presumed source of these cytokines (Feldmann et al. 1996; Gupta et al. 2001), and these cytokines lead to endothelial activation.

Increased vascular permeability due to loosening of the endothelial barrier is a normal and necessary physiologic function that allows cells and biomolecules to reach sites of inflammation, but widespread activation in many inflammatory diseases results in fluid movement that can be detrimental to the host. Clinical and laboratory findings in EVD, including tachypnea (with or without pulmonary edema), hypotension, oliguria, tachycardia, impaired distal perfusion, hypoalbuminemia, and hemoconcentration, are consistent with fluid extravasation into extravascular spaces secondary to increased vascular permeability (Rollin et al. 2007; Uyeki et al. 2016b; Hunt et al. 2015; Chertow et al. 2014). This constellation of clinical findings is thought ultimately to lead to hypovolemic shock in fatal cases.

In recent years, additional evidence that dysfunctional endothelia contribute to the disease process include the findings of increased levels of sICAM, thrombomodulin, PE-CAM, and P-selectin in patients with severe or fatal disease (McElroy et al. 2014a, b). All of these biomarkers, when released into the plasma, indicate an activated endothelium and/or breakdown of endothelial intercellular junctions. An activated endothelium is both pro-inflammatory and pro-coagulant, and likely contributes both to the ongoing inflammatory response that characterizes severe EVD and to the coagulopathy that has been observed in some patients (discussed in more detail below).

Also noteworthy is the sometimes conflicting body of evidence implicating the viral glycoprotein ($GP_{1,2}$) in endothelial dysfunction. The EBOV *GP* gene coding region produces two proteins based on a transcriptional editing site, the soluble GP (sGP) and the full-length structural $GP_{1,2}$ (Sanchez et al. 1996). The full-length $GP_{1,2}$ produced by pseudotyped retrovirus or virion-like particles (VLPs) can bind to and activate endothelial cells, leading to increased endothelial permeability (Wahl-Jensen et al. 2005; Yang et al. 1998). sGP has been detected in the plasma of infected individuals (Sanchez et al. 1999), and, in fact, inhibits TNF-mediated increases in vascular permeability in vitro, perhaps suggesting a compensatory mechanism to control virus-induced inflammation. A third form of the protein, known as shed GP, is shed from the surface of infected cells in vitro and increases permeability of cultured endothelial cells (Escudero-Perez et al. 2014). While shed GP was detected in infected guinea pigs, it has not yet been detected in vivo in humans. Finally, overexpression of $GP_{1,2}$ in explanted human, porcine, or NHP blood vessels leads to increased endothelial permeability and endothelial cytotoxicity mediated by the mucin domain of the protein (Yang et al. 2000). The relevance of this finding to EVD is unclear, since endothelial cells are infected long after endothelial function has already been compromised, and do not show cytopathic effects when infected in vitro (Geisbert et al. 2003c). Taken together, these data suggest that the various forms of EBOV GP may modulate endothelial function, but the precise role of the protein in human EVD pathogenesis is unclear.

8 Coagulopathy

The moniker "viral hemorrhagic fever" was applied to EBOV EVD during the first outbreak identified in 1976, and was appropriate because 78% of fatal cases had hemorrhagic manifestations, mostly melena (Ebola haemorrhagic fever in Zaire 1976/1978). This 1976 outbreak was unique, because the route of virus transmission was via injection in approximately one-third of the 288 patients, and this mode of entry could have contributed to the manifestations and severity of disease (the authors note that all patients who were infected by injection died). Notably, a concurrent outbreak of SUDV-caused EVD also had high frequencies (71%) of hemorrhagic manifestations. However, in several of the larger subsequent outbreaks where appropriate data were available, significantly fewer patients had hemorrhagic manifestations of disease: 41% (EBOV 1995), 30% (SUDV 2000), and 47% (BDBV 2007) (Okware et al. 2002; MacNeil et al. 2010; Bwaka et al. 1999). Additionally, in these three outbreaks, no association was observed between bleeding and death, arguing against the commonly held belief that hemorrhage equates to a fatal outcome. Furthermore, in the Western African outbreak, hemorrhagic manifestations were rarely reported; fewer than 15% of all patients from Liberia and Sierra Leone had any bleeding symptom recorded (Chertow et al. 2014; Schieffelin et al. 2014; Lado et al. 2015; Yan et al. 2015; Dallatomasina et al. 2015; Li et al. 2016; Qin et al. 2015), but two reports from a single center in Guinea reported bleeding in 51 and 26% of patients (Barry et al. 2014; Bah et al. 2015). Perhaps reported differences in hemorrhage frequency are related to genetic or nutritional factors that cannot be controlled for in observational reports. Regardless, hemorrhaging can occur during EVD, but is not the most prominent feature. In contrast, hemorrhage does seem to be a common (50–80% of patients) feature of MVD, based upon limited data from the two largest outbreaks to date (Colebunders et al. 2007; Roddy et al. 2010).

Hemorrhaging is a clinical sign that can be secondary to multiple types of hematologic disorders. In the simplest terms, two general categories of hematologic disorders manifest clinically as bleeding: low levels of platelets and coagulation factor deficiencies (Hunt 2014). Platelet counts have not been routinely measured in patients with EVD, but in one study of 150 patients with EVD during the West African outbreak, platelet counts were not especially low, ranging 119–247 $\times$ 10^9/L (normal range is 150–300) (Hunt et al. 2015). Interestingly, the same phenomenon was noted years ago in the NHP model, and while absolute platelet counts were not very low, platelet function was severely affected as a result of in vivo activation and degranulation (Fisher-Hoch et al. 1985). No measurements of platelet function have been reported to date in humans. However, elevated levels of sCD40L were observed in surviving patients with EVD caused by SUDV (McElroy et al. 2014a, b). Since platelets are the major source of sCD40L in the bloodstream (Henn et al. 2001), this finding suggests platelet activation in humans during EVD. These data suggest a process that consumes platelet functional activity in severe or fatal EVD; this process would be consistent with the finding that in lethal NHP studies of EVD, sCD40L levels are elevated initially, but decline to undetectable at the time of death (Ebihara et al. 2011).

MARV might be a bit different, since in the original report of the first outbreak in Europe in 1967, most patients had severe thrombocytopenia, sometimes less than 10 $\times$ 10^9/L (Martini 1973), coincident with significant hemorrhage in about half of the patients. Finally, the type of bleeding often described in EVD (and MVD) patients—epistaxis, conjunctival hemorrhages, bleeding into the GI tract and from the oral cavity—is mostly mucosal in nature, consistent with loss of platelet numbers or function.

The second general category of hematologic disorders that manifest as bleeding is deficiency in coagulation factors. Coagulation factors are quantitated clinically by measuring partial thromboplastin time (PTT) and prothrombin time (PT) to evaluate the intrinsic and extrinsic coagulation pathways. Unfortunately, these measurements have only been reported in case studies, and provide no consensus regarding the levels of PT and PTT during EVD. This is a clear information gap that needs to be addressed. An early report on MARV states that PTT and PT were measured in 10 patients, but the values obtained did not explain the observed severity of the hemorrhaging (Martini 1973).

DIC is often seen in critically ill patients, especially those with sepsis, and involves both low platelet counts and coagulation factor deficiencies. The bleeding seen in patients with EVD is often reported as due to DIC, although whether the criteria for DIC are met is unknown because the necessary laboratory tests are not routinely available. DIC laboratory features include thrombocytopenia, elevated fibrin split products, prolonged PT, and consumption of fibrinogen (Levi et al. 2009). As noted above, the level of thrombocytopenia seen in EVD patients rarely meets the criteria to assign a DIC score, but elevated fibrin split products (such as D-dimer) have been measured retrospectively, are elevated in EVD patients, and are associated with fatal outcomes (Rollin et al. 2007). PT measurements are normal in the few available case reports (Sueblinvong et al. 2015), and fibrinogen levels were not associated with outcome or hemorrhagic manifestations (McElroy et al. 2014a, b, 2016). Measurement of DIC markers is clearly an area that requires additional study for clarification.

Less conventional evaluations of factors involved in coagulation pathways have also been conducted. Thrombomodulin, a protein expressed on endothelial cells, has anticoagulant properties in the microenvironment of the cell surface. When present in the plasma, thrombomodulin can act more globally, as shown in one family with a genetic deficiency that results in elevated levels of free plasma thrombomodulin in association with a bleeding disorder (Langdown et al. 2014). Endothelial cells also release thrombomodulin when they become activated. Elevated plasma levels of thrombomodulin were associated with both hemorrhage and death in SUDV patients, and with more severe disease in a cohort of EVD patients (McElroy et al. 2014a, b, 2016), suggesting that loss of this protein from the endothelial surface exacerbates both endothelial dysfunction and coagulopathy during EVD. Additionally, tissue factor, which is implicated in coagulopathy observed in NHPs (Geisbert et al. 2003b), was also elevated in patients with severe EVD (McElroy et al. 2016). Von Willebrand factor (vWF), a protein that is present in both platelets and endothelial cells and mediates interactions between platelets and the damaged endothelium, was elevated both in SUDV-infected patients with hemorrhage and in pediatric SUDV-infected patients with fatal outcomes. It was also elevated in EBOV-infected patients with severe disease (McElroy et al. 2015b, 2016).

A complex interplay of activated endothelial cells, activated platelets, inflammation, and coagulopathy is clearly at work during EVD. How intervening in any one aspect of the network impacts human disease is still unknown. It would be invaluable to determine the effects on EVD outcome of readily available clinical products that affect aspects of these processes. Some compounds of interest are statins, which stabilize the endothelium; soluble GPIbα, which inhibits the interaction between platelets and vWF; and sCD40L, which appears to be consumed during severe disease.

9 The Roles of Co-infections, Co-morbidities, and Age

One key and largely unaddressed question is the role of co-infections and co-morbidities in EVD pathogenesis. Especially relevant to patient populations in the affected African nations are the possible contributions of malnutrition and malarial co-infection in the disease process. Malnutrition is prevalent in the regions affected by EVD (Wirth et al. 2016), and long-standing malnutrition leads to defects in both innate and adaptive cellular immune responses (Schaible and Kaufmann 2007). Malnutrition may contribute to the high case fatality rates observed during EVD outbreaks in Africa as compared to filoviral infections in patients repatriated to the US and Europe (Uyeki et al. 2016b; Martini 1973).

Malaria co-infection is likely to increase EVD-related mortality, although this has not been rigorously evaluated, a study of the effects of various antimalarial drugs has been conducted during the West African outbreak. Antimalarial drugs were routinely given to EVD-positive patients at the ETC in Foya, Liberia; during a time of artesunate-lumefantrine shortage, artesunate-amodiaquine was prescribed. While the amodiaquine preparation was associated with improved survival in malaria-negative patients, interestingly suggesting a direct antiviral effect, this effect was lost in the malaria-positive patients, suggesting that malaria and EVD co-infection lead to worse outcomes even when malaria is treated (Gignoux et al. 2016).

Also of potential consequence are co-infections with HIV or other hepatotropic viruses. HIV co-infection has only been examined in one study; during the SUDV outbreak in Gulu, 18% of the tested patients were HIV-1-positive by antibody testing. No differences in EVD outcome were observed based upon HIV status in this study (McElroy et al. 2014a, b), but no CD4 counts were obtained, so it is possible that all cases were newly acquired and the patients were not yet immune-compromised enough for HIV-1 infection to influence EVD outcome. One study evaluated publicly available next-generation sequencing data, and using a cohort of 49 patients, posited that co-infection with GB virus C (a common, clinically innocuous pegivirus infection) results in improved outcomes during EVD infection (Lauck et al. 2015). The results were somewhat confounded by the fact that age is a major determinant of both GB virus C infection and outcome during EVD.

A special mention must be made that early data regarding the effect of age on EBOV susceptibility (Dowell 1996) and outcome (Mupere et al. 2001; McElroy et al. 2014a, b) have been repeatedly observed in large cohorts during the West African outbreak (Team et al. 2015; Faye et al. 2015; Li et al. 2016; Schieffelin et al. 2014; Bower et al. 2016). Case fatality rates are high in children under 5, lowest in school-aged children, reaching a nadir around puberty, and increase again to peak in the elderly. This phenomenon has been seen in other infectious diseases in children, and suggests perhaps that school-aged children are in the perfect immunologic window of life, with a fully mature and functioning immune system without the alterations that occur secondary to the influence of sex hormones. To date, only one study has examined pediatric patients for laboratory evidence of this protective effect in EVD; this study demonstrated that pediatric patients have viral loads similar to adult patients (McElroy et al. 2014a, b) and thus do not appear to control the viremia better. However, higher levels of RANTES, a T cell chemokine, were associated with pediatric survival, an association not seen in adults. Thus, stronger immune responses in pediatric patients might contribute to better outcomes, but this remains to be proven definitively and will require additional research efforts.

10 Post-EVD Syndrome

Perhaps one of the most striking findings during the Western African EVD outbreak has been the identification of severe sequelae in EVD survivors long after recovery. These sequelae have important implications both for medical treatment and for public health. In 1975, one year before the identification of EBOV in Zaire, MARV was successfully isolated from the ocular fluid of a convalescent patient with uveitis (Gear et al. 1975). A later follow-up study described arthralgia, myalgia, and abdominal pain as common sequelae in EVD survivors of the Kikwit outbreak (Rowe et al. 1999). Similar findings in individuals long after recovery from BDBV infection (Clark et al. 2015) suggest that post-recovery sequelae may be common in filovirus infections.

To date, EBOV RNA has been detected in semen, ocular fluid, cerebrospinal fluid, breast milk, and other body fluids in EVD survivors for several weeks or even months after discharge (Green et al. 2016; Chughtai et al. 2016). Moreover, infectious virus has been isolated from semen (Uyeki et al. 2016a), ocular fluid (Varkey et al. 2015), saliva, and breast milk (Bausch et al. 2007), and epidemiological evidence of sexual EBOV transmission has been established (Mate et al. 2015). An important task will be determining the pathogenic potential of virus isolates from semen compared to those of blood from the same patient. Because sexual transmission seems to be uncommon (based on the large numbers of male survivors and few sexual transmission events), virus isolates from semen may be less infectious, either due to attenuating mutations or inactivation by EBOV-specific antibodies secreted at the mucosal surface. In general, the pathogenic features of EVD sequelae and their putative physiological mechanisms are poorly understood.

Importantly, many of the symptoms reported by EVD survivors, as well as some of the observed signs like uveitis and skin desquamation, suggest an inflammatory syndrome. Indeed, immune activation persists after the acute phase of EVD (McElroy et al. 2015a; Rowe et al. 1999), strongly suggesting continuous immune stimulation and postinfection autoimmunity. These hypotheses still need to be experimentally tested, but are consistent with virus persistence in immunoprivileged sites. Alternatively, sustained inflammation could be due to deposition of immune complexes in the joints or to viral antigen persistence, as seen commonly in alphavirus infections and influenza A virus infection respectively (Hoarau et al. 2010; Tamburini et al. 2014; Tappe et al. 2016).

Since some, but not all, EVD survivors suffer sequelae, the factors leading to post-EVD syndrome must be determined. Co-morbidities and co-infections are likely contributing to sequelae development, particularly those involving immune phenomena like bystander T cell activation (Fujinami et al. 2006). Also, a positive correlation has been described between viremia levels during the acute phase of EVD and increased risk of sequelae (Mattia et al. 2016). These findings suggest that during infection, EBOV may be confined to immunoprivileged sites by treatment or by the host immune response, leading to viral persistence, and that this phenomenon would be favored by high viral loads. Consistent with this hypothesis, one surviving patient with very high virus loads had virus persistence even within the central nervous system, followed by virus reactivation and meningoencephalitis (Jacobs et al. 2016).

In the future, the causes and molecular mechanisms of post-EVD syndrome must be determined. In particular, assessing whether reactivation of infectious virus is involved in sequelae development is a key. Due to the magnitude of the West African outbreak and the high number of surviving individuals, this is an important issue that needs to be addressed to establish adequate medical countermeasures and public health policies.

References

Adjemian J, Farnon EC, Tschioko F, Wamala JF, Byaruhanga E, Bwire GS, Kansiime E, Kagirita A, Ahimbisibwe S, Katunguka F, Jeffs B, Lutwama JJ, Downing R, Tappero JW, Formenty P, Amman B, Manning C, Towner J, Nichol ST, Rollin PE (2011) Outbreak of Marburg hemorrhagic fever among miners in Kamwenge and Ibanda Districts, Uganda, 2007. J Infect Dis 204(Suppl 3):S796–799. doi:10.1093/infdis/jir312

Agnandji ST, Huttner A, Zinser ME, Njuguna P, Dahlke C, Fernandes JF, Yerly S, Dayer JA, Kraehling V, Kasonta R, Adegnika AA, Altfeld M, Auderset F, Bache EB, Biedenkopf N, Borregaard S, Brosnahan JS, Burrow R, Combescure C, Desmeules J, Eickmann M, Fehling SK, Finckh A, Goncalves AR, Grobusch MP, Hooper J, Jambrecina A, Kabwende AL, Kaya G, Kimani D, Lell B, Lemaitre B, Lohse AW, Massinga-Loembe M, Matthey A, Mordmuller B, Nolting A, Ogwang C, Ramharter M, Schmidt-Chanasit J, Schmiedel S, Silvera P, Stahl FR, Staines HM, Strecker T, Stubbe HC, Tsofa B, Zaki S, Fast P, Moorthy V, Kaiser L, Krishna S, Becker S, Kieny MP, Bejon P, Kremsner PG, Addo MM, Siegrist CA (2016) Phase 1 Trials of rVSV Ebola Vaccine in Africa and Europe. N Engl J Med 374 (17):1647–1660. doi:10.1056/NEJMoa1502924

Alvarez CP, Lasala F, Carrillo J, Muniz O, Corbi AL, Delgado R (2002) C-type lectins DC-SIGN and L-SIGN mediate cellular entry by Ebola virus in cis and in trans. J Virol 76(13):6841–6844

Amman BR, Carroll SA, Reed ZD, Sealy TK, Balinandi S, Swanepoel R, Kemp A, Erickson BR, Comer JA, Campbell S, Cannon DL, Khristova ML, Atimnedi P, Paddock CD, Crockett RJ, Flietstra TD, Warfield KL, Unfer R, Katongole-Mbidde E, Downing R, Tappero JW, Zaki SR, Rollin PE, Ksiazek TG, Nichol ST, Towner JS (2012) Seasonal pulses of Marburg virus circulation in juvenile Rousettus aegyptiacus bats coincide with periods of increased risk of human infection. PLoS Pathog 8(10):e1002877. doi:10.1371/journal.ppat.1002877

Appay V, Dunbar PR, Callan M, Klenerman P, Gillespie GM, Papagno L, Ogg GS, King A, Lechner F, Spina CA, Little S, Havlir DV, Richman DD, Gruener N, Pape G, Waters A, Easterbrook P, Salio M, Cerundolo V, McMichael AJ, Rowland-Jones SL (2002) Memory CD8+ T cells vary in differentiation phenotype in different persistent virus infections. Nat Med 8(4):379–385. doi:10.1038/nm0402-379

Auffray C, Fogg D, Garfa M, Elain G, Join-Lambert O, Kayal S, Sarnacki S, Cumano A, Lauvau G, Geissmann F (2007) Monitoring of blood vessels and tissues by a population of monocytes with patrolling behavior. Science 317(5838):666–670. doi:10.1126/science.1142883

Bah EI, Lamah MC, Fletcher T, Jacob ST, Brett-Major DM, Sall AA, Shindo N, Fischer WA 2nd, Lamontagne F, Saliou SM, Bausch DG, Moumie B, Jagatic T, Sprecher A, Lawler JV, Mayet T, Jacquerioz FA, Mendez Baggi MF, Vallenas C, Clement C, Mardel S, Faye O, Faye O, Soropogui B, Magassouba N, Koivogui L, Pinto R, Fowler RA (2015) Clinical presentation of patients with Ebola virus disease in Conakry, Guinea. N Engl J Med 372(1):40–47. doi:10.1056/NEJMoa1411249

Baize S, Leroy EM, Georges AJ, Georges-Courbot MC, Capron M, Bedjabaga I, Lansoud-Soukate J, Mavoungou E (2002) Inflammatory responses in Ebola virus-infected patients. Clin Exp Immunol 128(1):163–168

Baize S, Leroy EM, Georges-Courbot MC, Capron M, Lansoud-Soukate J, Debre P, Fisher-Hoch SP, McCormick JB, Georges AJ (1999) Defective humoral responses and extensive intravascular apoptosis are associated with fatal outcome in Ebola virus-infected patients. Nat Med 5(4):423–426. doi:10.1038/7422

Banchereau J, Steinman RM (1998) Dendritic cells and the control of immunity. Nature 392 (6673):245–252. doi:10.1038/32588

Barry M, Traore FA, Sako FB, Kpamy DO, Bah EI, Poncin M, Keita S, Cisse M, Toure A (2014) Ebola outbreak in Conakry, Guinea: epidemiological, clinical, and outcome features. Med Mal Infect 44(11–12):491–494. doi:10.1016/j.medmal.2014.09.009

Bausch DG, Borchert M, Grein T, Roth C, Swanepoel R, Libande ML, Talarmin A, Bertherat E, Muyembe-Tamfum JJ, Tugume B, Colebunders R, Konde KM, Pirad P, Olinda LL, Rodier GR, Campbell P, Tomori O, Ksiazek TG, Rollin PE (2003) Risk factors for Marburg hemorrhagic fever, Democratic Republic of the Congo. Emerg Infect Dis 9(12):1531–1537. doi:10.3201/eid0912.030355

Bausch DG, Towner JS, Dowell SF, Kaducu F, Lukwiya M, Sanchez A, Nichol ST, Ksiazek TG, Rollin PE (2007) Assessment of the risk of Ebola virus transmission from bodily fluids and fomites. J Infect Dis 196(Suppl 2):S142–147. doi:10.1086/520545

Bornholdt ZA, Ndungo E, Fusco ML, Bale S, Flyak AI, Crowe JE Jr, Chandran K, Saphire EO (2016a) Host-primed Ebola Virus GP exposes a hydrophobic NPC1 receptor-binding pocket, revealing a target for broadly neutralizing antibodies. mBio 7(1):e02154–02115. doi:10.1128/mBio.02154-15

Bornholdt ZA, Turner HL, Murin CD, Li W, Sok D, Souders CA, Piper AE, Goff A, Shamblin JD, Wollen SE, Sprague TR, Fusco ML, Pommert KB, Cavacini LA, Smith HL, Klempner M, Reimann KA, Krauland E, Gerngross TU, Wittrup KD, Saphire EO, Burton DR, Glass PJ, Ward AB, Walker LM (2016b) Isolation of potent neutralizing antibodies from a survivor of the 2014 Ebola virus outbreak. Science 351(6277):1078–1083. doi:10.1126/science.aad5788

Bosio CM, Aman MJ, Grogan C, Hogan R, Ruthel G, Negley D, Mohamadzadeh M, Bavari S, Schmaljohn A (2003) Ebola and Marburg viruses replicate in monocyte-derived dendritic cells without inducing the production of cytokines and full maturation. J Infect Dis 188(11):1630–1638. doi:10.1086/379199

Bower H, Johnson S, Bangura MS, Kamara AJ, Kamara O, Mansaray SH, Sesay D, Turay C, Checchi F, Glynn JR (2016) Exposure-specific and age-specific attack rates for Ebola Virus disease in Ebola-affected households, Sierra Leone. Emerg Infect Dis 22(8). doi:10.3201/eid2208.160163

Brannan JM, Froude JW, Prugar LI, Bakken RR, Zak SE, Daye SP, Wilhelmsen CE, Dye JM (2015) Interferon alpha/beta receptor-deficient mice as a model for Ebola Virus disease. J Infect Dis 212(Suppl 2):S282–294. doi:10.1093/infdis/jiv215

Bray M, Raymond JL, Geisbert T, Baker RO (2002) 3-deazaneplanocin A induces massively increased interferon-alpha production in Ebola virus-infected mice. Antiviral Res 55(1):151–159

Buchbinder EI, Desai A (2016) CTLA-4 and PD-1 Pathways: Similarities, Differences, and Implications of Their Inhibition. Am J Clin Oncol 39(1):98–106. doi:10.1097/COC.0000000000000239

Bwaka MA, Bonnet MJ, Calain P, Colebunders R, De Roo A, Guimard Y, Katwiki KR, Kibadi K, Kipasa MA, Kuvula KJ, Mapanda BB, Massamba M, Mupapa KD, Muyembe-Tamfum JJ, Ndaberey E, Peters CJ, Rollin PE, Van den Enden E, Van den Enden E (1999) Ebola hemorrhagic fever in Kikwit, Democratic Republic of the Congo: clinical observations in 103 patients. J Infect Dis 179(Suppl 1):S1–7. doi:10.1086/514308

Carrion R Jr, Brasky K, Mansfield K, Johnson C, Gonzales M, Ticer A, Lukashevich I, Tardif S, Patterson J (2007) Lassa virus infection in experimentally infected marmosets: liver pathology and immunophenotypic alterations in target tissues. J Virol 81(12):6482–6490. doi:10.1128/JVI.02876-06

Centers for Disease C, Prevention (2009) Imported case of Marburg hemorrhagic fever—Colorado, 2008. MMWR Morb Mortal Wkly Rep 58(49):1377–1381

Chertow DS, Kleine C, Edwards JK, Scaini R, Giuliani R, Sprecher A (2014) Ebola virus disease in West Africa–clinical manifestations and management. N Engl J Med 371(22):2054–2057. doi:10.1056/NEJMp1413084

Chughtai AA, Barnes M, Macintyre CR (2016) Persistence of Ebola virus in various body fluids during convalescence: evidence and implications for disease transmission and control. Epidemiol Infect 144(8):1652–1660. doi:10.1017/S0950268816000054

Clark DV, Kibuuka H, Millard M, Wakabi S, Lukwago L, Taylor A, Eller MA, Eller LA, Michael NL, Honko AN, Olinger GG Jr, Schoepp RJ, Hepburn MJ, Hensley LE, Robb ML (2015) Long-term sequelae after Ebola virus disease in Bundibugyo, Uganda: a retrospective cohort study. Lancet Infect Dis 15(8):905–912. doi:10.1016/S1473-3099(15)70152-0

Colebunders R, Tshomba A, Van Kerkhove MD, Bausch DG, Campbell P, Libande M, Pirard P, Tshioko F, Mardel S, Mulangu S, Sleurs H, Rollin PE, Muyembe-Tamfum JJ, Jeffs B, Borchert M, International S, Technical Committee DRCWDMOIG (2007) Marburg hemorrhagic fever in Durba and Watsa, Democratic Republic of the Congo: clinical documentation, features of illness, and treatment. J Infect Dis 196(Suppl 2):S148–153. doi:10.1086/520543

Corti D, Misasi J, Mulangu S, Stanley DA, Kanekiyo M, Wollen S, Ploquin A, Doria-Rose NA, Staupe RP, Bailey M, Shi W, Choe M, Marcus H, Thompson EA, Cagigi A, Silacci C, Fernandez-Rodriguez B, Perez L, Sallusto F, Vanzetta F, Agatic G, Cameroni E, Kisalu N, Gordon I, Ledgerwood JE, Mascola JR, Graham BS, Muyembe-Tamfun JJ, Trefry JC, Lanzavecchia A, Sullivan NJ (2016) Protective monotherapy against lethal Ebola virus infection by a potently neutralizing antibody. Science 351(6279):1339–1342. doi:10.1126/science.aad5224

Cros J, Cagnard N, Woollard K, Patey N, Zhang SY, Senechal B, Puel A, Biswas SK, Moshous D, Picard C, Jais JP, D'Cruz D, Casanova JL, Trouillet C, Geissmann F (2010) Human CD14dim monocytes patrol and sense nucleic acids and viruses via TLR7 and TLR8 receptors. Immunity 33(3):375–386. doi:10.1016/j.immuni.2010.08.012

Cusi MG, Gandolfo C, Terrosi C, Gori Savellini G, Belmonte G, Miracco C (2016) Toscana virus infects dendritic and endothelial cells opening the way for the central nervous system. J Neurovirol 22(3):307–315. doi:10.1007/s13365-015-0395-2

Dallatomasina S, Crestani R, Sylvester Squire J, Declerk H, Caleo GM, Wolz A, Stinson K, Patten G, Brechard R, Gbabai OB, Spreicher A, Van Herp M, Zachariah R (2015) Ebola outbreak in rural West Africa: epidemiology, clinical features and outcomes. Trop Med Int Health TM & IH 20(4):448–454. doi:10.1111/tmi.12454

Deen GF, Knust B, Broutet N, Sesay FR, Formenty P, Ross C, Thorson AE, Massaquoi TA, Marrinan JE, Ervin E, Jambai A, McDonald SL, Bernstein K, Wurie AH, Dumbuya MS, Abad N, Idriss B, Wi T, Bennett SD, Davies T, Ebrahim FK, Meites E, Naidoo D, Smith S, Banerjee A, Erickson BR, Brault A, Durski KN, Winter J, Sealy T, Nichol ST, Lamunu M, Stroher U, Morgan O, Sahr F (2015) Ebola RNA Persistence in Semen of Ebola Virus disease survivors—preliminary report. N Engl J Med. doi:10.1056/NEJMoa1511410

Dowell SF (1996) Ebola hemorrhagic fever: why were children spared? Pediatr Infect Dis J 15(3):189–191

Dowell SF, Mukunu R, Ksiazek TG, Khan AS, Rollin PE, Peters CJ (1999) Transmission of Ebola hemorrhagic fever: a study of risk factors in family members, Kikwit, Democratic Republic of the Congo, 1995. Commission de Lutte contre les Epidemies a Kikwit. J Infect Dis 179(Suppl 1):S87–91. doi:10.1086/514284

Ebihara H, Rockx B, Marzi A, Feldmann F, Haddock E, Brining D, LaCasse RA, Gardner D, Feldmann H (2011) Host response dynamics following lethal infection of rhesus macaques with Zaire ebolavirus. J Infect Dis 204(Suppl 3):S991–999. doi:10.1093/infdis/jir336

Ebola haemorrhagic fever in Zaire, 1976 (1978). Bulletin of the World Health Organization 56 (2):271–293

Emond RT, Evans B, Bowen ET, Lloyd G (1977) A case of Ebola virus infection. Br Med J 2 (6086):541–544

Ersland K, Wuthrich M, Klein BS (2010) Dynamic interplay among monocyte-derived, dermal, and resident lymph node dendritic cells during the generation of vaccine immunity to fungi. Cell Host Microbe 7(6):474–487. doi:10.1016/j.chom.2010.05.010

Escudero-Perez B, Volchkova VA, Dolnik O, Lawrence P, Volchkov VE (2014) Shed GP of Ebola virus triggers immune activation and increased vascular permeability. PLoS Pathog 10(11): e1004509. doi:10.1371/journal.ppat.1004509

Ewer K, Rampling T, Venkatraman N, Bowyer G, Wright D, Lambe T, Imoukhuede EB, Payne R, Fehling SK, Strecker T, Biedenkopf N, Krahling V, Tully CM, Edwards NJ, Bentley EM, Samuel D, Labbe G, Jin J, Gibani M, Minhinnick A, Wilkie M, Poulton I, Lella N, Roberts R, Hartnell F, Bliss C, Sierra-Davidson K, Powlson J, Berrie E, Tedder R, Roman F, De Ryck I, Nicosia A, Sullivan NJ, Stanley DA, Mbaya OT, Ledgerwood JE, Schwartz RM, Siani L, Colloca S, Folgori A, Di Marco S, Cortese R, Wright E, Becker S, Graham BS, Koup RA, Levine MM, Volkmann A, Chaplin P, Pollard AJ, Draper SJ, Ballou WR, Lawrie A, Gilbert SC, Hill AV (2016) A monovalent Chimpanzee Adenovirus Ebola Vaccine boosted with MVA. N Engl J Med 374(17):1635–1646. doi:10.1056/NEJMoa1411627

Faye O, Andronico A, Faye O, Salje H, Boelle PY, Magassouba N, Bah EI, Koivogui L, Diallo B, Diallo AA, Keita S, Konde MK, Fowler R, Fall G, Cauchemez S, Sall AA (2015) Use of Viremia to evaluate the baseline case fatality ratio of Ebola Virus disease and inform treatment studies: a retrospective cohort study. PLoS Med 12(12):e1001908. doi:10.1371/journal.pmed.1001908

Feldmann H, Bugany H, Mahner F, Klenk HD, Drenckhahn D, Schnittler HJ (1996) Filovirus-induced endothelial leakage triggered by infected monocytes/macrophages. J Virol 70(4):2208–2214

Feldmann H, Geisbert TW (2011) Ebola haemorrhagic fever. Lancet 377(9768):849–862. doi:10.1016/S0140-6736(10)60667-8

Fischer RJ, Judson S, Miazgowicz K, Bushmaker T, Munster VJ (2016) Ebola Virus persistence in Semen Ex Vivo. Emerg Infect Dis 22(2):289–291. doi:10.3201/eid2202.151278

Fisher-Hoch SP, Platt GS, Neild GH, Southee T, Baskerville A, Raymond RT, Lloyd G, Simpson DI (1985) Pathophysiology of shock and hemorrhage in a fulminating viral infection (Ebola). J Infect Dis 152(5):887–894

Francesconi P, Yoti Z, Declich S, Onek PA, Fabiani M, Olango J, Andraghetti R, Rollin PE, Opira C, Greco D, Salmaso S (2003) Ebola hemorrhagic fever transmission and risk factors of contacts, Uganda. Emerg Infect Dis 9(11):1430–1437. doi:10.3201/eid0911.030339

Fujinami RS, von Herrath MG, Christen U, Whitton JL (2006) Molecular mimicry, bystander activation, or viral persistence: infections and autoimmune disease. Clin Microbiol Rev 19 (1):80–94. doi:10.1128/CMR.19.1.80-94.2006

Gear JS, Cassel GA, Gear AJ, Trappler B, Clausen L, Meyers AM, Kew MC, Bothwell TH, Sher R, Miller GB, Schneider J, Koornhof HJ, Gomperts ED, Isaacson M, Gear JH (1975) Outbreake of Marburg virus disease in Johannesburg. Br Med J 4(5995):489–493

Geisbert TW, Hensley LE, Larsen T, Young HA, Reed DS, Geisbert JB, Scott DP, Kagan E, Jahrling PB, Davis KJ (2003a) Pathogenesis of Ebola hemorrhagic fever in cynomolgus macaques: evidence that dendritic cells are early and sustained targets of infection. Am J Pathol 163(6):2347–2370. doi:10.1016/S0002-9440(10)63591-2

Geisbert TW, Young HA, Jahrling PB, Davis KJ, Kagan E, Hensley LE (2003b) Mechanisms underlying coagulation abnormalities in ebola hemorrhagic fever: overexpression of tissue factor in primate monocytes/macrophages is a key event. J Infect Dis 188(11):1618–1629. doi:10.1086/379724

Geisbert TW, Young HA, Jahrling PB, Davis KJ, Larsen T, Kagan E, Hensley LE (2003c) Pathogenesis of Ebola hemorrhagic fever in primate models: evidence that hemorrhage is not a direct effect of virus-induced cytolysis of endothelial cells. Am J Pathol 163(6):2371–2382. doi:10.1016/S0002-9440(10)63592-4

Gignoux E, Azman AS, de Smet M, Azuma P, Massaquoi M, Job D, Tiffany A, Petrucci R, Sterk E, Potet J, Suzuki M, Kurth A, Cannas A, Bocquin A, Strecker T, Logue C, Pottage T, Yue C, Cabrol JC, Serafini M, Ciglenecki I (2016) Effect of Artesunate-Amodiaquine on mortality related to Ebola Virus disease. N Engl J Med 374(1):23–32. doi:10.1056/NEJMoa1504605

Green E, Hunt L, Ross JC, Nissen NM, Curran T, Badhan A, Sutherland KA, Richards J, Lee JS, Allen SH, Laird S, Blackman M, Collacott I, Parker PA, Walbridge A, Phillips R, Sellu SJ, Dama A, Sheriff AK, Zombo J, Ngegba D, Wurie AH, Checchi F, Brooks TJ (2016) Viraemia and Ebola virus secretion in survivors of Ebola virus disease in Sierra Leone: a cross-sectional cohort study. Lancet Infect Dis. doi:10.1016/S1473-3099(16)30060-3

Gupta M, MacNeil A, Reed ZD, Rollin PE, Spiropoulou CF (2012) Serology and cytokine profiles in patients infected with the newly discovered Bundibugyo ebolavirus. Virology 423(2):119–124. doi:10.1016/j.virol.2011.11.027

Gupta M, Mahanty S, Ahmed R, Rollin PE (2001) Monocyte-derived human macrophages and peripheral blood mononuclear cells infected with ebola virus secrete MIP-1alpha and TNF-alpha and inhibit poly-IC-induced IFN-alpha in vitro. Virology 284(1):20–25. doi:10.1006/viro.2001.0836

Gupta M, Spiropoulou C, Rollin PE (2007) Ebola virus infection of human PBMCs causes massive death of macrophages, CD4 and CD8 T cell sub-populations in vitro. Virology 364(1): 45–54. doi:10.1016/j.virol.2007.02.017

Haniffa M, Collin M, Ginhoux F (2013) Ontogeny and functional specialization of dendritic cells in human and mouse. Adv Immunol 120:1–49. doi:10.1016/B978-0-12-417028-5.00001-6

Henn V, Steinbach S, Buchner K, Presek P, Kroczek RA (2001) The inflammatory action of CD40 ligand (CD154) expressed on activated human platelets is temporally limited by coexpressed CD40. Blood 98(4):1047–1054

Hoarau JJ, Jaffar Bandjee MC, Krejbich Trotot P, Das T, Li-Pat-Yuen G, Dassa B, Denizot M, Guichard E, Ribera A, Henni T, Tallet F, Moiton MP, Gauzere BA, Bruniquet S, Jaffar Bandjee Z, Morbidelli P, Martigny G, Jolivet M, Gay F, Grandadam M, Tolou H, Vieillard V, Debre P, Autran B, Gasque P (2010) Persistent chronic inflammation and infection by Chikungunya arthritogenic alphavirus in spite of a robust host immune response. J Immunol 184(10):5914–5927. doi:10.4049/jimmunol.0900255

Huang YW, Dryman BA, Li W, Meng XJ (2009) Porcine DC-SIGN: molecular cloning, gene structure, tissue distribution and binding characteristics. Dev Comp Immunol 33(4):464–480. doi:10.1016/j.dci.2008.09.010

Hunt BJ (2014) Bleeding and coagulopathies in critical care. N Engl J Med 370(22):2153. doi:10.1056/NEJMc1403768

Hunt L, Gupta-Wright A, Simms V, Tamba F, Knott V, Tamba K, Heisenberg-Mansaray S, Tamba E, Sheriff A, Conteh S, Smith T, Tobin S, Brooks T, Houlihan C, Cummings R, Fletcher T (2015) Clinical presentation, biochemical, and haematological parameters and their association with outcome in patients with Ebola virus disease: an observational cohort study. Lancet Infect Dis 15(11):1292–1299. doi:10.1016/S1473-3099(15)00144-9

Hutchinson KL, Rollin PE (2007) Cytokine and chemokine expression in humans infected with Sudan Ebola virus. J Infect Dis 196(Suppl 2):S357–363. doi:10.1086/520611

Ilinykh PA, Lubaki NM, Widen SG, Renn LA, Theisen TC, Rabin RL, Wood TG, Bukreyev A (2015) Different temporal effects of Ebola Virus VP35 and VP24 proteins on global gene expression in human dendritic cells. J Virol 89(15):7567–7583. doi:10.1128/JVI.00924-15

Jacobs M, Rodger A, Bell DJ, Bhagani S, Cropley I, Filipe A, Gifford RJ, Hopkins S, Hughes J, Jabeen F, Johannessen I, Karageorgopoulos D, Lackenby A, Lester R, Liu RS, MacConnachie A, Mahungu T, Martin D, Marshall N, Mepham S, Orton R, Palmarini M, Patel M, Perry C, Peters SE, Porter D, Ritchie D, Ritchie ND, Seaton RA, Sreenu VB, Templeton K, Warren S, Wilkie GS, Zambon M, Gopal R, Thomson EC (2016) Late Ebola virus relapse causing meningoencephalitis: a case report. Lancet. doi:10.1016/S0140-6736(16)30386-5

Jin H, Yan Z, Prabhakar BS, Feng Z, Ma Y, Verpooten D, Ganesh B, He B (2010) The VP35 protein of Ebola virus impairs dendritic cell maturation induced by virus and lipopolysaccharide. J Gen Virol 91(Pt 2):352–361. doi:10.1099/vir.0.017343-0

Klechevsky E, Banchereau J (2013) Human dendritic cells subsets as targets and vectors for therapy. Ann N Y Acad Sci 1284:24–30. doi:10.1111/nyas.12113

Kondratowicz AS, Lennemann NJ, Sinn PL, Davey RA, Hunt CL, Moller-Tank S, Meyerholz DK, Rennert P, Mullins RF, Brindley M, Sandersfeld LM, Quinn K, Weller M, McCray PB Jr, Chiorini J, Maury W (2011) T-cell immunoglobulin and mucin domain 1 (TIM-1) is a receptor for Zaire Ebolavirus and Lake Victoria Marburgvirus. Proc Natl Acad Sci U S A 108(20):8426–8431. doi:10.1073/pnas.1019030108

Kornbluth RS, Bot A (2012) In this issue: harnessing immunity against difficult targets, through the CD40-CD40L pathway. Int Rev Immunol 31(4):243–245. doi:10.3109/08830185.2012.702511

Kreuels B, Wichmann D, Emmerich P, Schmidt-Chanasit J, de Heer G, Kluge S, Sow A, Renne T, Gunther S, Lohse AW, Addo MM, Schmiedel S (2014) A case of severe Ebola virus infection complicated by gram-negative septicemia. N Engl J Med 371(25):2394–2401. doi:10.1056/NEJMoa1411677

Ksiazek TG, Rollin PE, Williams AJ, Bressler DS, Martin ML, Swanepoel R, Burt FJ, Leman PA, Khan AS, Rowe AK, Mukunu R, Sanchez A, Peters CJ (1999) Clinical virology of Ebola hemorrhagic fever (EHF): virus, virus antigen, and IgG and IgM antibody findings among EHF patients in Kikwit, Democratic Republic of the Congo, 1995. J Infect Dis 179(Suppl 1):S177–187. doi:10.1086/514321

Kvale EO, Dalgaard J, Lund-Johansen F, Rollag H, Farkas L, Midtvedt K, Jahnsen FL, Brinchmann JE, Olweus J (2006) CD11c+ dendritic cells and plasmacytoid DCs are activated by human cytomegalovirus and retain efficient T cell-stimulatory capability upon infection. Blood 107(5):2022–2029. doi:10.1182/blood-2005-05-2016

Lado M, Walker NF, Baker P, Haroon S, Brown CS, Youkee D, Studd N, Kessete Q, Maini R, Boyles T, Hanciles E, Wurie A, Kamara TB, Johnson O, Leather AJ (2015) Clinical features of patients isolated for suspected Ebola virus disease at Connaught Hospital, Freetown, Sierra Leone: a retrospective cohort study. Lancet Infect Dis 15(9):1024–1033. doi:10.1016/S1473-3099(15)00137-1

Langdown J, Luddington RJ, Huntington JA, Baglin TP (2014) A hereditary bleeding disorder resulting from a premature stop codon in thrombomodulin (p.Cys537Stop). Blood 124(12):1951–1956. doi:10.1182/blood-2014-02-557538

Lanini S, Portella G, Vairo F, Kobinger GP, Pesenti A, Langer M, Kabia S, Brogiato G, Amone J, Castilletti C, Miccio R, Zumla A, Capobianchi MR, Di Caro A, Strada G, Ippolito G, Group I-EESLS (2015) Blood kinetics of Ebola virus in survivors and nonsurvivors. J Clin Invest 125(12):4692–4698. doi:10.1172/JCI83111

Lauck M, Bailey AL, Andersen KG, Goldberg TL, Sabeti PC, O'Connor DH (2015) GB virus C coinfections in west African Ebola patients. J Virol 89(4):2425–2429. doi:10.1128/JVI.02752-14

Leon B, Lopez-Bravo M, Ardavin C (2007) Monocyte-derived dendritic cells formed at the infection site control the induction of protective T helper 1 responses against Leishmania. Immunity 26(4):519–531. doi:10.1016/j.immuni.2007.01.017

Leroy EM, Baize S, Debre P, Lansoud-Soukate J, Mavoungou E (2001) Early immune responses accompanying human asymptomatic Ebola infections. Clin Exp Immunol 124(3):453–460

Leroy EM, Baize S, Volchkov VE, Fisher-Hoch SP, Georges-Courbot MC, Lansoud-Soukate J, Capron M, Debre P, McCormick JB, Georges AJ (2000) Human asymptomatic Ebola infection and strong inflammatory response. Lancet 355(9222):2210–2215

Leroy EM, Epelboin A, Mondonge V, Pourrut X, Gonzalez JP, Muyembe-Tamfum JJ, Formenty P (2009) Human Ebola outbreak resulting from direct exposure to fruit bats in Luebo,

Democratic Republic of Congo, 2007. Vector Borne Zoonotic Dis 9(6):723–728. doi:10.1089/vbz.2008.0167

Leung LW, Martinez O, Reynard O, Volchkov VE, Basler CF (2011) Ebola virus failure to stimulate plasmacytoid dendritic cell interferon responses correlates with impaired cellular entry. J Infect Dis 204(Suppl 3):S973–977. doi:10.1093/infdis/jir331

Levi M, Toh CH, Thachil J, Watson HG (2009) Guidelines for the diagnosis and management of disseminated intravascular coagulation. British Committee for standards in haematology. Br J Haematol 145(1):24–33. doi:10.1111/j.1365-2141.2009.07600.x

Li J, Duan HJ, Chen HY, Ji YJ, Zhang X, Rong YH, Xu Z, Sun LJ, Zhang JY, Liu LM, Jin B, Zhang J, Du N, Su HB, Teng GJ, Yuan Y, Qin EQ, Jia HJ, Wang S, Guo TS, Wang Y, Mu JS, Yan T, Li ZW, Dong Z, Nie WM, Jiang TJ, Li C, Gao XD, Ji D, Zhuang YJ, Li L, Wang LF, Li WG, Duan XZ, Lu YY, Sun ZQ, Kanu AB, Koroma SM, Zhao M, Ji JS, Wang FS (2016) Age and Ebola viral load correlate with mortality and survival time in 288 Ebola virus disease patients. Int J Infect Dis IJID Official Publ Int Soc Infect Dis 42:34–39. doi:10.1016/j.ijid.2015.10.021

Lindgren T, Ahlm C, Mohamed N, Evander M, Ljunggren HG, Bjorkstrom NK (2011) Longitudinal analysis of the human T cell response during acute hantavirus infection. J Virol 85(19):10252–10260. doi:10.1128/JVI.05548-11

Liu L, Wei Q, Nishiura K, Peng J, Wang H, Midkiff C, Alvarez X, Qin C, Lackner A, Chen Z (2015) Spatiotemporal interplay of severe acute respiratory syndrome coronavirus and respiratory mucosal cells drives viral dissemination in rhesus macaques. Mucosal Immunol. doi:10.1038/mi.2015.127

Luan YY, Yao YM, Xiao XZ, Sheng ZY (2015) Insights into the apoptotic death of immune cells in sepsis. J Interferon Cytokine Res Official J Int Soc Interferon Cytokine Res 35(1):17–22. doi:10.1089/jir.2014.0069

Luczkowiak J, Arribas JR, Gomez S, Jimenez-Yuste V, de la Calle F, Viejo A, Delgado R (2016) Specific neutralizing response in plasma from convalescent patients of Ebola Virus disease against the West Africa Makona variant of Ebola virus. Virus Res 213:224–229. doi:10.1016/j.virusres.2015.12.019

Ludtke A, Ruibal P, Becker-Ziaja B, Rottstegge M, Wozniak DM, Cabeza-Cabrerizo M, Thorenz A, Weller R, Kerber R, Idoyaga J, Magassouba N, Gabriel M, Gunther S, Oestereich L, Munoz-Fontela C (2016) Ebola Virus disease is characterized by Poor activation and reduced levels of circulating CD16+ Monocytes. J Infect Dis 214(suppl 3):S275–S280. doi:10.1093/infdis/jiw260

MacNeil A, Farnon EC, Wamala J, Okware S, Cannon DL, Reed Z, Towner JS, Tappero JW, Lutwama J, Downing R, Nichol ST, Ksiazek TG, Rollin PE (2010) Proportion of deaths and clinical features in Bundibugyo Ebola virus infection, Uganda. Emerg Infect Dis 16(12):1969–1972. doi:10.3201/eid1612.100627

Mahanty S, Hutchinson K, Agarwal S, McRae M, Rollin PE, Pulendran B (2003) Cutting edge: impairment of dendritic cells and adaptive immunity by Ebola and Lassa viruses. J Immunol 170(6):2797–2801

Mari Saez A, Weiss S, Nowak K, Lapeyre V, Zimmermann F, Dux A, Kuhl HS, Kaba M, Regnaut S, Merkel K, Sachse A, Thiesen U, Villanyi L, Boesch C, Dabrowski PW, Radonic A, Nitsche A, Leendertz SA, Petterson S, Becker S, Krahling V, Couacy-Hymann E, Akoua-Koffi C, Weber N, Schaade L, Fahr J, Borchert M, Gogarten JF, Calvignac-Spencer S, Leendertz FH (2015) Investigating the zoonotic origin of the West African Ebola epidemic. EMBO Mol Med 7(1):17–23. doi:10.15252/emmm.201404792

Martines RB, Ng DL, Greer PW, Rollin PE, Zaki SR (2015) Tissue and cellular tropism, pathology and pathogenesis of Ebola and Marburg viruses. J Pathol 235(2):153–174. doi:10.1002/path.4456

Martinez O, Johnson JC, Honko A, Yen B, Shabman RS, Hensley LE, Olinger GG, Basler CF (2013) Ebola virus exploits a monocyte differentiation program to promote its entry. J Virol 87(7):3801–3814. doi:10.1128/JVI.02695-12

Martini GA (1973) Marburg virus disease. Postgrad Med J 49(574):542–546

Mate SE, Kugelman JR, Nyenswah TG, Ladner JT, Wiley MR, Cordier-Lassalle T, Christie A, Schroth GP, Gross SM, Davies-Wayne GJ, Shinde SA, Murugan R, Sieh SB, Badio M, Fakoli L, Taweh F, de Wit E, van Doremalen N, Munster VJ, Pettitt J, Prieto K, Humrighouse BW, Stroher U, DiClaro JW, Hensley LE, Schoepp RJ, Safronetz D, Fair J, Kuhn JH, Blackley DJ, Laney AS, Williams DE, Lo T, Gasasira A, Nichol ST, Formenty P, Kateh FN, De Cock KM, Bolay F, Sanchez-Lockhart M, Palacios G (2015) Molecular evidence of sexual transmission of Ebola Virus. N Engl J Med 373(25):2448–2454. doi:10.1056/NEJMoa1509773

Mattia JG, Vandy MJ, Chang JC, Platt DE, Dierberg K, Bausch DG, Brooks T, Conteh S, Crozier I, Fowler RA, Kamara AP, Kang C, Mahadevan S, Mansaray Y, Marcell L, McKay G, O'Dempsey T, Parris V, Pinto R, Rangel A, Salam AP, Shantha J, Wolfman V, Yeh S, Chan AK, Mishra S (2016) Early clinical sequelae of Ebola virus disease in Sierra Leone: a cross-sectional study. Lancet Infect Dis 16(3):331–338. doi:10.1016/S1473-3099(15)00489-2

McElroy AK, Erickson BR, Flietstra TD, Rollin PE, Nichol ST, Towner JS, Spiropoulou CF (2014a) Biomarker correlates of survival in pediatric patients with Ebola virus disease. Emerging infectious diseases

McElroy AK, Erickson BR, Flietstra TD, Rollin PE, Nichol ST, Towner JS, Spiropoulou CF (2014b) Ebola hemorrhagic Fever: novel biomarker correlates of clinical outcome. J Infect Dis 210(4):558–566. doi:10.1093/infdis/jiu088

McElroy AK, Akondy RS, Davis CW, Ellebedy AH, Mehta AK, Kraft CS, Lyon GM, Ribner BS, Varkey J, Sidney J, Sette A, Campbell S, Stroher U, Damon I, Nichol ST, Spiropoulou CF, Ahmed R (2015a) Human Ebola virus infection results in substantial immune activation. Proc Natl Acad Sci U S A 112(15):4719–4724. doi:10.1073/pnas.1502619112

McElroy AK, Erickson BR, Flietstra TD, Rollin PE, Towner JS, Nichol ST, Spiropoulou CF (2015b) Von Willebrand factor is elevated in individuals infected with Sudan virus and is associated with adverse clinical outcomes. Viral Immunol 28(1):71–73. doi:10.1089/vim.2014.0072

McElroy AK, Harmon JR, Flietstra TD, Campbell S, Mehta AK, Kraft CS, Lyon MG, Varkey JB, Ribner BS, Kratochvil CJ, Iwen PC, Smith PW, Ahmed R, Nichol ST, Spiropoulou CF (2016) Kinetic analysis of biomarkers in a Cohort of US patients with Ebola Virus disease. Clin Infect Dis Official Publ Infect Dis Soc Am 63(4):460–467. doi:10.1093/cid/ciw334

McNab F, Mayer-Barber K, Sher A, Wack A, O'Garra A (2015) Type I interferons in infectious disease. Nat Rev Immunol 15(2):87–103. doi:10.1038/nri3787

Merad M, Ginhoux F, Collin M (2008) Origin, homeostasis and function of Langerhans cells and other langerin-expressing dendritic cells. Nat Rev Immunol 8(12):935–947. doi:10.1038/nri2455

Mesman AW, de Vries RD, McQuaid S, Duprex WP, de Swart RL, Geijtenbeek TB (2012) A prominent role for DC-SIGN+ dendritic cells in initiation and dissemination of measles virus infection in non-human primates. PLoS ONE 7(12):e49573. doi:10.1371/journal.pone.0049573

Miller JD, van der Most RG, Akondy RS, Glidewell JT, Albott S, Masopust D, Murali-Krishna K, Mahar PL, Edupuganti S, Lalor S, Germon S, Del Rio C, Mulligan MJ, Staprans SI, Altman JD, Feinberg MB, Ahmed R (2008) Human effector and memory CD8+ T cell responses to smallpox and yellow fever vaccines. Immunity 28(5):710–722. doi:10.1016/j.immuni.2008.02.020

Misasi J, Gilman MS, Kanekiyo M, Gui M, Cagigi A, Mulangu S, Corti D, Ledgerwood JE, Lanzavecchia A, Cunningham J, Muyembe-Tamfun JJ, Baxa U, Graham BS, Xiang Y, Sullivan NJ, McLellan JS (2016) Structural and molecular basis for Ebola virus neutralization by protective human antibodies. Science 351(6279):1343–1346. doi:10.1126/science.aad6117

Mohamadzadeh M, Chen L, Schmaljohn AL (2007) How Ebola and Marburg viruses battle the immune system. Nat Rev Immunol 7(7):556–567. doi:10.1038/nri2098

Mohamadzadeh M, Coberley SS, Olinger GG, Kalina WV, Ruthel G, Fuller CL, Swenson DL, Pratt WD, Kuhns DB, Schmaljohn AL (2006) Activation of triggering receptor expressed on myeloid cells-1 on human neutrophils by marburg and ebola viruses. J Virol 80(14):7235–7244. doi:10.1128/JVI.00543-06

Mupere E, Kaducu OF, Yoti Z (2001) Ebola haemorrhagic fever among hospitalised children and adolescents in northern Uganda: epidemiologic and clinical observations. Afr Health Sci 1(2): 60–65

Murin CD, Fusco ML, Bornholdt ZA, Qiu X, Olinger GG, Zeitlin L, Kobinger GP, Ward AB, Saphire EO (2014) Structures of protective antibodies reveal sites of vulnerability on Ebola virus. Proc Natl Acad Sci U S A 111(48):17182–17187. doi:10.1073/pnas.1414164111

Okware SI, Omaswa FG, Zaramba S, Opio A, Lutwama JJ, Kamugisha J, Rwaguma EB, Kagwa P, Lamunu M (2002) An outbreak of Ebola in Uganda. Trop Med Int Health TM & IH 7 (12):1068–1075

Olal D, Kuehne AI, Bale S, Halfmann P, Hashiguchi T, Fusco ML, Lee JE, King LB, Kawaoka Y, Dye JM Jr, Saphire EO (2012) Structure of an antibody in complex with its mucin domain linear epitope that is protective against Ebola virus. J Virol 86(5):2809–2816. doi:10.1128/JVI.05549-11

Olinger GG Jr, Pettitt J, Kim D, Working C, Bohorov O, Bratcher B, Hiatt E, Hume SD, Johnson AK, Morton J, Pauly M, Whaley KJ, Lear CM, Biggins JE, Scully C, Hensley L, Zeitlin L (2012) Delayed treatment of Ebola virus infection with plant-derived monoclonal antibodies provides protection in rhesus macaques. Proc Natl Acad Sci U S A 109(44):18030–18035. doi:10.1073/pnas.1213709109

Onyango CO, Opoka ML, Ksiazek TG, Formenty P, Ahmed A, Tukei PM, Sang RC, Ofula VO, Konongoi SL, Coldren RL, Grein T, Legros D, Bell M, De Cock KM, Bellini WJ, Towner JS, Nichol ST, Rollin PE (2007) Laboratory diagnosis of Ebola hemorrhagic fever during an outbreak in Yambio, Sudan, 2004. J Infect Dis 196(Suppl 2):S193–198. doi:10.1086/520609

Qin E, Bi J, Zhao M, Wang Y, Guo T, Yan T, Li Z, Sun J, Zhang J, Chen S, Wu Y, Li J, Zhong Y (2015) Clinical Features of Patients With Ebola Virus Disease in Sierra Leone. Clin Infect Dis Official Publ Infect Dis Soc Am 61(4):491–495. doi:10.1093/cid/civ319

Rhein BA, Brouillette RB, Schaack GA, Chiorini JA, Maury W (2016) Characterization of human and murine T-cell immunoglobulin mucin domain 4 (TIM-4) IgV domain residues critical for Ebola Virus Entry. J Virol 90(13):6097–6111. doi:10.1128/JVI.00100-16

Rivera JA, McGuire TC (2005) Equine infectious anemia virus-infected dendritic cells retain antigen presentation capability. Virology 335(2):145–154. doi:10.1016/j.virol.2005.02.013

Roddy P, Thomas SL, Jeffs B, Nascimento Folo P, Pablo Palma P, Moco Henrique B, Villa L, Damiao Machado FP, Bernal O, Jones SM, Strong JE, Feldmann H, Borchert M (2010) Factors associated with Marburg hemorrhagic fever: analysis of patient data from Uige, Angola. J Infect Dis 201(12):1909–1918. doi:10.1086/652748

Rodriguez LL, De Roo A, Guimard Y, Trappier SG, Sanchez A, Bressler D, Williams AJ, Rowe AK, Bertolli J, Khan AS, Ksiazek TG, Peters CJ, Nichol ST (1999) Persistence and genetic stability of Ebola virus during the outbreak in Kikwit, Democratic Republic of the Congo, 1995. J Infect Dis 179(Suppl 1):S170–176. doi:10.1086/514291

Rollin PE, Bausch DG, Sanchez A (2007) Blood chemistry measurements and D-Dimer levels associated with fatal and nonfatal outcomes in humans infected with Sudan Ebola virus. J Infect Dis 196(Suppl 2):S364–371. doi:10.1086/520613

Rowe AK, Bertolli J, Khan AS, Mukunu R, Muyembe-Tamfum JJ, Bressler D, Williams AJ, Peters CJ, Rodriguez L, Feldmann H, Nichol ST, Rollin PE, Ksiazek TG (1999) Clinical, virologic, and immunologic follow-up of convalescent Ebola hemorrhagic fever patients and their household contacts, Kikwit, Democratic Republic of the Congo. Commission de Lutte contre les Epidemies a Kikwit. J Infect Dis 179(Suppl 1):S28–35. doi:10.1086/514318

Ruibal P, Oestereich L, Ludtke A, Becker-Ziaja B, Wozniak DM, Kerber R, Korva M, Cabeza-Cabrerizo M, Bore JA, Koundouno FR, Duraffour S, Weller R, Thorenz A, Cimini E, Viola D, Agrati C, Repits J, Afrough B, Cowley LA, Ngabo D, Hinzmann J, Mertens M, Vitoriano I, Logue CH, Boettcher JP, Pallasch E, Sachse A, Bah A, Nitzsche K, Kuisma E, Michel J, Holm T, Zekeng EG, Garcia-Dorival I, Wolfel R, Stoecker K, Fleischmann E, Strecker T, Di Caro A, Avsic-Zupanc T, Kurth A, Meschi S, Mely S, Newman E, Bocquin A, Kis Z, Kelterbaum A, Molkenthin P, Carletti F, Portmann J, Wolff S, Castilletti C, Schudt G, Fizet A, Ottowell LJ, Herker E, Jacobs T, Kretschmer B, Severi E, Ouedraogo N, Lago M,

Negredo A, Franco L, Anda P, Schmiedel S, Kreuels B, Wichmann D, Addo MM, Lohse AW, De Clerck H, Nanclares C, Jonckheere S, Van Herp M, Sprecher A, Xiaojiang G, Carrington M, Miranda O, Castro CM, Gabriel M, Drury P, Formenty P, Diallo B, Koivogui L, Magassouba N, Carroll MW, Gunther S, Munoz-Fontela C (2016) Unique human immune signature of Ebola virus disease in Guinea. Nature 533(7601):100–104. doi:10.1038/nature17949

Sanchez A, Geisbert T, Feldmann H (2007a) *Filoviridae:* Marburg and Ebola Viruses, vol 1. Fields' virology, 5th edn. Wolters kluwer/Lippincott Williams & Wilkins, Philadelphia

Sanchez A, Ksiazek TG, Rollin PE, Miranda ME, Trappier SG, Khan AS, Peters CJ, Nichol ST (1999) Detection and molecular characterization of Ebola viruses causing disease in human and nonhuman primates. J Infect Dis 179(Suppl 1):S164–169. doi:10.1086/514282

Sanchez A, Lukwiya M, Bausch D, Mahanty S, Sanchez AJ, Wagoner KD, Rollin PE (2004) Analysis of human peripheral blood samples from fatal and nonfatal cases of Ebola (Sudan) hemorrhagic fever: cellular responses, virus load, and nitric oxide levels. J Virol 78(19):10370–10377. doi:10.1128/JVI.78.19.10370-10377.2004

Sanchez A, Trappier SG, Mahy BW, Peters CJ, Nichol ST (1996) The virion glycoproteins of Ebola viruses are encoded in two reading frames and are expressed through transcriptional editing. Proc Natl Acad Sci U S A 93(8):3602–3607

Sanchez A, Wagoner KE, Rollin PE (2007b) Sequence-based human leukocyte antigen-B typing of patients infected with Ebola virus in Uganda in 2000: identification of alleles associated with fatal and nonfatal disease outcomes. J Infect Dis 196(Suppl 2):S329–336. doi:10.1086/520588

Schaible UE, Kaufmann SH (2007) Malnutrition and infection: complex mechanisms and global impacts. PLoS Med 4(5):e115. doi:10.1371/journal.pmed.0040115

Schieffelin JS, Shaffer JG, Goba A, Gbakie M, Gire SK, Colubri A, Sealfon RS, Kanneh L, Moigboi A, Momoh M, Fullah M, Moses LM, Brown BL, Andersen KG, Winnicki S, Schaffner SF, Park DJ, Yozwiak NL, Jiang PP, Kargbo D, Jalloh S, Fonnie M, Sinnah V, French I, Kovoma A, Kamara FK, Tucker V, Konuwa E, Sellu J, Mustapha I, Foday M, Yillah M, Kanneh F, Saffa S, Massally JL, Boisen ML, Branco LM, Vandi MA, Grant DS, Happi C, Gevao SM, Fletcher TE, Fowler RA, Bausch DG, Sabeti PC, Khan SH, Garry RF, Program KGHLF, Viral Hemorrhagic Fever C, Team WHOCR (2014) Clinical illness and outcomes in patients with Ebola in Sierra Leone. N Engl J Med 371(22):2092–2100. doi:10.1056/NEJMoa1411680

Schmaljohn A, Lewis GK (2016) Cell-targeting antibodies in immunity to Ebola. Pathog Dis 74 (4). doi:10.1093/femspd/ftw021

Schuh AJ, Amman BR, Jones ME, Sealy TK, Uebelhoer LS, Spengler JR, Martin BE, Coleman-McCray JA, Nichol ST, Towner JS (2017) Modelling filovirus mainteinance in nature by experimental transmission of Marburg virus between Egyptian rousette bats. Nat Commun 8:14446. doi:10.1038/ncomms14446

Schwartz AJ, Alvarez X, Lackner AA (2002) Distribution and immunophenotype of DC-SIGN-expressing cells in SIV-infected and uninfected macaques. AIDS Res Hum Retroviruses 18(14):1021–1029. doi:10.1089/08892220260235380

Sharma P, Allison JP (2015) Immune checkpoint targeting in cancer therapy: toward combination strategies with curative potential. Cell 161(2):205–214. doi:10.1016/j.cell.2015.03.030

Simmons G, Reeves JD, Grogan CC, Vandenberghe LH, Baribaud F, Whitbeck JC, Burke E, Buchmeier MJ, Soilleux EJ, Riley JL, Doms RW, Bates P, Pohlmann S (2003) DC-SIGN and DC-SIGNR bind ebola glycoproteins and enhance infection of macrophages and endothelial cells. Virology 305(1):115–123

Smith LM, Hensley LE, Geisbert TW, Johnson J, Stossel A, Honko A, Yen JY, Geisbert J, Paragas J, Fritz E, Olinger G, Young HA, Rubins KH, Karp CL (2013) Interferon-beta therapy prolongs survival in rhesus macaque models of Ebola and Marburg hemorrhagic fever. J Infect Dis 208(2):310–318. doi:10.1093/infdis/jis921

Sobarzo A, Perelman E, Groseth A, Dolnik O, Becker S, Lutwama JJ, Dye JM, Yavelsky V, Lobel L, Marks RS (2012) Profiling the native specific human humoral immune response to

Sudan Ebola virus strain Gulu by chemiluminescence enzyme-linked immunosorbent assay. Clin Vaccine Immunol CVI 19(11):1844–1852. doi:10.1128/CVI.00363-12

Sow MS, Etard JF, Baize S, Magassouba N, Faye O, Msellati P, Toure AI, Savane I, Barry M, Delaporte E, Postebogui Study G (2016) New evidence of long-lasting persistence of Ebola Virus genetic material in Semen of survivors. J Infect Dis. doi:10.1093/infdis/jiw078

Sueblinvong V, Johnson DW, Weinstein GL, Connor MJ Jr, Crozier I, Liddell AM, Franch HA, Wall BR, Kalil AC, Feldman M, Lisco SJ, Sevransky JE (2015) Critical care for multiple organ failure secondary to Ebola Virus disease in the United States. Crit Care Med 43(10):2066–2075. doi:10.1097/CCM.0000000000001197

Sundar K, Boesen A, Coico R (2007) Computational prediction and identification of HLA-A2.1-specific Ebola virus CTL epitopes. Virology 360(2):257–263. doi:10.1016/j.virol.2006.09.042

Takada A, Fujioka K, Tsuiji M, Morikawa A, Higashi N, Ebihara H, Kobasa D, Feldmann H, Irimura T, Kawaoka Y (2004) Human macrophage C-type lectin specific for galactose and N-acetylgalactosamine promotes filovirus entry. J Virol 78(6):2943–2947

Tamburini BA, Burchill MA, Kedl RM (2014) Antigen capture and archiving by lymphatic endothelial cells following vaccination or viral infection. Nat Commun 5:3989. doi:10.1038/ncomms4989

Tappe D, Perez-Giron JV, Just-Nubling G, Schuster G, Gomez-Medina S, Gunther S, Munoz-Fontela C, Schmidt-Chanasit J (2016) Sustained elevated cytokine levels during recovery phase of Mayaro Virus infection. Emerg Infect Dis 22(4):750–752. doi:10.3201/eid2204.151502

Team WHOER, Agua-Agum J, Ariyarajah A, Blake IM, Cori A, Donnelly CA, Dorigatti I, Dye C, Eckmanns T, Ferguson NM, Fowler RA, Fraser C, Garske T, Hinsley W, Jombart T, Mills HL, Murthy S, Nedjati Gilani G, Nouvellet P, Pelletier L, Riley S, Schumacher D, Shah A, Van Kerkhove MD (2015) Ebola virus disease among children in West Africa. N Engl J Med 372 (13):1274–1277. doi:10.1056/NEJMc1415318

Terajima M, Ennis FA (2011) T cells and pathogenesis of hantavirus cardiopulmonary syndrome and hemorrhagic fever with renal syndrome. Viruses 3(7):1059–1073. doi:10.3390/v3071059

Towner JS, Rollin PE, Bausch DG, Sanchez A, Crary SM, Vincent M, Lee WF, Spiropoulou CF, Ksiazek TG, Lukwiya M, Kaducu F, Downing R, Nichol ST (2004) Rapid diagnosis of Ebola hemorrhagic fever by reverse transcription-PCR in an outbreak setting and assessment of patient viral load as a predictor of outcome. J Virol 78(8):4330–4341

Uyeki TM, Erickson BR, Brown S, McElroy AK, Cannon D, Gibbons A, Sealy T, Kainulainen MH, Schuh AJ, Kraft CS, Mehta AK, Lyon GM 3rd, Varkey JB, Ribner BS, Ellison RT 3rd, Carmody E, Nau GJ, Spiropoulou C, Nichol ST, Stroher U (2016) Ebola Virus persistence in Semen of male survivors. Clin Infect Dis Official Publ Infect Dis Soc Am 62 (12):1552–1555. doi:10.1093/cid/ciw202

Uyeki TM, Mehta AK, Davey RT, Jr., Liddell AM, Wolf T, Vetter P, Schmiedel S, Grunewald T, Jacobs M, Arribas JR, Evans L, Hewlett AL, Brantsaeter AB, Ippolito G, Rapp C, Hoepelman AI, Gutman J, Working Group of the USECNoCMoEVDPitUS, Europe (2016) Clinical management of Ebola Virus disease in the United States and Europe. N Engl J Med 374(7):636–646. doi:10.1056/NEJMoa1504874

van Griensven J, Edwards T, de Lamballerie X, Semple MG, Gallian P, Baize S, Horby PW, Raoul H, Magassouba N, Antierens A, Lomas C, Faye O, Sall AA, Fransen K, Buyze J, Ravinetto R, Tiberghien P, Claeys Y, De Crop M, Lynen L, Bah EI, Smith PG, Delamou A, De Weggheleire A, Haba N, Ebola-Tx C (2016) Evaluation of convalescent plasma for Ebola Virus disease in Guinea. N Engl J Med 374(1):33–42. doi:10.1056/NEJMoa1511812

van Paassen J, Bauer MP, Arbous MS, Visser LG, Schmidt-Chanasit J, Schilling S, Olschlager S, Rieger T, Emmerich P, Schmetz C, van de Berkmortel F, van Hoek B, van Burgel ND, Osterhaus AD, Vossen AC, Gunther S, van Dissel JT (2012) Acute liver failure, multiorgan failure, cerebral oedema, and activation of proangiogenic and antiangiogenic factors in a case of Marburg haemorrhagic fever. Lancet Infect Dis 12(8):635–642. doi:10.1016/S1473-3099(12)70018-X

Varkey JB, Shantha JG, Crozier I, Kraft CS, Lyon GM, Mehta AK, Kumar G, Smith JR, Kainulainen MH, Whitmer S, Stroher U, Uyeki TM, Ribner BS, Yeh S (2015) Persistence of Ebola Virus in ocular fluid during convalescence. N Engl J Med 372(25):2423–2427. doi:10.1056/NEJMoa1500306

Villinger F, Rollin PE, Brar SS, Chikkala NF, Winter J, Sundstrom JB, Zaki SR, Swanepoel R, Ansari AA, Peters CJ (1999) Markedly elevated levels of interferon (IFN)-gamma, IFN-alpha, interleukin (IL)-2, IL-10, and tumor necrosis factor-alpha associated with fatal Ebola virus infection. J Infect Dis 179(Suppl 1):S188–191. doi:10.1086/514283

Wahid R, Cannon MJ, Chow M (2005) Dendritic cells and macrophages are productively infected by poliovirus. J Virol 79(1):401–409. doi:10.1128/JVI.79.1.401-409.2005

Wahl-Jensen VM, Afanasieva TA, Seebach J, Stroher U, Feldmann H, Schnittler HJ (2005) Effects of Ebola virus glycoproteins on endothelial cell activation and barrier function. J Virol 79 (16):10442–10450. doi:10.1128/JVI.79.16.10442-10450.2005

Wauquier N, Becquart P, Padilla C, Baize S, Leroy EM (2010) Human fatal zaire ebola virus infection is associated with an aberrant innate immunity and with massive lymphocyte apoptosis. PLoS Neglected Trop Dis 4(10). doi:10.1371/journal.pntd.0000837

Weiskopf D, Angelo MA, de Azeredo EL, Sidney J, Greenbaum JA, Fernando AN, Broadwater A, Kolla RV, De Silva AD, de Silva AM, Mattia KA, Doranz BJ, Grey HM, Shresta S, Peters B, Sette A (2013) Comprehensive analysis of dengue virus-specific responses supports an HLA-linked protective role for CD8+ T cells. Proc Natl Acad Sci U S A 110(22):E2046–2053. doi:10.1073/pnas.1305227110

Wherry EJ (2011) T cell exhaustion. Nat Immunol 12(6):492–499

Wilson JA, Hart MK (2001) Protection from Ebola virus mediated by cytotoxic T lymphocytes specific for the viral nucleoprotein. J Virol 75(6):2660–2664. doi:10.1128/JVI.75.6.2660-2664.2001

Wirth JP, Rohner F, Woodruff BA, Chiwile F, Yankson H, Koroma AS, Russel F, Sesay F, Dominguez E, Petry N, Shahab-Ferdows S, de Onis M, Hodges MH (2016) Anemia, micronutrient deficiencies, and Malaria in children and women in Sierra Leone prior to the Ebola outbreak—findings of a cross-sectional study. PLoS ONE 11(5):e0155031. doi:10.1371/journal.pone.0155031

Wolf T, Kann G, Becker S, Stephan C, Brodt HR, de Leuw P, Grunewald T, Vogl T, Kempf VA, Keppler OT, Zacharowski K (2015) Severe Ebola virus disease with vascular leakage and multiorgan failure: treatment of a patient in intensive care. Lancet 385(9976):1428–1435. doi:10.1016/S0140-6736(14)62384-9

Yan T, Mu J, Qin E, Wang Y, Liu L, Wu D, Jia H, Li Z, Guo T, Wang X, Qin Y, Li Y, Chen S, Zhang Y, Zhang J, Wu Y, Wang S, Li J (2015) Clinical characteristics of 154 patients suspected of having Ebola virus disease in the Ebola holding center of Jui Government Hospital in Sierra Leone during the 2014 Ebola outbreak. Eur J Clin Microbiol Infect Dis 34 (10):2089–2095. doi:10.1007/s10096-015-2457-z

Yang Z, Delgado R, Xu L, Todd RF, Nabel EG, Sanchez A, Nabel GJ (1998) Distinct cellular interactions of secreted and transmembrane Ebola virus glycoproteins. Science 279 (5353):1034–1037

Yang ZY, Duckers HJ, Sullivan NJ, Sanchez A, Nabel EG, Nabel GJ (2000) Identification of the Ebola virus glycoprotein as the main viral determinant of vascular cell cytotoxicity and injury. Nat Med 6(8):886–889. doi:10.1038/78645

Yen B, Mulder LC, Martinez O, Basler CF (2014) Molecular basis for ebolavirus VP35 suppression of human dendritic cell maturation. J Virol 88(21):12500–12510. doi:10.1128/JVI.02163-14

Yen JY, Garamszegi S, Geisbert JB, Rubins KH, Geisbert TW, Honko A, Xia Y, Connor JH, Hensley LE (2011) Therapeutics of ebola hemorrhagic Fever: whole-genome transcriptional analysis of successful disease mitigation. J Infect Dis 204(Suppl 3):S1043–1052. doi:10.1093/infdis/jir345

Zhang N, Bevan MJ (2011) CD8(+) T cells: foot soldiers of the immune system. Immunity 35 (2):161–168. doi:10.1016/j.immuni.2011.07.010

Zhu FC, Hou LH, Li JX, Wu SP, Liu P, Zhang GR, Hu YM, Meng FY, Xu JJ, Tang R, Zhang JL, Wang WJ, Duan L, Chu K, Liang Q, Hu JL, Luo L, Zhu T, Wang JZ, Chen W (2015) Safety and immunogenicity of a novel recombinant adenovirus type-5 vector-based Ebola vaccine in healthy adults in China: preliminary report of a randomised, double-blind, placebo-controlled, phase 1 trial. Lancet 385(9984):2272–2279. doi:10.1016/S0140-6736(15)60553-0

Nonhuman Primate Models of Ebola Virus Disease

Richard S. Bennett, Louis M. Huzella, Peter B. Jahrling, Laura Bollinger, Gene G. Olinger Jr and Lisa E. Hensley

Abstract Ebola virus disease (EVD) in humans is associated with four ebolaviruses: Ebola virus (EBOV), Sudan virus (SUDV), Bundibugyo virus (BDBV), and Taï Forest virus. To date, no documented cases of human disease have been associated with Reston virus. Here, we describe the nonhuman primate (NHP) models that currently serve as gold standards for testing ebolavirus vaccines and therapeutic agents and elucidating underlying mechanisms of pathogenesis. Although multiple models have been explored over the past 50 years, the predominance of published work has been performed in macaque models. This chapter will focus on the most commonly used models.

Contents

R.S. Bennett · L.M. Huzella · P.B. Jahrling · L. Bollinger · L.E. Hensley (✉)
Integrated Research Facility, National Institute of Allergy and Infectious Diseases, National Institutes of Health, Frederick, MD, USA
e-mail: lisa.hensley@nih.gov

G.G. Olinger Jr
Department of Medicine, National Emerging Infectious Diseases Laboratories, Boston University School of Medicine, Boston, MA, USA

Current Topics in Microbiology and Immunology (2017) 411:171–194
DOI 10.1007/82_2017_20

Published Online: 23 June 2017

1 Introduction

Following the largest Ebola virus disease (EVD) outbreak in history (2013–2016), the clinical picture of human EVD has been redefined. Prior to this epidemic, the average case fatality of EVD was 77% (Kuhn 2015). The high lethality observed in these past outbreaks humans is mirrored in nonhuman primate (NHP) models and argued as a strength of these models, and lethality (or meeting predetermined humane euthanasia criteria) is often a required study endpoint. However, during the 2013–2016 outbreak, the World Health Organization (WHO) reported a total of 28,616 confirmed, probable, and suspected cases in Western Africa with a case fatality of 39.5% (World Health Organization 2016). Although serological evidence of mild or asymptomatic infections was limited in past outbreaks, this outbreak differed as multiple mild EVD cases (coined "Ebola-light" cases) were noted (Richardson et al. 2016; Bellan et al. 2014; Leroy et al. 2001). During the outbreak, disease severity ranged from mild-to-severe disease. As EVD descriptions from the Ebola treatment clinics and from patients repatriated to Western hospitals differs from disease descriptions of past outbreaks, the filovirus research community is re-examining the fidelity of the existing animal models to recapitulate the full spectrum of what is known about human disease.

2 Background

Prior to the 2013–2016 outbreak in Western Africa, the public health and biodefense communities made a tremendous financial investment in both the development of animal models of EBOV infection and in medical countermeasures (MCMs). Given the sporadic nature of filovirus disease outbreaks and the limited number of filovirus disease cases that occurred prior to 2014, researchers previously assumed the advancement of any MCM would require approval under the "Animal Rule" (21 CFR 601.90–95 and 21 CFR 314.600–601) (U.S. Food and Drug Administration 2015). The animal efficacy rule was developed to facilitate approval of therapeutics for agents when human efficacy trials are not feasible or ethical. Under this guidance, the United States Food and Drug Administration may approve a product for which human safety is established, and adequate and well controlled animal studies were performed. To fulfill the criteria, selected animal models should accurately reflect human disease, and the benefit of the therapeutic agent under evaluation will likely translate into a similar effect in humans. Although multiple attempts to evaluate candidate MCMs occurred during the Western Africa EVD outbreak, most of these efforts were initiated at the later stages of the outbreak and were underpowered to generate sufficient data. Thus, most scientists, policy makers, and regulators believe that approval of any MCM will still require use of the "Animal Rule."

Historically, clinical disease and related pathology in NHPs infected with EBOV closely resemble the features observed in human EVD (Fig. 1). Primates of several species have been used to model EVD, including grivets or African green monkeys

Human sequelae observed from EVD outbreaks (1976–2016)

Blood / heart / lung issues

- Anemia ... 50%
- Cardiopathy / valvulopathy / tachycardia / palpitations ... 11-50%
- Shortness of breath ... 38%
- Arterial hypertension ... 7%
- Pericarditis ... 5%
- Myocarditis ... 1%

Central nervous system issues

- Insomnia / sleep disorder ... 18-75%
- Headache ... 22-68%
- Memory / concentration difficulties / confusion ... 2-43%

Co-infections

- Respiratory tract infection + otitis ... 27%
- Urinary tract infection / sexually transmitted infection ... 23%
- Malaria ... 14%
- Suppurative parotitis ... 5%

Eye / vision disorders

- All ocular disorders ... 42-57%
- Blurred vision ... 39-62%
- Uveitis ... 5-35%
- Retro-orbital pain ... 29%
- Yellowing of eyes ... 14%
- Conjunctivitis ... 11%
- Vision loss ... 5%

Gastrointestinal issues

- Anorexia ... 7-100%
- Difficulty swallowing ... 27%
- Gastritis / ulcer / gastroesophageal reflux ... 16%
- Constipation ... 14%
- Moderate acute malnutrition ... 11%
- Sore throat ... 7%
- Diarrhea / gastroenteritis ... 2-5%
- Severe acute malnutrition ... 1%

Pain / fatigue

- Arthralgia / joint stiffness ... 35-87%
- Fatigue / decreased exercise tolerance ... 8-77%
- Abdominal pain ... 12-54%
- Myalgia / muscle weakness ... 12-54%
- Back pain ... 33-46%
- Chest pain ... 31-32%

Reproductive issues

- Orchitis / testicular pain ... 5-27%
- Decreased libido ... 23%
- Sexual dysfunction ... 3-20%
- Amenorrhea ... 2-10%

Sensory issues

- Peripheral paresthesia or dysesthesia ... 38%
- Tinnitus / hearing loss ... 1-27%
- Dizziness / loss of balance ... 1-11%
- Change in taste ... 3%
- Change in smell ... 1%
- Numbness ... 1%

Skin / hair issues

- Alopecia ... 1-75%
- Skin disorders ... 27-49%

Social / mental health issues

- Stigma / rejection ... 32-97%
- Diminished work capacity ... 70%
- Depression or anxiety ... 1-50%
- Mood changes ... 33%
- Post-traumatic stress ... 21%

Fig. 1 Incidence of human sequelae observed during EVD outbreaks (1976–2016)

(AGMs) (*Chlorocebus sps*) (Bowen et al. 1978a, b; Davis et al. 1997; Fisher-Hoch et al. 1992; Ryabchikova et al. 1999a, b), cynomolgus (crab eating) macaques (*Macaca fascicularis*) (Fisher-Hoch et al. 1992; Geisbert et al. 2002, 2003b, c, d; Jahrling et al. 1996, 1999; Sullivan et al. 2000, 2003), rhesus monkeys (*Macaca mulatta*) (Bowen et al. 1978a, b; Fisher-Hoch et al. 1985; Geisbert et al. 2002, 2003a; Jaax et al. 1996), common marmosets (*Callithrix jacchus*) (Smither et al. 2013, 2015; Carrion et al. 2011), and hamadryas baboons (*Papio hamadryas*) (Chupurnov et al. 1995; Kudoyarova-Zubavichene et al. 1999; Mikhailov et al. 1994; Ryabchikova et al. 1999a, b).

Many the studies were conducted in either the rhesus or cynomolgus macaques using 1000 pfu of EBOV delivered via intramuscular (IM) route. While no published reference could be located, the selection of 1000 pfu IM challenge was based on models of a needle stick exposure from a late stage EVD case (Jahrling, personal communication). The rhesus monkey model is often selected for therapeutic evaluations since the disease progression is a bit slower than in cynomolgus macaques. The cynomolgus macaque model is selected more often for vaccine studies due to the slightly more compact disease course and consistent disease progression (Geisbert et al. 2003d; Rubins et al. 2007). Aerosol exposure, while not the natural route of infection, has been used to mimic accidental laboratory exposure or intentional virus release. In addition, investigators are exploring alternative exposure routes that may more closely mimic naturally acquired EBOV infections and are examining the impact of challenge agents given via alternative routes. (Alfson et al. 2015; Mire et al. 2016; Pratt et al. 2010; Reed et al. 2011; Sullivan et al. 2003; Johnson et al. 1995; Geisbert et al. 2008a; Willet et al. 2015; Jaax et al. 1996). To date, most NHP animal modeling studies have focused on EBOV exposure; studies evaluating BDBV or SUDV exposure are limited (Falzarano et al. 2011; Hensley et al. 2010; Pratt et al. 2010; Geisbert et al. 2009). A limited number of NHPs from other species have been evaluated as EVD models. Since a large proportion of studies have used rhesus and cynomolgus macaques by IM challenge route, this chapter will focus on these models.

3 Rhesus Monkeys

3.1 Intramuscular Challenge Models of Ebola Virus Infection

Rhesus monkeys have been used for the study of EBOV infection since the identification of EBOV in 1976 (Bowen et al. 1978a, b). Following IM inoculation with EBOV, animals become febrile on 2–4 days post-exposure (Fig. 2) and develop petechial hemorrhage by days 4–7 post-exposure (Fig. 3a) (Bowen et al. 1978a, b; Fisher-Hoch et al. 1985). Mild anorexia, dehydration, and decreased activity

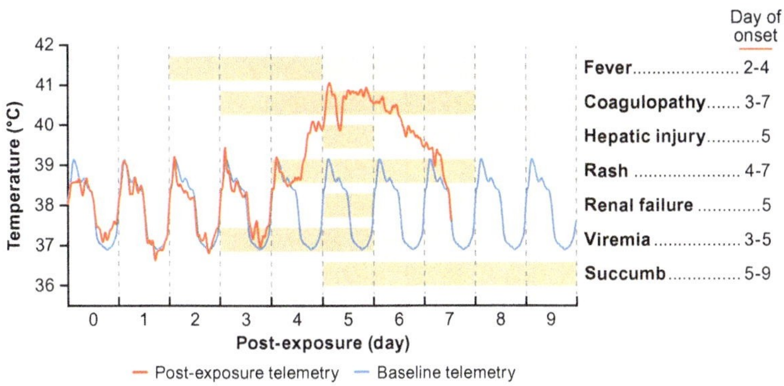

Fig. 2 Time course of clinical signs of Ebola virus infection following IM exposure in rhesus monkeys

typically occur 1–2 days after the onset of fever (Bowen et al. 1978a, b). Most animals succumb to infection on days 5–9 (average 6.5 days) post-inoculation depending on the isolate and virus preparation (Ebihara et al. 2011; Bowen et al. 1978a, b; Geisbert et al. 2002). Some but not all animals develop diarrhea, epistaxis, and or bleeding from venipuncture sites, gums, rectum, and vagina (Bowen et al. 1978a, b).

Monocytes, macrophages, and dendritic cells are the early target cells for viral infection, and seeding of the draining lymph nodes, spleen and liver are early events. Infected NHPs develop neutrophil-driven leukocytosis. An overall lymphopenia is manifested by a marked decrease in $CD8^+$ cells and NK cells from circulation, and widespread bystander apoptosis occurs both in peripheral blood mononuclear cells and lymphoid tissues (Geisbert et al. 2000).

Thrombocytopenia is observed, with the most marked drop often occurring on days 2–4 post-inoculation. Decreased hemoglobin and hematocrit are typical findings (Fisher-Hoch et al. 1983; Kortepeter et al. 2011). Significant drops in protein C activity and marked increases in fibrin degradation products, plasminogen activator inhibitor-1, and tissue-type plasminogen activator are noted from in-depth hematological studies (Geisbert et al. 2003c; Hensley et al. 2007). A decrease in protein C coagulation inhibitor activity occurs 1–2 days prior to an increase in coagulation times (Ebihara et al. 2011). During end-stage disease, animals experience a rapid reduction of plasma fibrinogen concentrations followed by an increase in intravascular fibrin deposition (Ebihara et al. 2011).

Changes in clinical chemistries during the early stages of disease are often unremarkable. Marked changes in the aspartate aminotransferase (AST), alanine aminotransferase (ALT), alkaline phosphatase (AP), and γ-glutamyltransferase concentrations occur in the later stages of disease. Marked increases in serum creatinine concentrations, rises in blood urea nitrogen (BUN) concentrations, and drops in total serum proteins are noted. Analysis of serum for circulating cytokines and chemokines reveals the development of a dysregulated state with the

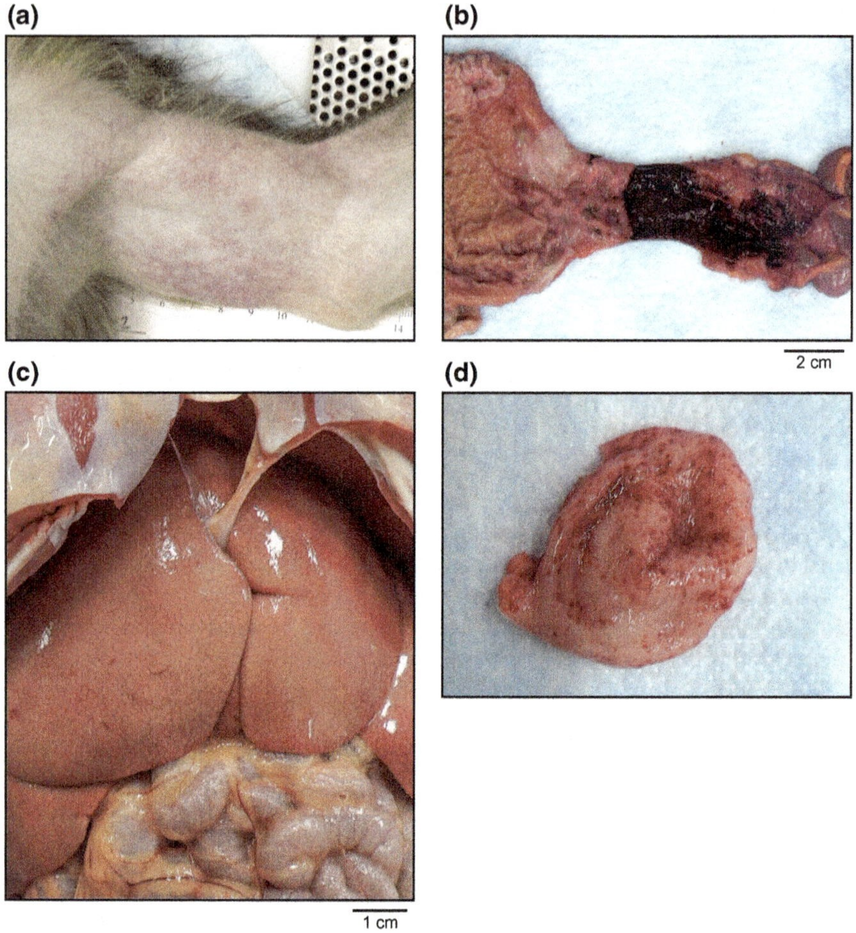

Fig. 3 Gross pathologic findings following IM Ebola virus exposure in rhesus monkeys. **a** NHP *left arm* with typical cutaneous petechial hemorrhage. **b** Stomach pyloric duodenal junction hemorrhage: Gastroduodenal junction with the pylorus to the *left* and the duodenum to the *right*. **c** Liver hepatomegaly and necrosis: Typical pallor and enlargement of the liver of an NHP, showing *rounded edges* and the typical reticulated pattern. **d** NHP urinary bladder with typical mucosal petechial hemorrhage

accumulation of many pro-inflammatory cytokines and chemokines. This accumulation commences as early as 3–4 days post-inoculation and includes interferon (IFN)-α, interleukin (IL)-1β, IL-6, macrophage inflammatory protein (MIP)-1α, IL-15, and IL-18 (Ebihara et al. 2011; Hensley et al. 2002).

Analysis of necropsied tissue reveals enlarged lymph nodes with the presence of congestion, hemorrhage, and edema. Marked congestion of the duodenum often with hemorrhage is observed occasionally at the gastroduodenal and ileocecal junction (Fig. 3b). Congestion and erythema of the gut-associated lymphoid tissues are also typical. The liver is often enlarged and friable with rounded edges and a reticulated pattern (Fig. 3c). Multifocal or coalescing hemorrhages may also be observed in the urinary bladder (Fig. 3d). Microscopically, a significant depletion of the germinal centers of the lymph nodes and widespread necrosis and apoptosis in the lymph nodes are noted (Fig. 4a). Similarly, in the spleen, widespread lymphoid depletion and the presence of numerous apoptotic cells and necrotic debris are observed. In addition, hemorrhage, congestion, and fibrin deposition are noted in the marginal zones of the spleen (Figs. 4b and 5). Similarly, fibrin and congestion are often observed in liver and kidneys. The sinusoids and spaces of Disse of the liver are typically expanded, and disruption of the sinusoidal endothelium may be observed. Hepatocytes are often degenerate or necrotic and contain variably sized intracytoplasmic viral inclusion bodies (Fig. 4c).

3.1.1 Alternative Routes of Ebola Virus Exposure

Exposure to aerosols containing EBOV is not believed to be a natural route of infection. The risk or belief that EBOV may be misused as a bioweapon has driven the development and characterization of EBOV small-particle aerosol exposure models (Cenciarelli et al. 2015; Twenhafel et al. 2013; Osterholm et al. 2015). Of interest is whether this route of exposure changes the disease course, if the amount of virus required to achieve a lethal infection differs from other routes, and if an aerosol exposure alters the efficacy of candidate MCMs. Attempts to establish lethal dose$_{50}$ (LD$_{50}$) or lethal dose$_{90}$ (LD$_{90}$) were unsuccessful, with lethal disease observed following a challenge dose of <10 pfu in rhesus, cynomolgus, and grivet monkeys (Reed et al. 2011). Despite these challenges, many scientists have persevered with pathogenesis and efficacy testing of MCMs. Results from the EBOV aerosol model sequential serial sampling study in rhesus monkeys reveal an early infection of the respiratory lymphoid tissues and early fibrin deposition in the splenic white pulp (Twenhafel et al. 2013). Twenhafel et al. propose that the pathology observed in the lung is specifically related to EBOV infection, unique to this route of infection, and is not seen in IM, intranasal, or intraperitoneal (IP) challenge models. In addition, perivasculitis and vasculitis in superficial dermal blood vessels of haired skin sampled from areas with visible rash were first noted in this study. Researchers are unclear if perivasculitis is linked to this particular model (Twenhafel et al. 2013).

Historically, conjunctival or oral inoculation of rhesus monkeys with EBOV produces a disease model that is consistent with IM challenge. Following conjunctival or oral inoculation with 5.2 $\log_{10}$ pfu of EBOV isolate Yambuku-Mayinga, 4/4 and 2/3 monkeys (one remained uninfected) succumbed to infection 7–8 days post-exposure (Jaax et al. 1996). On day 7 post-exposure, rash on the face, axillary, and inguinal areas was observed. However, the challenge dose used in this study is 100 times higher than a typical IM challenge (Jaax et al. 1996).

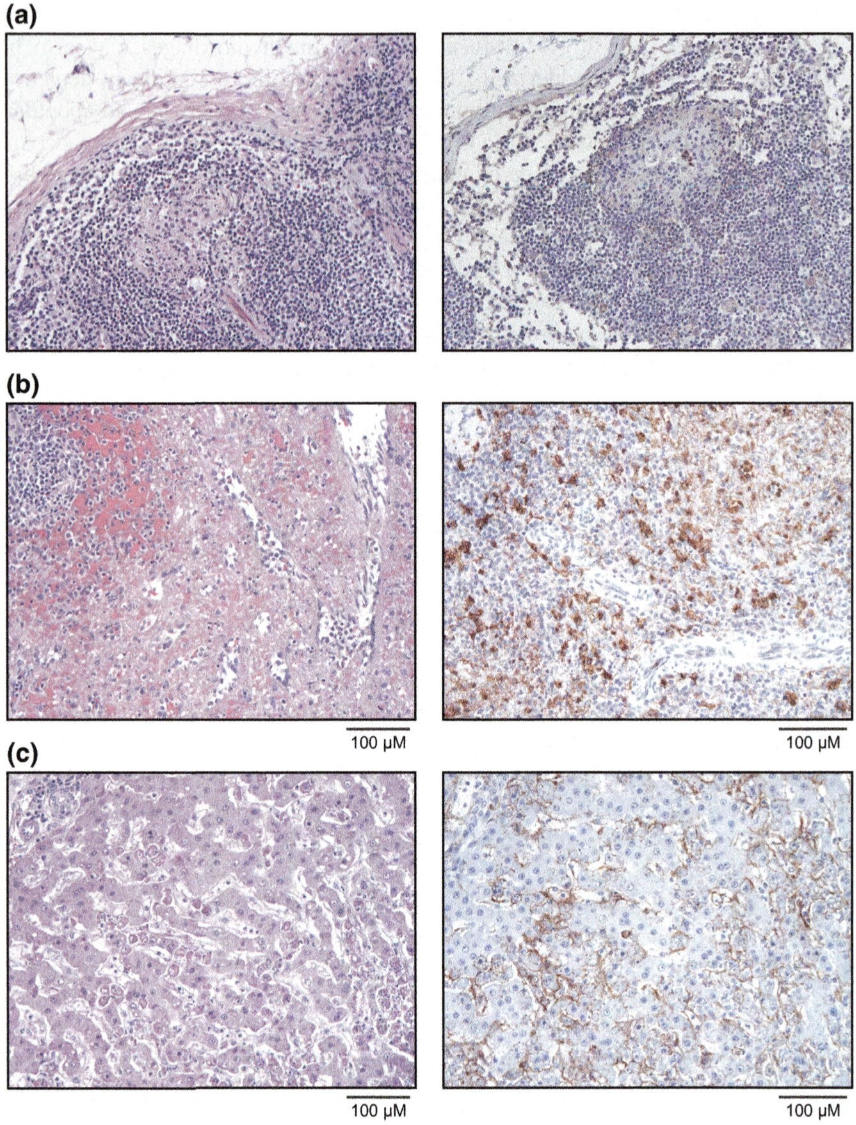
(a)
(b)
100 μM
100 μM
(c)
100 μM
100 μM

◂**Fig. 4** Hematoxylin and eosin (HE) and immunohistochemical (IHC) staining of common target organs in Ebola virus-infected rhesus monkeys. *Left column* is hematoxylin and eosin (HE) stain of target tissues. *Right column* is IHC stain of target tissues. **a** (*left*) Staining of an axillary lymph node at 20× magnification reveals focally extensive loss of lymphocytes, germinal center necrosis accompanied by fibrin deposition and edema. (*right*) IHC staining for EBOV-glycoprotein (GP) in a lymph node reveals positive staining in the cytoplasm of macrophages, endothelial cells, and fibrocytes (20× magnification). **b** (*left*) Staining of spleen periarteriolar lymphoid sheath at 20× magnification indicates marked depletion of lymphocytes. Approximately 80% of the normal architecture of the *red pulp* is replaced by dense, irregular sheets of fibrin. Remaining splenic cords and sinusoids are markedly congested with blood. *Upper left side* of figure is the periarteriolar lymphoid sheath. (*right*) IHC staining for EBOV-viral matrix protein (VP40) in the spleen at 20× magnification reveals positive serum staining in the splenic cords and sinusoids and within the cytoplasm of histiocytes (tissue macrophages). **c** (*left*) The normal sinusoidal architecture of the hepatic parenchyma is disrupted by moderate-to-marked diffuse hepatocellular swelling and multifocal degeneration and necrosis. The space of Disse is markedly expanded by diffuse edema. Numerous eosinophilic cytoplasmic inclusion bodies within hepatocytes are noted (20× magnification). (*right*) IHC staining for EBOV VP40 protein reveals positive serum staining in the hepatic sinusoids and multifocal, punctate, positive staining in individual hepatocytes (20× magnification)

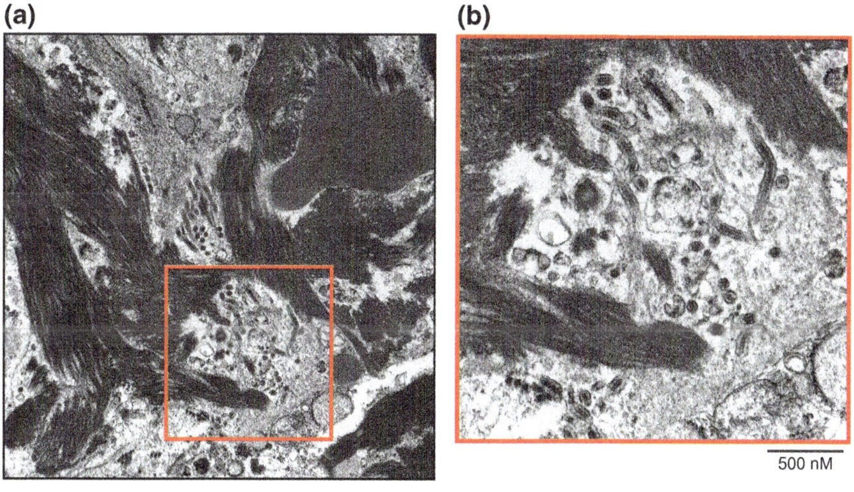

Fig. 5 Transmission electron microgram of a spleen of a rhesus monkey. **a** (*left*) Numerous mature virions are present within the *red pulp* with abundant fibrin (*dark material*) and a distorted red blood cell (24,200× magnification). (*right*) *Inset* of a portion of a (*left*, *red square*) at higher magnification. Red blood cell is not included in the image on *right*

3.1.2 Long-Term Ebola Virus Disease Progression

Over the last decade, results from a growing body of studies of EBOV-infected NHPs (treated with MCMs) describe delayed death or what is now referred to as long-term progression (LTP). First reported at the Negative Strand Virus meeting in 2003 (Hensley et al. 2003), the potential relevance and significance of these findings were for the most part overlooked until the 2013–2016 Western African EVD outbreak. NHPs that experience LTP may display or present with clinical signs not typically seen in classical disease models, including neurological manifestations and ocular involvement (Larsen et al. 2007; Alves et al. 2016). Although the development of a LTP phenotype is not well defined, the onset is usually not before 12 days post-exposure, (after the normal time-to-death in the rhesus monkey IM challenge model). From a summary of six animals experiencing LTP from several different treatment studies, an average time-to-death is 21.7 days (14–29 days) post-exposure, indicating that the phenomenon is not specific to any given candidate MCM (Larsen et al. 2007; Alves et al. 2016).

At necropsy, 5 of 6 animals experiencing LTP had glial nodules with EBOV antigens present in the brain suggestive of a meningoencephalitis or, at the minimum, an active inflammatory process accompanied by viral replication. Viral antigen was also detected in the pancreatic islets of Langerhans, pancreatic acini, the follicular and perifollicular areas of the thyroid, lungs, the cornea, retina, and or the pia/arachnoid mater surrounding the optic nerve of the animals. Surprisingly, the presence of viral antigen was not always associated with an inflammatory response. Although these animals were challenged by IM with EBOV, five of the six NHPs had interstitial pneumonia with multifocal areas of consolidation, alveolar inflammation, or pulmonary emboli. Examination of normal target organs, spleen, liver, and lymph nodes, suggests a viral clearance from these organs (Larsen et al. 2007). The small amount of available data suggest that late stage seeding of the brain, pancreas, thyroid, eye, and lung can occur during an elongated infection. However, in classic EBOV NHP models, this event occurs near the terminal stage and is most likely often overlooked or undetected. When the time-to-death window is extended slightly, the window for viral replication in these tissues is expanded, and the incidence of LTP is increased (Larsen et al. 2007; Alves et al. 2016).

Initially, EBOV targeting of eyes, brain, pancreas, thyroid, or lungs was referred to as potentially altered tropism. The presence of EBOV in these organs was not usually observed during early infection, and LTP was an infrequent event (Larsen et al. 2007). Results from current models suggest that infection of these tissues is more aptly referred to as a third phase or third wave. This phase requires a viremic state, breakdown of the blood–brain barrier, blood–testis barrier, access to other immune privileged compartments, and a disease course of sufficient length to detect viral replication in these tissues. The reports of prolonged EBOV persistence in the semen and vitreous fluid of human EVD survivors (Christie et al. 2015; Diallo et al. 2016; Mate et al. 2015; Crozier 2016) support the need to characterize the LTP phenotype and investigate the potential for the development of a model for EBOV persistence.

3.2 Intramuscular Challenge Models of Sudan Virus Infection

From limited information, rhesus monkeys challenged IM with 1000 pfu of SUDV intramuscularly had decreased activity and reduced appetite similar to that observed following challenge with EBOV. Cage-side observations also included the development of petechial rash, diarrhea, and epistaxis. Death occurred 7–10 days post-inoculation (Thi et al. 2015, 2016). Animals that succumbed to infection presented with splenic lymphoid depletion, expansion of the splenic red pulp with fibrin, multifocal necrotizing hepatitis with sinusoidal leukocytosis, and mild interstitial pneumonia (Thi et al. 2016). In a blinded pilot study comparing the SUDV variants Gulu and Yambio, similar lethality was observed following IM challenge of cynomolgus macaques, with one monkey surviving from each initial group of six (Johnson et al. 2015). From subsequent studies performed with the Gulu variant utilizing either IM or aerosol challenge, no animals survived (Wollen et al. 2015).

3.2.1 Alternative Routes of Sudan Virus Exposure

In a single aerosol challenge experiment, small groups of rhesus monkeys were exposed to 50 or 500 pfu of SUDV (Zumbrun et al. 2012). Five of six animals in the 50 pfu group and 6/6 in the 500 pfu group succumbed following challenge. Overt clinical signs of illness, including dyspnea (3/6), reduced stool (2/6), and fever (1/6), were observed on day 4 post-exposure in the higher dose group. The onset of fever occurred on day 5 post-exposure (3/6) in the lower challenge dose group. Reported hematological and changes in clinical chemistries were consistent with EBOV NHP models.

4 Cynomolgus Macaques

4.1 Intramuscular Challenge Models of Ebola Virus Infection

Following IM inoculation with 1000 pfu of EBOV, cynomolgus macaques develop a clinical picture of EVD that is nearly indistinguishable from the disease observed in rhesus monkeys. However, the disease course is compressed compared to that observed in rhesus monkeys with the normal disease course ranging from 6 to 7 days. Typically EBOV-infected cynomolgus macaques present with fever by day 3 post-exposure and develop a cutaneous rash by day 4 post-exposure (Geisbert et al. 2003b). As the disease progresses, leukocytosis and thrombocytopenia develop, fibrin degradation products (D-dimers) increase, and animals remaining with end-stage disease bleed from the nares or rectum (Geisbert et al. 2003b). EVD is almost uniformly lethal in this model. Significant gross lesions include macular rash on the face, inguinal, and axillary areas; pale and friable liver; enlarged, turgid spleen; pale discoloration of the kidneys; and enlarged lymph nodes (inguinal and axillary) (Trefry et al. 2015).

As also observed in rhesus monkeys, monocytes, macrophages, and dendritic cells are the primary sites of virus replication in cynomolgus macaques. In a sequential sampling study, the earliest EBOV-infected cells are macrophages and dendritic cells in the spleen and draining lymph nodes (Geisbert et al. 2003b). Kupffer cells in the liver are also targets of EBOV infection. Virus replication in endothelial cells, occurring in late stage disease, does not disrupt the architecture of the vasculature (Geisbert et al. 2003d).

A hallmark of EBOV infection is dysregulation of the normal host immune response with lymphopenia, bystander lymphocyte apoptosis, and destruction of lymphoid tissue (Hensley et al. 2002; Geisbert et al. 2000). Transcript analysis of circulating mononuclear cells reveals early and sustained IFN response detectable as early as day 2 post-inoculation prior to the onset of viremia and any detectable signs of illness (Rubins et al. 2007). In addition, strong changes are noted in transcripts in genes that regulate apoptosis, particularly tumor necrosis factor-related apoptosis inducing ligand (TRAIL), caspase-5, caspase-8, Fas-associated death domain protein (FADD), and the B-cell lymphoma 2 agonist of the death (BAD). From analysis of serum and plasma, a number of cytokines and chemokines accumulate, including IFN α and β, IL-6, IL-18, MIP-1α, and MIP-1β (Hensley et al. 2002).

The available transcript and protein data support the theory of a dysregulated host immune response. This response leads to an uncontrolled inflammatory state that is likely resembling the systemic inflammatory response syndrome (SIRS)-like state associated with septic shock (Rubins et al. 2007). Work in the macaque models previously demonstrated a strong upregulation of tissue factor, a trigger of the extrinsic arm of the coagulation pathway. EBOV-infected macrophages expressed tissue factor, and striking images of EBOV-infected macrophages encased in fibrin were seen in tissues collected from NHPs (Geisbert et al. 2003d). These findings support the hypothesis that the coagulation activation is triggered by the expression of tissue factor and not by the destruction of the endothelium (Geisbert et al. 2003d). The uncontrolled inflammatory responses may perpetuate the dysregulated host response amplifying the intertwined coagulation and inflammatory pathways.

At necropsy, cellular examination of other tissues besides the normally described target organs provide some unique potential insights that differ from that observed in rhesus monkeys. In late stage cynomolgus macaques, small foci of EBOV antigen-positive adrenal cortical cells were present in the zona glomerulosa, zona fasciculata, and zona reticularis of the cortices (Geisbert et al. 2003b). Antigen-positive germinal epithelial cells were infrequently noted in the tongue, esophagus, and lips in one of three animals at end-stage EBOV disease. However, no appreciable pathology was associated with these observations.

4.1.1 Alternative Routes of Ebola Virus Exposure

For cynomolgus macaques exposed via the aerosol route with ~1000 pfu of EBOV, onset of fever (detected with radiotelemetry devices) occurred as early as 3 days post-exposure; almost 2.5 days earlier than that observed in the rhesus monkey model (Reed et al. 2011). The onset of fever coincided with the detection of viremia. Changes in hematology were consistent with previously reported NHP models, and changes in both the prothrombin time and activated partial thromboplastin time were detected. Oral or conjunctival exposure of cynomolgus macaques to low-dose EBOV (10 pfu) failed to produce detectable disease or viremia (Mire et al. 2016). However, when the oral EBOV exposure dose was increased to 100 pfu, the animals became moribund and were euthanized on day 8 post-exposure (Mire et al. 2016). Conjunctival exposure to 100 pfu of EBOV resulted in sub-clinical disease with low viremia on day 14 post-exposure (Mire et al. 2016). Note that previous low-dose exposure experiments with other EBOV isolates using either aerosol, IM, or IP routes of administration resulted in fatalities (Sullivan et al. 2000, 2003). The small number of animals used in these studies limit interpretations.

4.2 Intramuscular Challenge Models of Sudan Virus Infection

Investigators have performed limited studies using the Nzara-Boniface SUDV isolate, derived from a fatal human case from the 1976 SUDV-caused outbreak, to test candidate MCMs against SUDV infection (Zumbrun et al. 2012; Warfield et al. 2015). The time-to-death following SUDV challenge was 6–10 days post-challenge, and clinical chemistries and hematological values were consistent with EBOV infection models (Warfield et al. 2015). Additional studies are needed to evaluate other SUDV isolates to characterize pathology and temporal progression of SUDV disease.

4.2.1 Alternative Routes of Sudan Virus Exposure

In a single study, NHPs challenged with either 50 or 500 pfu of SUDV Nzara-Boniface isolate developed fever, petechial rash, disruption in the diurnal changes in the heart rate, and an increase in the overall heart rate (Zumbrun et al. 2012). Most the animals developed dyspnea. Changes in hematology and clinical chemistries were consistent with that observed in previously reported models.

5 Additional Primate Models

5.1 African Green Monkeys

AGMs (genus *Chlorocebus*) refer to five to six monkey populations inhabiting most of sub-Saharan Africa (Grubb et al. 2003; Pfeifer 2017). The taxonomy has been revised over the years making confirmations of true species used in the limited number of older studies difficult. Thus, the term AGM will be used collectively, and we make no attempt to correct the reported taxonomy. Following aerosol exposure to a EBOV dose of 712 ± 442.7 pfu, fever onset began on day 5.2 ± 0.8 post-exposure, and animals were moribund on day 8.3 ± 0.5 post-exposure (Reed et al. 2011). The time-to-death in this study was 24–48 h longer than that observed with either rhesus or cynomolgus monkeys at the challenge dose used. However, the number of animals used for each of the study groups was small. Overall, clinical chemistry and hematological changes seen in the AGMs were consistent with that seen in rhesus and cynomolgus monkeys challenged with a comparable aerosol dose. However, platelet loss and increases in the prothrombin time and activated partial thromboplastin time were much more marked this model than that observed in either macaque model.

Two vervet monkeys (*Chlorocebus pygerythrus*) were exposed IP to 10^4 median guinea-pig infectious doses of EBOV passaged three times in guinea pigs. Animals survived 6 days post-inoculation. No rash was observed, but animals did develop diarrhea and intermittent melena. Gross lesions were similar to what was observed in rhesus monkeys (Baskerville et al. 1978). At necropsy, blood, heart, lung liver, spleen, adrenal glands, kidneys, mesenteric lymph nodes, and urine had viral titers ranging from 3.5 $\log_{10}$ guinea-pig infectious units/g or mL of tissue (urine) to 7.5 $\log_{10}$ guinea-pig infectious units/g of spleen (Bowen et al. 1978a, b).

Upon necropsy, EBOV infection in green monkeys presented as impairment of microcirculation by deposition of fibrin masses and thrombi in the organs (Ryabchikova et al. 1999a). However, the development of the classic petechial rash during the disease course was not consistently observed (Reed et al. 2011; Ryabchikova et al. 1999a; Baskerville et al. 1978; Bowen et al. 1978a, b). Monocytes and macrophages were the primary targets of EBOV infection (Ryabchikova et al. 1999a). In addition, lymphoid depletion in B-cell follicles and spleen and the absence of inflammatory infiltration were observed (Ryabchikova et al. 1999a). Involuted follicles were frequently congested with tangible body macrophages (Geisbert et al. 2000; Ryabchikova et al. 1999a). Necrosis in the liver, spleen, and kidneys was present (Ryabchikova et al. 1999a).

5.2 Baboons

Serological surveys of wild hamadryas baboons (*Papio hamadryas*) suggest that these animals are permissive to wild-type EBOV infection (Leroy et al. 2004; Johnson et al. 1982). Reports of experimental infection of baboons with EBOV (passaged two times in monkeys) or rodent-adapted EBOV are limited (Ryabchikova et al. 1999a; Ignatiev et al. 2000). In one experiment, hamadryas baboons were inoculated subcutaneously with EBOV passaged twice in monkeys with a dose of 20–50 mouse LD_{50} (Ryabchikova et al. 1999a). When infected with monkey-passaged virus, significant changes in blood chemistries (ALT, AST, bilirubin, urea, and creatinine) were observed starting on days 3–7 post-exposure. Up to 70% of the hamadryas baboons developed hemorrhage that progressed to hematemesis and bleeding from the rectum, vagina, skin, and mucosa (Ryabchikova et al. 1999a).

Of the animals infected with guinea-pig adapted virus, hemorrhagic syndrome also developed in one of three animals as evidenced by bleeding from the nose/mouth and hematochezia (Ignatiev et al. 2000). The other two animals bled from a blood draw puncture point. Viremia was detected on day 3 post-inoculation, 3 days prior to obvious clinical signs. Again, increases in ALT, AST, urea, and creatinine concentrations were observed. These animals underwent a period of initial hypercoagulation, followed by hypocoagulation starting on day 7 post-inoculation, and ending with no measurable clotting on day 9 post-inoculation (Ignatiev et al. 2000). As these studies used either monkey-passaged or rodent-adapted viruses, results may not be directly comparable results from other NHP studies that utilized cell culture-propagated EBOV.

Like humans, hamadryas baboons produce four isotypes of IgG antibodies, whereas IgG3 is not detected in macaques (Shearer et al. 1999). IgG3 is an activator of complement (Shearer et al. 1999). As immunologic markers of human EVD survival are identified, the baboon's complete humoral response may provide more opportunities to predict vaccine or therapeutic efficacy in humans than macaques.

5.3 Common Marmosets

Common marmosets (*Callithrix jacchus*) are small New World monkeys that have been used in limited filovirus studies (Smither et al. 2013, 2015; Carrion et al. 2011). Aerosol exposure of sexually mature common marmosets to 4–27 tissue culture infectious dose 50 of EBOV led to lethal disease. In this model, fever, hunched posture, unkempt fur, altered respiration, and a reluctance to move, eat, or drink were the initial clinical signs, with terminal disease occurring 6–8 days after virus challenge (Smither et al. 2015). No rash was observed in this animal model. However, thrombocytopenia, elevated ALT, BUN, and or creatinine, and increased blood clotting times were noted consistent with that observed with other NHP models and human disease (Smither et al. 2015). Gross and histological findings were consistent with other NHP models of EBOV infection, including lymphocyte depletion with evidence of apoptosis, and fibrin deposition in multiple organs.

Similarly, IM inoculation of marmosets with 10 or 1000 pfu of EBOV resulted in a severe systemic disease course, which resembled human disease course. The infection was characterized by fever, viremia, thrombocytopenia, neutrophilia, and disseminated intravascular coagulation leading to death 4–5 days post-inoculation. Similar to the aerosol model and other NHP models, severe lymphoid depletion and fibrin deposition were evident. Hematological changes and changes in the clinical chemistries were consistent with other NHP models of EBOV infection. The compact disease course combined with the monkey's small size, 320–450 g (roughly equivalent to guinea-pig weights for males and females 3–7 weeks of age) (Charles River Laboratories International 2016), may impact the utility of this model for interventional studies or studies requiring frequent blood collection. Alternatively, the small size may be an advantage for some BSL-4 containment laboratories with limited space for housing NHPs (Carrion et al. 2011).

6 Common Threads of Pathogenesis

Change in clinical chemistries, hematology and end-stage pathology are in general surprisingly consistent across the IM and aerosol NHP challenge models. Early changes in the lungs are one of the few differences in disease presentation with aerosol exposure compared to that observed following IM exposure (Twenhafel et al. 2013). For models with sufficient data, several commons threads of pathogenesis are readily noted: the development of a dysregulated host immune response characterized by a profound bystander lymphocyte apoptosis and a SIRS-like state; the activation of the coagulation cascade leading to widespread fibrin deposition; and the early infection of monocytes, macrophages, and dendritic cells.

EBOV infection of monocytes, macrophages, and dendritic cells is an early, critical event in EVD and leads to seeding of target organs. This early event promotes EBOV to co-opt the dendritic cells critical for the immune response against invading pathogens. EBOV infection of the dendritic cells has been suggested to be linked to both the lack of a robust immune response and a potential role in the development of bystander apoptosis seen in the lymph nodes of infected NHPs (Geisbert et al. 2003b; Ilinykh et al. 2015; Melanson et al. 2015; Yen et al. 2014; Bosio et al. 2003; Geisbert et al. 2000). Infection of monocytes and macrophages in vitro with EBOV leads to upregulation of TRAIL and multiple apoptosis-related genes, cytokines, chemokines, and tissue factor (Geisbert et al. 2000; Ludtke et al. 2016). Analysis of samples from infected NHPs has confirmed these findings and support that early infection of these target cells may be driving multiple critical downstream pathways. Data from both in vitro and in vivo experiments have hinted at just how complex these interactions are with activation of multiple apoptotic, inflammatory signaling, and coagulation pathways (Rubins et al. 2007; Geisbert et al. 2000, 2003a, b; Caballero et al. 2016).

The importance of the interaction between coagulation and inflammation as a response to severe infection has become increasingly appreciated. As such, coagulation and inflammation clearly go hand in hand. In fact, coagulation mediators are included as part of the spectrum known as SIRS, also referred to as a cytokine storm. Cytokines are key coordinators of inflammation and vascular dysfunction. Although the vascular endothelium may not be an early target of EBOV infection, the potential contribution of the vascular endothelium to EBOV pathogenesis cannot be overlooked. Cytokines and coagulation products can induce changes in endothelial cell structure that affect permeability, and they can also play a role in regulating the inflammatory response. Inflammatory mediators upregulate procoagulant factors such as tissue factor, inhibit fibrinolytic activity, and downregulate natural anticoagulant pathways (e.g., protein C anticoagulant pathway). Importantly, protein C activity rapidly declines in the EBOV models. Replacement with recombinant activated protein C was demonstrated to reduce morbidity and fatality (protected ≈20% of NHPs) and increase the mean time-to-death in an NHP experimental EBOV challenge model (Hensley et al. 2007). Not surprisingly, evaluation of treated NHPs that exhibited a beneficial effect had significant reductions in D-dimers and pro-inflammatory cytokines, including IL-6 and monocyte chemoattractant protein 1 (Hensley et al. 2007) A similar effect was seen in other intervention studies regardless of the class of intervention tested? sounds confusing (Geisbert et al. 2003a, b).

Other cytokines or chemokines may also be involved in modulating endothelial function during EBOV infections either directly or indirectly. For example, fibrin split products and thrombin can increase the production of pro-inflammatory cytokines such as IL-6. Production of IL-6 can in turn increase tissue factor production, activating the clotting cascade (Geisbert et al. 2003c). Theoretically, anti-cytokine therapies alone or in combination with anticoagulant strategies may aid in disrupting the downward spiral observed in filovirus infections. Previously, investigators demonstrated that treatment of NHP with a recombinant protein (i.e., nematode anticoagulant protein C2) targeting activation of the intrinsic pathway increased survival to 33% (Geisbert et al. 2003a). Results from this study demonstrate the potential for this compound and the benefit of targeting the clinical manifestations of the disease itself rather than direct viral replication.

7 Future Outlook

Currently, the most pressing questions facing the filovirus animal model community are: Does the current NHP model accurately reflect human disease? Does the model accurately predict the performance of any given countermeasure in humans? Are additional animal models needed? Some researchers question whether the current NHP lethal animal models are too stringent. The models may mimic severe EVD, but not the full spectrum of disease ranging from asymptomatic-to-severe infection, LTP, or viral persistence. Lethality in the current well-characterized NHP models is substantially higher than in human infections. Researchers are unclear if the measured case fatality (~40%) during the Western African 2013–2016 outbreak is directly a result of a large outbreak size leading to a more accurate outcome measurements, a product of improved supportive care, or reflection of genetic variation of the population or of the EBOV variant. To date, only limited attempts have incorporated aggressive supportive care in NHP EBOV studies (Jaax et al. 1996). As the field moves beyond the Western African EVD outbreak, researchers should step back and recognize and understand the limitations of the current models. Gaps in current EVD models include the inability of current models to mimic EVD sequelae and viral persistence. Addressing these gaps will likely require new or refinement of the existing models. The potential of NHP model(s) for these purposes is unknown, and pilot experiments should be initiated. Developing models of EBOV persistence as evidenced by LTP should be considered as separate, independent tasks and should not be confused with a need to determine if the existing models accurately reflect human EVD disease or predict MCM efficacy.

Acknowledgements We thank Donna Perry for providing the gross and histology images, John Bernbaum for providing the electron microscopy images, and Jiro Wada for figure preparation, respectively. The content of this publication does not necessarily reflect the views or policies of the US Department of Health and Human Services (DHHS) or of the institutions and companies affiliated with the authors. This work was funded in part through Battelle Memorial Institute's prime contract with the US National Institute of Allergy and Infectious Diseases (NIAID) under Contract No. HHSN272200700016I. R.S.B. and L.B. performed this work as employees of Battelle Memorial Institute. L.M.H. preformed this work as an employee of Charles River Laboratories, a subcontractor to Battelle Memorial Institute.

References

Alfson KJ, Avena LE, Beadles MW, Staples H, Nunneley JW, Ticer A et al (2015) Particle-to-PFU ratio of Ebola virus influences disease course and survival in cynomolgus macaques. J Virol 89(13):6773–6781. doi:10.1128/jvi.00649-15

Alves DA, Honko AN, Kortepeter MG, Sun M, Johnson JC, Lugo-Roman LA et al (2016) Necrotizing scleritis, conjunctivitis, and other pathologic findings in the left eye and brain of an Ebola virus-infected rhesus macaque (*Macaca mulatta*) with apparent recovery and a delayed time of death. J Infect Dis 213(1):57–60. doi:10.1093/infdis/jiv357

Baskerville A, Bowen ET, Platt GS, McArdell LB, Simpson DI (1978) The pathology of experimental Ebola virus infection in monkeys. J Pathol 125(3):131–138. doi:10.1002/path.1711250303

Bellan SE, Pulliam JR, Dushoff J, Meyers LA (2014) Ebola control: effect of asymptomatic infection and acquired immunity. Lancet 384(9953):1499–1500. doi:10.1016/s0140-6736(14)61839-0

Bosio CM, Aman MJ, Grogan C, Hogan R, Ruthel G, Negley D et al (2003) Ebola and Marburg viruses replicate in monocyte-derived dendritic cells without inducing the production of cytokines and full maturation. J Infect Dis 188(11):1630–1638. doi:10.1086/379199

Bowen ET, Platt GS, Simpson DI, McArdell LB, Raymond RT (1978a) Ebola haemorrhagic fever: experimental infection of monkeys. Trans R Soc Trop Med Hyg 72(2):188–191

Bowen ETW, Baskerville A, Cantell K, Mann GF, Simpson DIH, Zuckerman AJ (1978b) The effect of interferon on experimental Ebola virus infection in rhesus monkeys. In: Pattyn S (ed) Ebola virus haemorrhagic fever. Elsevier/North Holland, Amsterdam, pp 167–171

Caballero IS, Honko AN, Gire SK, Winnicki SM, Mele M, Gerhardinger C et al (2016) In vivo Ebola virus infection leads to a strong innate response in circulating immune cells. BMC Genom 17:707. doi:10.1186/s12864-016-3060-0

Carrion R Jr, Ro Y, Hoosien K, Ticer A, Brasky K, de la Garza M et al (2011) A small nonhuman primate model for filovirus-induced disease. Virology 420(2):117–124

Cenciarelli O, Gabbarini V, Pietropaoli S, Malizia A, Tamburrini A, Ludovici GM et al (2015) Viral bioterrorism: Learning the lesson of Ebola virus in West Africa 2013–2015. Virus Res 210:318–326. doi:10.1016/j.virusres.2015.09.002

Charles River Laboratories International. Hartley guinea pig growth chart. Charles River Laboratories International, Wilminton, MA. 2016. http://www.criver.com/products-services/basic-research/find-a-model/hartley-guinea-pig#pricing. Accessed 23 Dec 2016

Christie A, Davies-Wayne GJ, Cordier-Lassalle T, Blackley DJ, Laney AS, Williams DE et al (2015) Possible sexual transmission of Ebola virus—Liberia, 2015. MMWR Morb Mortal Wkly Rep 64(17):479–481

Chupurnov AA, Chernukhin IV, Ternovoi VA, Kudoiarova NM, Makhova NM, Azaev M et al (1995) Attempts to develop a vaccine against Ebola fever. Vopr Virusol 40(6):257–260

Crozier I (2016) Ebola virus RNA in the semen of male survivors of Ebola virus disease: the uncertain gravitas of a privileged persistence. J Infect Dis 214(10):1467–1469. doi:10.1093/infdis/jiw079

Davis KJ, Anderson AO, Geisbert TW, Steele KE, Geisbert JB, Vogel P et al (1997) Pathology of experimental Ebola virus infection in African green monkeys. Involvement of fibroblastic reticular cells. Arch Pathol Lab Med 121(8):805–819

Diallo B, Sissoko D, Loman NJ, Bah HA, Bah H, Worrell MC et al (2016) Resurgence of Ebola virus disease in Guinea linked to a survivor with virus persistence in seminal fluid for more than 500 days. Clin Infect Dis 63(10):1353–1356. doi:10.1093/cid/ciw601

Ebihara H, Rockx B, Marzi A, Feldmann F, Haddock E, Brining D et al (2011) Host response dynamics following lethal infection of rhesus macaques with Zaire ebolavirus. J Infect Dis 204 (Suppl 3):S991–S999. doi:10.1093/infdis/jir336

Falzarano D, Feldmann F, Grolla A, Leung A, Ebihara H, Strong JE et al (2011) Single immunization with a monovalent vesicular stomatitis virus-based vaccine protects nonhuman primates against heterologous challenge with Bundibugyo ebolavirus. J Infect Dis 204(Suppl 3):S1082–S1089. doi:10.1093/infdis/jir350

Fisher-Hoch SP, Brammer TL, Trappier SG, Hutwagner LC, Farrar BB, Ruo SL et al (1992) Pathogenic potential of filoviruses: role of geographic origin of primate host and virus strain. J Infect Dis 166(4):753–763

Fisher-Hoch SP, Platt GS, Lloyd G, Simpson DI, Neild GH, Barrett AJ (1983) Haematological and biochemical monitoring of Ebola infection in rhesus monkeys: implications for patient management. Lancet 2(8358):1055–1058. doi:10.1016/S0140-6736(83)91041-3 [pii]

Fisher-Hoch SP, Platt GS, Neild GH, Southee T, Baskerville A, Raymond RT et al (1985) Pathophysiology of shock and hemorrhage in a fulminating viral infection (Ebola). J Infect Dis 152(5):887–894

Geisbert TW, Daddario-Dicaprio KM, Geisbert JB, Reed DS, Feldmann F, Grolla A et al (2008a) Vesicular stomatitis virus-based vaccines protect nonhuman primates against aerosol challenge with Ebola and Marburg viruses. Vaccine 26(52):6894–6900

Geisbert TW, Daddario-DiCaprio KM, Williams KJ, Geisbert JB, Leung A, Feldmann F et al (2008b) Recombinant vesicular stomatitis virus vector mediates postexposure protection against Sudan Ebola hemorrhagic fever in nonhuman primates. J Virol 82(11):5664–5668. doi:10.1128/jvi.00456-08

Geisbert TW, Geisbert JB, Leung A, Daddario-DiCaprio KM, Hensley LE, Grolla A et al (2009) Single-injection vaccine protects nonhuman primates against infection with Marburg virus and three species of Ebola virus. J Virol 83(14):7296–7304. doi:10.1128/JVI.00561-09

Geisbert TW, Hensley LE, Gibb TR, Steele KE, Jaax NK, Jahrling PB (2000) Apoptosis induced in vitro and in vivo during infection by Ebola and Marburg viruses. Lab Invest 80(2):171–186

Geisbert TW, Hensley LE, Jahrling PB, Larsen T, Geisbert JB, Paragas J et al (2003a) Treatment of Ebola virus infection with a recombinant inhibitor of factor VIIa/tissue factor: a study in rhesus monkeys. Lancet 362(9400):1953–1958. doi:10.1016/s0140-6736(03)15012-x

Geisbert TW, Hensley LE, Larsen T, Young HA, Reed DS, Geisbert JB et al (2003b) Pathogenesis of Ebola hemorrhagic fever in cynomolgus macaques: evidence that dendritic cells are early and sustained targets of infection. Am J Pathol 163(6):2347–2370

Geisbert TW, Pushko P, Anderson K, Smith J, Davis KJ, Jahrling PB (2002) Evaluation in nonhuman primates of vaccines against Ebola virus. Emerg Infect Dis 8(5):503–507. doi:10.3201/eid0805.010284

Geisbert TW, Young HA, Jahrling PB, Davis KJ, Kagan E, Hensley LE (2003c) Mechanisms underlying coagulation abnormalities in Ebola hemorrhagic fever: overexpression of tissue factor in primate monocytes/macrophages is a key event. J Infect Dis 188(11):1618–1629

Geisbert TW, Young HA, Jahrling PB, Davis KJ, Larsen T, Kagan E et al (2003d) Pathogenesis of Ebola hemorrhagic fever in primate models: evidence that hemorrhage is not a direct effect of virus-induced cytolysis of endothelial cells. Am J Pathol 163(6):2371–2382

Grubb P, Butynski T, Oates J, Bearder S, Disotell T, Groves C et al (2003) Assessment of the diversity of African primates. Int J Primatol 24(6):1301–1357. doi:10.1023/B:IJOP.0000005994.86792.b9

Hensley L, Young H, Paragas J, Larsen T, Fisher R, Fritz E et al (2003) Pathogenesis of Ebola Hemorrhagic fever: the impact of viral infection of dendritic cells and other antigen presenting cells. Negative Strand Virus Meeting; Palazzo dei Congressi, Pisa, Italy

Hensley LE, Mulangu S, Asiedu C, Johnson J, Honko AN, Stanley D et al (2010) Demonstration of cross-protective vaccine immunity against an emerging pathogenic Ebola virus species. PLoS Pathog 6(5):e1000904. doi:10.1371/journal.ppat.1000904

Hensley LE, Stevens EL, Yan SB, Geisbert JB, Macias WL, Larsen T et al (2007) Recombinant human activated protein C for the postexposure treatment of Ebola hemorrhagic fever. J Infect Dis 196(Suppl 2):S390–S399

Hensley LE, Young HA, Jahrling PB, Geisbert TW (2002) Proinflammatory response during Ebola virus infection of primate models: possible involvement of the tumor necrosis factor receptor superfamily. Immunol Lett 80(3):169–179

Ignatiev GM, Dadaeva AA, Luchko SV, Chepurnov AA (2000) Immune and pathophysiological processes in baboons experimentally infected with Ebola virus adapted to guinea pigs. Immunol Lett 71(2):131–140. doi:10.1016/S0165-2478(99)00169-8

Ilinykh PA, Lubaki NM, Widen SG, Renn LA, Theisen TC, Rabin RL et al (2015) Different temporal effects of Ebola virus VP35 and VP24 proteins on global gene expression in human dendritic cells. J Virol 89(15):7567–7583. doi:10.1128/jvi.00924-15

Jaax NK, Davis KJ, Geisbert TJ, Vogel P, Jaax GP, Topper M et al (1996) Lethal experimental infection of rhesus monkeys with Ebola-Zaire (Mayinga) virus by the oral and conjunctival route of exposure. Arch Pathol Lab Med 120(2):140–155

Jahrling PB, Geisbert J, Swearengen JR, Jaax GP, Lewis T, Huggins JW et al (1996) Passive immunization of Ebola virus-infected cynomolgus monkeys with immunoglobulin from hyperimmune horses. Arch Virol Suppl 11:135–140

Jahrling PB, Geisbert TW, Geisbert JB, Swearengen JR, Bray M, Jaax NK et al (1999) Evaluation of immune globulin and recombinant interferon-alpha2b for treatment of experimental Ebola virus infections. J Infect Dis 179(Suppl 1):S224–S234. doi:10.1086/514310

Johnson BK, Gitau LG, Gichogo A, Tukei PM, Else JG, Suleman MA et al (1982) Marburg, Ebola and rift valley fever virus antibodies in East African primates. Trans R Soc Trop Med Hyg 76 (3):307–310

Johnson E, Jaax N, White J, Jahrling P (1995) Lethal experimental infections of rhesus monkeys by aerosolized Ebola virus. Int J Exp Pathol 76(4):227–236

Johnson J, Lewis B, Reed C, Mucker E, Hensley L, Honko A (2015) Comparison of Sudan virus variants Yambio and Gulu in cynomolgus macaques to support non-clinical testing of filovirus medical countermeasures. In: 7th international symposium on Filoviruses—Ebola West Africa and Recent Developments. Washington, DC 2015, p 058A

Kortepeter MG, Lawler JV, Honko A, Bray M, Johnson JC, Purcell BK et al (2011) Real-time monitoring of cardiovascular function in rhesus macaques infected with Zaire ebolavirus. J Infect Dis 204(Suppl 3):S1000–S1010. doi:10.1093/infdis/jir337

Kudoyarova-Zubavichene NM, Sergeyev NN, Chepurnov AA, Netesov SV (1999) Preparation and use of hyperimmune serum for prophylaxis and therapy of Ebola virus infections. J Infect Dis 179(Suppl 1):S218–S223. doi:10.1086/514294

Kuhn JH (2015) Ebolavirus and Marburgvirus infections. In: Kasper DL, Fauci AS, Hauser SL, Longo DL, Jameson JL (eds) Harrison's principles of internal medicine, 19 ed. McGraw-Hill Education, Columbus, pp 1323–9

Larsen T, Stevens EL, Davis KJ, Geisbert JB, Daddario-DiCaprio KM, Jahrling PB et al (2007) Pathologic findings associated with delayed death in nonhuman primates experimentally infected with Zaire Ebola virus. J Infect Dis 196(Suppl 2):S323–S328

Leroy EM, Baize S, Debre P, Lansoud-Soukate J, Mavoungou E (2001) Early immune responses accompanying human asymptomatic Ebola infections. Clin Exp Immunol 124(3):453–460

Leroy EM, Telfer P, Kumulungui B, Yaba P, Rouquet P, Roques P et al (2004) A serological survey of Ebola virus infection in central African nonhuman primates. J Infect Dis 190 (11):1895–1899. doi:10.1086/425421

Ludtke A, Ruibal P, Becker-Ziaja B, Rottstegge M, Wozniak DM, Cabeza-Cabrerizo M et al (2016) Ebola virus disease is characterized by poor activation and reduced levels of circulating CD16+ monocytes. J Infect Dis 214(suppl 3):S275–S280. doi:10.1093/infdis/jiw260

Mate SE, Kugelman JR, Nyenswah TG, Ladner JT, Wiley MR, Cordier-Lassalle T et al (2015) Molecular evidence of sexual transmission of Ebola virus. N Engl J Med 373(25):2448–2454. doi:10.1056/NEJMoa1509773

Melanson VR, Kalina WV, Williams P (2015) Ebola virus infection induces irregular dendritic cell gene expression. Viral Immunol 28(1):42–50. doi:10.1089/vim.2014.0091

Mikhailov VV, Borisevich IV, Chernikova NK, Potryvaeva NV, Krasnianskii VP (1994) The evaluation in hamadryas baboons of the possibility for the specific prevention of Ebola fever. Vopr Virusol 39(2):82–84

Mire CE, Geisbert JB, Agans KN, Deer DJ, Fenton KA, Geisbert TW (2016) Oral and conjunctival exposure of nonhuman primates to low doses of Ebola Makona virus. J Infect Dis 214(suppl 3): S263–S267. doi:10.1093/infdis/jiw149

Osterholm MT, Moore KA, Kelley NS, Brosseau LM, Wong G, Murphy FA et al (2015) Transmission of Ebola viruses: what we know and what we do not know. mBio 6(2):e00137. doi:10.1128/mBio.00137-15

Pfeifer S (2017) The demographic and adaptive history of the African green monkey. Mol Biol Evol. doi:10.1093/molbev/msx056

Pratt WD, Wang D, Nichols DK, Luo M, Woraratanadharm J, Dye JM et al (2010) Protection of nonhuman primates against two species of Ebola virus infection with a single complex adenovirus vector. Clin Vaccine Immunol 17(4):572–581

Reed DS, Lackemeyer MG, Garza NL, Sullivan LJ, Nichols DK (2011) Aerosol exposure to Zaire ebolavirus in three nonhuman primate species: differences in disease course and clinical pathology. Microbes Infect 13(11):930–936. doi:10.1016/j.micinf.2011.05.002

Richardson ET, Kelly JD, Barrie MB, Mesman AW, Karku S, Quiwa K et al (2016) Minimally symptomatic infection in an Ebola 'hotspot': a cross-sectional serorosurvey. PLoS Negl Trop Dis. 10(11):e0005087. doi:10.1371/journal.pntd.0005087

Rubins KH, Hensley LE, Wahl-Jensen V, Daddario DiCaprio KM, Young HA, Reed DS et al (2007) The temporal program of peripheral blood gene expression in the response of nonhuman primates to Ebola hemorrhagic fever. Genome Biol 8(8):R174. doi:10.1186/gb-2007-8-8-r174

Ryabchikova EI, Kolesnikova LV, Luchko SV (1999a) An analysis of features of pathogenesis in two animal models of Ebola virus infection. J Infect Dis 179(Suppl 1):S199–S202. doi:10.1086/514293

Ryabchikova EI, Kolesnikova LV, Netesov SV (1999b) Animal pathology of filoviral infections. Curr Top Microbiol Immunol 235:145–173

Shearer MH, Dark RD, Chodosh J, Kennedy RC (1999) Comparison and characterization of immunoglobulin G subclasses among primate species. Clin Diagn Lab Immunol 6(6):953–958

Smither SJ, Nelson M, Eastaugh L, Laws TR, Taylor C, Smith SA et al (2013) Experimental respiratory Marburg virus haemorrhagic fever infection in the common marmoset (Callithrix jacchus). Int J Exp Pathol 94(2):156–168. doi:10.1111/iep.12018

Smither SJ, Nelson M, Eastaugh L, Nunez A, Salguero FJ, Lever MS (2015) Experimental respiratory infection of marmosets (*Callithrix jacchus*) with Ebola virus Kikwit. J Infect Dis 212(Suppl 2):S336–S345. doi:10.1093/infdis/jiv371

Sullivan NJ, Geisbert TW, Geisbert JB, Xu L, Yang ZY, Roederer M et al (2003) Accelerated vaccination for Ebola virus haemorrhagic fever in non-human primates. Nature 424 (6949):681–684. doi:10.1038/nature01876

Sullivan NJ, Sanchez A, Rollin PE, Yang ZY, Nabel GJ (2000) Development of a preventive vaccine for Ebola virus infection in primates. Nature 408(6812):605–609. doi:10.1038/35046108

Thi EP, Lee AC, Geisbert JB, Ursic-Bedoya R, Agans KN, Robbins M et al (2016) Rescue of non-human primates from advanced Sudan ebolavirus infection with lipid encapsulated siRNA. Nat Microbiol. 1(10):16142. doi:10.1038/nmicrobiol.2016.142

Thi EP, Mire CE, Lee AC, Geisbert JB, Zhou JZ, Agans KN et al (2015) Lipid nanoparticle siRNA treatment of Ebola-virus-Makona-infected nonhuman primates. Nature 521(7552):362–365. doi:10.1038/nature14442

Trefry JC, Wollen SE, Nasar F, Shamblin JD, Kern SJ, Bearss JJ et al (2015) Ebola virus infections in nonhuman primates are temporally influenced by glycoprotein poly-u editing site populations in the exposure material. Viruses 7(12):6739–6754. doi:10.3390/v7122969

Twenhafel NA, Mattix ME, Johnson JC, Robinson CG, Pratt WD, Cashman KA et al (2013) Pathology of experimental aerosol Zaire ebolavirus infection in rhesus macaques. Vet Pathol 50(3):514–529. doi:10.1177/0300985812469636

U.S. Food and Drug Administration. Product development under the animal rule: guidance for industry. Center for Drug Evaluation and Research, Food and Drug Administration, Silver Spring, MD. 2015. http://www.fda.gov/downloads/drugs/guidancecomplianceregulatoryinformation/guidances/ucm399217.pdf. Accessed 20 December 2016

Warfield KL, Dye JM, Wells JB, Unfer RC, Holtsberg FW, Shulenin S et al (2015) Homologous and heterologous protection of nonhuman primates by Ebola and Sudan virus-like particles. PLoS ONE 10(3):e0118881. doi:10.1371/journal.pone.0118881

Willet M, Kurup D, Papaneri A, Wirblich C, Hooper JW, Kwilas SA et al (2015) Preclinical development of inactivated rabies virus-based polyvalent vaccine against rabies and filoviruses. J Infect Dis 212(Suppl 2):S414–S424. doi:10.1093/infdis/jiv251

Wollen SE, Trefry J, Pitt M, Honko A (2015) Confirmation of virulence of Sudan ebolavirus, Gulu variant, in cynomolgus macaques as a result of aerosol exposure. In: Pierson T, Bosio C (eds). ASM biodefense and emerging diseases research meeting; Washington Marriott Wardman Park Hotel, Washington DC. American Society for Microbiology, Washington, Poster #171A

World Health Organization. Situation report: Ebola virus disease—10 June 2016. World Health Organization, Geneva, CH. 2016. http://apps.who.int/iris/bitstream/10665/208883/1/ebolasitrep_10Jun2016_eng.pdf?ua=1. Accessed 20 Dec 2016

Yen B, Mulder LC, Martinez O, Basler CF (2014) Molecular basis for ebolavirus VP35 suppression of human dendritic cell maturation. J Virol 88(21):12500–12510. doi:10.1128/jvi.02163-14

Zumbrun EE, Bloomfield HA, Dye JM, Hunter TC, Dabisch PA, Garza NL et al (2012) A characterization of aerosolized Sudan virus infection in African green monkeys, cynomolgus macaques, and rhesus macaques. Viruses 4(10):2115–2136. doi:10.3390/v4102115

Small Animal Models for Studying Filovirus Pathogenesis

Satoko Yamaoka, Logan Banadyga, Mike Bray and Hideki Ebihara

Abstract Filovirus small animal disease models have so far been developed in laboratory mice, guinea pigs, and hamsters. Since immunocompetent rodents do not exhibit overt signs of disease following infection with wild-type filoviruses isolated from humans, rodent models have been established using adapted viruses produced through sequential passage in rodents. Rodent-adapted viruses target the same cells/tissues as the wild-type viruses, making rodents invaluable basic research tools for studying filovirus pathogenesis. Moreover, comparative analyses using wild-type and rodent-adapted viruses have provided beneficial insights into the molecular mechanisms of pathogenicity and acquisition of species-specific virulence. Additionally, wild-type filovirus infections in immunodeficient rodents have provided a better understanding of the host factors required for resistance to filovirus infection and of the immune response against the infection. This chapter provides comprehensive information on the filovirus rodent models and rodent-adapted filoviruses. Specifically, we summarize the clinical and pathological features of filovirus infections in all rodent models described to date, including the recently developed humanized and collaborative cross (CC) resource recombinant inbred (RI) intercrossed (CC-RIX) mouse models. We also cover the molecular determinants responsible for adaptation and virulence acquisition in a number of rodent-adapted filoviruses. This chapter clearly defines the characteristic and advantages/disadvantages of rodent models, helping to evaluate the practical use of rodent models in future filovirus studies.

S. Yamaoka · H. Ebihara (✉)
Department of Molecular Medicine, Mayo Clinic, 200 First Street SW, Rochester, MN 55905, USA
e-mail: Ebihara.Hideki@mayo.edu

L. Banadyga
Laboratory of Virology, Division of Intramural Research, National Institute of Allergy and Infectious Diseases, National Institutes of Health, Hamilton, MT 59840, USA

M. Bray
Division of Clinical Research, National Institute of Allergy and Infectious Diseases, National Institutes of Health, Bethesda, MD 20892, USA

Current Topics in Microbiology and Immunology (2017) 411:195–228
DOI 10.1007/82_2017_9

Published Online: 27 June 2017

Contents

1 Introduction

The recent and devastating West African Ebola virus disease (EVD) epidemic due to Ebola virus (EBOV) infection clearly highlighted the need for effective anti-filoviral countermeasures, including vaccines and postexposure therapeutics, to be deployed rapidly in an outbreak situation (Martins et al. 2016; Mendoza et al. 2016). Considering the development of such countermeasures, rodent models are most valuable for the initial testing of direct-acting antivirals and vaccines, prior to their final evaluation in nonhuman primates (NHPs) and eventual clinical trials. Furthermore, rodent models have been used extensively to understand molecular aspects of filovirus pathogenesis and host immune responses. In this chapter, we summarize the pathological features and molecular pathogenesis of filovirus infection in each rodent model, and we discuss the advantages and disadvantages of these models for understanding filoviral disease mechanisms.

2 Background

Ebola virus (EBOV) and Marburg virus (MARV) belong to the family *Filovirdae*, which includes three genera: *Ebolavirus*, *Marburgvirus*, and *Cuevavirus* (Bukreyev et al. 2014). The genus *Ebolavirus* includes of five species that each have a single type virus: *Zaire ebolavirus* with type virus Ebola virus (EBOV), *Sudan ebolavirus* with type virus Sudan virus (SUDV), *Bundibugyo ebolavirus* with type virus Bundibugyo virus (BDBV), *Taï Forest ebolavirus* with type virus Taï Forest virus (TAFV), and *Reston ebolavirus* with type virus Reston virus (RESTV). The genus *Marburgvirus* includes a single species, *Marburg marburgvirus*, which has two genetically closely related members, Marburg virus (MARV) and Ravn virus (RAVV). The recently established genus *Cuevavirus* also includes a single species, *Lloviu cuevavirus*, although the type virus Lloviu virus (LLOV) was only detected in bats and has not yet been isolated (Negredo et al. 2011; Bukreyev et al. 2014). Except for RESTV and LLOV, filoviruses are known to cause severe hemorrhagic fever in humans with high case-fatality rates (Sanchez et al. 2007; Banadyga and Ebihara 2015).

Filovirus particles are filamentous in shape, consisting of a nucleocapsid core surrounded by a viral matrix and a host-derived envelope studded with glycoprotein spikes. The filoviral single-stranded, negative-sense genome consists of a linear RNA molecule of approximately 19 kb that is composed of seven genes, encoding nine proteins for ebolaviruses and seven for marburgviruses (Sanchez et al. 2007; Brauburger et al. 2015). The termini of the genome comprise a 3' leader and a 5' trailer that contain replication/transcription promoters and genome packaging signals. Four of the gene products—the nucleoprotein (NP), virion protein 35 (VP35), VP30, and the RNA-dependent RNA polymerase L—comprise the ribonucleoprotein complex, which drives viral RNA synthesis, mRNA transcription, and replication of the RNA genome (Brauburger et al. 2015). The glycoprotein $GP_{1,2}$, which is expressed following transcriptional editing that leads to a +1 frame shift, mediates viral entry, including attachment to receptor molecules and membrane fusion (Jangra et al. 2015). EBOV, unlike MARV, also encodes soluble GP (sGP), which is expressed from the unedited GP transcript, and small soluble GP (ssGP), which is expressed following transcriptional editing that produces a +2 frame shift. The function of sGP has been postulated to modulate host immune responses and endothelial permeability (Mohan et al. 2012; Escudero-Pérez et al. 2014). VP40 functions as the viral matrix protein, which is essential for virion assembly and budding (Bornholdt et al. 2013; Oda et al. 2015), while VP24 is required for nucleocapsid assembly along with NP, VP30, and VP35 (Huang et al. 2002; Noda et al. 2005; Hoenen et al. 2006; Watt et al. 2014). VP35 also functions to inhibit the induction of a type I interferon (IFN) response, whereas ebolavirus VP24 and marburgvirus VP40 additionally act to inhibit the IFN signaling cascade directly (Basler et al. 2000; Reid et al. 2006; Valmas et al. 2010; Bale et al. 2012; Messaoudi et al. 2015; Oda et al. 2015).

3 General Pathological Features of Filovirus Infection in Humans and NHPs

Ebola virus disease (EVD) [also colloquially referred to as Ebola hemorrhagic fever, EHF] and Marburg virus disease (MVD) [also colloquially referred to as Marburg hemorrhagic fever, MHF] are among the most severe acute viral diseases. Severe/fatal filoviral hemorrhagic fevers are characterized by systemic viral replication; dysregulation of immune responses, including induction of uncontrolled pro-inflammatory responses and suppression of effective innate and acquired immune responses; hemorrhagic manifestations, including coagulation abnormalities, such as disseminated intravascular coagulation and a characteristic maculopapular/petechial rash; increased vascular permeability; multi-organ failure; and electrolyte imbalances and fluid redistribution that induce hypovolemic shock (Bray and Mahanty 2003; Geisbert and Jahrling 2004; Mahanty and Bray 2004; Mohamadzadeh et al. 2007; Messaoudi et al. 2015). Several clinical and pathological studies in infected humans and NHPs have demonstrated that cells of the mononuclear phagocytic system (MPS) (i.e., monocytes, macrophages) and dendritic cells are the first target cells (Geisbert et al. 2003a; Bray and Geisbert 2005), whereas infection of endothelial cells seems to occur at the terminal stage of infection (Geisbert et al. 2003b; Hensley and Geisbert 2005). Dendritic cells and cells of the MPS are usually located in peripheral tissues and circulate in target organs (e.g., liver, lymph nodes, spleen), resulting in the efficient transmission of the virus to noninfected cells and tissues. Filovirus infection in lymphoid tissues may contribute to the dysregulation of immune responses. For example, filovirus infection of dendritic cells is thought to paralyze early immune responses and impair the maturation of these cells and their ability to support T cell stimulation, which may be one of the mechanisms that lead to immune suppression during filovirus infection (Bosio et al. 2003; Mahanty et al. 2003b; Mohamadzadeh et al. 2007). Necrosis of hepatocytes in the liver caused by viral replication leads to the impairment of clotting factor synthesis, and a reduction in the synthesis of serum proteins (such as albumin) results in decreased plasma osmotic pressures and induces edema (Martines et al. 2015; Wolf et al. 2015). Recent studies suggest that filoviral hemorrhagic fevers, as well as some other viral hemorrhagic fevers, are at least in part driven by an uncontrolled release of cytokines known as a "cytokine storm," similar to what is seen in septic shock induced by gram-negative bacteria (Bray and Mahanty 2003; Bixler et al. 2015). Indeed, several reports support the idea that a cytokine storm is a major trigger that contributes to lymphocyte depletion, vascular leakage, and coagulation disorders (Wauquier et al. 2010; Ebihara et al. 2011).

4 Animal Disease Modeling in Filovirus Research

Animal models of filovirus infection have so far been developed in laboratory mice, guinea pigs, hamsters, ferrets, pigs, and NHPs. Of these models, NHPs are considered the "gold standard" since they are highly susceptible to filovirus infections and show hallmark pathological features similar to those seen in human infections. Viruses isolated from humans are lethal for NHPs without any adaptation, and virulence in NHPs is apparently not affected by amplification of virus in cell culture. NHP models have been developed with monkeys of several different species, including cynomologus macaques, rhesus macaques, grivets (African green monkeys), hamadryas baboons, and common squirrel monkeys, all of which display similar hemorrhagic fever syndromes but with species-specific differences in hemorrhagic manifestations (Kuhn et al. 2008).

Unlike NHPs, adult, immunocompetent rodents do not exhibit overt signs of disease following infection with wild-type filoviruses, and, as such, all filovirus rodent models described to date rely on either immunocompromised animals or rodent-adapted virus strains. Soon after the first recognized outbreaks of filovirus disease, attempts were made to isolate MARV and EBOV by inoculating specimens from patients into guinea pigs (Smith et al. 1967; Bowen et al. 1977), although such attempts caused only nonlethal, febrile illness in guinea pigs (Siegert et al. 1967; Smith et al. 1967). Additional early studies revealed that immunocompetent adult mice were completely resistant to filovirus-induced disease, although virus replication was detected. In contrast, newborn mice, which have incompletely developed immune systems, were susceptible to filovirus infection. Wild-type MARV (WT-MARV) caused lethal infection in newborn mice following intracerebral inoculation (Hofmann and Kunz 1970), whereas intracerebral or intraperitoneal inoculation of wild-type EBOV (WT-EBOV) was lethal in newborn mice up to 4 days old (Bowen et al. 1977; Van der Groen et al. 1979; Bray et al. 2001a). To develop rodent models that more accurately recapitulated filovirus virulence, immunocompromised animals, such as Severe combined immunodeficiency (SCID) mice were used, although wild-type viruses cause a prolonged wasting illness in SCID mice that bears no resemblance to filovirus disease in humans (Bray et al. 2001a; Warfield et al. 2007). In addition, and more commonly, viruses were adapted to rodents via sequential passage through guinea pigs or mice (Smith et al. 1967; Simpson et al. 1968; Bray et al. 1999; Volchkov et al. 2000; Warfield et al. 2009; Subbotina et al. 2010; Lofts et al. 2011; Qiu et al, 2014; Cross et al. 2015a). In addition, immunocompromised mouse models, including knockout mice, have been used to identify host factors required for resistance to infection and to elucidate the immunological responses to filovirus infections (Bray et al. 2001a; Gupta et al. 2001, 2004, 2005; Bray 2004; Panchal et al. 2009; Bradfute et al. 2010; de Wit et al. 2011; Raymond et al. 2011; Brannan et al. 2015; Hill-Batorski et al. 2015).

Table 1 Comparison of animal model systems for filovirus research

	Human	NHP	Ferret	Guinea pig	Hamster	Hu-BLT	CC-RIX	Mouse
Virus adaptation	No	No	No	Yes	Yes	No	Yes	Yes
Pathogenicity of MA-filovirus[a]	ND	+	ND	++	+++	ND	+++	+++
Pathogenicity of GPA-filovirus[a]	Yes[b]	+++[c]	ND	+++	–	ND	ND	–
Inoculation route	N/A	Various	in	ip/sc/ar	ip	ip	ip	ip/ar[d]/in[e]
Viremia	+++	+++	+++	++	++	+++	+++	+++
Tissue/cell tropism	+	+	+	+	+	+	+	+
Pyrexia	+++	+++	++	++	ND	ND	ND	–
Systemic inflammatory response	+++	+++	+/++	++	++	+++	ND	+
Hematology	+	+	+	+	+	+	ND	+
Blood chemistry	+	+	+	+	+	+	ND	+
Lymphocyte apoptosis	+	+	+	+	+	ND	+	+
Coagulation abnormalities	+++	+++	+++	+/++[f]	+++	ND	+++	+/–
Rash/hemorrhage	++	++	+	–/+	–/+	–	–	–/+
Availability	N/A	+	++	+++	+++	+	++	+++
Ease of handling	N/A	+	+	++	+++	+	++	+++
Cost	N/A	+++	++	+	+	+++	++	+
Large N	N/A	–	–	++	+++	–/+	+/++	+++
Tools and reagents	++	++	+/++	–/+	+/++	+++	+++	+++

[a]Only examined with rodent-adapted EBOV strains
[b]Laboratory accident with GPA-EBOV in Koltsovo, Russia, 2004
[c]Hamadryas Baboon (*Papio hamadryas*) (Ignatiev et al. 2000)
[d]BXD (C57BL/6 crossed with DBA/2) recombinant inbred strains and C57BL/6 mice (Zumbrun et al 2012)
[e]Partial lethality in mice infected with mouse-adapted MARV Angola variant strains (Qiu et al. 2014)
[f]Paramerters indicate coagulopathy were only examined in outbred guinea pigs
N/A not applicable; *ND* not determined; *ip* intraperitoneal; *sc* subcutaneous; *ar* aerosol; *in* intranasal

Compared to the NHP models, rodent models in general present several distinct advantages in the study of filovirus pathogenesis (Table 1). Their relatively small size and prolific reproduction make them inexpensive and easy to handle and house, especially under maximum containment conditions. For mice in particular, the availability of various strains with precisely defined genetic backgrounds, along with numerous reagents and experimental tools for work with mice and other rodents, make these animals appealing experimental models. Rodent-adapted filoviruses, specifically, target the same cells/tissues as wild-type filoviruses in humans and NHPs and have enabled the modeling of at least some of the pathogenic processes associated with disease in humans and NHPs. Moreover, comparative analyses using wild-type and rodent-adapted viruses have provided valuable insight into the molecular mechanisms of pathogenicity and acquisition of species-specific

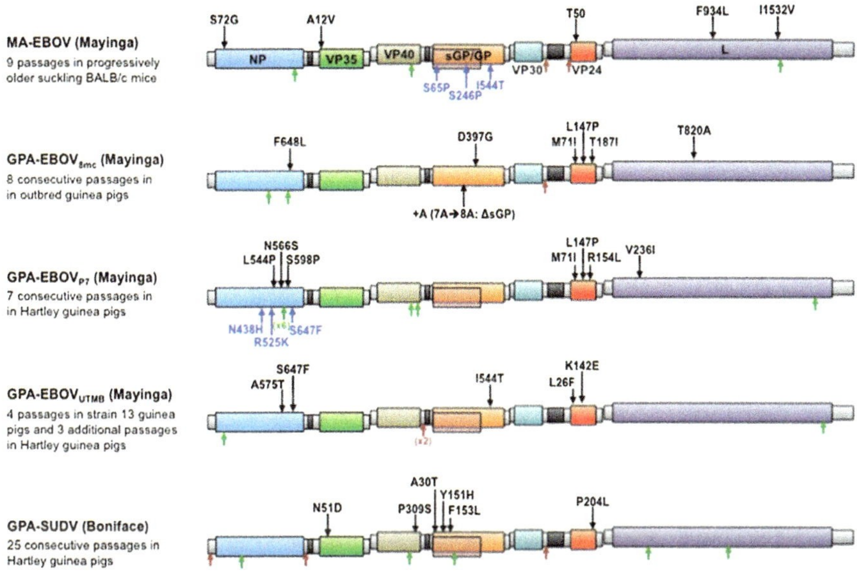

Fig. 1 Mutations in lethal rodent-adapted ebolavirus strains: Non-synonymous mutations are indicated with *black arrows*, and mutations are indicated using the single-letter amino acid code and the position of the residue substituted in each viral protein. *Blue arrows* with *blue text* indicate mutations that existed in the precursor virus used for rodent adaptation. *Green* and *brown arrows* indicate synonymous mutations and nucleotide substitutions in noncoding/untranslated regions, respectively

virulence. Several rodent-adapted lethal variants of filoviruses have been generated based on EBOV isolate Mayinga (EBOV-Mayinga), SUDV isolate Boniface (SUDV-Boniface), RAVV isolate Ravn, and several isolates of MARV (Figs. 1 and 2).

On the other hand, rodent models also present some disadvantages over NHP models. Although rodent-adapted filoviruses recapitulate many of the hallmark clinical features of disease, they often also display biological phenotypes different from the wild-type, parental viruses, leading to a disease course that does not always accurately reflect that seen in humans and NHPs. Accordingly, inconsistencies have been observed between the results obtained from the evaluation of vaccines and treatments in rodent versus NHP models (Geisbert et al. 2002). Nevertheless, rodent models are convenient and critical for modeling certain aspects of filovirus disease. Indeed, they have been used extensively to study aspects of pathogenicity and host immune responses, as well as to test the efficacy of post-exposure therapeutics and experimental vaccines prior to their final evaluation in NHPs.

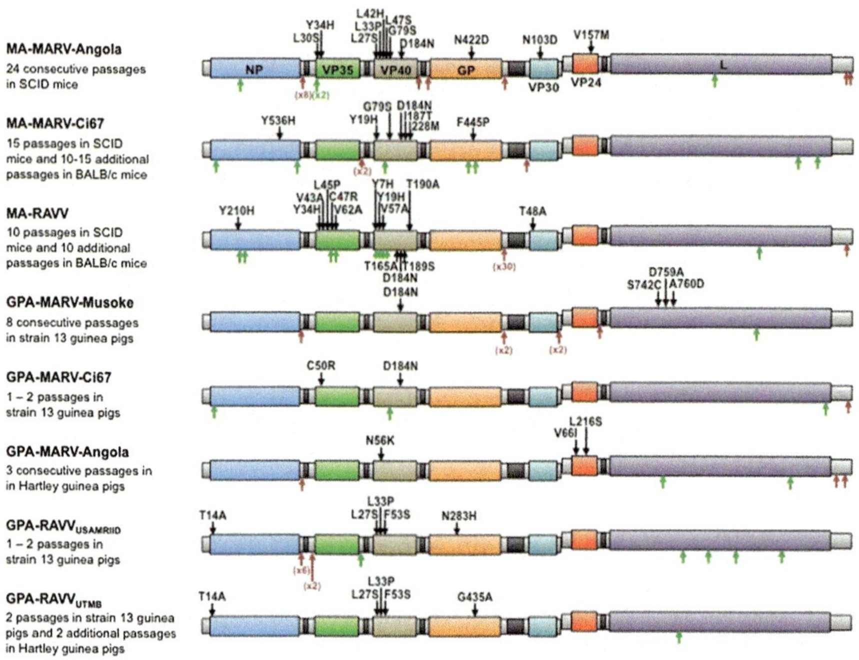

Fig. 2 Mutations in lethal rodent-adapted marburgvirus strains: Non-synonymous mutations are indicated with *black arrows*, and mutations are indicated using the single-letter amino acid code and the position of the residue substituted in each viral protein. *Blue arrows* with *blue text* indicate mutations that existed in the precursor virus used for rodent adaptation. *Green* and *brown arrows* indicate synonymous mutations and nucleotide substitutions in noncoding/untranslated regions, respectively

5 Mouse Models of Filovirus Infection

5.1 Immunocompetent Mouse Model with Mouse-Adapted Ebola Virus

WT-EBOV does not cause any clinical signs in adult mice, with neither intraperitoneal nor subcutaneous inoculation of WT-EBOV causing disease in mice over 8 days old (Bray et al. 2001a). The establishment of mouse-adapted EBOV (MA-EBOV), which was generated by repeated passage of EBOV-Mayinga (isolated from a patient during the initial outbreak of EBOV in Zaire 1976) in progressively older suckling BALB/c mice (Bray et al. 1999), enabled the investigation of EBOV pathogenesis in immunocompetent adult mice. An intraperitoneal dose of 1 or 100 plaque-forming units (pfu) of MA-EBOV caused lethal infection in BALB/c, C57BL/6, and CD-1 (ICR) mice ranging in age from 5 to 16 weeks old. All infected mice became visibly ill with ruffled fur and diminished activity, began to lose weight on day 3 postinfection, and died by day 7 (Bray et al. 1999, 2001a, b).

Elevated body temperatures were not detected in infected mice throughout the duration of illness (Bray et al. 2001b). Although all three strains of mice produced 100% lethality, CD-1 mice showed the highest susceptibility, with a time to death of 2–5 days post-infection, compared to a time to death of 4–7 and 6 days postinfection for BALB/c and C57BL/6 mice, respectively. Virus titers in the liver and spleen reached approximately 10^9 pfu/g on day 3, coincident with peak viremia in the sera, which was 7.5×10^7 pfu/ml. The livers and spleens of mice infected with MA-EBOV yielded virus titers more than a thousand times greater than the mean titer of WT-EBOV in these organs at the same time point (Bray et al. 1999). Subcutaneous inoculation of the virus did not cause symptomatic illness in 3-week-old mice (Bray et al. 2001a; Mahanty et al. 2003a). Interestingly, MA-EBOV also resulted in uniformly lethality and 30% lethality in BXD (C57BL/6 crossed with DBA/2) recombinant inbred and C57BL/6, respectively, following aerosol inoculation (Zumbrun et al. 2012).

Immunohistochemistry and in situ hybridization targeting viral RNA revealed that MA-EBOV initially infects macrophages and other cells of the MPS in the lymph nodes and spleen. Viral infection then spreads to Kupffer cells, hepatocytes, and sinusoidal endothelial cells in the liver; fibroblast-like cells (fibroblastic reticular cells) in the lymph nodes and spleen; and a number of organs, including the thymus, adrenal glands, lungs, gastrointestinal tract, kidneys, reproductive tract, and central nervous system (Bray et al. 1999; Gibb et al. 2001). Coincident with viral spread, severe histological lesions were observed in infected organs, especially in the spleen and liver by days 4 and 5 postinfection. Marginal zones of the spleen became almost indistinguishable because of extensive necrosis, and the liver developed disseminated coalescing degeneration and necrosis. Lymphoblastic cells were present in lymph nodes and spleen, concomitant with severe lymphocyte apoptosis in these organs as well as the thymus (Gibb et al. 2001; Bradfute et al. 2007, 2008). In contrast to other animal models (Geisbert et al. 1992; Jaax et al. 1996; Connolly et al. 1999), infrequent small deposits of fibrin were observed in the spleen and visceral vasculature of the mice (Bray et al. 1999; Gibb et al. 2001).

MA-EBOV-infected mice showed hemoconcentration with a rise in hemoglobin and hematocrit, leukocytosis, and severe thrombocytopenia. By day 5, mice inoculated with 100 pfu of virus exhibited a platelet count drop of approximately 60% from the baseline value (Bray et al. 2001b). Although hemorrhagic manifestations, such as spontaneous bleeding from the gastrointestinal tract or abdominal cavity, were occasionally observed in a small number of infected animals, no obvious hematological abnormalities suggestive of coagulopathy were observed in infected mice: the plasma prothrombin time (PT) and activated partial thromboplastin time (aPTT) remained normal throughout the duration of illness, and plasma fibrinogen concentrations showed no change from baseline, consistent with the lack of intensive fibrin deposition. Serum concentrations of liver-associated enzymes, especially aspartate aminotransferase (AST), were markedly elevated, reflecting severe liver damage (Bray et al. 2001b). Moreover, similar to humans and NHPs infected with EBOV, an increased inflammatory response, characterized by tumor necrosis factor (TNF)-α and monocyte chemotactic protein (MCP)-1 production, appeared early after infection and correlated with the increasing virus loads (Mahanty et al. 2003a).

5.2 Immunocompetent Mouse Model with Mouse-Adapted Marburg and Ravn Viruses

Three mouse-adapted marburgviruses have been generated based on RAVV (isolated from a fatal case in Kenya 1987), MARV isolate Cie67 (Ci67) (isolated in Marburg, West Germany 1967), and MARV variant Angola (isolated in Uíge, Angola 2005), each of which was generated by passage through immunodeficient SCID mice and, in the case of RAVV and MARV isolate Ci67, additional passaging through BALB/c mice. Twenty-four sequential passages of WT-RAVV—ten in SCID mice and fourteen in BALB/c mice—were required to generate mouse-adapted RAVV (MA-RAVV), which was uniformly lethal in adult immunocompetent BALB/c mice infected with 1000 or 100,000 pfu via the intraperitoneal route (Warfield et al. 2009). Mouse-adapted MARV variant Angola (MA-MARV-Angola) was generated by 24 passages in SCID mice (Qiu et al. 2014), whereas mouse-adapted MARV isolate Ci67 (MA-MARV-Ci67) required 15 passages in SCID mice followed by 14 in BALB/c mice (Lofts et al. 2011). Interestingly, whereas MA-RAVV killed both BALB/c and C57BL/6 mouse strains (by days 8 and 12 postinfection, respectively) when inoculated intraperitoneally, MA-MARV-Ci67 only killed 80% of infected BALB/c mice and 60% of infected C57BL/6 mice. Like MA-RAVV, MA-MARV-Angola caused 100% lethal disease in BALB/c mice following intraperitoneal inoculation; however, unlike MA-RAVV, MA-MARV-Angola also resulted in 80% lethality in mice following intranasal inoculation. None of the mouse-adapted marburgviruses caused disease following subcutaneous inoculation.

After intraperitoneal inoculation with 1000 pfu of MA-RAVV, reduction of body weight was observed from day 3 postinfection, and all infected mice died by day 8 without hemorrhagic manifestations. Viral titers in the blood of MA-RAVV-infected mice peaked on day 5, with titers over 10^5 pfu/ml. Infectious virus was detected in a number of organs, including the spleen, liver, lymph node, kidneys, lungs, brain, intestine, and gonads as early as day 3, and the highest titer was observed in the spleen on day 7, with values exceeding 10^5 pfu/g. In contrast to MA-RAVV infection, virus titers of WT-RAVV and WT-MARV-Ci67 in all mice tissues assayed were around the lower detection limit on day 7. Interestingly, MA-MARV-Ci67 grew to higher titer than MA-RAVV in almost all organs tested, despite the fact that MA-RAVV is more virulent in mice (Lofts et al. 2011).

Pathological and hematological features in mice infected with either MA-RAVV or MA-MARV-Angola are similar. Following infection with either virus, MARV antigen was detected 3–5 days postinfection in the spleen and liver, where it localized to macrophages and dendritic cells in the red pulp of the spleen and Kupffer cells and hepatocytes in the liver. Splenic lymphoid depletion and lymphocytosis were seen in follicles and periarteriolar lymphoid sheaths, accompanied by the appearance of large lymphoblastic cells in MA-RAVV-infected mice. The pathological changes in the liver, including hepatocellular degeneration and necrosis, were also prominent.

Hematologic analysis revealed a decrease in both peripheral white blood cells and lymphocyte numbers in MA-MARV-Angola and MA-RAVV-infected mice, as well as a remarkable reduction in platelets starting around days 3–5 postinfection. In addition, numbers of peripheral $CD4^+$, $CD8^+$, B, and NK cells decreased in MA-RAVV-infected mice. While elevated levels of D-dimers—a fibrin degradation product—were seen during the late stage of infection with MA-RAVV, significant fibrin deposition in tissues and prolonged PT and aPTT were not observed. Concentrations of liver-associated enzymes, including AST and alanine aminotransferase (ALT), increased over time, indicating diminution of liver function in infected mice. Moreover, a massive cytokine response occurred, with increased levels of numerous cytokines and chemokines, including TNF-α, IFN-γ, interleukin (IL)-5, and MCP-1. Notably, none of these changes were observed in mice infected with nonadapted wild-type viruses.

5.3 *Immunodeficient Mouse Models*

The mechanisms underlying the immune response to filovirus infection have been studied in mice with a variety of immune defects, including gene knockout mice. Bray et al. (2001a) reported that intraperitoneal inoculation of WT-EBOV or WT-MARV and subcutaneous inoculation of MA-EBOV, which produced no visible illness in immunocompetent mice, caused death in IFN-α/β receptor knockout ($IFNAR^{-/-}$) mice, and electron microscopy and immunohistochemistry confirmed large amounts of replicating virus and viral antigen in the livers and spleens of these mice. Additionally, mice lacking the transcription factor STAT1, which plays an important role in type I IFN signaling, were also highly susceptible to intraperitoneal inoculation of WT-EBOV or subcutaneous inoculation of MA-EBOV (Bray et al. 2001a). These reports prove that the type I IFN response is critical to resistance in mice, and they suggest its importance in primates.

Indeed, since type I IFN signaling-defective mouse models are highly susceptible to wild-type filoviruses, they have served as useful models for investigating the virulence and pathogenesis of filoviruses for which no rodent-adapted variants exist. Although WT-EBOV and SUDV-Boniface were capable of causing lethal infection in 129/Sv $IFNAR^{-/-}$ mice, EBOV variant Kikwit (EBOV-Kikwit), RESTV, and TAFV were unable to cause lethal illness (Bray et al. 2001a). A more recent study indicated that $IFNAR^{-/-}$ mice with a C57BL/6 background were more resistant to various ebolaviruses: SUDV-Boniface and SUDV variant Gulu were only partially lethal in this model, with 37 and 5% of mice succumbing via intraperitoneal inoculation, respectively, while BDBV, TAFV, and RESTV were incapable of causing lethal disease (Brannan et al. 2015). Interestingly, EBOV-Mayinga caused lethal illness in these IFNAR$^{-/-}$ mice, whereas EBOV-Kikwit did not, despite both viruses exhibiting similar virulence in NHP models. Infection of STAT1$^{-/-}$ mice with EBOV-Mayinga, EBOV-Kikwit, SUDV-Boniface, RESTV isolate AZ-1435, MARV isolate Musoke (MARV-Musoke), and RAVV produced visible illness in mice (Raymond et al. 2011), and a comparable study with RESTV variant Pennsylvania and Reston08-A demonstrated the same, with the Pennsylvania variant resulting in 50% lethality (de Wit et al. 2011). These studies suggest that the absence of STAT1, which would impact type I, II, and III IFN-mediated signaling, is more critical to filovirus infection in mice than the absence of IFNAR. Overall, nonadapted filoviruses that possess highly virulent phenotypes (EBOV, SUDV, MARV, and RAVV) in NHPs (and possibly in humans) tend to cause lethal or more severe illness than TAFV, BDBV, and RESTV in type I IFN signaling-defective mice.

For mouse models in which the adaptive immune system is compromised, it has been demonstrated that wild-type EBOV, MARV-Musoke, MARV-Ci67, and RAVV all cause lethal infection in SCID mice, although the course of illness was relatively long and infected mice did not die until 3–8 weeks after inoculation (Bray et al. 2001a; Warfield et al. 2007). These results again demonstrate the important role of innate immune responses, which can slow the progression of disease, but cannot prevent death. Recombination activating gene (Rag)-2 knockout mice resembled SCID mice in their susceptibility to infection with wild-type EBOV; however, nude mice, which lack T cells, developed disease only after intraperitoneal or subcutaneous inoculation with MA-EBOV, and not after inoculation with WT-EBOV. MA-EBOV caused lethal infection in beige mice, IFN-γ knockout mice, and TNF-α knockout mice, whereas WT-EBOV caused no apparent disease in those mice (Bray 2004).

Many other immunodeficient rodent models have been used for basic research to obtain a better understanding of the host immune response against filovirus infection in vivo. Gupta et al. (2001) demonstrated that the transfer of immune serum from EBOV-infected mice to SCID mice resulted in 100% survival after otherwise lethal challenge with EBOV, indicating that antibodies alone can protect mice from lethal infection. Moreover, mice transfused with CD8^{+} T cells from MA-EBOV-infected mice or mice vaccinated with recombinant Venezuelan equine encephalitis virus (VEEV) expressing NP survived after otherwise lethal EBOV challenge, indicating an important role for CD8^{+} T cells in protection against EBOV infection (Wilson et al. 2001; Bradfute et al. 2008). Further analyses using mice deficient in Fas, IFN-γ, or perforin revealed that the CD8^{+} T cell-mediated protection against EBOV infection was perforin-dependent (Gupta et al. 2005). Moreover, Panchal et al. (2009) established mice expressing differential levels of the leukocyte tyrosine phosphatase CD45, and the authors reported that mice expressing 11–77% CD45 levels were protected from lethal MA-EBOV challenge dependent on CD8^{+} T cells and IFN-γ, whereas mice expressing 0 or 100% CD45 were not. These data suggest that susceptibility to EBOV is influenced by a subtle balance of immune responses.

By using several kinds of immunodeficient mice, Gupta et al. (2004) demonstrated that $CD8^+$ T cells played an important role in protection against acute EBOV disease, while both $CD4^+$ T cells and antibodies were required for long-term protection. The results indicated that mice lacking B cells and depleted of $CD4^+$ T cells showed signs of illness about 50–70 days after infection, indicating the possibility of a persistent infection of EBOV under certain conditions of immunodeficiency. Moreover, Strong et al. (2008) used the mouse model with WT-EBOV to reveal that reactivation of EBOV from a state of persistent infection could be achieved by stimulating the Ras/MAPK pathway by phorbol-12-myristate treatment. Consistent with these findings, there is a report describing the development of meningoencephalitis, along with the detection of EBOV in cerebrospinal fluid, in a patient nine months after recovery from EVD (Jacobs et al. 2016), indicating that severe relapses of this disease can occur in humans.

Induction of uncontrolled pro-inflammatory responses by filovirus infection in MPS is thought to be one of the most significant drivers of the cytokine storm, coagulation abnormalities, increased vascular permeability, and multi-organ failure. Comparative analysis of cytokine/chemokine responses between mice infected with MA-EBOV intraperitoneally and mice infected with MA-EBOV subcutaneously (lethal and nonlethal infection, respectively) revealed that nonlethal subcutaneous inoculation was associated with an attenuated inflammatory response and early production of antiviral cytokines, particularly, IFN-α (Mahanty et al. 2003a). In a similar line of study, hypersecretion of interleukin-1 receptor antagonist (IL-1Ra) was associated with fatal human EBOV infection. Hence, to elucidate whether IL-1Ra had a pathogenic or protective role in fatal EBOV infection, Hill-Batorski et al. (2015) made use of IL-1Ra knockout mice. Their work suggests that IL-1Ra may have a protective effect by attenuating the intense pro-inflammatory response typically induced upon EBOV infection (Hill-Batorski et al. 2015). Additionally, several studies have suggested that lymphocyte apoptosis is one of the major factors underlying the immunosuppression observed in infected humans and NHPs. Several groups of transgenic mice defective in different apoptosis pathways were used to examine the mechanism underlying EBOV-induced lymphocyte death, and EBOV induces multiple proapoptotic stimuli (Bradfute et al. 2010). Moreover, it was demonstrated that there was no difference in survival after EBOV infection between wild-type mice and transgenic mice defective in apoptotic pathways, indicating that inhibiting lymphocyte apoptosis alone is not sufficient to improve survival in EBOV infection.

5.4 Recombinant Inbred Mouse Models

Recently, a collaborative cross (CC) resource recombinant inbred (RI) intercrossed (CC-RIX) mouse model with MA-EBOV was developed, which showed severe coagulopathy that was not evident in other conventional laboratory mouse strains, such as BALB/c, C57BL/6, and CD-1 (Rasmussen et al. 2014). CC-RI mouse lines were established by three generations of funnel breeding to incorporate eight mouse founder lines, including five inbred strains (C57BL/6 J, A/J, 129S1/SvImJ, NOD/ShiLtJ, and NZO/H1LuJ) and three wild-type-derived strains (CAST/EiJ, PWK/PhJ, and WSB/EiJ), followed by at least 20 generations of inbred mating. CC-RI strains derived from collaborative cross of the eight founders captures 90% of the most common genetic diversity across members of the three major *Mus musculus* subspecies, and different F_1 progeny of CC-RIX strains can be created by crossing CC-RI strains. Following MA-EBOV infection of the 8 CC-RI founders and 47 CC-RIX lines, several distinct pathological phenotypes became evident, classified as resistant, partially resistant, lethal with hepatitis, and lethal with hemorrhagic fever. In CC-RIX mouse lines classified as "lethal with hemorrhagic fever", mice infected with MA-EBOV had severe coagulopathy with prolongation of clotting times and late hypofibrinogemia. Transcriptomic profiling of these mice revealed a decrease in the expression levels of the endothelial tyrosine kinases *Tie1* and *Tek* on day 5 post-infection compared with mock-infected animals. In contrast, "resistant" CC-RIX mice exhibited constitutively upregulated *Tie1* and *Tek* expression. Tie1 and Tek positively regulate the activation of coagulation factors, and they therefore likely represent important host factors related to the hemorrhagic fever pathogenesis of EBOV infection. Combining the phenotype characterization and genome analysis of CC-RIX lines with filovirus infection will contribute to a better understanding of host factors that determine susceptibility, resistance, and activation of pathogenic processes in mice.

5.5 Humanized Mouse Model

In an effort to overcome the need to adapt filoviruses to mice, a humanized mouse model has recently been developed (Bird et al. 2016). Humanized BLT (Hu-BLT) mice, which are generated by engraftment of functional human macrophages, dendritic cells, T cells, B cells, and natural killer cells in NOD. Cg-Prkdc^{scid}Il2rg^{tmlWjl}/Sz (NSG) mice, are highly susceptible to intraperitoneal WT-EBOV infection, producing a 100% lethal outcome. Hu-BLT mice infected with WT-EBOV with dose range from 10^3 to 10^5 $TCID_{50}$ per animal died 6–15 days postinfection. Virus genome titers in target organs and blood were often greater than 10^7 $TCID_{50}$ genome equivalents. Cellular tropism of infection and histopathological changes in target organs were similar to those seen in other rodent models, NHPs, and humans. Immunological assays revealed that infected Hu-BLT mice induced extensive production of human pro-inflammatory cytokines and chemokines from human cells, with a profile similar to the uncontrolled pro-inflammatory responses seen in other animal models and humans. Interestingly, since susceptibility and/or fatality of EBOV-infected Hu-BLT mice was dependent on the donor of human hematopoietic stem cells, this mouse model may prove useful in elucidating human resistance factors against filovirus infection. It will also be worth examining the susceptibility of these mice to infection with other filoviruses, including RESTV and MARV.

6 Guinea Pig Models of Filovirus Infection

6.1 *Wild-Type Filovirus Infection in Guinea Pigs*

As previously described, guinea pigs were the first rodents used as a model for the isolation of filoviruses. Four to 10 days after inoculation with WT-MARV, guinea pigs lost weight and showed a mild febrile illness lasting 4–6 days (Smith et al. 1967; Simpson et al. 1968). Clinical manifestations, including anorexia, loss of excitability, diarrhea, and tufts of hair, were also observed (Ryabchikova et al. 2004). Similarly, infection with WT-EBOV produced only a febrile illness in adult outbred Hartley guinea pigs, whereas 20% of strain 13 inbred guinea pigs died, suggesting a higher susceptibility for the inbred animals. Although viral infection was found in the macrophages in the liver of guinea pigs, and the titer of EBOV in the liver was 1.1–1.3 $\times$ 10^5 PFU/ml on days 7–8 postinfection (Bray 2004 in *Ebola and Marburg virus*), spread of the virus from macrophages to hepatocytes was not observed (Mateo et al. 2011). The prominent histopathological feature in WT-EBOV-infected guinea pigs was the granuloma-like inflammatory foci, which were composed of monocytes/macrophages, in the liver (Ryabchikova et al. 1996). Widespread depletion of the lymphoid tissues and necrotic liver cells were also seen in guinea pigs infected with wild-type filoviruses.

6.2 *Guinea Pig-Adapted Ebolaviruses*

While wild-type filoviruses cause only transient febrile illness or partial lethality (in strain 13 guinea pigs), they acquired the ability to cause uniformly lethal infection in strain 13 inbred (Connolly et al. 1999; Bray et al. 2001b) and outbred Hartley guinea pigs as a consequence of serial passage in guinea pigs (Subbotina et al. 2010; Cross et al. 2015a; Cheresiz et al. 2016). Several guinea pig-adapted lethal viruses based on EBOV-Mayinga were generated (Fig. 1): GPA-EBOV$_{8mc}$ (Volchkov et al. 2000), GPA-EBOV$_{P7}$ (Subbotina et al. 2010), GPA-EBOV$_{USAMRIID}$ (Connolly et al. 1999), and GPA-EBOV$_{UTMB}$ (Cross et al. 2015a). Guinea pigs infected with these viruses became febrile 4–7 days postinfection and died on days 8–12 (Connolly et al. 1999; Bray et al. 2001b; Subbotina et al. 2010; Cross et al. 2015a). Notably although GPA-EBOV is highly adapted to guinea pigs, it did result in the lethal infection of a human following a laboratory accident, indicating that this virus retains virulence in humans (Table 1). Similarly, hamadryas baboons (*Papio hamadryas*) infected with

GPA-EBOV succumbed to an illness that displayed the clinical hallmarks of EVD, including coagulation abnormalities and increased serum concentrations of TNF-α and IFN-α at late stage of infection (Ignatiev et al. 2000).

Studies by Connolly et al. (1999) and Cross et al. (2015a) with GPA-EBOV$_{USAMRIID}$ and GPA-EBOV$_{UTMB}$ have provided thorough characterizations of GPA-EBOV pathogenesis in inbred strain 13 and outbred Hartley guinea pigs, respectively. Accordingly, we have based the majority of the following discussion on these two studies. In general, both inbred and outbred guinea pigs displayed similar clinical disease signs and pathological changes following infection with GPA-EBOV. Visible signs of hemorrhage were not usually observed, although some animals showed locally extensive hemorrhages in subcutaneous tissues of the distal limb and serosal hemorrhages in the stomach and other internal organs (Cross et al. 2015a). Viremia was seen within two days after inoculation and reached a peak on day 7, with values greater than 10^4 pfu/ml (Connolly et al. 1999; Jahrling et al. 1999; Subbotina et al. 2010; Cross et al. 2015a). High virus titers—approximately 10^5 pfu/g—were recorded in various tissues, including the spleen, liver, adrenal glands, lungs, pancreas, and kidneys.

In the liver, a number of infected Kupffer cells were detected beginning 1–2 days post-infection, followed by the detection of viral antigen in hepatocytes on day 3. Apoptotic hepatocytes and eosinophilic, intracytoplasmic viral inclusion bodies were detected from day 4 postinfection until the terminal stage of infection. Prominent histopathological changes in the liver during the middle to late stages of infection included progressive hepatocellular vacuolation, degeneration/necrosis, and sinusoidal leukocytosis. In the spleen, viral antigen was detected in mononuclear cells within the red pulp beginning one day post-infection. On day 3 postinfection, the number of antigen-positive mononuclear cells peaked within the red and white pulp. Beginning on day 4 postinfection and continuing to the terminal stage of disease, all infected animals displayed progressive lymphocyte depletion within the white pulp with a concomitant increase in macrophages containing phagocytosed apoptotic cells (known as tingible body macrophages). Within the splenic germinal centers, increasing numbers of apoptotic cells, which were mostly lymphocytes and cells of the MPS, were detected by TUNEL staining and peaked on day 5 postinfection. Extensive apoptosis in the spleen most likely contributed to lymphoid depletion and necrosis of the splenic follicles. Ultrastructurally, the red pulp and marginal zones of the spleen contained large fibrin deposits. Fibrin deposition was also prevalent in the sinusoids and in the subendothelial spaces of the liver where it was intermixed with virions and cellular debris (Connolly et al. 1999). In other lymphoid tissues, lymphocyte degeneration, and necrosis were observed in the mandibular lymph node and hemorrhagic lesions were observed in the subcapsular and medullary sinuses in inguinal lymph nodes. In addition, cortical medullary hemorrhage with cortical necrosis in the adrenal glands and interstitial pneumonia in the lungs were also observed.

Guinea pigs infected with GPA-EBOV showed developing leukocytosis due to an increasing neurotrophilia. As seen in other animal models, lymphopenia was remarkable in guinea pig models: lymphocyte counts declined from 7.5×10^3 cells/mm^3 on day 1 to 2.3×10^3 cells/mm^3 on day 9 postinfection. There was progressive and striking thrombocytopenia as pre-infection mean platelet counts declined from 500,000/μl to 50,000/μl on day 7 (Connolly et al. 1999). Elevations of blood urea nitrogen (BUN) and some serum enzymes, such as AST, ALT, and lactic acid dehydrogenase were observed in the late phase of disease. At late stages of infection, animals also showed signs of severe coagulation abnormalities, represented by prolongation of clotting factor times (PT and aPTT) and decreased protein C activity, as well as elevation of inflammatory mediators, such as TNF-α, IL-6, nitric oxide, and high-mobility group B1 (HMGB-1) in the serum (Cross et al. 2015a). Increased concentrations of von Willebrand factor were also detected at the late stage of disease. Moreover, concentrations of plasminogen activator inhibitor 1 was elevated and transiently spiked on day 4 postinfection. Notably, in the early stages of infection, tissue factor levels were also transiently increased, a phenomenon common to EBOV-infected NHPs and thought to be a key factor contributing to coagulation abnormalities (Geisbert et al. 2003c).

Interestingly, intraperitoneal infection of adult inbred strain 2 and 13 guinea pigs with MA-EBOV produced a lethal infection that was almost identical to that observed following infection with GPA-EBOV (Bray et al. 2001b). Infected guinea pigs began to lose weight 4–5 days postinfection, and they died on days 8–9, with viremic titers higher than those observed in guinea pigs infected with GPA-EBOV. Similar to guinea pigs infected with GPA-EBOV, but unlike mice infected with MA-EBOV, guinea pigs infected with MA-EBOV had prolonged clotting times throughout the course of disease (Bray et al. 2001b). It is also worth noting that, in contrast to the mouse model with MA-EBOV, subcutaneous inoculation of guinea pigs with MA-EBOV produced lethal infection, albeit with slower disease progression than guinea pigs infected intraperitoneally.

Most recently, a guinea pig model of lethal SUDV infection was developed based on SUDV-Boniface (Wong et al. 2015). Twenty-five serial passages in 6–8 week-old Hartley guinea pigs were required for the generation of GPA-SUDV, which caused 100% lethality in infected animals. Infected animals started losing weight on day 5 post-infection and died between days 9 and 14. The LD_{50} value was determined to be 5.3×10^{-2} $TCID_{50}$/animal. Guinea pigs infected intraperitoneally with a dose of 1000 LD_{50} had lymphocytopenia and thrombocytopenia as well as prolonged PT and aPTT, similar to what is seen in guinea pigs infected with GPA-EBOV. In addition, pathological changes in multiple organs during infection with GPA-SUDV were suggested based on the increased serum concentrations of albumin, alkaline phosphatase (ALP), ALT, blood urea nitrogen, creatinine, and globulin at the late and terminal stages of infection (10–11 days post-infection). The peak virus RNA titer in the liver, spleen, and blood reached approximately 10^6 genome equivalents/g or ml on day 5 when the animal started losing weight. Pathological findings were similar to the those found in guinea pigs infected with GPA-EBOV, and were described as multifocal necrotizing hepatitis and neutrophil infiltration in the liver and extensive neutrophil infiltration in the red pulp and multifocal cellular necrosis in the spleen.

Notably, Dowall et al. recently reported a catheterized guinea pig model of EBOV infection and demonstrated its usefulness for sequential blood sampling from the same group of animals (Dowall et al. 2013). Accordingly, this model will be valuable in assessing the hematological and biochemical changes that occur in filovirus-infected animals, because of its advantage of being able to track same group of animals being studied.

6.3 Guinea Pig-Adapted Marburgviruses

To date, GPA-RAVV and several GPA-MARV viruses (based on isolates Ci67 and Musoke and variant Angola) have been established by sequential passages of the viruses in either outbred Hartley or inbred strain 13 guinea pigs (Fig. 2). Following the original outbreak of MVD in 1967, attempts to adapt MARV in guinea pigs led to a 1–3 days decrease of the incubation period and development of a more severe disease (Simpson et al. 1968). It was reported that, by the 8th passage, all infected guinea pigs died on days 7–9, and their temperatures in the febrile stages often reached 41.1 °C (Smith et al. 1967; Simpson et al. 1968). Clinical signs in the animals also included bloated face and loss of appetite and weight. Since then, several guinea pig-adapted marburgviruses have been developed and used frequently as challenge viruses for examining the efficacy of potential vaccines (Hevey et al. 1997, 1998, 2001). Historically, detailed pathogenesis studies of the GPA-MARV/RAVV guinea pig model were not conducted, owing mainly to the lack of available reagents and tools, such as antibodies, quantitative reverse-transcription polymerase chain reaction (qRT-PCR) assays, and enzyme-linked immunosorbent assays (ELISA). However, recent work by Cross et al. (2015b) has characterized and compared the pathogenesis of GPA-RAVV$_{UTMB}$ and GPA-MARV-Angola infections in outbred guinea pigs, providing valuable information and a better understanding of the pathogenesis and disease progression of MARV infection in the guinea pig model.

A striking feature of GPA-MARV-infected guinea pigs is splenomegaly, with the spleen becoming enlarged up to three times its normal size. Lymphoid depletion and the appearance of lymphoblasts were also observed. Macroscopic abnormalities were also significant in the liver, which became soft and mottled light yellow to reddish brown in color (Smith et al. 1967; Simpson et al. 1968). Microscopically, disseminated single-cell and group necrosis of the liver with slight inflammation, early proliferation of Kupffer cells, and a marked fatty degeneration were observed (Korb and Slenczka 1971 in *Marburg Virus Disease*). Fibrinoid thrombi were deposited inside the sinusoids and Kupffer cells were enlarged. In the most recent study with GPA-RAVV and GPA-MARV-Angola, viremia was detected 3 days post-infection for both viruses, with similar titers. Peak viremia titers of GPA-MARV-Angola reached approximately 7.9 $\log_{10}$ PFU/ml by day 7 postinfection, which was about a log higher than the peak viremia titer of GPA-RAVV infection detected at terminal time points. Beginning on day 3 postinfection, virus replication was detected in various target organs, including the spleen, liver, pancreas, adrenal gland, kidney, and plasma from both GPA-MARV-Angola and GPA-RAVV-infected animals, with similar growth kinetics between the two viruses. Histopathology and immunohistochemistry demonstrated that viral antigen-positive cells were most apparent by day 3 postinfection, with the severity of the lesions and affected areas increasing over the duration of disease. In the liver, Kupffer cells and hepatocytes were early and major targets for infection, similar to GPA-EBOV infections. Progressive hepatocellular vacuolation, degeneration/necrosis with mineralization, and sinusoidal leukocytosis were observed from day 7 postinfection until the terminal stage of disease. In the spleen, GPA-MARV-Angola-infected animals had more extensive and scattered antigen-positive mononuclear cells than observed in GPA-RAVV-infected animals. From day 3 postinfection until the terminal stage of disease, progressive lymphocyte depletion with tingible body macrophages, hemorrhage, and fibrin deposition within the white pulp were observed in animals infected with both viruses. The numbers of apoptotic lymphocytes and cells of the MPS increased over time within the splenic germinal centers in GPA-RAVV and GPA-MARV-Angola-infected animals, as detected by TUNEL staining. Fibrin aggregates within vessels, along with the endothelium, and clusters that disperse into the adjacent red and white pulp were present only in guinea pigs infected with GPA-MARV-Angola. Lymphocyte depletion was observed in mandibular, axillary, and inguinal lymph nodes. In the late stage of infection, interstitial pneumonia with antigen-positive alveolar macrophages in the lung and hemorrhage at the corticomedullary junction with viral positive cells were also observed for both virus infections. Guinea pigs infected with GPA-RAVV or GPA-MARV-Angola developed leukocytosis due to an increasing neurotrophilia, lymphocytopenia, and thrombocytopenia, as seen in other animal models, including the EBOV guinea pig model. Elevations of some serum enzymes, such as AST, ALT, and ALP were observed in the late phase of disease for both the GPA-RAVV and GPA-MARV-Angola-infected animals.

The original studies in 1968 had already indicated evidence of severe coagulopathy in guinea pigs infected with MARV (Smith et al. 1967; Simpson et al. 1968). The concentrations of a number of clotting factors, as well as the thrombocyte count, dramatically decreased, and prolongation of thromboplastin time (PTT) was observed (Egbring et al. 1971 in *Marburg Virus disease*). The more recent study by Cross et al. (2015b) also demonstrated marked prolongation of clotting times (PT and aPTT) and increased levels of plasminogen activator inhibitor 1 and von Willebrand factor in both GPA-RAVV and GPA-MARV-Angola-infected animals at the late stage of disease. Conversely, circulating protein C activity and tissue factor concentrations progressively decreased until the time of death. Notably, the kinetics of tissue factor circulation differ between guinea pigs infected with GPA-EBOV and GPA-RAVV/MARV, mirroring similar differences observed in NHP infection with WT-EBOV and WT-MARV and suggesting that this protein plays a different role in the coagulopathy associated with these two filoviruses. Overall, these observations strongly suggest that both GPA-RAVV and

GPA-MARV-Angola infections lead to severe coagulation abnormalities in infected guinea pigs. In addition, activation of the kinin–kallikrein system, which regulates inflammation and coagulation, and perturbation of the fibrinolysis pathway were also observed, as evidenced by a depression in the metabolism of bradykinin, prekallikrein, and thrombin-activated fibrinolysis inhibitor (TAFI). Interestingly, GPA-RAVV-infected animals exhibited markedly increased TAFI concentrations 3 days postinfection, whereas GPA-MARV-Angola-infected animals did not exhibit increased TAFI concentrations by the late stage of infection. Furthermore, serum concentrations of inflammatory mediators, including TNF-α, IL-6, nitric oxide, and HMGB-1, increased during the late stages of infection. Moreover, in GPA-RAVV-infected animals, an earlier and marked increased in expression of HMGB-1 in Kupffer cells and hepatocytes was detected compared with GPA-MARV-Angola-infected animals and mock-infected animals. Intriguingly, HMGB-1 concentrations are increased in patients with severe cases of Crimean-Congo hemorrhagic fever and hemorrhagic fever with renal syndrome (Resman Rus et al. 2016). Thus, the pathobiological significance of these biological markers, including TAFI and HMGB-1, in filovirus disease in guinea pigs warrants further investigation.

7 Syrian Golden Hamster Models of Filovirus Infection

7.1 The Hamster Model with Mouse-Adapted Ebola Virus

As is the case for mice and outbred guinea pigs, adult hamsters are resistant to wild-type filovirus infection and do not develop illness upon virus inoculation (Simpson et al. 1969; Zlotnik and Simpson 1969; Ebihara et al. 2013). Accordingly, a Syrian golden hamster model of EBOV was established by infection of 6-week-old hamsters with MA-EBOV (Ebihara et al. 2013). Hamsters intraperitoneally inoculated with MA-EBOV (Bray et al. 1999) developed signs of disease, such as ruffled fur and decreased activity, beginning on day 3 postinfection, and all animals died by days 4 and 5 (Ebihara et al. 2013). Viremia was detected on day 2 and reached approximately 10^8 ffu/ml by day 4. Virus replication was also systemic, with high virus titers detected in the spleen, liver, kidneys, heart, lungs, and brain. As with the mouse model, subcutaneous inoculation with MA-EBOV did not cause visible illness in hamsters.

In infected hamsters, rapid and increased distribution of viral antigen was immunohistochemically observed in the spleen and liver, the main target organs for infection (Ebihara et al. 2013). In the spleen, virus mainly replicated in macrophages and marginal reticular-like cells in the red pulp and marginal zone. In the liver, Kupffer cells were the first target cells for both WT-EBOV and MA-EBOV, which then spread to hepatocytes starting at day 2 postinfection. Conversely, only a few antigen-positive hepatocytes were detected in WT-EBOV-infected hamsters, even after 4 days postinfection. Virus replication was also observed in MPS cells in the mesenteric lymph node, accompanied with inflammation represented by histiocytosis and neutrophilia. Extensive lymphocytolysis induced by apoptosis was also observed in the mesenteric lymph node, as well as the spleen. In the liver, hepatitis associated with hemorrhage and fibrin deposition was shown in ill hamsters.

The most important feature of the hamster model of EBOV infection is severe coagulopathy, which is a hallmark of filovirus infection but typically absent in other rodent models. MA-EBOV-infected hamsters had significantly prolonged PT, aPTT, and TT during the late stage of infection (Ebihara et al. 2013). Concentrations of plasma fibrinogen increased and peaked on day 4 postinfection, and suddenly dropped below normal levels prior to death. Other factors, including thrombocytopenia and decreased protein C concentrations, were also observed in the infected hamsters. Moreover, in contrast to WT-EBOV, MA-EBOV infection induced the expression of a number of cytokine/chemokine genes, including IL-1b, IL-6, TNF-a, IL-12p35, IP-10, and IL-10 in the spleen and liver. In the blood of terminal MA-EBOV-infected hamsters, intense elevation of cytokine/chemokine gene expression was observed, indicating an uncontrolled immune response in moribund animals. In contrast, the type I IFN response—represented by STAT1, PKR, and Mx2 gene expression—was severely impaired at the early stage of infection in target organs of MA-EBOV-infected hamsters, likely contributing to the efficient replication of this virus. Conversely, intense upregulation of these type I IFN related genes was detected in hamsters infected with WT-EBOV. Together, these data suggest that acquisition of the ability to evade the host type I IFN response in may be one of the mechanisms for adaptation in hamsters.

7.2 The Hamster Model with Marburg Virus

Although a recent model has yet to be established for MARV infection of hamsters, historical work offers some insight into the pathogenesis of this virus in hamsters. Zlotnik (1971 in *Marburg Virus Disease*) demonstrated that only suckling hamsters were susceptible to WT-MARV infection by either intracerebral or intraperitoneal inoculation, although only 40–80% of infected animals developed disease. Inoculation of hamsters with MARV that had been passaged nine times through suckling hamsters caused lethal infection in 90% of 5–6 week-old hamsters. Pathological findings in organs of infected hamsters were basically similar to those demonstrated in other animal models, but encephalitis, which was observed in hamsters after intracerebral inoculation, was a characteristic feature of this model (Simpson et al. 1969; Zlotnik and Simpson 1969).

8 Comparison Between Rodent and NHP Models

As is the case with humans, NHPs (e.g., cynomolgus macaques, rhesus macaques, grivets (African green monkeys), hamadryas baboons) are susceptible to lethal illness following inoculation with wild-type filoviruses isolated from human patients, without any process of adaptation. Although the viral strain, dose, route, and species of NHP appears to influence the severity of disease, the clinical signs, and pathological/hematological features observed in NHPs experimentally infected with filoviruses are quite similar to those of patients with filoviral hemorrhagic fever. Therefore, the NHP model is recognized as the "gold standard" for studying filovirus pathogenesis and for evaluating the efficacy of medical countermeasures (Table 1).

Despite the fact that serial passaging of wild-type filoviruses in rodents has resulted in the successful establishment of lethal rodent-adapted strains, some aspects of the pathogenic findings in rodent models are still different from those of NHP models. Route-dependent inoculation susceptibility is not observed in guinea pigs and NHPs, but it is seen in mice and hamsters (Table 1). Whereas guinea pigs and NHPs develop severe or lethal illness following inoculation of adapted (GPs) or unadapted (NHPs) by any route, including intraperitoneal, subcutaneous, and aerosol routes, mice and hamsters do not develop symptomatic disease after subcutaneous inoculation of MA-EBOV. Other than SCID mice, which developed a prolonged illness, the progression of disease is nearly identical between rodents and NHPs: infected animals become acutely ill around days 3–7 postinfection, show weight loss, anorexia, and decreased activity, and die around day 7 (Bray et al. 1999, 2001b; Connolly et al. 1999; Subbotina et al. 2010; Hensley et al. 2011; Ebihara et al. 2013). Pyrexia, an important clinical sign of disease in NHPs, is observed in guinea pigs but not in mice. While maculopapular skin rashes appear on the forehead, fore and hind limbs, and chest in macaques on days 4–6 postinfection, none of the rodent models exhibit such a rash. Obvious and extensive hemorrhagic manifestations are usually seen consistently only in NHP models, although sporadic instances of hemorrhage are observed in infected mice, guinea pigs, and hamsters (Table 2) (Wahl-Jensen et al. 2012). Viremia is detectable within 2–3 days after infection in all animal models. The maximum virus titer in the blood of terminal mice and hamsters is approximately 10^8 pfu/ml, similar to that in macaques (Jahrling et al. 1999; Jones et al. 2005; Geisbert et al. 2008; Hensley et al. 2011), whereas guinea pig models tend to display lower viremic and organ virus titers (Table 2). Target organs and cells are the same among all animal models, with the main target organs being the spleen and liver, although viral replication is observed systemically (e.g., lymph node, kidney, adrenal, lung, brain) (Jahrling et al. 1999; Geisbert et al. 2008). The cells of the MPS, such as monocytes, macrophages, and dendritic cells in the lymph nodes and spleen, are the primary and preferred replication sites of both the wild-type virus in macaques and the adapted viruses in rodents. Viral antigen is also detected in Kupffer cells and cells lining the sinusoids in the liver during the early stages of infection, followed by detection mainly in hepatocytes during later stages. Pathological changes, including inflammatory cell infiltration and parenchymal cell necrosis, in lymphoid organs and the liver are similar among all animal models. Fibrin deposition in tissues is observed in guinea pigs and hamsters, but not as marked as that shown in NHPs, and it is infrequently seen in mice (Table 2).

Table 2 Comparison of clinical and pathological features in different animal models of Ebola virus infection[a–j]

	NHP[a,b,j]	Guinea pig[c,d,e,j]	Hamster[f,j]	Mouse[g,h,i,j]
Average time to death	6–8 days	5–12 days	4–5 days	4–7 days
First detection of viremia (mean titre[k])	Day 3 (2.5–3.5)	Day 2 (1.6–3.0)	Day 2 (3.5)	Day 1 (1.5)
Mean peak viremia titre[k] (day pi)	6.0–8.0 (Day 6)	5.2 (Day 3–7)	>7.0 (Day 5)	>7.5 (Day 3)
Spleen pathology				
Lymphoid depletion, necrosis, and apoptosis	Diffuse, severe	Diffuse, severe	Diffuse, moderate to severe	Multifocal, mild
Inflammatory lesions	Neutrophlic	Neutrophlic	Neutrophlic	Neutrophlic
Mean peak virus titre[k] (day pi)	>8.0 (day 5)[a]	6.4 (day 9)[c] 3.7 (day 3)[d]	>7.0 (day 5)	>8.0 (day 3)
Liver pathology				
Hepatocellular degeneration and necrosis	Diffuse, random	Diffuse, random	Diffuse, midzonal	Diffuse, random
Inflammatory lesions	Neutrophilic	Neutrophilic and histiocytic	Neutrophilic	Neutrophilic
Mean peak virus titre[k] (day pi)	>8.0 (day 5)[a]	4.3–5.5 (day 8–9)	>6.5 (day 5)	>8.0 (day 5)
Coagulopathy				
Prolonged PT	++	++/+++[d,f]	++	–
Prolonged aPTT	+++	++[d,f]	+++	–
Prolonged TT	++	–[d,f]	++	ND
Late hypofibinogenemia	+++	–/+[d,f]	+++	–/+
Decreased protein C activity %	++	+	++	+
D-dimer detection	yes	ND	ND	ND
Thrombocytopenia	++	++	++	++
Tissue factor detection	+	+[d]	ND	ND
Fibrin deposition in spleen and liver	+++	++	+	+
Hemorrhagic manifestations				
Rash	++	–	–	–
Other hemorrhage (internal and subcutaneous tissues)	++	++	++	+

[a]Geisbert et al. (2003a): cynomolgus macaques infected with WT-EBOV (dose: 1000 pfu/animal)
[b]Ebihara et al. (2011): rhesus macaques infected with WT-EBOV (dose: 1000 pfu/animal)
[c]Connolly et al. (1999): inbred strain 13 guinea pigs infected with GPA-EBOV$_{USAMRIID}$ (dose: $10^{3.8}$ pfu/animal)
[d]Cross et al. (2015a): outbred Hartley guinea pigs infected with GPA-EBOV$_{UTMB}$ (dose: 5000 pfu/animal)
[e]Cheresiz et al. (2016): outbred Hartley guinea pigs infected with several GPA-EBOV strains including 8 mc (dose: not available)
[f]Ebihara et al. (2013): Syrian golden hamsters infected with MA-EBOV (dose: 1000 pfu/animal)
[g]Bray et al. (1999): inbred BALB/c, C57BL/6, and outbred ICR (CD-1) mice infected with MA-EBOV (dose: 1–300 pfu/animal)
[h]Bray et al. (2001b): BALB/c mice infected with MA-EBOV (dose: 100 pfu/animal)
[i]Gibb et al. (2001): BALB/c mice infected with MA-EBOV (dose: 100 pfu/animal)
[j]Wahl-Jensen et al. (2012): Comparative review of clinical and pathological changes in EBOV animal models
[k]Log_{10} plaque-forming units (focus-forming units)/ml

During filovirus infection, NHP models exhibit overt coagulopathy indicated by prolongation of PT, aPTT and TT, as well as thrombocytopenia, increased fibrinogen and D-dimer concentrations, and decreased protein C activity (Geisbert et al. 2007; Hensley et al. 2011). In contrast, mice infected with MA-EBOV do not consistently display coagulation abnormalities. Compared to mice, guinea pigs had coagulopathy but no hypofibinogenemia was observed, which is in contrast to what is observed in NHPs and hamsters (Wahl-Jensen et al. 2012). Notably, the course of coagulopathy in hamsters resembles that seen in macaques, indicating the adequacy of this model for studying filovirus disease (Table 2). Massive expression of a number of inflammatory cytokines and chemokines—another hallmark of filovirus disease in humans and NHPs—has been demonstrated in infected mice (Mahanty et al. 2003a), guinea pigs (Cross et al. 2015a, b) and hamsters by monitoring innate immunological gene expressions using qRT-PCR (Ebihara et al. 2013).

9 Genomic Analysis

To identify the molecular determinants and mechanisms of filovirus adaptation and the acquisition of virulence in rodent models, reverse genetics has been used to generate and characterize a variety of recombinant viruses containing various subsets of the mutations found in rodent-adapted filoviruses. In the case of EBOV, comparative genomic analysis between WT-EBOV-Mayinga (GenBank accession number AF086833), MA-EBOV (GenBank accession number AF499101) (Bray et al. 1999), GPA-EBOV$_{8mc}$ (Genbank accession number AF272001) (Volchkov et al. 2000), GPA-EBOV$_{P7}$ (Genbank accession number EU224440) (Subbotina et al. 2010), and GPA-EBOV$_{UTMB}$ (Cross et al. 2015a GenBank accession # is not available) revealed the presence of 9 to 19 nucleotide substitutions among the rodent-adapted EBOV variants (Volchkov et al. 2000; Ebihara et al. 2006; Subbotina et al. 2010; Cross et al. 2015a). Notably, almost all of the amino acid changes in these viruses are found in NP and VP24, suggesting important roles for these proteins in the acquisition of virulence in mice and guinea pigs, although the exact sites of the mutations in GPA-EBOV differ from those in MA-EBOV (Fig. 1) (Banadyga et al. 2016). For MA-EBOV, studies using reverse genetics demonstrated that the amino acid mutations in NP (S72G) and VP24 (T50I) were primarily responsible for the acquisition of virulence in mice, although all of the other mutations contributed to some extent to the virulence (Ebihara et al. 2006). Further analysis demonstrated that this recombinant virus grew to higher titers compared to WT-EBOV in IFN-stimulated mouse macrophages, suggesting that NP and VP24 play an important role in counteracting the IFN-mediated antiviral state. VP24 is known to function as an IFN antagonist by binding karyopherin-α1, 5, and 6 and blocking nuclear accumulation of STAT1. However, Reid et al. (2007) showed that there was no difference in the ability of wild-type or mouse-adapted VP24 to inhibit STAT1 nuclear translocation in human or mouse fibroblasts (NIH3T3 cells). Although it remains unclear how NP is involved in the mouse adaptation process, the interaction of NP with VP24 in the efficient formation of nucleocapsids in target cells (macrophages) might be critical. Indeed, for GPA-EBOV$_{8mc}$, a reverse genetics approach revealed that the three amino acid mutations in VP24 (M71I, L147P, and T187I) were the minimum requirements for a fully virulent phenotype in guinea pigs. VP24 derived from both WT-EBOV and GPA-EBOV$_{8mc}$ showed similar abilities to inhibit the IFN response in human and guinea pig cell lines, suggesting that the mechanism of EBOV adaptation to guinea pigs is not linked to evasion of the type I IFN response. Instead, evidence suggests that the guinea pig adaptation of EBOV is linked to an enhancement of viral replication in target cells. Recombinant EBOV possessing the three guinea pig-adapted mutations in VP24 was better able to replicate and form viral inclusion bodies in guinea pig primary macrophages than WT-EBOV. Interestingly, GPA-EBOV$_{8mc}$ and GPA-EBOV$_{P7}$ share two identical amino acid substitutions in VP24 (M71I and L147P), strongly implying that this protein is critical to guinea pig adaptation of EBOV.

For GPA-SUDV, full-length sequence determination revealed 16 nucleotide substitutions, of which six resulted in amino acid changes: one in VP35, one in VP40, three in GP, and one in VP24. The effects of these nucleotide/amino acid changes on SUDV virulence in guinea pigs remain to be elucidated.

Genome sequence information has so far been made available for eight rodent-adapted marburgviruses: MA-RAVV (GenBank accession number EU500826), MA-MARV-Ci67 (Genbank accession GQ433351), MA-MARV-Angola (GenBank accession number KM2611523), GPA-MARV-Musoke (GenBank accession number AY430365), GPA-MARV-Ci67 (GenBank accession number AY430365), GPA-MARV-Angola (GenBank accession number DQ447653.1), GPA-RAVV$_{USAMRIID}$ (GenBank accession number EF446131), and GPA-RAVV$_{UTMB}$ (Hevey et al. 1997; Lofts et al. 2007, 2011; Cross et al. 2015a; Banadyga et al. 2016). Notably, all rodent-adapted MARV and RAVV variants possess amino acid substitutions in VP40, which is known to play an important role in virus particle assembly and budding, as well as transcription regulation and IFN signaling antagonism (Fig. 2). In contrast to EBOV, which uses VP24 to inhibit type I IFN signaling, MARV and RAVV VP40 inhibits type I IFN signaling by targeting janus kinase 1 (JAK1), an upstream signaling molecule in the IFN response cascade (Valmas et al. 2010). Interestingly, some of the mutations in MA-RAVV and MA-MARV-Ci67 VP40 have been shown to confer the ability to inhibit IFN signaling in mouse cells and alter the susceptibility of VP40 to human but not murine tetherin (Valmas et al. 2011; Feagins and Basler 2014; Feagins et al. 2015). Together, these data suggest the host innate/antiviral response represents a barrier to marburgvirus infection in rodents that can be overcome by amino acid changes in VP40. Another significant amino acid substitution is D184 N in VP40, which is found in all mouse-adapted RAVV and MARV viruses, as well as GPA-MARV-Musoke and GPA-MARV-Ci67. This mutation has no effect on the ability of VP40 to inhibit the IFN response in either guinea pig or human cells—instead it contributes to the enhancement of VP40's role as the viral matrix protein by improving viral budding and viral RNA synthesis. Indeed, reverse genetics studies demonstrated that MARV containing only the D184 N substitution in VP40 outgrows WT-MARV in guinea pig cells (Koehler et al. 2015). The roles of other rodent-adapted mutations in NP, VP35, GP, VP30, VP24, and L remain to be elucidated.

10 Future Perspectives

Rodent models are invaluable basic research tools for better understanding filovirus pathogenesis. In particular, the combination of rodent models and filovirus reverse genetics systems has been, and will continue to be, a very powerful research strategy for elucidating the molecular basis of filovirus replication and pathogenesis in vivo. Indeed, this approach has already been applied to identify the molecular determinants responsible for adaptation and virulence acquisition in a number of rodent-adapted filoviruses. Understanding the molecular mechanisms that underlie filovirus rodent adaptation has provided valuable insight into the roles that viral proteins play in pathogenic processes in vivo, which is a crucial to understanding how filoviruses cause such a unique, complicated, and devastating hemorrhagic fever syndrome. For instance, during filovirus adaptation in rodents, viruses gain the ability to (i) replicate in target cells/tissues, (ii) evade the host type I IFN-mediated antiviral response, (iii) decrease the local inflammatory response, and (iv) induce uncontrolled systemic inflammatory responses, all of which contribute to filovirus disease pathogenesis. It is noteworthy that MA-EBOV appeared to be less virulent for NHPs than WT-EBOV, as 2 of 3 animals survived infection (Bray et al. 2001b). The identification of the specific attenuating mutations, in addition to adaptive mutations, will also provide important information for understanding the pathogenicity of filoviruses.

Reverse genetics systems for rodent-adapted filoviruses will also allow us to further elucidate the functions of viral proteins in vivo in a way that would not otherwise be possible. Before reverse genetics systems for the generation of rodent-adapted filoviruses were available, the role of viral proteins and their functional motifs in viral replication and pathogenicity could not be easily determined in infected animals, with studies having to rely on NHPs or immunodeficient mice. However, the use of recombinant rodent-adapted EBOV mutants has enabled unprecedented investigations into viral protein functions and pathogenic processes. For example, using MA-EBOV and GPA-EBOV, it was demonstrated that disruption of the interferon inhibitory domain of VP35 rendered the virus completely attenuated in mice and guinea pigs (Hartman et al. 2008; Prins et al. 2010). Additionally, recombinant GPA-EBOV mutants that do not express sGP were used to elucidate the role of this protein in pathogenesis in guinea pigs (Hoenen et al. 2015; Volchkova et al. 2015). Indeed, the availability of rodent-adapted filovirus reverse genetics will improve our ability to understand viral protein function and the molecular processes leading to filoviral hemorrhagic fever.

Numerous small animal models have been developed for use in filovirus research. The use of recently developed humanized and CC-RIX mouse models

(Bird et al. 2016; Rasmussen et al. 2014) will be able to shed more light on how genetic and immunological backgrounds contribute to resistance of filovirus infection and recovery from disease—invaluable information for the development of effective therapeutic schemes. Moreover, lethal ferret models for BDBV, SUDV, and EBOV infection using nonadapted viruses (Cross et al. 2016) may offer a convenient alternative to the NHP model (Table 1). In summary, filovirus rodent models will continue to promote basic as well as translational research, contributing to a more detailed and better molecular understanding of filovirus virulence and host responses that will be crucial to improving our ability to control infections and future outbreaks.

Acknowledgements L. B. is supported by the Division of Intramural Research, NIAID, NIH. Opinions, interpretations, conclusions, and recommendations are those of the authors and are not necessarily endorsed by the NIAID and NIH.

References

Bale S, Julien JP, Bornholdt ZA, Kimberlin CR, Halfmann P, Zandonatti MA, Kunert J, Kroon GJ, Kawaoka Y, MacRae IJ, Wilson IA, Saphire EO (2012) Marburg virus VP35 can both fully coat the backbone and cap the ends of dsRNA for interferon antagonism. PLoS Pathog 8(9): e1002916. doi:10.1371/journal.ppat.1002916

Banadyga L, Ebihara H (2015) Epidemiology and Pathogenesis of Filovirus Infections. In: Pattnaik AK, Whitt MA (eds) Biology and Pathogenesis of Rhabdo-and Filoviruses. World Scientific Publishing Co. Pte. Ltd., Singapore, pp 453–486

Banadyga L, Dolan MA, Ebihara H (2016) Rodent-Adapted Filoviruses and the Molecular Basis of Pathogenesis. J Mol Biol 428(17):3449–3466. doi:10.1016/j.jmb.2016.05.008

Basler CF, Wang X, Mühlberger E, Volchkov V, Paragas J, Klenk HD, García-Sastre A, Palese P (2000) The Ebola virus VP35 protein functions as a type I IFN antagonist. Proc Natl Acad Sci U S A 97(22):12289–12294

Bird BH, Spengler JR, Chakrabarti AK, Khristova ML, Sealy TK, Coleman-McCray JD, Martin BE, Dodd KA, Goldsmith CS, Sanders J, Zaki SR, Nichol ST, Spiropoulou CF (2016) Humanized mouse model of Ebola virus disease mimics the immune responses in human disease. J Infect Dis 213:703–711. doi:10.1093/infdis/jiv538

Bixler SL, Goff AJ (2015) The role of cytokines and chemokines in filovirus infection. Viruses 7 (10):5489–5507. doi:10.3390/v7102892

Bosio CM, Aman MJ, Grogan C, Hogan R, Ruthel G, Negley D, Mohamadzadeh M, Bavari S, Schmaljohn A (2003) Ebola and Marburg viruses replicate in monocyte derived dendritic cells without inducing the production of cytokines and full maturation. J Infect Dis 188(11):1630–1638

Bornholdt ZA, Noda T, Abelson DM, Halfmann P, Wood MR, Kawaoka Y, Saphire EO (2013) Structural rearrangement of ebola virus VP40 begets multiple functions in the virus life cycle. Cell 154(4):763–774. doi:10.1016/j.cell.2013.07.015

Bowen ETW, Lloyd G, Harris WJ, Platt GS, Baskerville A, Vella EE (1977) Viral haemorrhagic fever in southern Sudan and northern Zaire: preliminary studies on the aetiological agent. Lancet 1(8011):571–573. doi:10.1016/S0140-6736(77)92001 3

Bradfute SB, Braun DR, Shamblin JD, Geisbert JB, Paragas J, Garrison A, Hensley LE, Geisbert TW (2007) Lymphocyte death in a mouse model of Ebola virus infection. J Infect Dis 196:S296–S304. doi:10.1086/520602

Bradfute SB, Warfield KL, Bavari S (2008) Functional CD8+ T cell responses in lethal Ebola virus infection. J Immunol 180:4058–4066. doi:10.4049/jimmunol.180.6.4058

Bradfute SB, Swanson PE, Smith MA, Watanabe E, McDunn JE, Hotchkiss RS, Bavari S (2010) Mechanisms and consequences of ebolavirus-induced lymphocyte apoptosis. J Immunol 184:327–335. doi:10.4049/jimmunol.0901231

Brannan JM, Froude JW1, Prugar LI, Bakken RR, Zak SE, Daye SP, Wilhelmsen CE, Dye JM (2015) Interferon α/β Receptor-Deficient Mice as a Model for Ebola Virus Disease. J Infect Dis. 212:S282–94. doi:10.1093/infdis/jiv215

Brauburger K, Deflubé LR, Mühlberger E (2015) Filovirus Transcription and Replication. In: Pattnaik AK, Whitt MA (eds) Biology and pathogenesis of Rhabdo and Filoviruses. World Scientific Publishing Co. Pte. Ltd., Singapore, pp 515–555

Bray M, Davis K, Geisbert T, Schmaljohn C, Huggins J (1999) A mouse model for evaluation of prophylaxis and therapy of Ebola hemorrhagic fever. J Infect Dis 179:S248–S258. doi:10.1086/514292

Bray M (2001) The role of the type I interferon response in the resistance of mice to filovirus infection. J Gen Virol 82:1365–1373. doi:10.1099/0022-1317-82-6-1365

Bray M, Hatfill S, Hensley L, Huggins JW (2001) Haematological, biochemical and coagulation changes in mice, guinea-pigs and monkeys infected with a mouse-adapted variant of Ebola Zaire virus. J Comp Pathol 125:243–253. doi:10.1053/jcpa.2001.0503

Bray M, Mahanty S (2003) Ebola hemorrhagic fever and septic shock. J Infect Dis 188(11):1613–1617

Bray M (2004) Pathogenesis of filovirus infection in Mice. In: Klenk HD, Feldmann H (eds) Ebola and Marburg viruses—molecular and cellular biology. Horizon Bioscience, Wymondham, Norfolk, United Kingdom, pp 255–277

Bray M, Geisbert TW (2005) Ebola virus: the role of macrophages and dendritic cells in the pathogenesis of Ebola hemorrhagic fever. Int J Biochem Cell Biol 37(8):1560–1566

Bukreyev AA, Chandran K, Dolnik O, Dye JM, Ebihara H, Leroy EM, Mühlberger E, Netesov SV, Patterson JL, Paweska JT, Saphire EO, Smither SJ, Takada A, Towner JS, Volchkov VE, Warren TK, Kuhn JH (2014) Discussions and decisions of the 2012–2014 International Committee on Taxonomy of Viruses (ICTV) *Filoviridae* Study Group, January 2012–June 2013. Arch Virol 159(4):821–830

Cheresiz SV, Semenova EA, Chepurnov AA (2016) Adapted lethality: what we can learn from guinea pig-adapted Ebola virus infection model. Adv Virol 2016:8059607. doi:10.1155/2016/8059607

Connolly BM, Steele KE, Davis KJ, Geisbert TW, Kell WM, Jaax NK, Jahrling PB (1999) Pathogenesis of experimental Ebola virus infection in guinea pigs. J Infect Dis 179:S203–S217. doi:10.1086/514305

Cross RW, Fenton KA, Geisbert JB, Mire CE, Geisbert TW (2015a) Modeling the disease course of Zaire ebolavirus infection in the outbred guinea pig. J Infect Dis 212:S305–S315. doi:10.1093/infdis/jiv237

Cross RW, Fenton KA, Geisbert JB, Ebihara H, Mire CE, Geisbert TW (2015b) Comparison of the pathogenesis of the Angola and Ravn strains of Marburg virus in the outbred guinea pig model. J Infect Dis 212:S258–S270. doi:10.1093/infdis/jiv182

Cross RW, Mire CE, Borisevich V, Geisbert JB, Fenton KA, Geisbert TW (2016) The domestic ferret (*Mustela putorius furo*) as a lethal infection model for three different species of *Ebolavirus*. J Infect Dis 214(4):565–569. doi:10.1093/infdis/jiw209

de Wit E, Munster VJ, Metwally SA, Feldmann H (2011) Assessment of rodents as animal models for Reston ebolavirus. J Infect Dis 204:S968–S972. doi:10.1093/infdis/jir330

Dowall S, Taylor I, Yeates P, Smith L, Rule A, Easterbrook L, Bruce C, Cook N, Corbin-Lickfett K, Empig C, Schlunegger K, Graham V, Dennis M, Hewson R (2013) Catheterized guinea pigs infected with Ebola Zaire virus allows safer sequential sampling to determine the pharmacokinetics profile of a phosphatidylserine-targeting monoclonal antibody. Antiviral Res 97(2):108–111. doi:10.1016/j.antiviral.2012.11.003

Ebihara H, Takada A, Kobasa D, Jones S, Neumann G, Theriault S, Bray M, Feldmann H, Kawaoka Y (2006) Molecular determinants of Ebola virus virulence in mice. PLoS Pathog 2: e73. doi:10.1371/journal.ppat.0020073

Ebihara H, Rockx B, Marzi A, Feldmann F, Haddock E, Brining D, LaCasse RA, Gardner D, Feldmann H (2011) Host response dynamics following lethal infection of rhesus macaques with Zaire ebolavirus. J Infect Dis 204(Suppl 3):S991–S999. doi:10.1093/infdis/jir336

Ebihara H, Zivcec M, Gardner D, Falzarano D, LaCasse R, Rosenke R, Long D, Haddock E, Fischer E, Kawaoka Y, Feldmann H (2013) A Syrian golden hamster model recapitulating Ebola hemorrhagic fever. J Infect Dis 207:306–318. doi:10.1093/infdis/jis626

Egbring R, Slenczka W, Baltzer G (1971) Clinical manifestations and mechanism of the haemorrhagic diathesis in Marburg virus disease. In: Martini GA, Siegert R (eds) Marburg virus disease. Springer, New York, pp 41–49

Escudero-Pérez B, Volchkova VA, Dolnik O, Lawrence P, Volchkov VE (2014) Shed GP of Ebola virus triggers immune activation and increased vascular permeability. PLoS Pathog 10(11): e1004509. doi:10.1371/journal.ppat.1004509

Feagins AR, Basler CF (2014) The VP40 protein of Marburg virus exhibits impaired budding and increased sensitivity to human tetherin following mouse adaptation. J Virol 88:14440–14450. doi:10.1128/JVI.02069-14

Feagins AR, Basler CF (2015) Amino acid residue at position 79 of Marburg virus vp40 confers interferon antagonism in mouse cells. J Infect Dis 212:S219–S225. doi:10.1093/infdis/jiv010

Geisbert TW, Jahrling PB, Hanes MA, Zack PM (1992) Association of Ebola-related Reston virus particles and antigen with tissue lesions of monkeys imported to the United States. J Comp Pathol 106:137–152. doi:10.1016/0021-9975(92)90043-T

Geisbert TW, Pushko P, Anderson K, Smith J, Davis KJ, Jahrling PB (2002) Evaluation in nonhuman primates of vaccines against Ebola virus. Emerg Infect Dis 8(5):503–507

Geisbert TW, Hensley LE, Larsen T, Young HA, Reed DS, Geisbert JB, Scott DP, Kagan E, Jahrling PB, Davis KJ (2003a) Pathogenesis of Ebola hemorrhagic fever in cynomolgus macaques: evidence that dendritic cells are early and sustained targets of infection. Am J Pathol 163(6):2347–2370

Geisbert TW, Young HA, Jahrling PB, Davis KJ, Larsen T, Kagan E, Hensley LE (2003b) Pathogenesis of Ebola hemorrhagic fever in primate models: evidence that hemorrhage is not a direct effect of virus-induced cytolysis of endothelial cells. Am J Pathol 163(6):2371–2382

Geisbert TW, Young HA, Jahrling PB, Davis KJ, Kagan E, Hensley LE (2003c) Mechanisms underlying coagulation abnormalities in Ebola hemorrhagic fever: overexpression of tissue factor in primate monocytes/macrophages is a key event. J Infect Dis 188:1618–1629. doi:10.1086/379724

Geisbert TW, Jahrling PB (2004) Exotic emerging viral diseases: progress and challenges. Nat Med 10(12 Suppl):S110–S121

Geisbert TW, Daddario-DiCaprio KM, Geisbert JB, Young HA, Formenty P, Fritz EA, Larsen T, Hensley LE (2007) Marburg virus Angola infection of rhesus macaques: pathogenesis and treatment with recombinant nematode anticoagulant protein c2. J Infect Dis 196:S372–S381. doi:10.1086/520608

Geisbert TW, Daddario-Dicaprio KM, Geisbert JB, Reed DS, Feldmann F, Grolla A, Ströher U, Fritz EA, Hensley LE, Jones SM, Feldmann H (2008) Vesicular stomatitis virus-based vaccines protect nonhuman primates against aerosol challenge with Ebola and Marburg viruses. Vaccine. 26:6894–6900. doi:10.1016/j.vaccine.2008.09.082

Gibb TR, Bray M, Geisbert TW, Steele KE, Kell WM, Davis KJ, Jaax NK (2001) Pathogenesis of experimental Ebola Zaire virus infection in BALB/c mice. J Comp Pathol 125:233–242. doi:10.1053/jcpa.2001.0502

Gupta M, Mahanty S, Bray M, Ahmed R, Rollin PE (2001) Passive transfer of antibodies protects immunocompetent and immunodeficient mice against lethal Ebola virus infection without complete inhibition of viral replication. J Virol 75:4649–4654. doi:10.1128/JVI.75.10.4649-4654.2001

Gupta M, Mahanty S, Greer P, Towner JS, Shieh WJ, Zaki SR, Ahmed R, Rollin PE (2004) Persistent infection with Ebola virus under conditions of partial immunity. J Virol 78:958–967. doi:10.1128/JVI.78.2.958-967.2004

Gupta M, Greer P, Mahanty S, Shieh WJ, Zaki SR, Ahmed R, Rollin PE (2005) CD8-mediated protection against Ebola virus infection is perforin dependent. J Immunol 174:4198–4202. doi:10.4049/jimmunol.174.7.4198

Hartman AL, Bird BH, Towner JS, Antoniadou ZA, Zaki SR, Nichol ST (2008) Inhibition of IRF-3 activation by VP35 is critical for the high level of virulence of Ebola virus. J Virol 82:2699–2704. doi:10.1128/JVI.02344-07

Hensley LE, Geisbert TW (2005) The contribution of the endothelium to the development of coagulation disorders that characterize Ebola hemorrhagic fever in primates. Thromb Haemost 94(2):254–261

Hensley LE, Alves DA, Geisbert JB, Fritz EA, Reed C, Larsen T, Geisbert TW (2011) Pathogenesis of Marburg hemorrhagic fever in cynomolgus macaques. J Infect Dis 204:S1021–S1031. doi:10.1093/infdis/jir339

Hevey M, Negley D, Geisbert J, Jahrling P, Schmaljohn A (1997) Antigenicity and vaccine potential of Marburg virus glycoprotein expressed by baculovirus recombinants. Virology 239:206–216. doi:10.1006/viro.1997.8883

Hevey M, Negley D, Pushko P, Smith J, Schmaljohn A (1998) Marburg virus vaccines based upon alphavirus replicons protect guinea pigs and nonhuman primates. Virology 251:28–37. doi:10.1006/viro.1998.9367

Hevey M, Negley D, VanderZanden L, Tammariello RF, Geisbert J, Schmaljohn C, Smith JF, Jahrling PB, Schmaljohn AL (2001) Marburg virus vaccines: comparing classical and new approaches. Vaccine 20:586–593. doi:10.1016/S0264-410X(01)00353-X

Hill-Batorski L, Halfmann P, Marzi A, Lopes TJ, Neumann G, Feldmann H, Kawaoka Y (2015) Loss of interleukin 1 receptor antagonist enhances susceptibility to Ebola virus infection. J Infect Dis 212:S329–S335. doi:10.1093/infdis/jiv335

Hoenen T, Groseth A, Kolesnikova L, Theriault S, Ebihara H, Hartlieb B, Bamberg S, Feldmann H, Ströher U, Becker S (2006) Infection of naive target cells with virus-like particles: implications for the function of Ebola virus VP24. J Virol 80(14):7260–7264

Hoenen T, Marzi A, Scott DP, Feldmann F, Callison J, Safronetz D, Ebihara H, Feldmann H (2015) Soluble glycoprotein is not required for Ebola virus virulence in guinea pigs. J Infect Dis 212:S242–S246. doi:10.1093/infdis/jiv111

Hofmann H, Kunz C (1970) Ein mauspathogener stamm des Marburg virus (A strain of "Marburg virus" (Rhabdovirus simiae) pathogenic to mice). Arch Gesamte Virusforsch 32:244–248

Ignatiev GM, Dadaeva AA, Luchko SV, Chepurnov AA (2000) Immune and pathophysiological processes in baboons experimentally infected with Ebola virus adapted to guinea pigs. Immunol Lett 71(2):131–140

Huang Y, Xu L, Sun Y, Nabel GJ (2002) The assembly of Ebola virus nucleocapsid requires virion-associated proteins 35 and 24 and posttranslational modification of nucleoprotein. Mol Cell 10(2):307–316

Jaax NK, Davis KJ, Geisbert TJ, Vogel P, Jaax GP, Topper M, Jahrling PB (1996) Lethal experimental infection of rhesus monkeys with Ebola-Zaire (Mayinga) virus by the oral and conjunctival route of exposure. Arch Pathol Lab Med 120:140–155

Jacobs M, Rodger A, Bell DJ, Bhagani S, Cropley I, Filipe A, Gifford RJ, Hopkins S, Hughes J, Jabeen F et al (2016) Late Ebola virus relapse causing meningoencephalitis: a case report. Lancet 388(10043):498–503. doi:10.1016/S0140-6736(16)30386-5

Jahrling PB, Geisbert TW, Geisbert JB, Swearengen JR, Bray M, Jaax NK, Huggins JW, LeDuc JW, Peters CJ (1999) Evaluation of immune globulin and recombinant interferon-α2b for treatment of experimental Ebola virus infections. J Infect Dis 179:S224–S234. doi:10.1086/514310

Jangra RK, Mittler E, Chandran K (2015) Filovirus entry into susceptible cells. In: Pattnaik AK, Whitt MA (eds) Biology and pathogenesis of Rhabdo-and filoviruses. World Scientific Publishing Co. Pte. Ltd., Singapore, pp 487–514

Jones SM, Feldmann H, Ströher U, Geisbert JB, Fernando L, Grolla A, Klenk HD, Sullivan NJ, Volchkov VE, Fritz EA, Daddario KM, Hensley LE, Jahrling PB, Geisbert TW (2005) Live attenuated recombinant vaccine protects nonhuman primates against Ebola and Marburg viruses. Nat Med 11:786–790. doi:10.1038/nm1258

Korb G, Slenczka W (1971) Histologic findings in livers and spleens of guinea pigs after infection by the Marburg virus. In: Martini GA, Siegert R (eds) Marburg virus disease. Springer-Verlag, New York, pp 123–124

Koehler A, Kolesnikova L, Welzel U, Schudt G, Herwig A, Becker S (2015) A single amino acid change in the Marburg virus matrix protein VP40 provides a replicative advantage in a species-specific manner. J Virol 90(3):1444–1454. doi:10.1128/JVI.02670-15

Kuhn JH (2008) Clinical presentation of filoviral disease. In: Calisher CH (eds) Filoviruses—A compendium of 40 years of epidemiological, clinical, and laboratory studies. Springer, New York, pp 100–120

Lofts LL, Ibrahim MS, Negley DL, Hevey MC, Schmaljohn AL (2007) Genomic differences between guinea pig lethal and nonlethal Marburg virus variants. J Infect Dis 196:S305–S312. doi:10.1086/520585

Lofts LL, Wells JB, Bavari S, Warfield KL (2011) Key genomic changes necessary for an in vivo lethal mouse marburgvirus variant selection process. J Virol 85:3905–3917. doi:10.1128/JVI.02372-10

Mahanty S, Gupta M, Paragas J, Bray M, Ahmed R, Rollin PE (2003a) Protection from lethal infection is determined by innate immune responses in a mouse model of Ebola virus infection. Virology 312:415–424. doi:10.1016/S0042-6822(03)00233-2

Mahanty S, Hutchinson K, Agarwal S, McRae M, Rollin PE, Pulendran B (2003b) Cutting edge: impairment of dendritic cells and adaptive immunity by Ebola and Lassa viruses. J Immunol 6:2797–2801

Mahanty S, Bray M (2004) Pathogenesis of filoviral haemorrhagic fevers. Lancet Infect Dis 4 (8):487–498

Martines RB, Ng DL, Greer PW, Rollin PE, Zaki SR (2015) Tissue and cellular tropism, pathology and pathogenesis of Ebola and Marburg viruses. J Pathol 235(2):153–174. doi:10.1002/path.4456

Martins KA, Jahrling PB, Bavari S, Kuhn JH (2016) Ebola virus disease candidate vaccines under evaluation in clinical trials. Expert Rev Vaccines 27:1–12

Mateo M, Carbonnelle C, Reynard O, Kolesnikova L, Nemirov K, Page A, Volchkova VA, Volchkov VE (2011) VP24 is a molecular determinant of Ebola virus virulence in guinea pigs. J Infect Dis 204:S1011–S1020. doi:10.1093/infdis/jir338

Mendoza EJ, Qiu X, Kobinger GP (2016) Progression of Ebola therapeutics during the 2014–2015 outbreak. Trends Mol Med 22(2):164–73. doi: 10.1016/j.molmed.2015.12.005

Messaoudi I, Amarasinghe GK, Basler CF (2015) Filovirus pathogenesis and immune evasion: insights from Ebola virus and Marburg virus. Nat Rev Microbiol 13(11):663–676. doi:10.1038/nrmicro3524

Mohamadzadeh M, Chen L, Schmaljohn AL (2007) How Ebola and Marburg viruses battle the immune system. Nat Rev Immunol 7:556–567. doi:10.1038/nri2098

Mohan GS, Li W, Ye L, Compans RW, Yang C (2012) Antigenic subversion: a novel mechanism of host immune evasion by Ebola virus. PLoS Pathog 8(12):e1003065. doi:10.1371/journal.ppat.1003065

Negredo A, Palacios G, Vázquez-Morón S, González F, Dopazo H, Molero F, Juste J, Quetglas J, Savji N, de la Cruz Martínez M, Herrera JE, Pizarro M, Hutchison SK, Echevarría JE, Lipkin WI, Tenorio A (2011) Discovery of an ebolavirus-like filovirus in Europe. PLoS Pathog 7(10):e1002304

Noda T, Aoyama K, Sagara H, Kida H, Kawaoka Y (2005) Nucleocapsid-like structures of Ebola virus reconstructed using electron tomography. J Vet Med Sci 67(3):325–328

Panchal RG, Bradfute SB, Peyser BD, Warfield KL, Ruthel G, Lane D, Kenny TA, Anderson AO, Raschke WC, Bavari S (2009) Reduced levels of protein tyrosine phosphatase CD45 protect mice from the lethal effects of Ebola virus infection. Cell Host Microbe 6:162–173. doi:10.1016/j.chom.2009.07.003

Prins KC, Delpeut S, Leung DW, Reynard O, Volchkova VA, Reid SP, Ramanan P, Cárdenas WB, Amarasinghe GK, Volchkov VE, Basler CF (2010) Mutations abrogating VP35 interaction with double-stranded RNA render Ebola virus avirulent in guinea pigs. J Virol 84:3004–3015. doi:10.1128/JVI.02459-09

Oda S, Noda T, Wijesinghe KJ, Halfmann P, Bornholdt ZA, Abelson DM, Armbrust T, Stahelin RV, Kawaoka Y, Saphire EO (2015) Crystal structure of Marburg virus vp40 reveals a broad, basic patch for matrix assembly and a requirement of the N-terminal domain for immunosuppression. J Virol 90(4):1839–1848. doi:10.1128/JVI.01597-15

Qiu X, Wong G, Audet J, Cutts T, Niu Y, Booth S, Kobinger GP (2014) Establishment and characterization of a lethal mouse model for the Angola strain of Marburg virus. J Virol 88:12703–12714. doi:10.1128/JVI.01643-14

Rasmussen AL, Okumura A, Ferris MT, Green R, Feldmann F, Kelly SM, Scott DP, Safronetz D, Haddock E, LaCasse R, Thomas MJ, Sova P, Carter VS, Weiss JM, Miller DR, Shaw GD, Korth MJ, Heise MT, Baric RS, de Villena FP, Feldmann H, Katze MG (2014) Host genetic diversity enables Ebola hemorrhagic fever pathogenesis and resistance. Science 346:987–991. doi:10.1126/science.1259595

Raymond J, Bradfute S, Bray M (2011) Filovirus infection of STAT-1 knockout mice. J Infect Dis 204:S986–S990. doi:10.1093/infdis/jir335

Reid SP, Leung LW, Hartman AL, Martinez O, Shaw ML, Carbonnelle C, Volchkov VE, Nichol ST, Basler CF (2006) Ebola virus VP24 binds karyopherin α1 and blocks STAT1 nuclear accumulation. J Virol 80(11):5156–5167

Reid SP, Valmas C, Martinez O, Sanchez FM, Basler CF (2007) Ebola virus VP24 proteins inhibit the interaction of NPI-1 subfamily karyopherin alpha proteins with activated STAT1. J Virol 81:13469–13477. doi:10.1128/JVI.01097-07

Resman Rus K, Fajs L, Korva M, Avšič-Županc T (2016) HMGB1 is a potential biomarker for severe viral hemorrhagic fevers. PLoS Negl Trop Dis 10(6):e0004804. doi:10.1371/journal.pntd.0004804

Ryabchikova E, Kolesnikova L, Smolina M, Tkachev V, Pereboeva L, Baranova S, Grazhdantseva A, Rassadkin Y (1996) Ebola virus infection in guinea pigs: presumable role of granulomatous inflammation in pathogenesis. Arch Virol 141:909–921

Ryabchikova E, Smolina M, Grajdantseva A, Rassadkin J (2004) Ebola virus infection in the guinea pig. In: Klenk HD, Feldmann H (eds) Ebola and Marburg viruses—molecular and cellular biology. Horizon Bioscience, Wymondham, Norfolk, United Kingdom, pp 239–254

Sanchez A, Geisbert TW, Feldmann H, Knipe DM et al. (2007) Fields virology. Lippincott Williams and Wilkins, pp 1410–1448

Siegert R, Shu Hsin-Lu, Slenczka W, Peters D, Müller G (1967) Zur Ätiologie einer unbekannten, von Affen ausgegangenen menschlichen Infektionskrankheit (On the cause of a previously unknown human infection transmitted from monkeys). Dtsch Med Wochenschr 92:2341–2343. doi:10.1055/s-0028-1106144

Simpson DI, Zlotnik I, Rutter DA (1968) Vervet monkey disease. Experiment infection of guinea pigs and monkeys with the causative agent. Br J Exp Pathol 49:458–464

Simpson DI (1969) Vervet monkey disease. Transmission to the hamster. Br J Exp Pathol 50: 389–392

Smith CE, Simpson DIH, Bowen ETW, Zlotnik I (1967) Fatal human disease from vervet monkeys. Lancet 290:1119–1121. doi:10.1016/S0140-6736(67)90621-6

Strong JE, Wong G, Jones SE, Grolla A, Theriault S, Kobinger GP, Feldmann H (2008) Stimulation of Ebola virus production from persistent infection through activation of the Ras/MAPK pathway. Proc Natl Acad Sci U S A 105:17982–17987. doi:10.1073/pnas.0809698105

Subbotina E, Dadaeva A, Kachko A, Chepurnov A (2010) Genetic factors of Ebola virus virulence in guinea pigs. Virus Res 153:121–133. doi:10.1016/j.virusres.2010.07.015

Valmas C, Grosch MN, Schümann M, Olejnik J, Martinez O, Best SM, Krähling V, Basler CF, Mühlberger E (2010) Marburg virus evades interferon responses by a mechanism distinct from ebola virus. PLoS Pathog 6:e1000721. doi:10.1371/journal.ppat.1000721

Valmas C, Basler CF (2011) Marburg virus VP40 antagonizes interferon signaling in a species-specific manner. J Virol 85:4309–4317. doi:10.1128/JVI.02575-10

Van der Groen G, Jacob W, Pattyn SR (1979) Ebola virus virulence for newborn mice. J Med Virol 4:239–240. doi:10.1002/jmv.1890040309

Volchkov VE, Chepurnov AA, Volchkova VA, Ternovoj VA, Klenk HD (2000) Molecular characterization of guinea pig-adapted variants of Ebola virus. Virology 277:147–155. doi:10.1006/viro.2000.0572

Volchkova VA, Dolnik O, Martinez MJ, Reynard O, Volchkov VE (2015) RNA editing of the GP gene of Ebola virus is an important pathogenicity factor. J Infect Dis 212(Suppl 2):S226–S233. doi:10.1093/infdis/jiv309

Warfield KL, Alves DA, Bradfute SB, Reed DK, VanTongeren S, Kalina WV, Olinger GG, Bavari S (2007) Development of a model for marburgvirus based on severe-combined immunodeficiency mice. Virol J 4:108. doi:10.1186/1743-422X-4-108

Warfield KL, Bradfute SB, Wells J, Lofts L, Cooper MT, Alves DA, Reed DK, VanTongeren SA, Mech CA, Bavari S (2009) Development and characterization of a mouse model for Marburg hemorrhagic fever. J Virol 83:6404–6415. doi:10.1128/JVI.00126-09

Wahl-Jensen V, Bollinger L, Safronetz D, de Kok-Mercado F, Scott DP, Ebihara H (2012) Use of the Syrian hamster as a new model of Ebola virus disease and other viral hemorrhagic fevers. Viruses. 4(12):3754–3784. doi:10.3390/v4123754

Watt A, Moukambi F, Banadyga L, Groseth A, Callison J, Herwig A, Ebihara H, Feldmann H, Hoenen T (2014) A novel life cycle modeling system for Ebola virus shows a genome length-dependent role of VP24 in virus infectivity. J Virol 88(18):10511–10524. doi:10.1128/JVI.01272-14

Wauquier N, Becquart P, Padilla C, Baize S, Leroy EM (2010) Human fatal Zaire Ebola virus infection is associated with an aberrant innate immunity and with massive lymphocyte apoptosis. PLoS Negl Trop Dis. 4:e837. doi:10.1371/journal.pntd.0000837

Wilson JA, Hart MK (2001) Protection from Ebola virus mediated by cytotoxic T lymphocytes specific for the viral nucleoprotein. J Virol 75:2660–2664. doi:10.1128/JVI.75.6.2660-2664.2001

Wolf T, Kann G, Becker S, Stephan C, Brodt HR, de Leuw P, Grünewald T, Vogl T, Kempf VA, Keppler OT, Zacharowski K (2015) Severe Ebola virus disease with vascular leakage and multiorgan failure: treatment of a patient in intensive care. Lancet 385(9976):1428–1435. doi:10.1016/S0140-6736(14)62384-9

Wong G, He S, Wei H, Kroeker A, Audet J, Leung A, Cutts T, Graham J, Kobasa D, Embury-Hyatt C, Kobinger GP, Qiu X (2015) Development and characterization of a guinea pig-adapted Sudan virus. J Virol 90:392–399. doi:10.1128/JVI.02331-15

Zlotnik I, Simpson DI (1969) The pathology of experimental vervet monkey disease in hamsters. Br J Exp Pathol 50:393–399

Zlotnik I (1971) "Marburg Disease" The pathology of experimentally infected hamsters. In: Martini GA, Siegert R (eds) Marburg virus disease. Springer, New York, pp 129–135

Zumbrun EE, Abdeltawab NF, Bloomfield HA, Chance TB, Nichols DK, Harrison PE, Kotb M, Nalca A (2012) Development of a murine model for aerosolized ebolavirus infection using a panel of recombinant inbred mice. Viruses 4(12):3468–3493. doi:10.3390/v4123468

Accelerating Vaccine Development During the 2013–2016 West African Ebola Virus Disease Outbreak

Elizabeth S. Higgs, Sheri A. Dubey, Beth A.G. Coller, Jakub K. Simon, Laura Bollinger, Robert A. Sorenson, Barthalomew Wilson, Martha C. Nason and Lisa E. Hensley

Abstract The Ebola virus disease outbreak that began in Western Africa in December 2013 was unprecedented in both scope and spread, and the global response was slower and less coherent than was optimal given the scale and pace of the epidemic. Past experience with limited localized outbreaks, lack of licensed medical countermeasures, reluctance by first responders to direct scarce resources to

---

E.S. Higgs (✉) · R.A. Sorenson
National Institute of Allergy and Infectious Diseases, National Institutes of Health, Bethesda, MD, USA
e-mail: ehiggs@niaid.nih.gov

R.A. Sorenson
e-mail: robert.sorenson@nih.gov

L. Bollinger · L.E. Hensley
Integrated Research Facility, National Institute of Allergy and Infectious Diseases, National Institutes of Health, Frederick, MD, USA
e-mail: bollingerl@niaid.nih.gov

L.E. Hensley
e-mail: lisa.hensley@nih.gov

S.A. Dubey · B.A.G. Coller · J.K. Simon
Merck and Co. Inc, Kenilworth, NJ, USA
e-mail: sheri.dubey@merck.com

B.A.G. Coller
e-mail: beth-ann.coller@merck.com

J.K. Simon
e-mail: jakub.simon@merck.com

M.C. Nason
Biostatistics Research Branch, National Institute of Allergy and Infectious Diseases, National Institutes of Health, Bethesda, MD, USA
e-mail: mnason@niaid.nih.gov

B. Wilson
Partnership for Research on Ebola Virus in Liberia, Monrovia, Liberia
e-mail: barthalomeww@liberiaerp.com

Current Topics in Microbiology and Immunology (2017) 411:229–262
DOI 10.1007/82_2017_53

Published Online: 17 September 2017

clinical research, community resistance to outside interventions, and lack of local infrastructure were among the factors delaying clinical research during the outbreak. Despite these hurdles, the global health community succeeded in accelerating Ebola virus vaccine development, in a 5-month interval initiating phase I trials in humans in September 2014 and initiating phase II/III trails in February 2015. Each of the three Ebola virus disease-affected countries, Sierra Leone, Guinea, and Liberia, conducted a phase II/III Ebola virus vaccine trial. Only one of these trials evaluating recombinant vesicular stomatitis virus expressing Ebola virus glycoprotein demonstrated vaccine efficacy using an innovative mobile ring vaccination trial design based on a ring vaccination strategy responsible for eradicating smallpox that reached areas of new outbreaks. Thoughtful and intensive community engagement in each country enabled the critical community partnership and acceptance of the phase II/III in each country. Due to the delayed clinical trial initiation, relative to the epidemiologic peak of the outbreak in the three countries, vaccine interventions may or may not have played a major role in bringing the epidemic under control. Having demonstrated that clinical trials can be performed during a large outbreak, the global research community can now build on the experience to implement trials more rapidly and efficiently in future outbreaks. Incorporating clinical research needs into planning for future health emergencies and understanding what kind of trial designs is needed for reliable results in an epidemic of limited duration should improve global response to future infectious disease outbreaks.

Contents

1 Introduction

Prior to 2014 Ebola virus disease (EVD) outbreak, EVD outbreaks occurred primarily in rural Central Africa, involved less than 500 persons, and were brought under control through infection control practices. When an EVD outbreak began in the Guéckédou prefecture, Guinea, in late 2013 (Baize et al. 2014; Dudas et al. 2017), the outbreak was not identified for several months (World Health Organization 2014a). By the time the initial EVD cases were confirmed in March 2014 (Nyenswah et al. 2014), cases had spread across local and national borders and into urban centers, ultimately leading to sustained urban person-to-person transmission in Liberia, Sierra Leone, and Guinea, with importation into Senegal and Mali, Nigeria, Europe, and the US (Bray and Chertow DS 2016; World Health Organization 2015b). When the World Health Organization (WHO) declared a public health emergency of international concern (PHEIC) on August 8, 2014, the weekly incidence already exceeded over 500 EVD cases per week in the three most affected countries—Liberia, Guinea, and Sierra Leone—sparking fears of EVD around the world (World Health Organization 2014d). The 2014–2016 West African EVD outbreak ultimately resulted in over 28,000 cases and 11,000 deaths (World Health Organization 2016a).

Before the West African EVD outbreak, the need for clinical research during an international emerging infectious disease emergency was neither fully realized nor integrated into organizational policies, planning, and practices. Recognition of the critical role of clinical research to inform clinical care, diagnostics, and new vaccines and therapeutics for emerging and re-emerging infectious diseases was largely absent. Beyond the lack of recognition of the potential of clinical research to aid in curbing the spread of epidemics, emergency clinical research was viewed by many as a potential distraction from classic public health strategies and practices or a dilution of critical resources. This end result was a global community ill prepared for a coordinated response. Even the Global Health Security Agenda (GHSA), based on the pillars of "Prevent, Detect, Respond" did not have the vision to include clinical research as part of the response (Frieden et al. 2014). Thus, some confusion and disagreements were inevitable in the midst of a public health emergency. During the early stages of the epidemic, results of some models predicted up to 1 million EVD cases.

By August 2014, most global stakeholders understood the need for and were working toward accelerated development of a safe and effective Ebola virus (EBOV) vaccine. At the time of the PHEIC declaration, only two EBOV vaccine studies in humans had been completed, both investigating a DNA-based vaccine which ultimately failed to elicit high antibody titers (Kibuuka et al. 2015; Martin et al. 2006). The two lead experimental vaccine candidates in August 2014, both expressing the surface EBOV glycoprotein (GP), were the a) recombinant vesicular stomatitis virus (rVSVΔG-ZEBOV-GP) vaccine developed by the Public Health Agency of Canada (PHAC) and NewLink Genetics, then licensed to Merck & Co., Inc. (Kenilworth, NJ USA) for further development and b) recombinant DNA chimpanzee adenovirus type 3 (rChAd3-EBO Z) vaccine developed by the Vaccine Resource Center at the NIH and Glaxo Smith Kline (GSK), respectively (Grady D 2014). These vaccines had yet to enter phase I clinical trials.

The WHO used its convening authority throughout the PHEIC to bring global researchers together to accelerate and facilitate Ebola vaccine development. The first-in-human study of rChAd3-EBO Z sponsored by the NIH began on September 2, 2014 at the NIH Clinical Center in Bethesda, Maryland (Ledgerwood et al. 2017; Regules et al. 2017). Additional Phase I studies in the US, Europe, and Africa for both the rVSVΔG-ZEBOV-GP and the rChAd3-EBO Z vaccines provided data for both safety and dose selection for Phase II/III studies. By June 2015, three large vaccine studies were underway in the EVD-affected countries (Henao-Restrepo et al. 2015; Widdowson et al. 2016; Doe-Anderson et al. 2016). As the outbreak progressed, other candidate vaccines entered clinical trials including adenovirus platforms (Ad5) (Li et al. 2017; Zhu et al. 2015; Dolzhikova et al. 2017). By October 2015, another phase II trial (NCT02509494) evaluating replication incompetent adenovirus-vectored vaccine (Ad26.ZEBOV) expressing the EBOV Mayinga variant GP (Janssen Vaccines and Prevention BV, Leiden, The Netherlands) with modified vaccinia Ankara virus (MVA-BN-Filo) expressing GP of EBOV, Sudan virus (formerly known as Ebola virus Sudan), Marburg Virus Musoke variant, and the nucleoprotein of Taï forest virus (formerly known as Cote d'Ivoire Ebola virus) was initiated (Keusch et al. 2017) in Sierra Leone. In 2017, the largest phase II vaccine trial to date, Partnership for Research on Ebola Vaccinations (PREVAC) also referred to as PREVAIL V in Liberia, was initiated across the three countries (NCT02876328). This randomized placebo-controlled clinical trial will provide data for three vaccine strategies on the safety and durability of immune responses for the Ad26.ZEBOV (rHAd26) vaccine with an MVA-BN-Filo boost, and the rVSVΔG-ZEBOV-GP vaccine with or without boosting in both children and adults.

Though a full account of the accelerated vaccine development is beyond the scope of this chapter, it provides an overview of some critical elements and principles that may be useful for future efforts. These elements include preclinical data leading to initial candidate selection; global efforts to coordinate and collaborate on vaccine studies; manufacturers' actions to accelerate production and development; innovative trial designs; and the critical elements of social mobilization, community engagement, and communications. The phase II/III vaccine studies conducted in Western Africa required all these elements. The chapter concludes with lessons learned and recommendations for accelerated emergency research response.

1.1 Identification of Leading Vaccine Candidates

At the time of the outbreak, a number of candidate vaccine platforms had demonstrated efficacy in preclinical animal models. In addition to the rVSVΔG-ZEBOV-GP and the rChAd-EBO Z vaccines, multiple vaccines were under development including vaccines based on other replication-deficient adenoviruses (Geisbert et al. 2011), a rabies vaccine vector (Johnson et al. 2016), a virus-like particle (Warfield et al. 2007), and a Venezuelan equine encephalitis vector (Herbert et al. 2013). Down selection from the list of available vaccines to the rVSVΔG-ZEBOV-GP and the rChAd-EBO Z vaccines was heavily based on preclinical efficacy and safety, durability, availability of clinical-grade supplies, development state, and proposed vaccine dosing schedule.

The rVSVΔG-ZEBOV-GP and the rChAd-EBO Z vaccines demonstrated immunogenicity and efficacy in nonhuman primate studies (Geisbert et al. 2008a; Marzi et al. 2013; Qiu et al. 2009; Stanley et al. 2014), which positioned both vaccines as lead experimental candidates for accelerated efforts. The rVSVΔG-ZEBOV-GP vaccine is a live attenuated chimeric virus consisting of a single recombinant vesicular stomatitis virus (rVSV) (variant Indiana) in which the gene for the VSV G glycoprotein was deleted and replaced with the gene for the Ebola virus glycoprotein (ZEBOV GP), Kikwit variant. These changes result in a replication-competent chimeric virus with a VSV backbone and ZEBOV GP envelope (Widdowson et al. 2016). As observed for similar chimeric vaccines, the substitution of the native virus GP genes with a heterologous GP leads to significant attenuation (Geisbert et al. 2008b). The expression of the heterologous ZEBOV GP on the surface of the virus does not perturb the proper assembly of the recombinant virus particle, which resembles the native VSV bullet-like shape (Fig. 1). However, ZEBOV GP expression is thought to narrow the cell tropism of the recombinant virus and eliminate neurovirulence associated with wild-type (WT) VSV (Geisbert et al. 2008b; Mire et al. 2012).

Results from published preclinical studies evaluating rVSVΔG-ZEBOV-GP in rodents and nonhuman primates (NHPs) demonstrated robust humoral immunogenicity following a single intramuscular (IM) vaccination, and high efficacy (100% survival) against IM EBOV challenge 4–6 weeks after vaccination (Geisbert and Feldmann 2011; Jones et al. 2007; Geisbert et al. 2009; Jones et al. 2005; Qiu et al. 2009). Other studies demonstrated that the rVSVΔG-ZEBOV-GP vaccine was protective and well tolerated, lacking neurovirulent properties, in simian–human immunodeficiency virus-infected NHPs (Geisbert et al. 2008b).

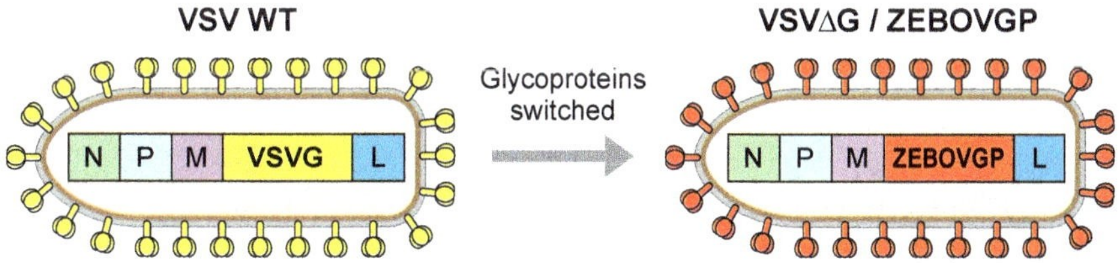

Fig. 1 Design of the recombinant vesicular stomatitis virus expressing Ebola virus glycoprotein (rVSVΔG-ZEBOV-GP) vaccine. This vaccine consists of a recombinant VSV backbone in which the surface glycoprotein (G) of wild-type (WT) VSV has been completely replaced by the Ebola virus Kikwit variant surface glycoprotein

In parallel to the start of the human Phase I trials, a dose-ranging study in NHPs was conducted at the US Army Medical Research Institute for Infectious Diseases (USAMRIID) to assess immunogenicity and efficacy of rVSVΔG-ZEBOV-GP vaccine at doses similar to those used in the Phase I studies (Trefry et al., unpublished data). This study demonstrated 100% survival upon IM challenge with approximately 1000 plaque-forming units (pfu) of WT EBOV, Kikwit variant, 42 days after a single IM immunization of rVSVΔG-ZEBOV-GP at a nominal dose of 2×10^7 or 1×10^8 pfu. Administration of a 3×10^6 pfu dose of rVSVΔG-ZEBOV-GP vaccine protected 7 out of 8 animals challenged with EBOV. All vaccinated animals developed IgG to EBOV GP as measured by Filovirus Animal Nonclinical Group GP-enzyme-linked immunosorbent assay (FANG GP-ELISA) prior to challenge.

The first vaccine to demonstrate efficacy in nonhuman primates (NHP) utilized a prime-boost strategy with DNA plasmid expressing the GP and Ad5 vector replication defective vector expressing GP. The original Ad5 vector contained both a deletion to the E1 in addition to a deletion/substitution of E3 (Sullivan et al. 2000). Follow-on studies demonstrated the use of a single dose of the Ad5-GP vaccine to be sufficient to elicit a protective immune response in NHPs (Sullivan et al. 2003). However, subsequent studies showed that the presence of pre-existing immunity to the adenovirus type 5 (Ad5) vector reduced the efficacy of the Ad5-GP vaccine alone in NHPs. These findings drove the investigation of alternative adenovirus vectors primarily derived from rare serotypes and animals. Down selection of candidate vectors was based on multiple factors including the existing seroprevalence, ability to circumvent Ad5 immunity, the efficiency of production of the vector, and their immunogenicity.

Characterization of several novel Ad vectors derived from rare serotypes suggested the Ad26 vector was most immunogenic of the vectors evaluated in NHPs. However, neither the antibody responses nor cellular responses reached the levels observed with the Ad5 vectors (Abbink et al. 2007). Evaluation of this vector in NHPs protected 75% of the NHPs using a single-dose strategy and required a prime-boost strategy to achieve 100% protection (Geisbert 2011).

The ChAd vectors were evaluated to identify an alternative to human adenovirus (HuAd) vectors that could avoid the challenges of pre-existing immunity while rapidly inducing a robust protective immune response with a single dose of vaccine. These vectors were explored in human clinical trials for other indications including malaria, human immunodeficiency virus 1 infection (HIV), and hepatitis C and had been demonstrated to be highly immunogenic. The observed T cell responses were reported to be comparable to those observed with HuAd5 vectors and such responses were of high magnitude and long lived (Capone et al. 2013). Furthermore, ChAd vectors were demonstrated to be 100% efficacious for EBOV infection in NHPs by multiple laboratories (Kobinger et al. 2006; Stanley et al. 2014).

Despite the number of studies performed by various laboratories using various adenovirus vectors, limited preclinical data were available on the efficacy and safety of candidate rChAd-EBO Z vaccine prior to advancement into Phase I trials.

(Widdowson et al. 2016; World Health Organization 2014c; Keusch G et al. 2017; World Health Organization 2015a). In a multi-arm, dose-ranging study, cynomolgus macaques received either monovalent ChAd3 vaccine expressing EBOV GP, monovalent ChAd63 expressing EBOV GP (De Santis et al. 2016), a bivalent ChAd3 vaccine expressing EBOV GP and Sudan virus GP, or were placed into one of the prime-boost groups to receive ChAd3 followed by either ChAd3, ChAd63, or MVA (Stanley et al. 2014). All eight NHPs that received monovalent ChAd3 vaccine (1×10^{10} or 1×10^{11} particle units [pu]) 5 weeks prior to challenge with 1000 pfu of EBOV were completely protected against EVD. Surprisingly, significant breakthrough was observed in the ChAd63 group. Comparison of the survival data from the dose–response study of the ChAd3 demonstrated comparable potency to that previously observed with HuAd5 studies in NHPs (Stanley et al. 2014). This study on the monovalent ChAd3 EBO Z vaccine formed the basis for Phase I trials in humans.

1.2 *Efforts to Coordinate Ebola Virus Vaccine Studies*

The first WHO Consultation on Potential Ebola Therapies and Vaccines meeting September 4–5, 2014 (World Health Organization 2014c) convened 29 countries and broadly addressed the need to accelerate research and development (Maurice 2014). Participants expressed substantial differences on the acceptability of various study design elements, such as randomization and the use of placebo. Some participants from both nongovernmental organizations and research organizations (e.g., Médecins Sans Frontières [MSF]) opposed randomized controlled studies. These participants argued that diverting scarce resources from patient care or withholding any potential treatment or vaccine would be unethical in view of bed shortages and fatality rates close to 90% in some facilities (Calain 2016). Other participants countered that rigorous clinical research was essential to demonstrate both safety and efficacy, and hence essential to both avoid harming research participants and improving patient outcomes.

The United Nations Security Council (UNSC) responded to transport restrictions by airlines, shipping companies, and other countries; government lockdowns in Sierra Leone; quarantines in Liberia; and concerns that the EVD outbreak threatened gains in the fragile postwar states of Sierra Leone and Liberia. On September 18, 2014, the UNSC approved an EVD resolution urging member states to end travel bans, deploy medical assets, and educate the public (United Nations Security Council 2014). The UNSC resolution, coupled with dire modeling predictions of explosive EVD spread, raised public awareness and injected a global sense of urgency into the quest for EBOV vaccines.

On October 23, 2014, WHO Director General Margaret Chan launched what would become a series of high-level meetings to accelerate vaccine development. Margaret Chan opened the first "High-level Meeting on Ebola Vaccines: Access and Financing," with a challenge "to accomplish within a matter of months, work that normally takes years, with no compromise whatsoever of international standards for safety and efficacy…to give the African people and their health authorities the best product that the world scientists, working collectively, have to offer." At this meeting, the Liberia-US Joint Clinical Research Partnership, newly established to accelerate development of EBOV vaccines and therapeutics, shared plans for a randomized, placebo-controlled phase II/III trial. This trial would include 28,000 participants in Liberia, comparing each of the two lead experimental vaccines to placebo (Keusch et al. 2017; World Health Organization 2014e). In addition, the US Centers for Disease Control and Prevention (CDC) presented a plan to conduct a step-wedge design trial in Sierra Leone. Shortly afterward, the WHO convened parties interested in developing a phase III trial in Guinea, which would later become known as the "Ring Study" (Henao-Restrepo et al. 2017).

2 Overview of Phase I Studies of Two Virus Vector Vaccines

The goal of any clinical vaccine development program is to evaluate safety, immunogenicity, and efficacy to support product licensure, starting with Phase I clinical trials in healthy adult volunteers across a range of vaccine doses. It is not unusual for vaccine development and licensure to take 10 years or more. Through loosely coordinated and intensely focused efforts, the global health research community accelerated the clinical development of the two leading Ebola vaccines candidates, rVSVΔG-ZEBOV-GP and ChAd3-EBO Z, progressing in 5 months from phase I first-in-human studies to phase II/III studies in the EVD outbreak setting. The accelerated development of both vaccines benefited from the experience of large pharmaceutical companies.

2.1 Vesicular Stomatitis Virus Expressing Ebola Virus Glycoprotein

Starting in October 2014, eight phase I clinical trials assessing rVSVΔG-ZEBOV-GP at doses ranging from 3×10^3 to 1×10^8 plaque-forming units (pfu) were launched (Table 1). These studies were conducted in healthy volunteers in EVD non-endemic countries in North America, Europe, and Africa. Overall, nearly 800 adults and 40 pediatric subjects received rVSVΔG-ZEBOV-GP in the phase I program, and preliminary results from some trials have been published (Regules et al. 2017; Agnandji et al. 2016; Huttner et al. 2015).

Preliminary safety and immunogenicity data from the North American trial sites were analyzed in January 2015 for selection of the rVSVΔG-ZEBOV-GP dose for

Table 1 Phase I clinical trials of rVSVΔG-ZEBOV GP vaccine

Protocol number and site	N	Trial Description	Dose-Escalation Regimens (pfu)
V920-001-00 (NLG 0207; WRAIR 2163); US, ClinicalTrials.gov: NCT02269423	39	Randomized, single-center, double-blind, placebo-controlled, dose-escalation study of 3 sequential cohorts	3×10^6, 2×10^7, 1×10^8 (each, $n = 10$), or placebo ($n = 9$)
V920-002-00 (NLG 0407; NIH (NIAID) 15-I-0001); US, ClinicalTrials.gov: NCT02280408	39	Randomized, double-blind, placebo-controlled, dose-escalation study of a prime (day 0) then boost (day 28) regimen (2 doses total)	3×10^6, 2×10^7, 1×10^8 (each, $n = 10$), or placebo ($n = 9$)
V920-003-00 (#CI 1401); Halifax, CA, ClinicalTrials.gov: NCT02374385	40	Randomized, single-center, double-blind controlled, dose-ranging study	1×10^5, 5×10^5, 3×10^6 (each, $n = 10$), Placebo ($n = 10$)
V920-004-00 (NLG 0507); US, ClinicalTrials.gov: NCT02314923	512	Randomized, multi-center, double-blind, placebo-controlled, dose–response study	3×10^3, 3×10^4, 3×10^5 (each, $n = 64$), 3×10^6 ($n = 84$), 9×10^6, 2×10^7 (each $n = 47$), 1×10^8 (each, $n = 48$), placebo ($n = 94$)
V920-005-00 (WHO); Geneva, ClinicalTrials.gov: NCT02287480	115	Dose-finding, randomized, single-center, double-blind[a], placebo-controlled study	3×10^5 ($n = 50$), 1×10^7 ($n = 35$), 5×10^7 ($n = 15$), placebo ($n = 15$)
V920-006-00 (WHO); Hamburg, ClinicalTrials.gov: NCT02283099	30	Open-label, single-center, dose-escalation study	3×10^5, 3×10^6, 2×10^7 (each, $n = 10$)
V920-007-00 (WHO); Gabon, ClinicalTrials.gov: not registered	115[b]	Randomized, open-label, dose-escalation study	3×10^3 ($n = 20$), 3×10^4 ($n = 20$), 3×10^5 ($n = 20$), 3×10^6 ($n = 39$), 2×10^7 ($n = 16$)
V920-008-00 (WHO); Kenya, ClinicalTrials.gov: NCT02296983	40	Open-label, dose-escalation study	3×10^6, 1×10^7 (each, $n = 20$)

[a]Excluding 19 run-in subjects and those in the deployable group (entire group received rVSV-ΔG-ZEBOV-GP)
[b]Additional 40 pediatric subjects 6–17 years of age also included

phase II/III trials. The vaccine was generally well tolerated across all doses. Common systemic adverse events were consistent with those observed following administration of other live virus vaccines, including headache, fatigue, myalgia, subjective fever, chills, and arthralgia (Regules et al. 2017). These signs and symptoms were generally mild (grades 1–2) and transient, and such effects resolved by day 4 postvaccination.

Several cases of arthritis were initially observed in the first cohort of subjects in the second week following study initiation at the Geneva site (Huttner et al. 2015). The study was temporarily halted to more fully investigate these observations. These events appeared to be self-limited and resolved generally within 1–3 weeks or longer. The recovery of the vaccine viral RNA from the joints, when joint fluid was available for testing, suggests a virally mediated (rather than immune-mediated) process. Such arthritis has been seen after the administration of other live virus vaccines, particularly the rubella vaccine in adult women (Institute of Medicine and Committee to Review Adverse Effects of Vaccines 2011; Institute of Medicine and Committee to Review the Adverse Consequences of Pertussis and Rubella Vaccines 1991). Following a review of the data on the cases in Geneva, and noting the absence of similar events in other trial sites, the Data and Safety Monitoring Board (DSMB) determined that dosing could resume at the Geneva site, at a dose of 1×10^5 pfu, and the trial was reinitiated on January 5, 2015. Arthralgia and arthritis were subsequently added to solicited adverse events for other protocols including those in Western Africa. Across the phase I program, arthritis has been observed in <5% of subjects with the exception of the Geneva trial.

In addition to the safety data, immunogenicity data that were available in January 2015 were pooled (study days 0 [day of vaccination] and 7, 14, and 28 postvaccination) from the three North American phase I studies (Table 1) and assessed for dose selection. The immune responses were measured by an ELISA developed by the Filovirus Animal Model Nonclinical Working Group, also referred to as the FANG ELISA at USAMRIID. Geometric mean FANG GP-ELISA titers for this cohort on day 28 postvaccination were 857 ELISA units (EU)/ml at 1×10^5 pfu dose ($n = 10$), 800 EU/ml at 5×10^5 pfu dose ($n = 10$), 1361 EU/ml at 3×10^6 pfu dose ($n = 30$), and 4079 EU/ml at 2×10^7 pfu dose ($n = 20$).

Responses at the 2×10^7 pfu dose were statistically significantly higher ($p < 0.001$) for all pairwise comparisons between 2×10^7 pfu and lower doses, and a dose of 2×10^7 pfu was selected enabling the initiation of the first phase II/III EBOV vaccine trial in Liberia. The phase I studies, and subsequent phase II/III clinical trials, were conducted in collaboration with many partners including regulators, academic institutions, international public health agencies, nongovernmental organizations, the US military, and other US and non-US governmental agencies,[1] which enabled Merck & Co., Inc. and NewLink Genetics to move this

[1]Governmental agencies (e.g., PHAC, US NIH, US CDC, US Biomedical Advanced Research and Development Authority [BARDA], US Defense Treat Reduction Agency [DTRA], Norwegian Institute of Public Health]), clinical researchers (e.g., WHO-led VSV Ebola Consortium, Canadian Center for Vaccinology, Walter Reed Army Institute of Research [WRAIR], Sierra Leone Medical School, Liberia-US Clinical Research Partnership), funding organizations (The Wellcome Trust, Joint Vaccine Acquisition Program), and regulatory agencies/Ministries of Health (e.g., US Food and Drug Administration [FDA], European Medicines Agency, Health Canada, Ministry of Health

vaccine forward at an unprecedented pace. Furthermore, frequent engagement with regulatory agencies and their rapid feedback was a critical part of the accelerated progress.

2.2 *Chimpanzee Type 3 Adenovirus Expressing Ebola Virus Glycoprotein*

Prior to the EVD outbreak, the NIH, National Institute of Allergy and Infectious Diseases (NIAID) Vaccine Research Center and GSK were developing ChAd3-EBO Z vaccine. Through the GSK-NIAID partnership, the first phase I clinical trial in adults for safety and immunogenicity was rapidly initiated at the NIH Clinical Center in Bethesda MD on September 2, 2014. Numerous other phase I clinical trials evaluated the safety and immunogenicity of the ChAd3-EBO Z vaccine in a variety of populations: healthy adults in Switzerland, United Kingdom (UK), Cameroon, Mali, Nigeria, and Senegal, and healthy children in Mali and Senegal, Table 2 (Ledgerwood et al. 2017; De Santis et al. 2016; Tapia et al. 2016; U.S. National Institutes of Health 2017a, b; Ewer et al. 2016). Participants from the US, Mali, UK, and Switzerland enrolled in dose-escalation phase I studies received ChAd3-EBO Z doses ranging from 1×10^{10} to 1×10^{11} pu with or without a boost from MVA-BN-Filo. (Tapia et al. 2016; Ewer et al. 2016; De Santis et al. 2016). These populations were chosen to show the range of responses from volunteers of different genetic backgrounds.

A single dose of the ChAd3-EBO Z vaccine was well tolerated with mild-to-moderate adverse events in numerous phase I studies, including in African populations in Mali using the ChAd3-EBO Z with MVA-BN-Filo boost (Tapia et al. 2016), the Switzerland trial (monovalent ChAd3-EBO Z without boost) (De Santis et al. 2016), US trial (ChAd3 EBO Z without boost) (Tapia et al. 2016), and the UK trial (monovalent ChAd3-EBO Z with MVA-BN-Filo boost) (Ewer et al. 2016). The highest dose and the broadest range of doses studied were in the Mali trials (Tapia et al. 2016). Some evidence from the US trial at the highest dose tested (1×10^{11} pu) and another dose-escalation trial at lower doses suggest that the incidence and severity of adverse effects are dose-related (Tapia et al. 2016; Ewer et al. 2016). The placebo-controlled results from the Switzerland trial indicated that more local and systemic adverse effects (e.g., local pain, fatigue, malaise, muscular-articular pain, chills, fever, headache) occurred in the active arm than in the placebo-treated arm (De Santis et al. 2016). Data from these dose-ranging trials suggested 1 or 2×10^{11} pu of ChAd3-EBO Z as the dose range for additional phase II–III trials (Ledgerwood et al. 2017; Tapia et al. 2016; Kennedy et al. 2017).

(Footnote 1 continued)

and Social Welfare of Liberia, Ministry of Health and Public Hygiene of Guinea, Ministry of Health and Sanitation of Sierra Leone).

Table 2 Phase I clinical trials of ChAd3 EBO Z vaccine

Trial location Study identifier Vaccine	IM Dosage Escalation Regimen	Most common adverse effects	Immune response
• Switzerland (De Santis et al. 2016) • NCT02289027 • ChAd3-EBO Z Mayinga variant	Placebo-controlled, dose-escalation trial of two groups of potentially deployed healthcare workers or nondeployed workers receiving low dose: 2.5 × 10^{10} vp (9 potentially deployed, 42 nondeployed) or high dose: 5 × 10^{10} vp (9 potentially deployed, 40 nondeployed) or placebo (n = 20)	Local pain, fatigue, malaise	• GMC of antibodies Low: 44.9 μg/mL, High: 51 μg/ml; Placebo: 5.2 μg/ml • CD4 response Low: 61%, High: 57% of volunteers • CD8 response Low: 69%, High: 67% of volunteers
• United Kingdom (Ewer et al. 2016) • NCT02240875 • ChAd3-EBO Z Mayinga variant with MVA-BN-Filo boost	Open-label study in adults primed with • 1 × 10^{10} vp (n = 20), • 2.5 × 10^{10} vp (n = 20), or • 5 × 10^{10} vp (n = 20), then MVA-BN-Filo	Mild local pain, headache, myalgia, fatigue, lymphopenia, anemia prior to boost	• Virus-specific GMT at 4 weeks: 752 in primed adults only that persisted through 6 months • Neutralizing antibodies GMT at 4 weeks: 14.9 • T cell response peaked at 14 days post-prime, 633 spot-forming cells
• Mali (Tapia et al. 2016) • NCT02267109 (Mali) • ChAd3-EBO Z then MVA-BN-Filo boost	Open-label study in healthcare workers primed with • 1 × 10^{10} pu (n = 10) • 2.5 × 10^{10} pu (n = 20) • 5 × 10^{10} pu (n = 20), or • 1 × 10^{11} pu (n = 11), then double-blind randomized study of subsequent boost Double-blind randomized study in adults primed with • 2.5 × 10^{10} pu (n = 15) • 5 × 10^{10} pu (n = 15), no boost	Mild local pain, mild headache, fatigue, fever, lymphopenia	GMT at 4 weeks after prime: 220–1446.9 that persisted for 259 days; dose-dependent CD4 response: 22% after prime that persisted through endpoint CD8 response: 16% after prime that persisted through endpoint
• US (Tapia et al. 2016) • NCT02231866 • ChAd3-EBO Z	Single-blind study in adults receiving: 1 × 10^{10} pu (n = 10) or 1 x10^{11} pu (n = 10)	Mild headache, local pain, fatigue, lymphopenia	GMT: 531.5 low dose; 1255.9 high dose

Antibody titers as measured by ELISA and neutralization assays peaked at 1 month after vaccination and were maintained to study endpoint or until boost with multivalent vaccine (4–48 weeks postvaccination) (Ledgerwood et al. 2017; Ewer et al. 2016; Wang et al. 2017; Tapia et al. 2016). Development of EBOV GP antibodies was dose-related, and antibody responses were sustained to study endpoint (Tapia et al. 2016). In the Oxford study (monovalent ChAd3 EBo Z with MVA-BN-Filo boost), seropositivity (defined as a geometric mean titer >166 EU) declined from 82.6% at 28 days to 25% at 6-month postvaccination.

T cell response to vaccination was measured by flow cytometry and intracellular cytokine (interferon, interleukin 2, tumor necrosis factor) staining of peripheral blood mononuclear cells (Tapia et al. 2016). Another study measured T cell responses ex vivo by interferon-γ enzyme-linked immunospot (ELISPOT) (Ewer et al. 2016). T cell responses to ChAd3-EBO Z vaccination were slight in Malian adults; 31% of participants mounted memory CD4 or CD8 responses following exposure of T cells to peptides from EBOV Kikwit variant glycoprotein (Tapia et al. 2016). In participants who received priming with ChAd3-EBO Z vaccine and a placebo booster at 11–16 weeks after priming, the slight T cell (CD4 AND CD8) responses noted during the priming were stable through study endpoint (day 28 after boost). Among the participants boosted with MVA-BN-Filo vaccine, 85% of participants mounted strong CD4 and CD8 responses. T cell responses, as measured ex vivo by interferon-γ ELISPOT assay, peaked at day 14 postvaccination and returned to baseline around day 90 postvaccination. Similar to results in Malian adults (Tapia et al. 2016), boosting with MVA-BN-Filo in pediatric subjects markedly increased virus-specific antibodies by a factor of 12 and GP-specific T cells by a factor of 5 compared to priming with ChAd3-EBO Z only (Ewer et al. 2016).

3 Overview of EBOV Vaccine Efficacy from Phase II/III Studies

Three Phase II/III clinical trials evaluating ChAd3-EBO Z and rVSVΔG-ZEBOV-GP vaccines began in the EVD epidemic region in February through April 2015. The PREVAIL I study sponsored by the NIH and conducted by the Liberian-NIH Partnership started in February 2015 (Kennedy et al. 2017). The "Ebola ça Suffit" ring vaccination trial, sponsored by the WHO and partners (Henao-Restrepo et al. 2015, 2017), began in March 2015 in Guinea. Concurrently, WHO and MSF conducted a safety and immunogenicity study in frontline workers (FLW) in Guinea (PACTR201503001057193). In April 2015, the Sierra Leone Trial to Introduce a Vaccine against Ebola (STRIVE) study was initiated by the CDC and the Sierra Leone Medical School with support from BARDA (Widdowson et al. 2016). In addition to these studies in the outbreak regions, Merck & Co., Inc. started a phase III study in August 2015 in North America and Europe to generate safety and manufacturing lot consistency data important for licensure of the rVSVΔG-ZEBOV-GP vaccine (Table 3).

Table 3 Phase II and III clinical trials of Ebola virus vaccines

Sponsor Trial Name Trial Identifier	Location	Vaccine(s)	Study Objectives/Comments
• NIH (NIAID) • PREVAIL • ClinicalTrials.gov: NCT02344407	Liberia	• rVSVΔG-ZEBOV-GP • ChAd3 EBO Z • Placebo	• Safety and efficacy of each vaccine compared to placebo (3 arms; n = 9000 per arm planned)—canceled due to declining number of cases • Phase II safety and immunogenicity sub-study (n ~1500)
• WHO/NIPH/MSF • Ebola ça suffit • Pan African Clinical Trials: • PACTR201503001057193	Guinea	rVSVΔG-ZEBOV-GP	• Safety and efficacy of immediate vs. delayed ring vaccination (n ~10,000); delayed vaccination arm 21 days later • Parallel safety and immunogenicity study of front line workers (n ~1800)
• CDC • STRIVE • ClinicalTrials.gov: NCT02378753	Sierra Leone	rVSVΔG-ZEBOV	• Safety and efficacy in immediate vs. delayed vaccination groups; delayed arm 18–24 weeks later • Study focused on high-risk workers including healthcare workers (n ~8000) • Safety sub-study (n ~400) • Immunogenicity sub-study (n ~500)
• Merck • V920-012 • ClinicalTrials.gov: NCT02503202	US, Canada, Spain	• rVSVΔG-ZEBOV-GP • Placebo	• Safety through day 42 postvaccination, including detailed work-ups of joint and skin symptoms (n ~1200) • Immunogenicity assessments at days 28 and 180 postvaccination including lot consistency assessment at day 28 • Subset of subjects extended follow-up out to 2 years postvaccination

3.1 PREVAIL I Trial

The initial design of this phase III study called for approximately 28,000 healthy men and women to be randomized to either ChAD3-EBO Z, rVSVΔG-ZEBOV-GP, or matching placebo in equal numbers (Kennedy et al. 2017). This three-arm, double-blind, randomized, controlled trial was designethe safety data are reassuringd to assess the safety and efficacy of each of the two candidate vaccines against a pooled-saline placebo more efficiently than two separate trials. The willingness of the pharmaceutical companies to join attests to the shared recognition of the need for expedience in the PHEIC. The trial design stipulated that 600 participants were to be enrolled into a phase II sub-study to collect safety and immunogenicity data with a DSMB review to follow, without a pause in enrollment. As the EVD epidemic waned in Liberia (Fig. 2), the ability to collect sufficient data to assess efficacy also waned (Keusch et al. 2017). Following the DSMB advice, the phase III arm of the trial was eventually dropped. The Phase II arm was expanded to include safety and immunogenicity assessments and long-term follow-up on 1500 volunteers total, with 500 volunteers in each vaccine group and in the placebo group. Volunteers were followed up at 1 week, 1 month, 2 months, and then every 2 months through 12 months and then yearly thereafter for safety and longevity of immune responses. Both vaccines were found safe and immunogenic at 1 month and 1 year (Bolay 2016; Kennedy et al. 2017). Injection site reactions, headache, muscle pain, fever, and fatigue were significantly greater during week one in both

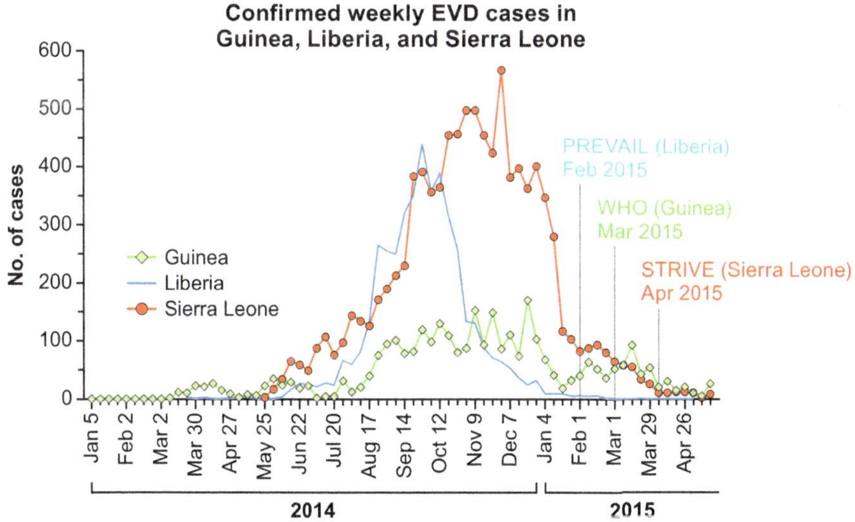

Fig. 2 Initiation of phase II–II clinical trials and incidence of EVD cases in Guinea, Liberia, and Sierra Leone. Adapted from World Health Organization Ebola data and statistics (World Health Organization 2016b, c, d)

vaccine groups compared to placebo (p <0.001), but the incidence of these adverse effects declined to incidence observed with placebo at 1 month postvaccination. The incidence of cumulative serious adverse events (SAEs) at 12 months did not differ significantly between the vaccine arms and placebo. In the ChAD3-EBO Z arm, 40 (8.0%) participants experienced SAEs compared to 47 (9.4%) in rVSVΔG-ZEBOV-G arm and 59 (11.8%) in the placebo arm. Some 71% of SAEs were due to malaria (Kennedy et al. 2017). A 2-week postvaccination active assessment of joint problems failed to identify a difference between the vaccines and placebo (Kennedy et al. 2017).

Immunogenicity was assessed by IgG levels against the EBOV GP using the FANG GP-ELISA assay (Kennedy et al. 2017). A positive vaccine response is defined as a $\log_{10}$ titer increase or more than 2 standard deviations above the placebo change in the absence of elevated baseline antibodies. In both vaccines, antibody responses peaked at 1 month and declined at 6 and 12 months. At 1 month, 70.8% of the participants receiving ChAd3 EBO Z vaccine and 83.7% of the participants receiving rVSVΔG-ZEBOV GP responded to vaccination, respectively (p <0.001 for each comparison to placebo). Antibody responses were largely maintained at 1 year in 63.5% of the ChAd3 vaccinees, 79.5% of the rVSVΔG-ZEBOV vaccinees, and 6.8% of placebo recipients, respectively (p <0.001 for each vaccine compared to placebo) (Kennedy et al. 2017).

3.2 STRIVE Trial

The CDC partnered with the College of Medicine and Allied Health Sciences, University of Sierra Leone, and the Sierra Leone Ministry of Health and Sanitation to design and implement the Sierra Leone Trial to Introduce a Vaccine against Ebola (STRIVE, NCT02378753, PACTR201502001037220). The study enrolled healthcare and other response workers caring for EVD patients or managing the deceased beginning in April 2015. The trial was originally designed to use a step-wedge design, in which participants were randomized to receive the rVSVΔG-ZEBOV-GP vaccine with varying periods of delay. At the time this clinical study started, the EVD epidemic was slowing (Fig. 2), and data were again insufficient to assess efficacy. STRIVE proceeded as a phase II study (n = 8673) with a safety sub-study (n = 453) and an immunogenicity sub-study (n = 539). Participants were individually randomized to either immediate (within a week) vaccination with rVSVΔG-ZEBOV-GP or vaccination 6 months after enrollment (Widdowson et al. 2016).

Although complete results have not been published, some preliminary findings were reported by Widdowson et al. No confirmed cases of EVD or vaccine-related SAEs were reported as of April 2016, 1 year after enrollment. The report by the National Academies of Sciences, Engineering, and Medicine concluded that "Although still incomplete, the safety data are reassuring" (Keusch et al. 2017).

3.3 *Ebola ça Suffit! Trial*

A third Phase III EVD vaccine trial took place in Guinea under the leadership and regulatory sponsorship of the WHO. This trial has the distinction of being the only one of the three West African trials to accrue enough participants with EVD infections from ongoing transmission chains for the investigators to draw conclusions about vaccine efficacy. Modeled after smallpox eradication campaigns, the open-label, cluster-randomized Ebola ça Suffit! trial (PACTR201503001057193) focused on identifying clusters (rings) of at-risk adults who had had contact with an EVD index case or who were contacts of contacts (Henao-Restrepo et al. 2015; Keusch et al. 2017). Upon identification of a new laboratory-confirmed EVD index case, individuals in each ring were enumerated based on reported contact and social or residential proximity. Then, the entire ring was randomized 1:1 as a single unit to receive either immediate or delayed, defined as 21 days later, vaccination with rVSVΔG-ZEBOV-GP (Keusch et al. 2017).

The primary endpoint of this trial focused on EVD cases that occurred between 10 and 31 days post-randomization. EVD cases diagnosed on days 0–9 postvaccination were excluded from the primary analysis to exclude subjects who were likely infected pre-vaccination and to allow time for an adaptive immune response to develop following vaccination. In the rings assigned to delayed vaccination, the teams returned to vaccinate consenting and eligible individuals on day 21 post-randomization. Mobile research vaccination teams were deployed to areas within Guinea with newly identified EVD cases or transmission chains. The trial was planned to have approximately 190 rings (Henao-Restrepo et al. 2015). Assuming an average ring size of 50 participants and an intra-class correlation of 0.05, the trial would have power to detect vaccine efficacy of 70% or more.

With the novel design of the trial using a mobile research team that focused on high-risk contacts within rings, 41 cases of EVD were identified, 16 of which were between days 10 and 31 postvaccination. An interim analysis was conducted in July 2015 (Henao-Restrepo et al. 2015). Following the interim analysis, the DSMB recommended discontinuing randomization to the delayed vaccination arm and offering the vaccine immediately to subsequent rings, including children ages 6–17 (Henao-Restrepo et al. 2017; Keusch et al. 2017).

At the final analysis, 98 rings were randomized to either immediate or delayed vaccination, and 19 rings were offered immediate vaccination without randomization (Henao-Restrepo et al. 2017). No cases of EVD were seen 10 or more days postvaccination in any subject who received vaccine. When individuals vaccinated in the immediate arm were compared to individuals in the delayed vaccination group who were eligible and consented on day 0, the estimated vaccine efficacy was 100%, with a 95% confidence interval from 63.5 to 100% ($p = 0.0471$) (Table 4). In the analysis of all eligible participants in 7 rings that had been randomized to delayed vaccination, including those who did not consent or did not receive vaccination, 16 EVD cases occurred (Keusch et al. 2017).

Table 4 Results of the Ebola ça suffit phase II trial (Henao-Restrepo et al. 2017)

	Four randomized comparisons			
	1	2	3	4
Immediate vaccination arm	All vaccinated	All vaccinated	All eligible	All
#clusters (#individuals):	51 (2108)	51 (2108)	51 (3212)	51 (4513)
#clusters with 1+ case (#cases)	0 (0)	0 (0)	4 (7)	5 (10)
Delayed vaccination arm	All consented on Day 0	All eligible	All eligible	All
#clusters (#individuals):	46 (1429)	47 (3075)	47 (3075)	47 (4529)
#clusters with 1 + case (#cases)	4 (10)	7 (16)	7 (16)	8 (22)
Estimated vaccine efficacy[a]	100%	100%	64.6%	64.6%
***p*-value**	0.0471	0.0045	0.344	0.3761

[a]Note that, as is appropriate in cluster-randomized trials, rings were randomized in their entirety so that all individuals in a ring would get the same treatment assignment. The vaccine efficacy estimates and p values are computed based on the number of rings with one or more cases, not on the number of individual cases

Based on published critiques (Zhang et al. 2016; Krause 2015; Kieny et al. 2016; Keusch et al. 2017), the authors conducted additional analyses (eight different comparisons) that were included in the final data set (Henao-Restrepo et al. 2017). Four of these comparisons include only information from the randomized part of the trial, and therefore provide the most rigorous assessment of vaccine efficacy (Table 4). To partially address the criticism that these two groups may not be comparable, all immediately vaccinated versus all those eligible who consented on day 0 are compared in column 2 of Table 4. If the study had been blinded, these groups would likely be close to ideal comparator groups that led to an estimated vaccine efficacy of 100%, with a similar 95% confidence interval of 68.9–100% (p = 0.0471). In this unblinded study, however, researchers have difficulty knowing whether the two sets of individuals are comparable, and whether exposure rates would be influenced by the fact that they were or were not vaccinated.

In the last two columns, comparisons that include all eligible individuals in both arms (comparison 3) or all individuals identified as belonging to the ring (comparison 4) are also depicted. These comparisons are consistent with the traditional intent-to-treat approach and subject to the same criticisms as the other arms, since researchers included the same groups of individuals for both arms. However, the inclusion of individuals who did not receive vaccination in the immediate arm yielded lower estimates of vaccine efficacy (64.6% for both comparisons), falling short of statistical significance (Henao-Restrepo et al. 2017). Importantly, the high efficacy and incidence of EVD cases before elimination of randomization suggest that the Ebola ça Suffit trial itself may have had some contribution to foreshortening the epidemic of EVD in Guinea by direct and indirect aversion of cases.

The trial in Guinea also included an open-label parallel study of the rVSVΔG-ZEBOV-GP in healthcare workers that was conducted by MSF and sponsored by WHO. In this study, over 1800 healthcare workers were vaccinated, and safety and immunogenicity data were collected. This study provided an important contribution both to public health, by providing potential protection for people at risk, and to scientific understanding of the vaccine's profile. Since no samples for evaluation of immunogenicity were collected in the ring study itself, the MSF/WHO healthcare worker study, along with the STRIVE and PREVAIL trials, will provide important data on the immunogenicity of the vaccine in at-risk populations in Guinea, Sierra Leone, and Liberia, respectively, and may contribute to understanding correlates of protection.

3.4 Summary

Three large Ebola vaccine efficacy trials were planned in each of the three primary EVD-affected countries and sponsored by public partners (NIH, CDC, and WHO). The three studies were started 6–8 months after declaration of the PHEIC. Two of the three studies were unable to assess efficacy due to fortunate diminution of the outbreak in Liberia and Sierra Leone. Only investigators of the PREVAIL study used the ChAD3-EBO Z vaccine, vaccinating 500 adults. The ChAD3-EBO Z vaccine was found to be safe and immunogenic compared to placebo and demonstrated a similar safety profile to that found in the phase I studies. From over 17,000 subjects who received rVSVΔG-ZEBOV in twelve Phase I, II, and III trials since late 2014 (Tables 1 and 3) enough data were accrued to proceed with a licensing application to regulatory agencies. Preliminary safety data in healthy, nonpregnant adults suggest that the vaccine is generally well tolerated, and the safety profile observed in phase II and III trials is consistent with the results summarized previously for the phase I studies. Based on vaccine efficacy as reported in the Guinea ring vaccination (Ebola ça Suffit!) trial and safety assessed to date in all clinical trials, the rVSVΔG-ZEBOV GP vaccine appears to have a positive benefit-risk ratio for adults (Regules et al. 2017; Huttner et al. 2015; Agnandji et al. 2016; Henao-Restrepo et al. 2015, 2017). Common systemic adverse events included headache, pyrexia, fatigue, and myalgia, most which were mild-to-moderate and of short duration (Regules et al. 2017; Huttner et al. 2015; Agnandji et al. 2016).

4 Accelerated Ebola Virus Vaccine Development: A Pharmaceutical Company's Response

The global mobilization against the EVD epidemic and the partnerships it engendered were pivotal for rapid progress in development, compressing work that typically takes decades to less than 5 years in the case of the rVSVΔG-ZEBOV-GP vaccine. Merck & Co., Inc. contributed by putting its expertise and resources behind development of the rVSVΔG-ZEBOV-GP vaccine, joining the effort with NewLink pharmaceuticals in the fall of 2014, during the height of the EVD outbreak. Preclinical data generated by PHAC demonstrating the efficacy and safety of the vaccine in relevant animal models, and availability of clinical-grade vaccine supplies prepared by PHAC in 2013 were critical for rapid advancement.

Cumulatively, the phase I, II, and III studies are expected to provide a solid body of clinical data supporting the safety and efficacy of the rVSVΔG-ZEBOV-GP vaccine. These studies are part of an accelerated development program that also includes nonclinical safety and efficacy data and manufacturing process validation. A licensure package is under preparation at the time of writing. To further accelerate licensure and availability of the vaccine to those at risk, Merck & Co., Inc. applied for and received Breakthrough Therapy Designation from the US FDA and Priority Medicines status from the European Medicines Agency in June of 2016. With these designations, ongoing interactions with key regulatory bodies ensure alignment of product filing packages with regulatory requirements in support of licensure. To access the vaccine if needed prior to licensure, Merck also filed an Emergency Use Assessment and Listing (EUAL) application with the WHO. If approved, EUAL would expedite the availability of the vaccine outside clinical trials in a public health emergency. Expanded access clinical protocols have also been developed for vaccination of contacts of EVD cases in countries most likely to experience a future EVD outbreak.

5 Social Mobilization and Community Engagement for Vaccine Studies in Liberia

During the initial phase of the EVD epidemic, government-enlisted partners, including WHO, were overwhelmed by medical and logistical demands, exacerbated by cultural and geographic challenges. Outside organizations mounting a response to the epidemic lacked experience in the Liberian cultural context, and Liberians mistrusted their own existing healthcare system and political leaders. EBOV spread rapidly through social networks in a culture that stresses compassionate, hands-on care for the ill, and ceremonial rites for the deceased. The rapid spread of a disease new to Western Africa generated fears and many misconceptions that spread through communities and society by word of mouth (World Health Organization 2014b), triggering societal disruption. The government was overwhelmed, admittedly lacking the capacity to respond to a severe, sustained, and unprecedented public health crisis. Overflowing healthcare facilities turned away new patients (World Health Organization 2015a). Inadequate infrastructure and understanding of transmission and variable disease manifestations and sequelae created a volatile situation that undermined conventional control measures and constantly delivered surprises.

Amid the outbreak, Liberians still had hope, courage, commitment, and the resilience of the local communities to combat this crisis. Community leaders set up response teams (EVD task forces) to carry out contact tracing, case investigation and reporting, and surveillance (Fallah et al. 2017). The community-based Ebola task force also instituted quarantine measures and provided food and water to those in quarantine. The Ministry of Health understood the successes of these community-based efforts and formally supported them. The community-based initiative (CBI) program was instituted to support volunteers who had been sacrificing their time and meager resources.

The success of the people and the government of Liberia in overcoming EVD was not primarily the result of international support, but rather of community resilience and cooperation to awaken from the "nightmare" called Ebola. As several experts noted, when technical interventions violate entrenched cultural practices, cultural norms will prevail. Control efforts must work within the culture to accommodate or change attitudes, not against long-standing norms (Hankins 2016; Keusch et al. 2017).

Before the EVD outbreak in Liberia, little clinical research had been conducted in the country, and most citizens were "research naïve." In this context, the Liberian Ministry of Health and Social Welfare requested that the U.S. Department of Health and Human Services collaborate in a clinical research effort to rapidly develop EBOV vaccines and therapeutics (also known as PREVAIL). Social mobilization and community engagement were quickly identified as essential to the success of the randomized, placebo-controlled Ebola vaccine trial (PREVAIL I). The PREVAIL partnership established the Social Mobilization, Communication, and Community Engagement (SMC) team to promote acceptance of clinical trials, address misconceptions, and counteract fears hindering cooperation with outside organizations and local government agencies. In addition, the team provided social support for fellow community members enrolled in the trials (Doe-Anderson et al. 2016; Webmaster Administration 2017). The SMC team strategy was to leverage existing community structures in harnessing trust and to build collaboration and sustainable partnerships with inhabitants of targeted communities.

Discussing and understanding the basic clinical research concepts with all stakeholders are essential to full community participation in clinical research, and the PREVAIL SMC team strategy was designed to ensure community engagement. Social mobilization for the PREVAIL I trial had four distinct but coordinated pillars: (1) advocacy, (2) community engagement, (3) communication, and (4) monitoring and evaluation. These four pillars provided the basic structure for confidence and trust in the research team. The SMC strategy created a space for dialogue between researchers and communities to share their perspectives on study designs, facilitating the recruitment and retention of study participants.

5.1 Advocacy

The goal of the advocacy pillar is to ensure that key formal and informal decision-makers, opinion shapers, political leaders, and religious and traditional leaders have a comprehensive understanding of the study and its potential risks and benefits for the participants and their communities. Advocacy begins with ethnographic community mapping to identify major stakeholders and ensure proper community entry. Fruitful engagement with national stakeholders and local leaders and community members (gate keepers) requires respect for their cultural values and principles. Meetings with government officials and other national stakeholders prepare the way for political support. Dialogue with gatekeepers smooths community entry through better understanding of cultural, religious, and traditional norms and their interplay with community concerns.

5.2 Community Engagement

Community engagement included outreach to targeted populations, including community leaders, through community meetings, posters, flyers, brochures, and other educational materials covering, for example, frequently asked questions. Community leaders organized engagement meetings to build sustainable partnerships through existing structures. Community perspectives noted during the meetings were used to craft corresponding messages. Through these meetings, the research team also began to understand issues affecting the community, cultural values to consider, and methods of clear communication regarding research issues. Effective entry into the community required meeting formal and informal authorities to solicit approval before engaging individual community members. Not cultivating community leadership could imply disrespect for the community, thereby leading to rejection or noncompliance from the community.

5.3 Communication

The communication pillar is the major information dissemination arm of the SMC team. To avoid myths and misconceptions, accurate information was disseminated to targeted communities through flyers, posters, radio shows covering frequently asked questions, press statements/releases, media conferences, interpersonal communication, drama, music, and jingles. In Liberia, songs and drama are living means of communication, though they may be disregarded by outsiders who do not understand the fabric of Liberian society. Researchers must build understanding of the cultural context surrounding research to incorporate community norms and bring universal ethical standards to bear upon their specific local actions.

5.4 *Monitoring and Evaluation*

The SMC team established joint monitoring and evaluation mechanisms with community leaders, governments, and partners through focus group discussions, qualitative questionnaires, and interviews. These dialogues helped to ensure that the impacts of the study interventions were well documented and consistent with ethical and protocol guidelines of standardized clinical research. Monitoring and evaluation with full involvement of the community may necessitate reprogramming and redesigning activities. For example, the team reviewed messages and communication channels to enhance effective communication and engagements with all stakeholders.

5.5 *Impact of Social Mobilization and Community Engagement*

The PREVAIL SMC team faced many challenges in communicating to the community information about an EBOV vaccine when the name "Ebola" by itself evoked horror, suspicion, and disgust in the minds of many Liberians. Wild rumors were rife: EVD was brought to Africa by westerners who wanted to make money through pharmaceutical sales. Others believed that the EBOV vaccine was a reinforcement mechanism to infect more people with the virus. Some community dwellers thought that participants' blood specimens would be used for commercial purposes. One SMC team goal was to dispel the plethora of conspiracy theories floating around Western Africa and the world beyond.

After 3 months of intense engagement with community leaders and members, 1500 volunteers consented to enroll in the vaccine trial. The overall retention rate of study participants was quite impressive, with over 98% of participants returning for their follow-up visits within the first 6 months of enrollment. This phenomenal achievement would not have been possible without the support of formal and informal community leaders, who in effect championed the research. Their contributions began with rectification of names: "Trial" became "Study" to enhance understanding. Moreover, community members were recruited from within the research area to follow participants enrolled in the study. These trackers provided another channel of communication, ensuring adherence to follow-up schedules, and reporting issues affecting participation. Their role in enhancing trust and sustainable collaboration cannot be overemphasized.

In a context of fear and societal disruption, community involvement must be considered an essential ethical norm by the research community, particularly in public health emergencies. Experiences from the vaccine study in Liberia demonstrate that the prevention and control of a public health emergency such as EVD outbreak can only be advanced from a social cultural perspective informed by early, genuine engagement with communities. This is a crucial lesson learned from both the EVD outbreak response and the vaccine study in Liberia.

6 Lessons Learned for Epidemic Preparedness

A great deal of thought has gone into reviewing the EVD epidemic response and how to improve international readiness for the next infectious disease crisis (d'Harcourt 2016). With hindsight, many of the responders who were skeptical about clinical research in 2014 now understand that an earlier focus on clinical research in the EVD crisis may have brought the epidemic to a close sooner. If expertise on clinical research and clinical care had been incorporated into the initial assessment team in March 2014, the need for accelerated Ebola virus vaccine and therapeutic research agenda could have been identified. Had the accelerated EBOV vaccine development timeline begun then, and a phase II/III vaccine study would have begun in August 2014 during the peak transmission period, more data on safety and efficacy would have been likely.

In the United States, both the Department of Health and Human Services and the National Security Council formally recognized the need for clinical research as an integral part of infectious disease emergency planning (U.S. Department of Health and Human Services 2016, 2017). A National Academy of Sciences report (Keusch et al. 2017) focuses on improvements in three main areas: strengthening capacity, engaging communities, and facilitating international coordination and collaboration. Better health and research systems in the affected countries, memorandums of agreement on research needs such as data and sample sharing, and capacity for rapid ethical review within affected countries would help speed implementation of needed research.

Lessons from the trials themselves are valuable as well. Ideally, phase I safety and dose-ranging studies for known emerging infectious pathogens with epidemic potential will be conducted prior to outbreaks. Conducting a randomized controlled clinical trial during a public health emergency was both feasible and ethical, incorporating a social cultural perspective informed by early, genuine engagement with communities and stakeholders according to Good Participatory Practices. From a design perspective, the collaboration within PREVAIL I trial of two large pharmaceutical companies was both laudable and efficient.

Despite the successful completion of a phase III efficacy study, the ring study correlates of immunity remain unknown. The ring study was not able to obtain blood samples from participants which may have contributed to this knowledge. Preparedness plans for future vaccine studies should include acquisition of blood samples. The innovation of using mobile teams to bring the ring study to areas of high geographic incidence and identification of individuals at highest epidemiologic risk enabled completion of an efficacy trial comparing immediate to delayed groups that may lead to vaccine licensure. However, elements of this ring vaccination design comparison have been questioned (Zhang et al. 2016; Krause 2015; Kieny et al. 2016; Keusch et al. 2017). For example, individuals who were eligible for vaccination but did not consent might have different risk factors, exposures, or behaviors than those who were vaccinated. An additional concern is that the contacts in the immediate and delayed arms could be handled differently in such a design. Finally, recent information about "super-shedders or transmitters" (Lau et al. 2017) is a reminder that exposure risk can be variable in infectious disease transmission. Such criticisms combined with the success of the novel ring study design in capturing EVD endpoints during an outbreak can inform future vaccine study design. Vaccine study designs for future outbreak should consider leveraging mobile research teams to high incidence areas, identification of high-risk individuals, combined with concomitant individual randomization within clusters to active versus control vaccine to mitigate confounding factors and maximize efficiency. Discussion of such designs is ongoing.

7 Conclusions

In recognition of the need for better preparedness for emerging pathogens such as EBOV and Zika virus, national governments and the international public health community have undertaken a number of initiatives. The WHO is undertaking multifaceted efforts to improve its performance in emergencies. In addition to the formation of a new Emergency Program (World Health Organization 2016e), WHO has produced a blueprint to prevent epidemics which "underlines the importance of research as an integral element of the response to any epidemic" (World Health Organization 2016e). These publications are a clear mandate for better preparing countries and the international community for the next infectious disease emergency and for integrating clinical research into both preparation and response.

Other programs include U.S. initiatives led by the National Security Council and Department of Health and Human Services (U.S. Department of Health and Human Services 2016, 2017) and the recently established Coalition for Epidemic Preparedness and Innovations (CEPI) (Rottingen et al. 2017; Brende et al. 2017; Plotkin 2017). CEPI will focus on ensuring that medical countermeasures for pathogens likely to cause an outbreak are in the development pipeline. These endeavors incorporate the understanding that clinical research can play a vital role in ending ongoing epidemics, not only in preventing or preparing for future ones.

As highlighted previously, focusing on community engagement is critical to the success of clinical research in the context of an epidemic. Preparation, including the identification of promising medicines and vaccines in advance of an outbreak and the availability of preclinical data, phase I studies, and clinical supplies to support rapid clinical trial implementation, is also key. Finally, the response to the 2014–2016 EVD outbreak highlights that public–private partnerships can help bring experience and expertise from all sectors to bear, and produce extraordinary achievements in the face of a daunting public health challenge. Significant clinical research capacity has been established in the West African region for future preparedness. Much work remains, but stakeholders have broad agreement for moving forward.

Aside from the stark demonstration of the imperative role of clinical research, the EVD epidemic provided invaluable experience on how to accelerate clinical research. In the context of a broken healthcare system in a resource-poor environment, efficient trial design, collaboration, and robust community engagement are essential (Keusch et al. 2017). The need for rapid clinical research and accelerated development of medical countermeasures is now more widely recognized as critical to a global health response (Keusch et al. 2017; U.S. Department of Health and Human Services 2017). Those working on EBOV vaccine trials during the epidemic were learning how to do emergency research. It is incumbent upon researchers and policymakers to ensure that what they learned is incorporated in preparedness planning and brought to bear on the next emerging infectious disease emergency.

8 Funding

Funding for research on rVSVΔG-ZEBOV-GP was provided by Médecins sans Frontières, Merck & Co., Inc., Kenilworth, NJ, USA, NewLink Genetics, the Public Health Agency of Canada, the US Department of Health and Human Services (Biomedical Advanced Research and Development Authority and the National Institutes of Health), the US Department of Defense (Defense Threat Reduction Agency and the Joint Vaccine Acquisition Program), the Wellcome Trust, and the World Health Organization. This work was supported in part under Battelle Memorial Institute contract (Contract No. HHSN272200700016I) with NIAID.

Acknowledgements The authors thank Jiro Wada, Integrated Research Facility, for figure preparation.

References

Abbink P, Lemckert AA, Ewald BA, Lynch DM, Denholtz M, Smits S, Holterman L, Damen I, Vogels R, Thorner AR, O'Brien KL, Carville A, Mansfield KG, Goudsmit J, Havenga MJ, Barouch DH (2007) Comparative seroprevalence and immunogenicity of six rare serotype recombinant adenovirus vaccine vectors from subgroups B and D. J Virol 81(9):4654–4663. doi:10.1128/jvi.02696-06

Agnandji ST, Huttner A, Zinser ME, Njuguna P, Dahlke C, Fernandes JF, Yerly S, Dayer JA, Kraehling V, Kasonta R, Adegnika AA, Altfeld M, Auderset F, Bache EB, Biedenkopf N, Borregaard S, Brosnahan JS, Burrow R, Combescure C, Desmeules J, Eickmann M, Fehling SK, Finckh A, Goncalves AR, Grobusch MP, Hooper J, Jambrecina A, Kabwende AL, Kaya G, Kimani D, Lell B, Lemaitre B, Lohse AW, Massinga-Loembe M, Matthey A, Mordmuller B, Nolting A, Ogwang C, Ramharter M, Schmidt-Chanasit J, Schmiedel S,

Silvera P, Stahl FR, Staines HM, Strecker T, Stubbe HC, Tsofa B, Zaki S, Fast P, Moorthy V, Kaiser L, Krishna S, Becker S, Kieny MP, Bejon P, Kremsner PG, Addo MM, Siegrist CA (2016) Phase 1 trials of rVSV Ebola vaccine in Africa and Europe. N Engl J Med 374(17):1647–1660. doi:10.1056/NEJMoa1502924

Baize S, Pannetier D, Oestereich L, Rieger T, Koivogui L, Magassouba N, Soropogui B, Sow MS, Keita S, De Clerck H, Tiffany A, Dominguez G, Loua M, Traore A, Kolie M, Malano ER, Heleze E, Bocquin A, Mely S, Raoul H, Caro V, Cadar D, Gabriel M, Pahlmann M, Tappe D, Schmidt-Chanasit J, Impouma B, Diallo AK, Formenty P, Van Herp M, Gunther S (2014) Emergence of Zaire Ebola virus disease in Guinea. N Engl J Med 371(15):1418–1425. doi:10.1056/NEJMoa1404505

Bolay F (2016) A randomized controlled trial of the safety and immunogenicity of two Ebola vaccines. In: Abstracts of the 23rd Conference on Retroviruses and Opportunistic Infections, John B Hynes Veterans Memorial Convention Center, Boston, 22–25 Feb 2016. Abstract 76LB. CROI Foundation/International Antiviral Society-USA, San Francisco, CA. http://www.croiconference.org/sessions/randomized-controlled-trial-safety-and-immunogenicity-two-ebola-vaccines. Accessed 7 Apr 2017

Bray M, Chertow DS (2016) Epidemiology and pathogenesis of Ebola virus disease. UpToDate, Wolters Kluwer. https://www.uptodate.com/contents/epidemiology-and-pathogenesis-of-ebola-virus-disease?source=search_result&search=bray%20ebola&selectedTitle=3 ~ 23. Accessed 10 Apr 2017

Brende B, Farrar J, Gashumba D, Moedas C, Mundel T, Shiozaki Y, Vardhan H, Wanka J, Rottingen JA (2017) CEPI-a new global R&D organisation for epidemic preparedness and response. Lancet 389(10066):233–235. doi:10.1016/s0140-6736(17)30131-9

Calain P (2016) The Ebola clinical trials: a precedent for research ethics in disasters. J Med Ethics. doi:10.1136/medethics-2016-103474

Capone S, D'Alise AM, Ammendola V, Colloca S, Cortese R, Nicosia A, Folgori A (2013) Development of chimpanzee adenoviruses as vaccine vectors: challenges and successes emerging from clinical trials. Expert Rev Vaccin 12(4):379–393. doi:10.1586/erv.13.15

De Santis O, Audran R, Pothin E, Warpelin-Decrausaz L, Vallotton L, Wuerzner G, Cochet C, Estoppey D, Steiner-Monard V, Lonchampt S, Thierry AC, Mayor C, Bailer RT, Mbaya OT, Zhou Y, Ploquin A, Sullivan NJ, Graham BS, Roman F, De Ryck I, Ballou WR, Kieny MP, Moorthy V, Spertini F, Genton B (2016) Safety and immunogenicity of a chimpanzee adenovirus-vectored Ebola vaccine in healthy adults: a randomised, double-blind, placebo-controlled, dose-finding, phase 1/2a study. Lancet Infect Dis 16(3):311–320. doi:10.1016/s1473-3099(15)00486-7

d'Harcourt E (2016) The Ebola lessons reader: what's being said, what's missing and why it matters. International Rescue Committee. https://www.rescue.org/sites/default/files/document/563/theebolalessonsreaderlowres.pdf. Accessed 3 May 2017

Doe-Anderson J, Baseler B, Driscoll P, Johnson M, Lysander J, McNay L, Njoh WS, Smolskis M, Wehrlen L, Zuckerman J (2016) Beating the odds: successful establishment of a phase II/III clinical research trial in resource-poor Liberia during the largest-ever Ebola outbreak. Contemp Clin Trials Commun 4:68–73. doi:10.1016/j.conctc.2016.06.008

Dolzhikova IV, Zubkova OV, Tukhvatulin AI, Dzharullaeva AS, Tukhvatulina NM, Shcheblyakov DV, Shmarov MM, Tokarskaya EA, Simakova YV, Egorova DA, Scherbinin DN, Tutykhina IL, Lysenko AA, Kostarnoy AV, Gancheva PG, Ozharovskaya TA, Belugin BV, Kolobukhina LV, Pantyukhov VB, Syromyatnikova SI, Shatokhina IV, Sizikova TV, Rumyantseva IG, Andrus AF, Boyarskaya NV, Voytyuk AN, Babira VF, Volchikhina SV, Kutaev DA, Bel'skih AN, Zhdanov KV, Zakharenko SM, Borisevich SV, Logunov DY, Naroditsky BS, Gintsburg AL (2017) Safety and immunogenicity of GamEvac-Combi, a heterologous VSV- and Ad5-vectored Ebola vaccine: an open phase I/II trial in healthy adults in Russia. Hum Vaccin Immunother 13(3):613–620. doi:10.1080/21645515.2016.1238535

Dudas G, Carvalho LM, Bedford T, Tatem AJ, Baele G, Faria NR, Park DJ, Ladner JT, Arias A, Asogun D, Bielejec F, Caddy SL, Cotten M, D'Ambrozio J, Dellicour S, Caro AD, Diclaro JW, Duraffour S, Elmore MJ, Fakoli LS, Faye O, Gilbert ML, Gevao SM, Gire S, Gladden-Young A, Gnirke A, Goba A, Grant DS, Haagmans BL, Hiscox JA, Jah U, Kugelman JR, Liu D, Lu J, Malboeuf CM, Mate S, Matthews DA, Matranga CB, Meredith LW, Qu J, Quick J, Pas SD, Phan MVT, Pollakis G, Reusken CB, Sanchez-Lockhart M, Schaffner SF, Schieffelin JS, Sealfon RS, Simon-Loriere E, Smits SL, Stoecker K, Thorne L, Tobin EA, Vandi MA, Watson SJ, West K, Whitmer S, Wiley MR, Winnicki SM, Wohl S, Wölfel R, Yozwiak NL, Andersen KG, Blyden SO, Bolay F, Carroll MW, Dahn B, Diallo B, Formenty P, Fraser C, Gao GF, Garry RF, Goodfellow I, Günther S, Happi CT, Holmes EC, Kargbo B, Keïta S, Kellam P, Koopmans MPG, Kuhn JH, Loman NJ, Magassouba NF, Naidoo D, Nichol ST, Nyenswah T, Palacios G, Pybus OG, Sabeti PC, Sall A, Ströher U, Wurie I, Suchard MA, Lemey P, Rambaut A (2017) Virus genomes reveal factors that spread and sustained the Ebola epidemic. Nature. doi:10.1038/nature22040. http://www.nature.com/nature/journal/vaop/ncurrent/abs/nature22040.html#supplementary-information. Accessed 13 Apr 2017

Ewer K, Rampling T, Venkatraman N, Bowyer G, Wright D, Lambe T, Imoukhuede EB, Payne R, Fehling SK, Strecker T, Biedenkopf N, Krahling V, Tully CM, Edwards NJ, Bentley EM, Samuel D, Labbe G, Jin J, Gibani M, Minhinnick A, Wilkie M, Poulton I, Lella N, Roberts R, Hartnell F, Bliss C, Sierra-Davidson K, Powlson J, Berrie E, Tedder R, Roman F, De Ryck I, Nicosia A, Sullivan NJ, Stanley DA, Mbaya OT, Ledgerwood JE, Schwartz RM, Siani L, Colloca S, Folgori A, Di Marco S, Cortese R, Wright E, Becker S, Graham BS, Koup RA, Levine MM, Volkmann A, Chaplin P, Pollard AJ, Draper SJ, Ballou WR, Lawrie A, Gilbert SC, Hill AV (2016) A monovalent chimpanzee adenovirus Ebola vaccine boosted with MVA. N Engl J Med 374(17):1635–1646. doi:10.1056/NEJMoa1411627

Fallah MP, Skrip LA, Raftery P, Kullie M, Borbor W, Laney AS, Blackley DJ, Christie A, Dokubo EK, Lo TQ, Coulter S, Baller A, Vonhm BT, Bemah P, Lomax S, Yeiah A, Wapoe-Sackie Y, Mann J, Clement P, Davies-Wayne G, Hamblion E, Wolfe C, Williams D, Gasasira A, Kateh F, Nyenswah TG, Galvani AP (2017) Bolstering community cooperation in Ebola resurgence protocols: combining field blood draw and point-of-care diagnosis. PLoS Med 14(1):e1002227. doi:10.1371/journal.pmed.1002227

Frieden TR, Tappero JW, Dowell SF, Hien NT, Guillaume FD, Aceng JR (2014) Safer countries through global health security. Lancet 383(9919):764–766. doi:10.1016/s0140-6736(14)60189-6

Grady D (2014) Ebola vaccine, ready for test, sat on the shelf. N Y Times, 23 Oct 2014. https://www.nytimes.com/2014/10/24/health/without-lucrative-market-potential-ebola-vaccine-was-shelved-for-years.html?_r=1. Accessed 27 Apr, 2017

Geisbert TW, Feldmann H (2011) Recombinant vesicular stomatitis virus-based vaccines against Ebola and Marburg virus infections. J Infect Dis 204(Suppl 3):S1075–S1081. doi:10.1093/infdis/jir349

Geisbert TW, Daddario-Dicaprio KM, Geisbert JB, Reed DS, Feldmann F, Grolla A, Ströher U, Fritz EA, Hensley LE, Jones SM, Feldmann H (2008a) Vesicular stomatitis virus-based vaccines protect nonhuman primates against aerosol challenge with Ebola and Marburg viruses. Vaccine 26(52):6894–6900. doi:10.1016/j.vaccine.2008.09.082

Geisbert TW, Daddario-Dicaprio KM, Lewis MG, Geisbert JB, Grolla A, Leung A, Paragas J, Matthias L, Smith MA, Jones SM, Hensley LE, Feldmann H, Jahrling PB (2008b) Vesicular stomatitis virus-based Ebola vaccine is well-tolerated and protects immunocompromised nonhuman primates. PLoS Pathog 4(11):e1000225. doi:10.1371/journal.ppat.1000225

Geisbert TW, Geisbert JB, Leung A, Daddario-DiCaprio KM, Hensley LE, Grolla A, Feldmann H (2009) Single-injection vaccine protects nonhuman primates against infection with Marburg virus and three species of Ebola virus. J Virol 83(14):7296–7304. doi:10.1128/jvi.00561-09

Geisbert TW, Bailey M, Hensley L, Asiedu C, Geisbert J, Stanley D, Honko A, Johnson J, Mulangu S, Pau MG, Custers J, Vellinga J, Hendriks J, Jahrling P, Roederer M, Goudsmit J, Koup R, Sullivan NJ (2011) Recombinant adenovirus serotype 26 (Ad26) and Ad35 vaccine vectors bypass immunity to Ad5 and protect nonhuman primates against Ebolavirus challenge. J Virol 85(9):4222–4233. doi:10.1128/jvi.02407-10

Hankins C (2016) Good participatory practice guidelines for trials of emerging (and re-emerging) pathogens that are likely to cause severe outbreaks in the near future and for which few or no medical countermeasures exist (GPP-EP). Outcome document of the consultative process. World Health Organization. http://www.who.int/csr/research-and-development/documents/GPP-EP-Dec-2016.pdf?ua=1. Accessed 26 Apr 2017

Henao-Restrepo AM, Longini IM, Egger M, Dean NE, Edmunds WJ, Camacho A, Carroll MW, Doumbia M, Draguez B, Duraffour S, Enwere G, Grais R, Gunther S, Hossmann S, Konde MK, Kone S, Kuisma E, Levine MM, Mandal S, Norheim G, Riveros X, Soumah A, Trelle S, Vicari AS, Watson CH, Keita S, Kieny MP, Rottingen JA (2015) Efficacy and effectiveness of an rVSV-vectored vaccine expressing Ebola surface glycoprotein: interim results from the Guinea ring vaccination cluster-randomised trial. Lancet 386(9996):857–866. doi:10.1016/s0140-6736(15)61117-5

Henao-Restrepo AM, Camacho A, Longini IM, Watson CH, Edmunds WJ, Egger M, Carroll MW, Dean NE, Diatta I, Doumbia M, Draguez B, Duraffour S, Enwere G, Grais R, Gunther S, Gsell PS, Hossmann S, Watle SV, Konde MK, Keita S, Kone S, Kuisma E, Levine MM, Mandal S, Mauget T, Norheim G, Riveros X, Soumah A, Trelle S, Vicari AS, Rottingen JA, Kieny MP (2017) Efficacy and effectiveness of an rVSV-vectored vaccine in preventing Ebola virus disease: final results from the Guinea ring vaccination, open-label, cluster-randomised trial (Ebola Ca Suffit!). Lancet 389(10068):505–518. doi:10.1016/s0140-6736(16)32621-6

Herbert AS, Kuehne AI, Barth JF, Ortiz RA, Nichols DK, Zak SE, Stonier SW, Muhammad MA, Bakken RR, Prugar LI, Olinger GG, Groebner JL, Lee JS, Pratt WD, Custer M, Kamrud KI, Smith JF, Hart MK, Dye JM (2013) Venezuelan equine encephalitis virus replicon particle vaccine protects nonhuman primates from intramuscular and aerosol challenge with ebolavirus. J Virol 87(9):4952–4964. doi:10.1128/jvi.03361-12

Huttner A, Dayer JA, Yerly S, Combescure C, Auderset F, Desmeules J, Eickmann M, Finckh A, Goncalves AR, Hooper JW, Kaya G, Krahling V, Kwilas S, Lemaitre B, Matthey A, Silvera P, Becker S, Fast PE, Moorthy V, Kieny MP, Kaiser L, Siegrist CA (2015) The effect of dose on the safety and immunogenicity of the VSV Ebola candidate vaccine: a randomised double-blind, placebo-controlled phase 1/2 trial. Lancet Infect Dis 15(10):1156–1166. doi:10.1016/s1473-3099(15)00154-1

Institute of Medicine, Committee to Review Adverse Effects of Vaccines (2011) Adverse effects of vaccines: evidence and causality. National Academies Press, Washington, DC

Institute of Medicine, Committee to Review the Adverse Consequences of Pertussis and Rubella Vaccines (1991) Adverse effects of pertussis and rubella vaccines: a report of the committee to review the adverse consequences of pertussis and rubella vaccines. National Academy Press, Washington, DC

Johnson RF, Kurup D, Hagen KR, Fisher C, Keshwara R, Papaneri A, Perry DL, Cooper K, Jahrling PB, Wang JT, Ter Meulen J, Wirblich C, Schnell MJ (2016) An inactivated rabies virus based Ebola vaccine, FILORAB1, adjuvanted With glucopyranosyl lipid A in stable emulsion confers complete protection in nonhuman primate challenge models. J Infect Dis 214 (suppl 3):S342–S354. doi:10.1093/infdis/jiw231

Jones SM, Feldmann H, Stroher U, Geisbert JB, Fernando L, Grolla A, Klenk HD, Sullivan NJ, Volchkov VE, Fritz EA, Daddario KM, Hensley LE, Jahrling PB, Geisbert TW (2005) Live attenuated recombinant vaccine protects nonhuman primates against Ebola and Marburg viruses. Nat Med 11(7):786–790. doi:10.1038/nm1258

Jones SM, Stroher U, Fernando L, Qiu X, Alimonti J, Melito P, Bray M, Klenk HD, Feldmann H (2007) Assessment of a vesicular stomatitis virus-based vaccine by use of the mouse model of Ebola virus hemorrhagic fever. J Infect Dis 196(Suppl 2):S404–S412. doi:10.1086/520591

Kennedy S, Bolay F, Kieh M, Grandits G, Badio M, Ballou R, Eckes R, Feinberg M, Follmann D, Grund B, S G, Hensley L, Higgs E, Janosko K, Johnson M, Kateh F, Logue J, Marchand J, Monath T, Nason M, Nyenswah T, Roman F, Stavale E, Wolfson J, Neaton J, Lane H (2017) Phase 2 placebo-controlled trial of 2 vaccines to prevent Ebola in Liberia. N Engl J Med (in press)

Keusch G, McAdam K, Cuff P, Mancher M, Busta ER (2017) Integrating clinical research into epidemic response: the Ebola experience. National Academies Press. https://www.nap.edu/catalog/24739/integrating-clinical-research-into-epidemic-response-the-ebola-experience. Accessed 26 Apr 2017

Kibuuka H, Berkowitz NM, Millard M, Enama ME, Tindikahwa A, Sekiziyivu AB, Costner P, Sitar S, Glover D, Hu Z, Joshi G, Stanley D, Kunchai M, Eller LA, Bailer RT, Koup RA, Nabel GJ, Mascola JR, Sullivan NJ, Graham BS, Roederer M, Michael NL, Robb ML, Ledgerwood JE (2015) Safety and immunogenicity of Ebola virus and Marburg virus glycoprotein DNA vaccines assessed separately and concomitantly in healthy Ugandan adults: a phase 1b, randomised, double-blind, placebo-controlled clinical trial. Lancet 385 (9977):1545–1554. doi:10.1016/s0140-6736(14)62385-0

Kieny MP, Longini IM, Henao-Restrepo AM, Watson CH, Egger M, Edmunds WJ (2016) Changes in the primary outcome in Ebola vaccine trial–Authors' reply. Lancet 387 (10027):1509–1510. doi:10.1016/s0140-6736(16)00686-3

Kobinger GP, Feldmann H, Zhi Y, Schumer G, Gao G, Feldmann F, Jones S, Wilson JM (2006) Chimpanzee adenovirus vaccine protects against Zaire Ebola virus. Virology 346(2):394–401. doi:10.1016/j.virol.2005.10.042

Krause PR (2015) Interim results from a phase 3 Ebola vaccine study in Guinea. Lancet 386 (9996):831–833. doi:10.1016/s0140-6736(15)00002-1

Lau MS, Dalziel BD, Funk S, McClelland A, Tiffany A, Riley S, Metcalf CJ, Grenfell BT (2017) Spatial and temporal dynamics of superspreading events in the 2014–2015 West Africa Ebola epidemic. Proc Natl Acad Sci U S A 114(9):2337–2342. doi:10.1073/pnas.1614595114

Ledgerwood JE, DeZure AD, Stanley DA, Coates EE, Novik L, Enama ME, Berkowitz NM, Hu Z, Joshi G, Ploquin A, Sitar S, Gordon IJ, Plummer SA, Holman LA, Hendel CS, Yamshchikov G, Roman F, Nicosia A, Colloca S, Cortese R, Bailer RT, Schwartz RM, Roederer M, Mascola JR, Koup RA, Sullivan NJ, Graham BS (2017) Chimpanzee adenovirus vector Ebola vaccine. N Engl J Med 376(10):928–938. doi:10.1056/NEJMoa1410863

Li JX, Hou LH, Meng FY, Wu SP, Hu YM, Liang Q, Chu K, Zhang Z, Xu JJ, Tang R, Wang WJ, Liu P, Hu JL, Luo L, Jiang R, Zhu FC, Chen W (2017) Immunity duration of a recombinant adenovirus type-5 vector-based Ebola vaccine and a homologous prime-boost immunisation in healthy adults in China: final report of a randomised, double-blind, placebo-controlled, phase 1 trial. Lancet Glob Health 5(3):e324–e334. doi:10.1016/s2214-109x(16)30367-9

Martin JE, Sullivan NJ, Enama ME, Gordon IJ, Roederer M, Koup RA, Bailer RT, Chakrabarti BK, Bailey MA, Gomez PL, Andrews CA, Moodie Z, Gu L, Stein JA, Nabel GJ, Graham BS (2006) A DNA vaccine for Ebola virus is safe and immunogenic in a phase I clinical trial. Clin Vaccine Immunol 13(11):1267–1277. doi:10.1128/cvi.00162-06

Marzi A, Engelmann F, Feldmann F, Haberthur K, Shupert WL, Brining D, Scott DP, Geisbert TW, Kawaoka Y, Katze MG, Feldmann H, Messaoudi I (2013) Antibodies are necessary for rVSV/ZEBOV-GP-mediated protection against lethal Ebola virus challenge in nonhuman primates. Proc Natl Acad Sci U S A 110(5):1893–1898. doi:10.1073/pnas.1209591110

Maurice J (2014) WHO meeting chooses untried interventions to defeat Ebola. Lancet 384(9948): e45–e46. doi:10.1016/S0140-6736(14)61411-2

Mire CE, Miller AD, Carville A, Westmoreland SV, Geisbert JB, Mansfield KG, Feldmann H, Hensley LE, Geisbert TW (2012) Recombinant vesicular stomatitis virus vaccine vectors expressing filovirus glycoproteins lack neurovirulence in nonhuman primates. PLoS. doi: https://doi.org/10.1371/journal.pntd.0001567

Nyenswah T, Fahnbulleh M, Massaquoi M, Nagbe T, Bawo L, Falla JD, Kohar H, Gasasira A, Nabeth P, Yett S, Gergonne B, Casey S, Espinosa B, McCoy A, Feldman H, Hensley L, Baily M, Fields B, Lo T, Lindblade K, Mott J, Boulanger L, Christie A, Wang S, Montgomery J, Mahoney F (2014) Ebola epidemic—Liberia, March–October 2014. MMWR Morb Mortal Wkly Rep 63(46):1082–1086

Plotkin SA (2017) Vaccines for epidemic infections and the role of CEPI. Hum Vaccin Immunother:1–8. doi:10.1080/21645515.2017.1306615

Qiu X, Fernando L, Alimonti JB, Melito PL, Feldmann F, Dick D, Stroher U, Feldmann H, Jones SM (2009) Mucosal immunization of cynomolgus macaques with the VSVDeltaG/ZEBOVGP vaccine stimulates strong Ebola GP-specific immune responses. PLoS ONE 4(5):e5547. doi:10.1371/journal.pone.0005547

Regules JA, Beigel JH, Paolino KM, Voell J, Castellano AR, Hu Z, Munoz P, Moon JE, Ruck RC, Bennett JW, Twomey PS, Gutierrez RL, Remich SA, Hack HR, Wisniewski ML, Josleyn MD, Kwilas SA, Van Deusen N, Mbaya OT, Zhou Y, Stanley DA, Jing W, Smith KS, Shi M, Ledgerwood JE, Graham BS, Sullivan NJ, Jagodzinski LL, Peel SA, Alimonti JB, Hooper JW, Silvera PM, Martin BK, Monath TP, Ramsey WJ, Link CJ, Lane HC, Michael NL, Davey RT Jr, Thomas SJ (2017) A recombinant vesicular stomatitis virus Ebola vaccine. N Engl J Med 376(4):330–341. doi:10.1056/NEJMoa1414216

Rottingen JA, Gouglas D, Feinberg M, Plotkin S, Raghavan KV, Witty A, Draghia-Akli R, Stoffels P, Piot P (2017) New vaccines against epidemic infectious diseases. N Engl J Med 376 (7):610–613. doi:10.1056/NEJMp1613577

Stanley DA, Honko AN, Asiedu C, Trefry JC, Lau-Kilby AW, Johnson JC, Hensley L, Ammendola V, Abbate A, Grazioli F, Foulds KE, Cheng C, Wang L, Donaldson MM, Colloca S, Folgori A, Roederer M, Nabel GJ, Mascola J, Nicosia A, Cortese R, Koup RA, Sullivan NJ (2014) Chimpanzee adenovirus vaccine generates acute and durable protective immunity against ebolavirus challenge. Nat Med 20(10):1126–1129. doi:10.1038/nm.3702

Sullivan NJ, Sanchez A, Rollin PE, Yang ZY, Nabel GJ (2000) Development of a preventive vaccine for Ebola virus infection in primates. Nature 408(6812):605–609. doi:10.1038/35046108

Sullivan NJ, Geisbert TW, Geisbert JB, Xu L, Yang ZY, Roederer M, Koup RA, Jahrling PB, Nabel GJ (2003) Accelerated vaccination for Ebola virus haemorrhagic fever in non-human primates. Nature 424(6949):681–684. doi:10.1038/nature01876

Tapia MD, Sow SO, Lyke KE, Haidara FC, Diallo F, Doumbia M, Traore A, Coulibaly F, Kodio M, Onwuchekwa U, Sztein MB, Wahid R, Campbell JD, Kieny MP, Moorthy V, Imoukhuede EB, Rampling T, Roman F, De Ryck I, Bellamy AR, Dally L, Mbaya OT, Ploquin A, Zhou Y, Stanley DA, Bailer R, Koup RA, Roederer M, Ledgerwood J, Hill AV, Ballou WR, Sullivan N, Graham B, Levine MM (2016) Use of ChAd3-EBO-Z Ebola virus vaccine in Malian and US adults, and boosting of Malian adults with MVA-BN-Filo: a phase 1, single-blind, randomised trial, a phase 1b, open-label and double-blind, dose-escalation trial, and a nested, randomised, double-blind, placebo-controlled trial. Lancet Infect Dis 16(1):31–42. doi:10.1016/s1473-3099(15)00362-x

United Nations Security Council (2014) Resolution 2176 (2014) adopted by the Security Council at its 7263rd meeting, on 15 September 2014. United Nations Security Council. http://undocs.org/S/RES/2176(2014). Accessed 28 Mar 2017

U.S. Department of Health and Human Services (2016) Ebola response improvement plan based on lessons learned from the 2014–2016 Ebola epidemic. US Department of Health and Human Services. https://www.phe.gov/Preparedness/responders/ebola/Documents/EbolaIP.pdf. Accessed 3 May 2017, Washington, DC

U.S. Department of Health and Human Services (2017) Ebola response improvement plan progress report. U.S. Department of Health and Human Services. https://www.phe.gov/Preparedness/responders/ebola/Documents/erip-report-2017.pdf. Accessed 3 May 2017, Washington, DC

U.S. National Institutes of Health (2017a) A study to evaluate the safety and immunogenicity of a candidate Ebola vaccine in adults. https://clinicaltrials.gov/ct2/show/NCT02485301?term=nct02485301&rank=1. Accessed 24 Mar 2017

U.S. National Institutes of Health (2017b) A study to evaluate the safety and immunogenicity of a candidate Ebola vaccine in children. https://clinicaltrials.gov/ct2/show/NCT02548078?term=nct02548078&rank=1. Accessed 24 Mar 2017

Wang Y, Li J, Hu Y, Liang Q, Wei M, Zhu F (2017) Ebola vaccines in clinical trial: the promising candidates. Hum Vaccin Immunother 13(1):153–168. doi:10.1080/21645515.2016.1225637

Warfield KL, Swenson DL, Olinger GG, Kalina WV, Aman MJ, Bavari S (2007) Ebola virus-like particle-based vaccine protects nonhuman primates against lethal Ebola virus challenge. J Infect Dis 196(Suppl 2):S430–S437. doi:10.1086/520583

Webmaster Administration (2017) PREVAIL SMC ends March 2017 on high note. Daily Observer, Monrovia, LR, 6 Apr 2017. https://www.liberianobserver.com/news/prevail-smc-ends-March-2017-on-high-note/. Access 13 Apr 2017

Widdowson MA, Schrag SJ, Carter RJ, Carr W, Legardy-Williams J, Gibson L, Lisk DR, Jalloh MI, Bash-Taqi DA, Kargbo SA, Idriss A, Deen GF, Russell JB, McDonald W, Albert AP, Basket M, Callis A, Carter VM, Ogunsanya KR, Gee J, Pinner R, Mahon BE, Goldstein ST, Seward JF, Samai M, Schuchat A (2016) Implementing an Ebola vaccine study —Sierra Leone. MMWR Suppl 65(3):98–106. doi:10.15585/mmwr.su6503a14 https://www.cdc.gov/mmwr/volumes/65/su/su6503a14.htm?s_cid=su6503a14_w. Accessed 29 June 2017

World Health Organization (2014a) Ebola virus disease in Guinea. World Health Organization. http://www.afro.who.int/en/disease-outbreaks/outbreak-news/4063-ebola-virus-disease-in-guinea.html. Accessed 13 Apr 2017

World Health Organization (2014b) Liberia: working with communities is the key to stopping Ebola. World Health Organization. http://www.who.int/features/2014/liberia-stopping-ebola/en/. Accessed 27 Mar 2017

World Health Organization (2014c) Meeting summary of the WHO consultation on potential Ebola therapies and vaccines. World Health Organization. http://apps.who.int/iris/bitstream/10665/136103/1/WHO_EVD_Meet_EMP_14.1_eng.pdf?ua=1. Accessed 27 Apr 2017

World Health Organization (2014d) Statement on the 1st meeting of the IHR Emergency Committee on the 2014 Ebola outbreak in West Africa. World Health Organization. http://www.who.int/mediacentre/news/statements/2014/ebola-20140808/en/. Accessed 28 Mar 2017

World Health Organization (2014e) WHO high-level meeting on Ebola vaccines access and financing: summary report. World Health Organization. http://apps.who.int/iris/bitstream/10665/137184/1/WHO_EVD_Meet_EMP_14.2_eng.pdf?ua=1. Accessed 12 May 2017

World Health Organization (2015a) Factors that contributed to undetected spread of the Ebola virus and impeded rapid containment. One year into the Ebola epidemic. World Health Organization. http://www.who.int/csr/disease/ebola/one-year-report/factors/en/. Accessed 27 Mar 2017

World Health Organization (2015b) One year into the Ebola epidemic: a deadly tenacious and unforgiving virus. World Health Organization. http://www.who.int/csr/disease/ebola/one-year-report/introduction/en/ Accessed 14 Apr 2017

World Health Organization (2016a) Ebola virus disease: situation report. World Health Organization. http://apps.who.int/iris/bitstream/10665/208883/1/ebolasitrep_10Jun2016_eng.pdf?ua=1. Accessed 28 Mar 2017

World Health Organization (2016b) Graph of new confirmed cases per epi per week for Guinea. World Health Organization. http://apps.who.int/gho/data/view.ebola-sitrep.ebola-country-GIN-20160511-graph?lang=en. Accessed 11 July 2017

World Health Organization (2016c) Graph of new confirmed cases per epi per week for Sierra Leone. World Health Organization. http://apps.who.int/gho/data/view.ebola-sitrep.ebola-country-SLE-20160511-graph. Accessed 11 July 2017

World Health Organization (2016d) Graph of new confirmed caseswe per epi week for Liberia. World Health Organization. http://apps.who.int/gho/data/view.ebola-sitrep.ebola-country-LBR-20160511-graph. Accessed 11 July 2017

World Health Organization (2016e) Progress report on the development of the WHO health emergencies programme, 30March 2016, WHO, Geneva. http://www.who.int/about/who_reform/emergencycapacities/emergency-programme/en/. Accessed 30 August 2017

Zhang Y, Feng S, Cowling BJ (2016) Changes in the primary outcome in Ebola vaccine trial. Lancet 387(10027):1509. doi:10.1016/s0140-6736(16)30127-1

Zhu FC, Hou LH, Li JX, Wu SP, Liu P, Zhang GR, Hu YM, Meng FY, Xu JJ, Tang R, Zhang JL, Wang WJ, Duan L, Chu K, Liang Q, Hu JL, Luo L, Zhu T, Wang JZ, Chen W (2015) Safety and immunogenicity of a novel recombinant adenovirus type-5 vector-based Ebola vaccine in healthy adults in China: preliminary report of a randomised, double-blind, placebo-controlled, phase 1 trial. Lancet 385(9984):2272–2279. doi:10.1016/s0140-6736(15)60553-0

Therapeutics Against Filovirus Infection

John Connor, Gary Kobinger and Gene Olinger

Abstract Therapies for filovirus infections are urgently needed. The paradoxical issue facing therapies is the need for rigorous safety and efficacy testing, adhering to the principle tenant of medicine to do no harm, while responding to the extreme for a treatment option during an outbreak. Supportive care remains a primary goal for infected patients. Years of research into filoviruses has provided possible medical interventions ranging from direct antivirals, host-factor supportive approaches, and passive immunity. As more basic research is directed toward understanding these pathogens and their impact on the host, effective approaches to treat patients during infection will be identified. The ability to manage outbreaks with medical interventions beyond supportive care will require clinical trial design that will balance the benefits of the patient and scientific community.

Contents

J. Connor (✉)
Department of Microbiology, National Emerging Infectious Diseases Laboratories, Boston University School of Medicine, 620 Albany Street, Boston, MA 02118, USA
e-mail: jhconnor@bu.edu

G. Kobinger
Department of Microbiology, Immunology and Infectious Diseases, Faculty of Medicine, Universite Laval, 2705 Boulevard Laurier, RC-709, Ville de Québec QC G1V 4G2, Canada
e-mail: Gary.Kobinger@crchudequebec.ulaval.ca

G. Olinger
Department of Medicine, National Emerging Infectious Diseases Laboratories, Boston University School of Medicine, 620 Albaney Street, Boston, MA 02118, USA
e-mail: golinger@bu.edu; golinger@mriglobal.org

Current Topics in Microbiology and Immunology (2017) 411:263–290
DOI 10.1007/82_2017_12

Published Online: 27 June 2017

1 Introduction

The 2013–2016 Ebola virus disease (EVD) outbreak in West Africa serves as an example of the dramatic health consequences of severe viral disease transmissions at a time when there are few effective therapies. The appearance and rapid spread of Ebola virus (EBOV) in Guinea, Sierra Leone, and Liberia was unprecedented, and the world's response was off balance and somewhat disorganized at many stages. Much needed potential postexposure therapies were not available in significant supply and guidelines for their use were absent or created on-the-fly (Jacobs et al. 2015). Difficulties in organization and deployment led to few effective antiviral trials in humans (Cohen and Enserink 2016). These difficulties were encountered and occurred despite a longstanding research effort to develop countermeasures against filoviruses that produced many promising candidates. Sadly, the effort to deploy therapies was inadequate to impact the magnitude of disease observed in this outbreak, which should serve as a reminder of the importance of continuing development of these approaches.

This review focuses on postexposure therapies that have been identified and developed for filoviruses, with a heavy focus on those for EBOV. As is applicable, we discuss their development stages following the West African EVD outbreak. This represents a single waypoint in what is hopefully a continuing effort to steadily improve and broaden the therapy and prophylactic options to treat this explosive disease. We first discuss approaches that have sought to block virus replication by directly targeting virus components, followed by a discussion of approaches that target host-factors. Antibody therapeutics are presented as a separate approach because of the extensive issues associated with their testing and development. We also discuss the role of animal models in developing effective therapies and the need for preparedness to deploy potential therapies quickly in future outbreaks.

It is important to note that postexposure therapeutics were not deployed in filovirus outbreaks before the West African outbreak. Earlier outbreaks were halted by epidemiological approaches that effectively detected and isolated cases and case contacts, thus breaking chains of transmission. In these outbreaks, the course of disease was not usually significantly impacted by medical intervention. Healthcare options have been limited during most outbreaks. This has meant that highly desired basic supportive treatment such as provision of replacement fluids, electrolytes, and control of pain and other symptoms was often not achieved. Against this background, the ability to provide effective and targeted postexposure therapies to individuals with filovirus disease has been an obvious priority.

The existing pipeline of potential therapies was a result of years of investment in researching both the pathology of filoviruses and the replication cycle of the viruses themselves. The research has been slow due to the high containment required when working with filoviruses (Biosafety Level 4 maximum containment) and the resulting higher costs to research these high consequence pathogens. Most efforts to develop medical countermeasures have begun with the in vitro and small animal models followed by a staged development into larger animal models.

The regulations 21 CFR 314.600 through 314.650 (drugs) or 21 CFR 601.90 through 601.95 (biological products), commonly referred to as the FDA Animal Efficacy Rule has been an important consideration in the development of medical countermeasures. While focused on allowing intervention approval when human efficacy studies are not ethical and feasible, the development pathway has limited progress of the various interventions. In some cases, lack of resources, limited human clinical data, clear path for development, and lack of agreement on appropriate animal models hampered promising candidate therapies. Linking animal model data to those of human disease is difficult and often does not fully encompass the various clinical symptomology and outcomes observed. Uniformly lethal models of disease have been prioritized due to logistical and resource constraints, which has likely limited potential interventions that could have clinical benefit in humans.

Much of the early success in identifying potential postexposure therapies has come through approaches that seek to block the virus before and/or during cellular infection. Understanding virus replication has been paramount in this aspect of antiviral target discovery. The different life cycle steps of virus entry, replication, assembly and egress from the infected cell have yielded many targets amenable to inhibition and thus are targets for antivirals. Small molecules, small inhibitory RNAs (siRNA) approaches, and antibody approaches have all been used as approaches to inhibit virus replication.

Additionally, our increasing understanding of the pathology of filovirus infection has offered intervention strategies. The understanding of the pathology induced by the virus in hosts, the protective innate and adaptive immune responses that are created or blocked, and basic virus–host interactions have provided great insight into the disease and approaches to alter disease symptoms, morbidity, and mortality. Small molecules, recombinant proteins, including host proteins have been used to disrupt virus-induced and maladaptive host responses. Following the early discovery and development of these interventions, the focus has been to determine the tolerability of interventions within the animal model(s) and efficacy of these interventions unaccompanied by supportive care or combination approaches. As for most viruses, a combination of therapies may be most advantageous in a clinical setting and the search for complimentary and synergistic therapies will be a focus for future research.

Currently there are a variety of approaches that have demonstrated safety and efficacy in animal models. The development of the various potential therapies will rely on both animal models and human safety studies. During outbreaks, clinical studies demonstrating efficacy and safety in humans are possible. However, the animal model remains a key development pathway for preclinical and clinical development. During outbreaks, clinical trials are fraught with ethical, logistical, and political issues that must be considered and managed. Conducting such studies during an outbreak is an enormous challenge and often hastily and poorly designed given the colossal demand by patients and their families. There is a strong pressure to conduct studies under very challenging conditions. Proper study design, adequate data collection, and sharing results are all critical to advancing treatments and to fully assess the safety and efficacy, moreover the continuation of human testing. The past EVD outbreak in West Africa offered considerable insight into this for the current and future development of therapies for filoviruses and other emerging pathogens.

2 Antivirals—Virus Life Cycle Targets

A focus of research into antivirals that block EBOV infection has been the identification of small molecules that directly target specific aspects of the viral life cycle. Figure 1 illustrates the life cycle of EBOV. Like other viruses, EBOV begins its life cycle through attachment of its virions to the surface of a cell through GP/cell surface receptor interactions. The virion is taken up into intracellular vesicles where the GP is proteolytically cleaved to allow additional receptor binding and entry into the cell cytoplasm. Entry into the cytoplasm is marked by primary mRNA transcription and translation, followed by genome amplification and further mRNA transcription. The virion assembly process occurs continuously through the virus life cycle as components become available (Feldmann 2013). To date, small molecules have been identified that target each step of the virus life cycle.

2.1 Small Molecules Targeting EBOV Entry

Many molecules that have some antiviral activity against EBOV target the entry process. Significant among these are small molecules that alter cholesterol biosynthesis or intracellular trafficking of cholesterol. These include cationic amphipathic drugs such as clomiphene and toremiphene (Zhao et al. 2016). These molecules block entry in vitro and are efficacious in an animal model of disease (Johansen et al. 2013, 2015). Molecules such as U18666A that target cholesterol transport and the Neimann-Pick type C 1 protein (NPC1) that is the entry receptor

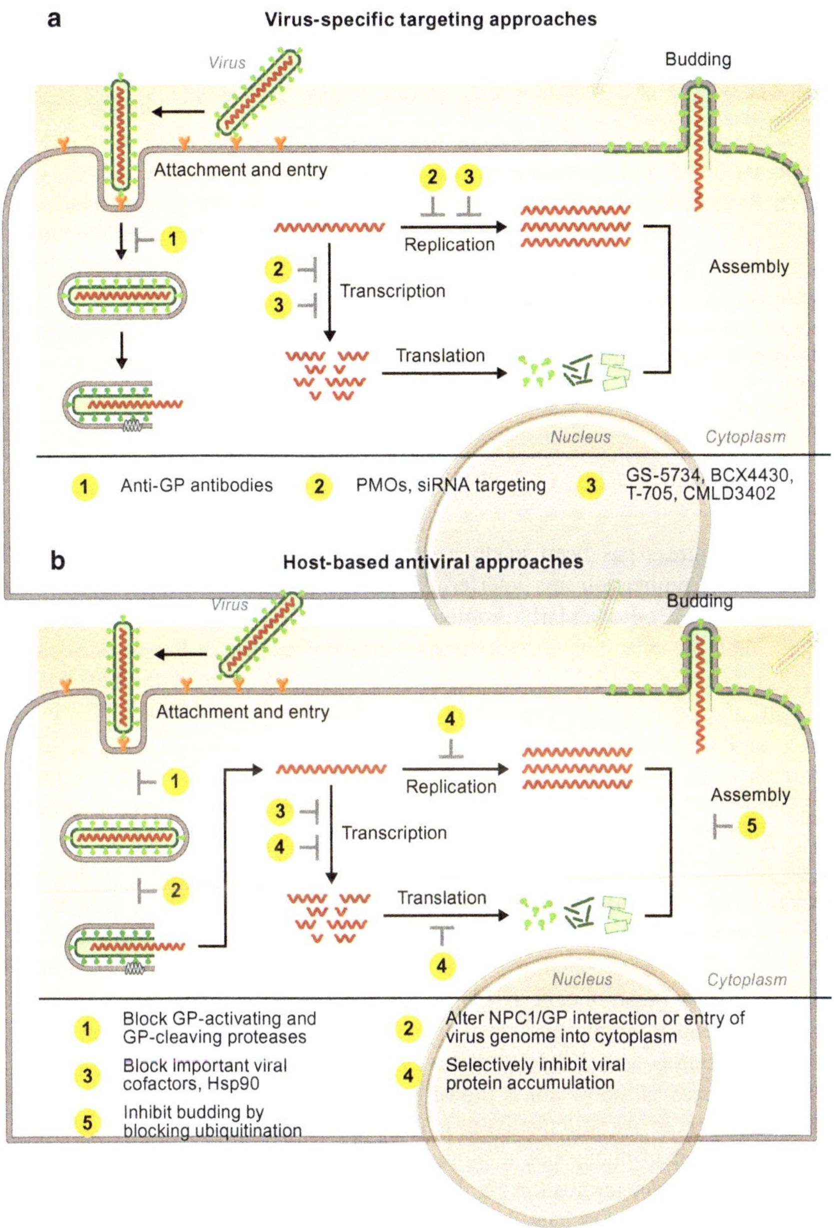

Fig. 1 **a** Virus-specific targeting approaches. **b** Host-based antiviral approaches

for EBOV have been shown to block EBOV replication in vitro. Additionally, several molecules such as compound 3.4 and MBX2254 and MBX2270, disrupt the interaction between EBOV and NPC1 and block virus infection in vitro (Basu et al.

2015; Cote et al. 2011). These compounds highlight the multifaceted potential for targeting the EBOV GP/NPC1 interaction as an effective means of blocking infection in vitro, suggesting potential for these cholesterol manipulation molecules to have an antiviral effect against EBOV infection.

In addition to the effectiveness of these compounds, a number of additional compounds have been identified that block EBOV replication at the entry step. These include molecules such as arbidol (Pecheur et al. 2016), a molecule with broad-spectrum antiviral activity that appears to target entry at the membrane fusion step. It also includes G-protein coupled receptor antagonists (Cheng et al. 2015), small molecule inhibitors of ErbB kinases and PI3 K pathway kinases (Saeed et al. 2008). Inhibitors of cathepsin and cathepsin-like proteases have been shown to block EBOV infection in vitro (Misasi et al. 2012; Zhou et al. 2015).

2.2 *Small Molecules Targeting EBOV Replication*

Significant progress has been made in identifying small molecules that directly inhibit EBOV replication by interfering with the viral RNA-dependent RNA polymerase (RdRP). Particularly prominent in this area are nucleoside analogs that are believed to be substrates for polymerization for viral RdRPs but not for cellular RNA polymerases. Favipiravir, a nucleoside analog developed initially for the treatment of influenza, has been shown to effectively limit EBOV infection in aerosol and intramuscular animal models of infection (Smither et al. 2014; Oestereich et al. 2014). Favipiravir is currently clinically used to treat influenza infection in Japan (Yen 2016) and was utilized as a treatment for EBOV infection during the 2013–2016 outbreak. Due to the rapid implementation of the trial and chaotic nature of the situation the trial was unable to provide strong evidence on associated benefits or risks (Sissoko et al. 2016b). In addition to favipiravir, two other nucleoside analogs had promising results in primate models of EVD. BCX4430, a nucleoside analog blocks EBOV infection in mouse and primate models of disease (Warren et al. 2014). A different purine analog, GS-5734 also has the ability to reduce viremia and promote survival of primates infected with EBOV, confirming strong potential as an anti-EBOV treatment in vivo (Warren et al. 2016). These compounds all show some promise but show different PK parameters that may influence their overall effectiveness (Madelain et al. 2016).

Aside from direct nucleoside analogs, additional compounds have been identified that target virus replication at the level of replication. CMLDBU3402, an indoline alkaloid that has activity against multiple viruses. Experiments using minigenome and full-length virus approaches showed that the molecule blocks EBOV through limiting virus-dependent transcription (Filone et al. 2013). It is possible that this molecule could be combined with nucleoside analogs or other treatments to provide synergistic protection against infection by targeting different aspects of polymerase function.

2.3 *Antisense Approaches Targeting Viral RNA Products*

Antisense RNA has also been used as an approach for limiting the replication of EBOV. Phosphorodiamidate morpholino oligomers (PMOs), which inhibit gene translation by steric blockage of ribosomal assembly, have been one successful approach. A combination of EBOV-specific PMOs targeting sequences of viral mRNAs for the VP24, VP35, and RdRP (L) proteins protected rodents in both pre- and postexposure therapeutic regimens (Warfield et al. 2006). In rhesus macaque models, treatment with a combination of the PMOs of VP24, VP35, and L from 2 days prior to EBOV challenge through day 9 of the infection protected 3 of 4 (75%) rhesus macaques against lethal infection (Warfield et al. 2006). Furthermore, it was demonstrated that the antiviral potency of PMOs could be enhanced by chemical modification, either by conjugating PMOs with peptides or by introducing positive charge to the PMOs (PMO*plus*™, Avi BioPharma, Inc. (Swenson et al. 2009). Subsequently, PMO*plus* targeting EBOV VP24 and VP35 or MARV Musoke VP24 and NP showed significant protection of mice and guinea pigs against lethal challenge with EBOV and MARV Musoke, respectively (Warren et al. 2010). AVI-6002 PMO*plus* against both EBOV VP24 and VP35, and AVI-6003 PMO*plus* against MARV VP24 and NP, were developed and tested for treatment efficacy using NHP models. These PMOs, delivered 30–60 min postexposure, protected 62.5% of rhesus macaques against otherwise lethal EBOV infection and 100% of cynomolgus macaques against MARV Musoke infection (Warren et al. 2010).

Small-interfering RNA (siRNA) also can effectively limit EBOV replication. siRNA targeting the EBOV RdRp formulated in stable nucleic acid-lipid particles (SNALPs) completely protected guinea pigs when administered shortly after an otherwise lethal EBOV infection (Geisbert et al. 2006). In rhesus macaques, a combination of siRNA targeting the EBOV L, VP24, and VP35 were formulated in SNALPs and administrated intravenously. Two of three macaques, which were treated four times with siRNA at 30 min, 1, 3, and 5 days after challenge, survived lethal infection. Furthermore, all four animals treated seven times at 30 min, 1–6 days after challenge survived (Geisbert et al. 2010). Later studies with a re-formulated siRNA to target EBOV-Makona provided further evidence of protection in the NHP model of disease (Thi et al. 2015).

3 Exploiting Host-Factors as Therapeutic Targets

In addition to directly targeting viral factors in the replication process, there has also been success in attacking EBOV replication by targeting host-factors that are essential for Ebola virus replication. Small molecules that are likely targeting host-factors required for EBOV entry have been discussed in the entry section above. In addition to these factors, it has been shown that p38 MAPK inhibitors can also block virus entry (Johnson et al. 2014), as can inhibitors of the PI3 K/Akt signaling pathway (Saeed et al. 2008).

Success has been seen through the indirect targeting of the EBOV RdRP through the host cell chaperone Hsp90. Studies have shown that compounds that limit Hsp90 function will block EBOV replication in vitro, presumably through destabilizing the L RdRP subunit and decreasing viral transcription and replication (Smith et al. 2010). Heme oxygenase activators inhibit EBOV replication, thought the stage is not well established at this point (Hill-Batorski et al. 2013).

Preemptive activation of innate immune responses could also be an effective approach for limiting EBOV infection Pretreatment of cells with interferon beta or gamma prior to EBOV infection has a strong antiviral effect on cells in tissue culture (McCarthy et al. 2016; Rhein et al. 2015). Treatment of cells with RLR agonist has a similar effect on suppressing EBOV replication (Pattabhi et al. 2016). These approaches set off a series of host responses to infection, including the increased transcription of hundreds if not antiviral genes. Understanding the exact mechanism of how innate immune activators limit infection is difficult to know a priori. This potential for multiple mechanisms of action may be advantageous. At least one study supports the idea that an interferon treatment approach could have an effect in vivo (Smith et al. 2013).

Coagulation abnormalities are one of the most prominent hallmarks of filovirus infection. Tissue factor may play an important role in triggering the hemorrhagic complications in NHPs infected with filoviruses (Geisbert et al. 2003a). Overexpression of tissue factor, which performs as the primary cellular inhibitor of the coagulation protease cascades, is one of the causes of DIC and thrombosis-related organ failure. Recombinant nematode anticoagulant protein c2 (rNAPc2), which directly inhibits factor VII and tissue factor, provided partial postexposure protection to rhesus macaques infected with ebolavirus (Geisbert et al. 2003a, b; Hensley et al. 2007). In rNAPc2-treated rhesus macaques, the mean survival time (11.7 days) was longer than that in untreated control monkeys (8.3 days) and 33% of EBOV-infected macaques survived. In MARV Angola-infected rhesus macaques treated with rNAPc2, 1 of 6 (17%) monkeys survived and the mean time to death for five animals was significantly prolonged compared with those of the untreated control monkeys. rNAPc2 demonstrated a clear improvement in survival rate and an increase in mean survival time in a normally 100% lethal model of filovirus infection.

Activated protein C (APC) is generated from protein C, which is a vitamin K-dependent plasma protein, and inactivates factors V and VIII to down-regulate thrombin generation. Circulating levels of protein C were rapidly and significantly reduced in cynomolgus macaques and rhesus macaques during EBOV infections, because protein C is be produced in the liver, which is a main target of filovirus infection (Geisbert et al. 2003b). In rhesus macaques, administration of recombinant human APC (rhAPC) at 30–60 min after challenge and continuing for 7 days, protected 2 of 11 (18%) monkeys against otherwise lethal EBOV infection (Hensley et al. 2007). The mean survival time in the rhAPC-treated monkeys was prolonged compared with the untreated monkeys (Hensley et al. 2007).

During the 2014–2015 outbreak, a long list of proposed therapies ranging from novel small molecules to cardiac modulators to silver particles, and green tea were proposed as therapies. Limited capacity within BSL-4 laboratories did not allow for rapid in vitro and in vivo assessments. For those that were prioritized, most were unsuccessful in initial studies and were abandoned (Glisic et al. 2015; Haque et al. 2015). Some candidates progressed to animal model testing with various levels of efficacy. These and newly developed small molecules are keeping the drug pipeline filled with candidates.

4 Antibodies as Therapies

Defense against pathogens and toxins is provided at least in part by the humoral immune response. Some of the earliest treatments for infection were based on passive immunity, the process of providing serum from previously infected humans or animals to those suffering from infection. In fact, the first Nobel Prize in medicine was awarded for the development of antiserum to combat diphtheria. The power of this approach was self-evident and led to the hypothesis of a "magic bullet" approach, an ideal therapeutic approach, which selectively targets and destroys a disease-causing organism or toxin. While antiserum treatments continued to be develop for other diseases and for intoxinations; the full realization of the "magic bullet" would take another 70 years to be realized with the development of monoclonal antibodies. The ability to utilize antibody therapeutic efforts for infectious diseases has been fraught with complex issues and conflicting observations (Graham et al. 2015). In contrast, the development of small molecules has been simpler and cheaper leading to a preference for antibiotics and eventual abandonment of passive immunity approaches. The promise of both passive immunity with blood products and monoclonal antibodies for treatment of infectious diseases, specifically viral diseases, has been limited because of the higher cost and the learning curve to understand antibody function and antibody characteristics associated with potency and efficacy in vivo. One exception is plaivizumab (Synagis brand name for MedImmune), which has been effectively developed and clinically used to treat human respiratory syncytial virus (HRSV) infections in infants since 1998 (Sandritter 1999). In the past two decades, both serum treatments using traditional and novel technologies, as well as monoclonal antibodies have been rediscovered and shown to have efficacy in animal models of filoviruses infection. Recently, the use of monoclonal antibodies, but not immune sera, has shown promises in the treatment and cure of humans infected with Ebola virus and appear to be a potential therapeutic option. Previous efforts focused on blocking virus entry by targeting the virus glycoprotein. More recently, expanding on this approach, engineered antibody methods and targeting host proteins have been considered.

4.1 Convalescent Whole Blood and Plasma

Passive immunity can be achieved before or after exposure to a pathogen by administering antiviral antibodies to patients from others that have been infected and survived the infection. For Junín and Lassa virus infections, passive transfer of sera collected from survivors has proven effective when treatments were initiated soon after infection (Enria et al. 1984; Jahrling and Peters 1984; Jahrling et al. 1985). The approach is often a salvage therapy approach when no other option exists and can raise numerous concerns during clinical management of a patient. Paramount is the potential for doing harm to the recipient from blood matching requirements and possible transmission of other advantageous pathogens. The need to match blood types (ABO) and antigen (Rh negative/positive) lessens its clinical utility in resource-limited situations. In most emergency uses, the inability to pretest for quality including potency and the absence of other blood borne pathogens in donor blood creates a hurdle for implementation in austere clinical settings (van Griensven et al. 2016a, b, c). Despite these limitations, the approach has been attempted during filovirus outbreaks and laboratory exposures.

In the face of an outbreak and limited options, the age-old approach of immune human convalescent plasma transfers for EVD infection was first reported by Emond et al. (1977). During the 1995 Zaire (now Democratic Republic of the Congo) EVD outbreak, blood transfusions from convalescent patients were once again attempted with eight EVD infected patients (Mupapa et al. 1999). Unfortunately, a clinical benefit was unclear and was impeded by a variety of complications including the study design, the low number of cases, and the possible impact of improved supportive care provided to the treated patients.

More recently during the 2013–2016 EVD outbreak, the approach was attempted and the outcome was uncertain. For clinical assessment, 99 patients diagnosed with EVD were enrolled in a nonrandomized, comparative study who received two consecutive transfusions of 200–250 ml of ABO-compatible convalescent plasma obtained from separate convalescent donor (van Griensven et al. 2016a, b, c). No significant survival benefit was observed with the passive transfer to confirmed EVD patients. Most notable, the antiviral quality of the immune plasma provided was not known prior to administration and it is likely that there were large differences in titer and antiviral activity. In contrast to earlier studies with EVD in NHPs during which protection was observed with pooled convalescent sera after concentration of the total IgG antibodies (Dye et al. 2012), the administration of convalescent serum from macaques infected with EBOV-Makona was not protective in challenged NHPs (Mire et al. 2016). Combined, these findings suggest that unconcentrated convalescent sera have limited value to protect against established infection. While the NHP data suggest that the approach may have clinical benefit, there are many aspects that likely have contributed to the clinical failure of the approach. First and foremost, the overall quality and quantity of anti-EBOV antibodies in each treatment is an important confounding factor. To date, there is no potency assay for serum or plasma, and both virus-specific neutralizing and non-neutralizing antibodies can provide protection in animal models. The antiviral potency between donors can't be assumed, and thus every transfer is confounded by the lack of a potency determination. In some cases, the specific EBOV-neutralizing capacity and overall antibody titer may be analyzed following the study to assess the characteristics of the preparation but clinical use did not permit this assessment during the outbreak. In vitro neutralization assessments may be of little benefit since it is unclear if neutralization activity is required for protective efficacy. Effective dosing, epitope specificity, non-neutralizing protective antibodies with alternative in vivo functions are yet to be understood to develop a potency assay for adequate assessment. As was learned more than a century ago with diphtheria antitoxin, a potency assay is necessary to interpret the effectiveness of convalescent treatment. It is likely that a clinical benefit was observed but only with specific donors with a unique, yet undefined quality of antibody response contained in the serum. The in vitro neutralization and titer of patients is highly variable in survivors (Sobarzo et al. 2013; van Griensven et al. 2016a, b, c). Patients received plasma with anti-EBOV IgG antibodies, but levels of neutralizing antibodies were low in many donations. Knowledge gaps about potency and quality of the donor transfers remains a hurdle for this classical clinical therapy for filoviruses. More research is needed to clarify if convalescent passive immunity will be efficacious for filoviruses. Compounding this technical issue, sources of human immune plasma are scarce early in outbreaks; therefore, antiviral preparations using animals as a source of immune plasma had to be pursued.

4.2 Heterologous Animal-Derived Hyperimmune Polyclonal Antibodies

After the MVD and EVD outbreaks in 1967 and 1976, there was an evident need for a prophylactic and therapeutic option for infected people that could be stockpiled and prepared before new cases were identified. With scarce options and limited human survivors, heterologous hyperimmune serum was an obvious approach. Control over quantity and quality of the preparation is often a benefit of this approach. Traditionally, non-susceptible animals can be immunized with a pathogen to induce a protective antiserum. Alternatively, inactivated virus can be used to generate hyperimmune serum in animals. Hyperimmune serum was one of the first perceived successful therapies for Ebola and Marburg virus infection following accidental laboratory exposures. Sheep, goats, and horse serum were used to generate hyperimmune serum that could protect in both rodent and nonhuman primate animal models of EVD and MVD (Borisevich et al. 1995, 1996; Krasnianskii et al. 1995; Krasnianskii et al. 1994; Kudoyarova-Zubavichene et al. 1999; Markin et al. 1997; Mikhailov et al. 1994; Dowall et al. 2016; Borisevich et al. 2008). Further, the earlier efforts demonstrated the first logical approach to placing an emergency-use drug in a better position for clinical use by conducting safety assessments in healthy human volunteers. Later studies conducted by others using macaque infection models suggested these candidates did not have protective efficacy in other nonhuman primate models of EVD, thus lessening the enthusiasm for the products (Jahrling et al. 1996).

In addition to the conflicting animal model data observed, cost, storage issues, and immunological complications associated with the use of heterologous sera in humans, such as serum sickness and immediate hypersensitivity significantly limited progress in using heterologous hyperimmune serum (Casadevall and Scharff 1995). Traditionally, antibodies from hyperimmune serum can be fractionated into smaller F(ab) or $F(ab')_2$ fragments to avoid the serious side effects observed in human treatments. Theoretically, this would severely impact potency since it could reduce the antibody half-life and in vivo potency. One approach that builds on this approach is the use of hyperimmune serum product derived from a transchromosomic (Tc) bovine in which the bovine immunoglobulin genes are knocked out and the human immunoglobulin genes have been inserted and are expressed. Immunized animals develop a fully human antiviral antibodies that can be collected and purified to treat infected patients. Early animal protection studies demonstrate the efficacy of the EBOV-specific hyperimmune serum (Dye et al. 2016). Combined with the experience of the approach with other viral diseases, the predicted cost and scalability projections, and ongoing human safety studies, the approach is a leading method to respond to emerging outbreaks, including EVD and MVD (Garraud 2017 #165).

4.3 Monoclonal Approaches

As mentioned previously, classical passive immunity approaches such as human intravenous immunoglobulin (IVIG), heterologous animal-derived polyclonal antibodies (pAbs), have a variety of inherent limitations. Monoclonal antibodies (mAb) offer the full realization of the "magic bullet" approach sought in the early days of passive immunity. Paradoxically, while overcoming many of the issues related to pAbs, identification of the optimal antibody clone or clones in the complexity of the enormous and diverse repertoire of antigen specific and nonspecific antibody responses is like finding a needle in a haystack. Years of brute force efforts were required to realize the utility of this approach for filoviruses (Hiatt et al. 2015). However, technological developments, such as improved purification techniques and the ability to engineer humanized mAbs, have greatly lessened the antibody panning process, allowing for increased specificity, expanded the range of possible target and selection (Jin and Simmons 2016). Antibody therapy approaches have focused on the filoviral glycoprotein or on host specific targets that are necessary for filovirus to function.

4.4 Anti-GP MAb-Based Therapies

The selection of the viral glycoprotein as the target for monoclonal antibodies coincided with the development of the vaccines and the early understanding of the role of antibodies in filovirus infection. For EBOV and MARV, vaccine and adoptive transfer studies in rodents revealed a protective role for antibodies against the viral glycoprotein. Strategies to identify GP-specific antibodies focused on either vaccination of animals or human survivors. Numerous therapeutic anti-GP full-length mAbs have been described (Moekotte et al. 2016). These mAbs target different epitopes (some linear, but most of them conformational) of the GP EBOV protein.

However, passive immunity offered limited clinical value in most animal models. As an example, an early GP-specific antibody, a human monoclonal antibody (mAbs) derived from an EVD survivor with strong virus neutralizing capacity was identified. Studies in guinea pigs demonstrated protective efficacy (Parren et al. 2002). Furthermore, passive therapy with convalescent phase blood from ebolavirus immune macaques failed to protect naïve primates (Jahrling et al. 2007). Similarly, passive immune therapy with hyperimmune EVD equine serum failed to demonstrate clinical benefits in the macaque model of infection (Jahrling et al. 1996). For more than 15 years, attempts to achieve passive immune therapy in primates were unsuccessful in the U.S. laboratories. In contrast, baboon models of EVD and MVD demonstrated the utility of the approach (Borisevich et al. 1995; Borisevich et al. 2008). The inability to replicate the successful studies with baboon models of disease, U.S. researchers became dogmatic about the utility for passive immunity as a efficacious treatment for EVD (Jahrling et al. 2007).

The first set of experiments that documented a successful use of passive immunization for EBOV infection in macaques was published in 2012 (Dye et al. 2012; Qiu et al. 2012a, b; Pettitt et al. 2013; Olinger et al. 2012). In these studies, the use of polyclonal antibodies and later monoclonal antibodies protected NHPs from EBOV infection. The pAb that was directly recovered and concentrated from NHPs that survived EBOV or MARV infection protected naïve animals when infected 48 h after infection. The first successful use of anti-GP (EBOV) mAb was based on a mixture of monoclonal antibodies (Qiu et al. 2012a; Olinger et al. 2012). Nearly simultaneously, MB003 and ZMAb, each consisting of three unique monoclonal antibodies against the viral glycoprotein protected NHPs infected with EBOV with a few limitations. An optimized product consisting of the two of the neutralizing antibodies from ZMAb and one antibody from MB003 was developed as ZMapp. The cocktail of three murine antibodies were humanized and produced in transgenic tobacco leaves. ZMapp demonstrated a high level of protection in NHPs when given at 3–5 days after lethal challenge (Qiu et al. 2014). Subsequently during the 2013–2016 Western African EVD outbreak, ZMAPP or convalescent plasma and ZMapp were administered to 25 laboratory confirmed cases of EVD on compassionate grounds. Early use was complicated by logistical and production limitations as well as confounded by the lack of control arm(s) and variability of additional supportive care approaches, however 22 of the 25 people survived (24 of the 25 people receiving at least 2 doses of ZMApp survived) (Group and Multi-National 2016). Ultimately, a randomized clinical trial was initiated toward the conclusion of the outbreak for ZMapp. As stated at the conclusion of the study, ZMapp appeared to be beneficial compared to standard of care, but lacked the prespecified statistical threshold for efficacy set at 97.5% (Group and Multi-National 2016). The inability to reach the statistical threshold was in part due to the lack of new cases as the outbreak ended. Therefore, an early termination of the trial occured after less than half of predicted enrollees had been recruited (72 out of a predicted 200).

Currently, mAb-based therapies have proven to be the most efficient strategy to reverse the progression of a lethal EBOV challenge in animal models of disease in the face of very limited therapeutic options. Thus, numerous teams have led research efforts on the development of monoclonal antibodies directed toward the viral glycoprotein. Antibodies with greater potency, broader reactivity, and bi-functional capabilities are desired. Isolating antibodies from human survivors offers a unique opportunity to obtain protective antibodies. While ZMapp requires multiple antibodies for optimal efficacy others have identified antibodies with potency that only requires a single antibody for protection in the NHP model (Corti et al. 2016). The monoclonal antibodies mAb100 and mAb114 isolated from a survivor in 2006 provide protective capacity and avoid the generation of virus escape mutations (Kugelman et al. 2015). In preliminary NHP studies, mAb100 and mAb114 in combination or mAb114 monotherapy protected NHPs from a lethal infection. More extensive in vitro and in vivo studies are necessary to ensure that viral escape mutants will not be induced in monotherapy approaches. Ultimately, an immune therapy consisting of fewer antibodies, less frequent treatments, and lower overall concentration of antibodies required could offer lower costs and clinical utility.

Another approach has focused on finding broadly reactive antibodies that can recognize a broader range of filoviruses. Most vaccines induce a protective response that is limited to either EBOV or Sudan virus (SUDV). Multiple pan-ebola virus antibodies were identified that react to the EBOV, SUDV, Bundibugyo virus (BDBV), and Reston virus (RESTV). It is rare that the vaccination with one ebolavirus provides cross protective (pan-ebolavirus) immunity. The ideal vaccine or therapeutic would provide protection to ebolaviruses and marburgviruses (pan-filovirus). Despite the challenge, several pan-ebolavirus and pan-filovirus monoclonal antibodies have been identified (Holtsberg et al. 2015; Fusco et al. 2015). Moreover, the cross-reactive monoclonal antibodies offer protection in postexposure murine models of infection demonstrating systemic utility. More research is needed to understand these novel epitopes and the utility of these monoclonal antibodies in NHP models of infection.

Recently, antibody engineering has been used to combine knowledge about GP-binding epitopes and knowledge about the virus entry mechanism to further the use of monoclonal antibodies as a therapeutic. Bispecific antibodies (bsMab) are artificially engineered monoclonal antibodies (MAbs) that consist of two distinct binding sites and are capable of binding two different antigens noncovalently. The approach has been demonstrated with two types of bispecific antibodies, each

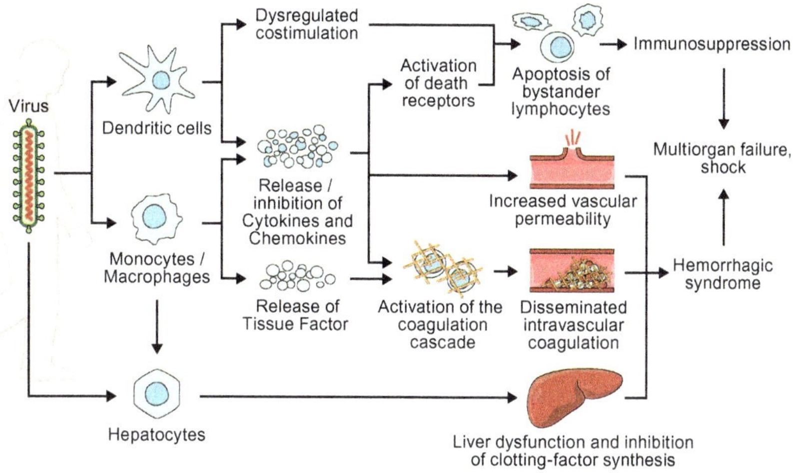

Fig. 2 Virus dissemination during infection and sequelae. Cartoon illustrates that EBOV initially infects immune cells, and from there disseminates to hepatocytes. Continued replication of the virus leads to several downstream responses including the alteration of dendritic cell function and the release of cytokines and tissue factor. These events can contribute to increased vascular permeability, cell death, and the activation of clotting cascades that contribute to overall pathology

consisting of two monoclonal antibodies combined into one antibody structure. One bispecific antibody was devised to neutralize the viral glycoprotein epitope that binds to NPC1, a critical step in the entry processs, whereas the other variable region targets the host NPC1 (Wec et al. 2016). The antibody FVM09, which binds to the surface glycoproteins of all ebola viruses was used to target the virus, offering the potential of broader virus specificity. The bispecific function is provided by the antibody MR72 which targets the NPC1-binding site in glycoprotein which becomes accessible by all filoviruses in lysosomes. Within the lysosome, the antibody MR72 was combined with FVM09 in a bispecific mAb. With one bispecific antibody targeting the "lock" (NPC1) and the other targeting the "key" (the virus's NPC1-binding protein), both have the potential for preventing Ebola virus from interacting with NPC1 and escaping from the lysosome into the cytoplasm (Fig. 2).

4.5 *Targeting Critical Virus–Host Interactions with Antibodies*

A major focus has been focused on targeting the virus by selecting antibodies that recognize and incapacitate the virus. An alternative approach is to target virus critical host proteins or proteins that are exposed because of virus infection. Special consideration is necessary as the antibodies are targeting a host protein, process or tissue, resulting in a delicate balance between stopping the virus and doing no harm to the host. Critical to these targets is the basic research that has revealed the critical host proteins usurped by the virus. One advantage of targeting host proteins is to limit viral escape mutations. The aforementioned entry mechanism utilizing the host protein NPC1, revealed a conserved entry mechanism for host interactions for all filoviruses (Aman 2016). Similarly, several studies have focused on targeting critical host proteins with monoclonal antibodies. One such approach has been the monoclonal human-mouse chimeric antibody that targets phospholipid phosphatidylserine (PS). Bavituximab (PGN401) is a monoclonal antibody to PS that was identified for oncology drug development (Thorpe 2010). Normally, PS is restricted to the inner side of the cell plasma membrane, however, under stress conditions, including infection by filoviruses, PS is flipped to the outer membrane. As a result, cells that are undergoing stress either by infection and cancer can be targeted. The drug candidate offers broad-spectrum utility since PGN401 can recognize PS change in cells infected with hepatitis C virus, Pichindé virus, influenza A virus, vaccina virus, vesicular stomatitis Indiana virus and murine cytomegalovirus (Soares et al. 2008). Similarly, the ability to target ebola virus-infected cells was demonstrated (Dowall et al. 2013, 2015). The mechanism of action for PGN401 is likely more complex than simply targeting infected cells for clearance by antibody clearance mechanisms. Changes in PS are known to suppress immune responses and inflammatory responses by binding to macrophage PS receptors (Gerber et al. 2016). The systemic impact of these changes in cytokines, inflammation, and potential tissue destruction during filovirus infection remains unknown.

5 Challenges of Discovery, Animal Models, Preclinical and Clinical Development

Prior to the outbreak in Western Africa, most of the knowledge about filovirus disease was established on the animal models and less on human infections. Candidate antivirals were often discovered with in vitro studies, followed by rodent and nonhuman primate animal protection assessment (Fig. 3). The animal models or filoviruses, in rodents and nonhuman primates induce an acute infection with uniform lethality in less than 10 days of exposure. The lethal dose often is low as measured by in vitro assays. NHPs models and have been traditionally the primary development hurdle for potential medical countermeasures to proceed to preclinical assessments. Although survival has been the primary measure; morbidity, delay to death, and reduction in viremia are secondary measures of effectiveness. A strong weight has been placed on examining models that most completely replicate the most severe syndromic outcome of disease that exists in the human population, that of hemorrhagic complication or death.

In humans, EVD and MVD are not uniformly lethal. Our understanding of viral immunity suggests that there will be a continuum of outcomes ranging from resistance to infection as observed with other viral diseases (Rasmussen et al. 2014),

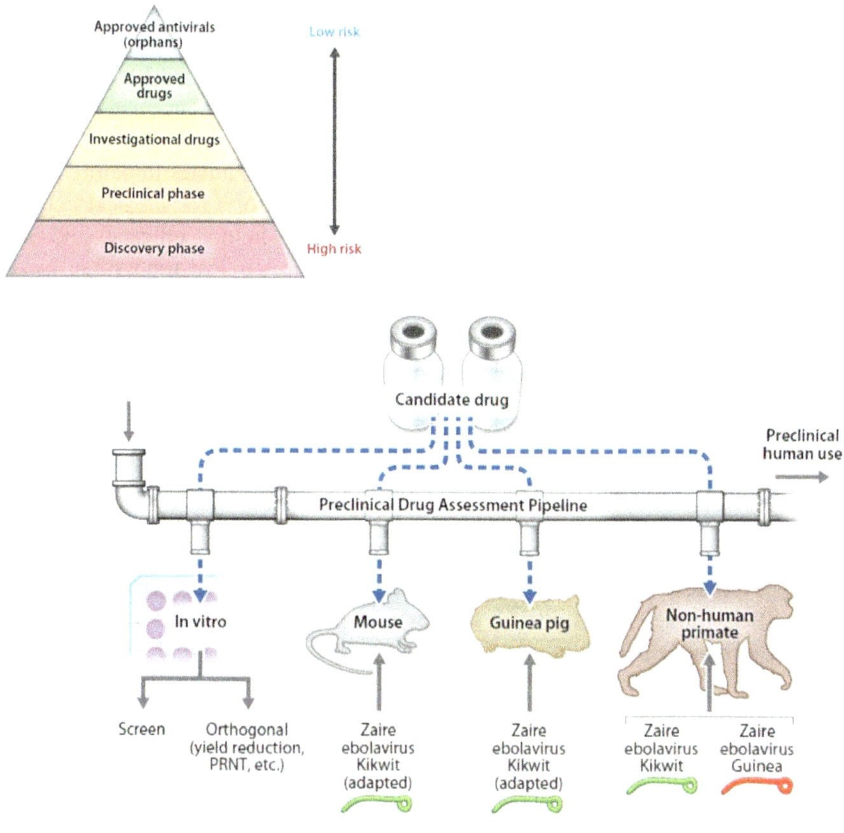

Fig. 3 Therapeutic hierarchy and development pipeline

to asymptomatic cases (Sissoko et al. 2016a), to varying degrees of symptomatic clinical manifestations disease (Xu et al. 2016; Leligdowicz et al. 2016; Mattia et al. 2016). Diarrhea was observed in West African EBOV outbreak, a clinical feature that is rare in the earlier NHP models of disease (Chertow et al. 2014). More recent data with the EBOV-Makona variant suggest that 10–20% of NHPs exhibit diarrhea (Wong et al. 2016), which was also observed in the most recent outbreak (Hunt et al. 2015). In humans, survival is common and clinical sequela is observed with moderate to severe long-term impacts on the patient. In some survivors, virus has been detected in reproductive, eye, and CNS tissue in humans long after symptoms subside and the patient is declared virus free (Brainard et al. 2016a, b). Some of these clinical features have been observed in NHPs (Larsen et al. 2007) but remain poorly characterized. Thus, the current knowledge gained from the animal models is of the most acute and lethal outcome. Further efforts recapitulating more of the human experience with animal models will be necessary to understand the effectiveness of medical interventions in preventing long-term sequelae.

As the efforts to progress toward improved filovirus treatment continue, the ability to compare and contrast the human and animal model data will be essential. Opportunities for phase 1 or 2 clinical trials in Africa on sustained EVD outbreaks are fraught with ethical, moral, regulatory, and logistical constraints. Therefore, it is very difficult to compare human and animal model datasets, allowing for the understanding of what models to use to test therapeutic candidates in before human use. There were a number of advanced development efforts for medical countermeasures, which began the regulatory steps toward a human efficacy studies in Western Africa (Cardile et al. 2016a, b). These human clinical trials often lacked the predictive power due to limited patient enrollments at the end of the outbreak and/or declining case fatality rates in the associated ETUs. Assessment in animal models and human clinical safety assessment will advance candidates for any future outbreak.

Unfortunately, there was no clear clinical evidence of a miracle post exposure therapeutic that is efficacious against EBOV, although positive results on the development of a vaccine were obtained (Agnandji et al. 2016; Huttner et al. 2015). The clinical experience suggests that obtaining optimal bedside supportive care has an impact on patient survival rates although this was also never formally evaluated with appropriate control arms and thus remain speculative in essence (Sueblinvong et al. 2015; Liddell et al. 2015). There were several studies that suggest that therapeutic benefit beyond supportive care was not observed and a few that had trends toward showing benefit (Cardile et al. 2016a, b).

Therapeutic clinical options became paramount to clinicians, politicians and patients as cases continued to rise in Western Africa. Most drugs tested in the clinical trial conducted in Western Africa, Europe, and the U.S. had shown benefit in the animal models of EVD. The NHP model remains the most stringent and acceptable model to demonstrate efficacy against an otherwise lethal challenge and was considered the ideal threshold for consideration for advancement to human use. Examples of the various levels of preclinical assessment include ZMapp, Brincidofovir, TKM-130803, and Favipiravir (T-705) (Dunning et al. 2016a, b; Group and Multi-National 2016; Bai et al. 2016). TKM-130803, Favipiravir, and ZMapp, demonstrated efficacy in animal models of infection. Brincidofovir had demonstrated in vitro efficacy under specific in vitro conditions, however efficacy for EVD in an animal had not been conducted before human-use studies in Western Africa (McMullan et al. 2016). Both Brincidofivir and Favipiravir had preexisting human clinical safety data. In some cases, with limited in vitro or human safety data clinical application of antivirals were initiated at various Ebola Treatment Units throughout Liberia, Sierra Leone, and Guinea (Richardson et al. 2017). Uncontrolled studies and even rogue treatments by clinicians lead to a continuum of drugs being assessed, often without any or limited preliminary data from in vitro or in vivo models. In these sporadic and compassionate attempts to provide care to patients, the ability to ascertain efficacy compared to "standard of care" was nearly impossible due to study design. These studies further highlighted the need to link basic, applied, and preclinical data for candidate antivirals to the clinical application.

6 Timely and Appropriate Therapeutic Approaches of Clinical Intervention During an Outbreak

The experience with therapeutics for EVD have illustrated the difficulty of developing therapeutics in the absence and then during an outbreak. The inability to have candidates in Phase I safety trials impeded the ability to proceed with compassionate use of clinical trials. The lack of demonstrated safety profiles and a regulatory approach to perform clinical studies during an outbreak in resource-limited settings significantly limited the ability to respond and to implement ethical studies that could provide a definitive assess the value of intervention. The experience and the knowledge gained from understanding the linkage between the research and the clinical experience offers a unique opportunity to prepare the global response network to prepare for the next outbreak response.

Therapeutics must advance beyond preclinical studies and need to be supported to the point of demonstrating safety in healthy populations before their entry into an emergency clinical trial. The value of randomized or adaptive clinical trial designs should be discussed globally to determine the appropriate approach for the situation, culture and population. In the middle of an outbreak, this discussion becomes difficult, if not impossible. Establishment of regulatory framework in areas where filovirus may impact the population will be a necessary step. Evaluating the true contribution of supportive care and its various algorithms will be critical in assessing any co-administered therapeutic. Along with fluid management, enhanced levels of clinical assessment and diagnostic testing may be ways to further improve survival beyond what has been seen elsewhere and in previous outbreaks of EVD.

Once supportive care is established, preplanned a coordinated clinical protocol will be necessary to consider clinical efficacy studies. Next, as the therapeutics options change due to discovery research, the ranking of therapeutic options will vary and should be discussed at least annually to be better prepared for any future outbreak. A unified and well-planned approach before an outbreak may avoid the chaos observed in the West African EVD outbreak, lessen confusion and rogue efforts to treat patients in the future. A positive note resides to effectively control the next outbreak caused by EBOV. There is now a vaccine (rVSV-EBOV) that was shown to be protective in an advanced human trial prophylactically and ZMapp, which was reported by NIH to be a promising therapeutic and is currently available through the FDA emergency rule protocol and continue to advance toward licensure also using the animal rule (Huttner et al. 2015; Agnandji et al. 2016; Group and Multi-National 2016).

Marrying the knowledge from the past, understanding the gaps, and developing new knowledge including improved animal models will be critical for future responses especially to filoviruses such as Marburg virus (MARV), SUDV, BDBV, and possibly new emerging viruses. Adequate funding must be provided to place clinical options ready to be used in an outbreak, including the much dreaded "valley of death" where drugs suffer from high attrition rates as they move from successful laboratory testing and begin to fail in preclinical and clinical stages of development

(Butler 2008). Adequate funding during preclinical and early clinical safety assessments of promising therapies is an essential step in development that is poorly funded. In a concurrent approach, the refined animal models that best recapitulate the human disease syndromes, new therapeutic options can be adequately assessed and prepared for emergency compassionate use in any future outbreaks.

7 Future and Outlook

The pipeline to drug development is mature and filled with promising candidates. Basic research reveals continuous opportunities to inhibit virus disease in humans. Once discovered, applied and translational studies move the candidate therapy to human safety studies. The steady research efforts of the numerous decades provided the knowledge and the countermeasure options that were used in the 2014–2015 EVD outbreak. Unfortunately, many of these promising candidates were hampered by the lack of funding to proceed from preclinical to safety clinical assessments to reduce the risk associated with emergency clinical use. Fortunately, research continues and the pipeline is robust and promising. A clearer regulatory pathway is being applied to the candidate products that can assist during outbreaks and beyond (Russek-Cohen et al. 2016). Continued funding and the subsequent research will provide additional options for therapies for filoviruses. Some of these options are broad-spectrum and may have utility against all filoviruses including SUDV and marburgviruses; and there are equal efforts to obtain virus-specific and host-directed therapies. In conclusion, the pipeline offers a wealth of emerging options and our experience during the past three years, demonstrates the gaps and the importance of this research effort.

Acknowledgements The authors would like to thank Jiro Wada for figure preparation, respectively. The content of this publication does not necessarily reflect the views or policies of the US Department of Health and Human Services (DHHS) or of the institutions and companies affiliated with the authors. JHC acknowledges support from NIH R21 AI121933, NIH RO1 AI1096159-04. This work was funded in part through Battelle Memorial Institute's prime contract with the US National Institute of Allergy and Infectious Diseases (NIAID) under Contract No. HHSN272200700016I. J.W. performed this work as employees of Battelle Memorial Institute.

References

Agnandji ST, Huttner A, Zinser ME, Njuguna P, Dahlke C, Fernandes JF, Yerly S, Dayer JA, Kraehling V, Kasonta R, Adegnika AA, Altfeld M, Auderset F, Bache EB, Biedenkopf N, Borregaard S, Brosnahan JS, Burrow R, Combescure C, Desmeules J, Eickmann M, Fehling SK, Finckh A, Goncalves AR, Grobusch MP, Hooper J, Jambrecina A, Kabwende AL, Kaya G, Kimani D, Lell B, Lemaitre B, Lohse AW, Massinga-Loembe M, Matthey A, Mordmuller B, Nolting A, Ogwang C, Ramharter M, Schmidt-Chanasit J, Schmiedel S, Silvera P, Stahl FR, Staines HM, Strecker T, Stubbe HC, Tsofa B, Zaki S, Fast P, Moorthy V,

Kaiser L, Krishna S, Becker S, Kieny MP, Bejon P, Kremsner PG, Addo MM, Siegrist CA (2016) Phase 1 Trials of rVSV Ebola Vaccine in Africa and Europe. New Engl J Med 374 (17):1647–1660. doi:10.1056/NEJMoa1502924

Aman MJ (2016) Chasing Ebola through the Endosomal Labyrinth. MBio 7(2), e00346. doi:10.1128/mBio.00346-16

Bai CQ, Mu JS, Kargbo D, Song YB, Niu WK, Nie WM, Kanu A, Liu WW, Wang YP, Dafae F, Yan T, Hu Y, Deng YQ, Lu HJ, Yang F, Zhang XG, Sun Y, Cao YX, Su HX, Sun Y, Liu WS, Wang CY, Qian J, Liu L, Wang H, Tong YG, Liu ZY, Chen YS, Wang HQ, Kargbo B, Gao GF, Jiang JF (2016) Clinical and Virological Characteristics of Ebola Virus Disease Patients Treated With Favipiravir (T-705)-Sierra Leone 2014. Clin Infect Dis : Official publication Infect Dis Soc Am 63(10):1288–1294. doi:10.1093/cid/ciw571

Basu A, Mills DM, Mitchell D, Ndungo E, Williams JD, Herbert AS, Dye JM, Moir DT, Chandran K, Patterson JL, Rong L, Bowlin TL (2015) Novel Small Molecule Entry Inhibitors of Ebola Virus. J Infect Dis 212(Suppl 2):425–434. doi:10.1093/infdis/jiv223

Borisevich IV, Mikhailov VV, Krasnianskii VP, Gradoboev VN, Lebedinskaia EV, Potryvaeva NV, Timan'kova GD (1995) Development and study of the properties of immunoglobulin against Ebola fever. Vopr Virusol 40(6):270–273

Borisevich IV, Mikhailov VV, Potryvaeva NV, Malinkin IuN, Kirillov AP, Krasnianskii VP, Markov VI, Makhlai AA, Lebedinskaia EV (1996) Development of the immunoenzyme test-system for detection of Ebola virus antigen. Vopr Virusol 41(5):232–234

Borisevich IV, Potryvaeva NV, Mel'nikov SA, Evseev AA, Krasnianskii VP, Maksimov VA (2008) Design of equine serum-based Marburg virus immunoglobulin. Vopr Virusol 53(1): 39–41

Brainard J, Hooper L, Pond K, Edmunds K, Hunter PR (2016a) Risk factors for transmission of Ebola or Marburg virus disease: a systematic review and meta-analysis. Int J Epidemiol 45 (1):102–116. doi:10.1093/ije/dyv307

Brainard J, Pond K, Hooper L, Edmunds K, Hunter P (2016b) Presence and persistence of ebola or marburg virus in patients and survivors: a rapid systematic review. PLoS Neglected Tropical Dis 10(2):e0004475. doi:10.1371/journal.pntd.0004475

Butler D (2008) Gabon centre refocuses on emerging diseases. Nature 455(7217), 1156–1157. doi:10.1038/4551156b

Cardile AP, Downey LG, Wiseman PD, Warren TK, Bavari S (2016a) Antiviral therapeutics for the treatment of Ebola virus infection. Curr Opin Pharmacol 30:138–143. doi:10.1016/j.coph.2016.08.016

Cardile AP, Warren TK, Martins KA, Reisler RB, Bavari S (2016b) Will there be a cure for Ebola? Annu Rev Pharmacol Toxicol. doi:10.1146/annurev-pharmtox-010716-105055

Casadevall A, Scharff MD (1995) Return to the past: the case for antibody-based therapies in infectious diseases. Clin Infect Dis Official Publication Infect Dis Soc Am 21(1):150–161

Cheng H, Lear-Rooney CM, Johansen L, Varhegyi E, Chen ZW, Olinger GG, Rong L (2015) Inhibition of Ebola and Marburg virus entry by G Protein-coupled receptor Antagonists. J Virol 89(19):9932–9938. doi:10.1128/jvi.01337-15

Chertow DS, Kleine C, Edwards JK, Scaini R, Giuliani R, Sprecher A (2014) Ebola virus disease in West Africa–clinical manifestations and management. New Engl J Med 371(22):2054–2057. doi:10.1056/NEJMp1413084

Cohen J, Enserink M (2016) Infectious Disease. As Ebola epidemic draws to a close, a thin scientific harvest. Science 351(6268):12–13. doi:10.1126/science.351.6268.12

Corti D, Misasi J, Mulangu S, Stanley DA, Kanekiyo M, Wollen S, Ploquin A, Doria-Rose NA, Staupe RP, Bailey M, Shi W, Choe M, Marcus H, Thompson EA, Cagigi A, Silacci C, Fernandez-Rodriguez B, Perez L, Sallusto F, Vanzetta F, Agatic G, Cameroni E, Kisalu N, Gordon I, Ledgerwood JE, Mascola JR, Graham BS, Muyembe-Tamfun JJ, Trefry JC, Lanzavecchia A, Sullivan NJ (2016) Protective monotherapy against lethal Ebola virus infection by a potently neutralizing antibody. Science 351(6279):1339–1342. doi:10.1126/science.aad5224

Cote M, Misasi J, Ren T, Bruchez A, Lee K, Filone CM, Hensley L, Li Q, Ory D, Chandran K, Cunningham J (2011) Small molecule inhibitors reveal Niemann-Pick C1 is essential for Ebola virus infection. Nature 477(7364):344–348. doi:10.1038/nature10380

Dowall S, Taylor I, Yeates P, Smith L, Rule A, Easterbrook L, Bruce C, Cook N, Corbin-Lickfett K, Empig C, Schlunegger K, Graham V, Dennis M, Hewson R (2013) Catheterized guinea pigs infected with Ebola Zaire virus allows safer sequential sampling to determine the pharmacokinetic profile of a phosphatidylserine-targeting monoclonal antibody. Antiviral Res 97(2):108–111. doi:10.1016/j.antiviral.2012.11.003

Dowall SD, Bosworth A, Rayner E, Taylor I, Landon J, Cameron I, Coxon R, Al Abdulla I, Graham VA, Hall G, Kobinger G, Hewson R, Carroll MW (2016) Post-exposure treatment of Ebola virus disease in guinea pigs using EBOTAb, an ovine antibody-based therapeutic. Sci Rep 6:30497. doi:10.1038/srep30497

Dowall SD, Graham VA, Corbin-Lickfett K, Empig C, Schlunegger K, Bruce CB, Easterbrook L, Hewson R (2015) Effective binding of a phosphatidylserine-targeting antibody to Ebola virus infected cells and purified virions. J Immunol Res 2015:347903. doi:10.1155/2015/347903

Dunning J, Kennedy SB, Antierens A, Whitehead J, Ciglenecki I, Carson G, Kanapathipillai R, Castle L, Howell-Jones R, Pardinaz-Solis R, Grove J, Scott J, Lang T, Olliaro P, Horby PW (2016a) Experimental treatment of ebola virus disease with Brincidofovir. PLoS ONE 11(9): e0162199. doi:10.1371/journal.pone.0162199

Dunning J, Sahr F, Rojek A, Gannon F, Carson G, Idriss B, Massaquoi T, Gandi R, Joseph S, Osman HK, Brooks TJ, Simpson AJ, Goodfellow I, Thorne L, Arias A, Merson L, Castle L, Howell-Jones R, Pardinaz-Solis R, Hope-Gill B, Ferri M, Grove J, Kowalski M, Stepniewska K, Lang T, Whitehead J, Olliaro P, Samai M, Horby PW (2016b) Experimental treatment of Ebola virus disease with TKM-130803: a Single-Arm Phase 2 Clinical Trial. PLoS Med 13(4):e1001997. doi:10.1371/journal.pmed.1001997

Dye JM, Herbert AS, Kuehne AI, Barth JF, Muhammad MA, Zak SE, Ortiz RA, Prugar LI, Pratt WD (2012) Postexposure antibody prophylaxis protects nonhuman primates from filovirus disease. Proc Natl Acad Sci USA 109(13):5034–5039. doi:10.1073/pnas.1200409109

Dye JM, Wu H, Hooper JW, Khurana S, Kuehne AI, Coyle EM, Ortiz RA, Fuentes S, Herbert AS, Golding H, Bakken RA, Brannan JM, Kwilas SA, Sullivan EJ, Luke TC, Smith G, Glenn G, Li W, Ye L, Yang C, Compans RW, Tripp RA, Jiao JA (2016) Production of Potent Fully Human Polyclonal Antibodies against Ebola Zaire Virus in Transchromosomal Cattle. Sci Rep 6:24897. doi:10.1038/srep24897

Emond RT, Evans B, Bowen ET, Lloyd G (1977) A case of Ebola virus infection. BMJ 2 (6086):541–544

Enria D, Fernandez N, Briggiler A, Levis S, Maiztegui J (1984) Importance of dose of neutralising antibodies in treatment of Argentine haemorrhagic fever with immune plasma. The Lancet 324 (8397):255–256

Feldmann HSa, Geisbert TW (2013) Filoviridae: Marburg and Ebola viruses. In: Knipe DM, Howley PM (eds) Fields Firology, 6th edn. Lippincott Williams and Wilkins, Philadelphia, pp 923–956

Filone CM, Hodges EN, Honeyman B, Bushkin GG, Boyd K, Platt A, Ni F, Strom K, Hensley L, Snyder JK, Connor JH (2013) Identification of a broad-spectrum inhibitor of viral RNA synthesis: validation of a prototype virus-based approach. Chem Biol 20(3):424–433. doi:10.1016/j.chembiol.2013.02.011

Fusco ML, Hashiguchi T, Cassan R, Biggins JE, Murin CD, Warfield KL, Li S, Holtsberg FW, Shulenin S, Vu H, Olinger GG, Kim DH, Whaley KJ, Zeitlin L, Ward AB, Nykiforuk C, Aman MJ, Berry JD, Saphire EO (2015) Protective mAbs and Cross-Reactive mAbs Raised by Immunization with Engineered Marburg Virus GPs. PLoS Pathog 11(6):e1005016. doi:10.1371/journal.ppat.1005016

Geisbert TW, Hensley LE, Jahrling PB, Larsen T, Geisbert JB, Paragas J, Young HA, Fredeking TM, Rote WE, Vlasuk GP (2003a) Treatment of Ebola virus infection with a recombinant inhibitor of factor VIIa/tissue factor: a study in rhesus monkeys. Lancet 362 (9400):1953–1958. doi:10.1016/s0140-6736(03)15012-x

Geisbert TW, Hensley LE, Kagan E, Yu EZ, Geisbert JB, Daddario-DiCaprio K, Fritz EA, Jahrling PB, McClintock K, Phelps JR, Lee AC, Judge A, Jeffs LB, MacLachlan I (2006) Postexposure protection of guinea pigs against a lethal ebola virus challenge is conferred by RNA interference. J Infect Dis 193(12):1650–1657. doi:10.1086/504267

Geisbert TW, Lee AC, Robbins M, Geisbert JB, Honko AN, Sood V, Johnson JC, de Jong S, Tavakoli I, Judge A, Hensley LE, Maclachlan I (2010) Postexposure protection of non-human primates against a lethal Ebola virus challenge with RNA interference: a proof-of-concept study. Lancet 375(9729):1896–1905. doi:10.1016/s0140-6736(10)60357-1

Geisbert TW, Young HA, Jahrling PB, Davis KJ, Kagan E, Hensley LE (2003b) Mechanisms underlying coagulation abnormalities in ebola hemorrhagic fever: overexpression of tissue factor in primate monocytes/macrophages is a key event. J Infect Dis 188(11):1618–1629. doi:10.1086/379724

Gerber DE, Spigel DR, Giorgadze D, Shtivelband M, Ponomarova OV, Shan JS, Menander KB, Belani CP (2016) Docetaxel combined With Bavituximab in previously treated, advanced Nonsquamous non-small-cell lung Cancer. Clin Lung Cancer 17(3):169–176. doi:10.1016/j.cllc.2016.02.003

Glisic S, Paessler S, Veljkovic N, Perovic VR, Prljic J, Veljkovic V (2015) Improving attrition rates in Ebola virus drug discovery. Expert Opin Drug Discov 10(9), 1025–1032. doi:10.1517/17460441.2015.1062872

Group PIW, Multi-National PIIST (2016) A randomized, controlled trial of ZMapp for Ebola virus infection. New Engl J Med 375(15):1448–1456. doi:10.1056/NEJMoa1604330

Haque A, Hober D, Blondiaux J (2015) Addressing therapeutic options for ebola virus infection in current and future outbreaks. Antimicrob Agents Chemot 59(10), 5892-5902. doi:10.1128/Aac.01105-15

Hensley LE, Stevens EL, Yan SB, Geisbert JB, Macias WL, Larsen T, Daddario-DiCaprio KM, Cassell GH, Jahrling PB, Geisbert TW (2007) Recombinant human activated protein C for the postexposure treatment of Ebola hemorrhagic fever. J Infect Dis 196(Suppl 2):390–399. doi:10.1086/520598

Hiatt A, Pauly M, Whaley K, Qiu X, Kobinger G, Zeitlin L (2015) The emergence of antibody therapies for Ebola. Hum Antibodies 23(3–4), 49–56. doi:10.3233/HAB-150284

Hill-Batorski L, Halfmann P, Neumann G, Kawaoka Y (2013) The cytoprotective enzyme heme oxygenase-1 suppresses Ebola virus replication. J Virol 87(24):13795–13802. doi:10.1128/jvi.02422-13

Holtsberg FW, Shulenin S, Vu H, Howell KA, Patel SJ, Gunn B, Karim M, Lai JR, Frei JC, Nyakatura EK, Zeitlin L, Douglas R, Fusco ML, Froude JW, Saphire EO, Herbert AS, Wirchnianski AS, Lear-Rooney CM, Alter G, Dye JM, Glass PJ, Warfield KL, Aman MJ (2015) Pan-ebolavirus and Pan-filovirus Mouse Monoclonal Antibodies: Protection against Ebola and Sudan Viruses. J Virol 90(1):266–278. doi:10.1128/JVI.02171-15

Hunt L, Gupta-Wright A, Simms V, Tamba F, Knott V, Tamba K, Heisenberg-Mansaray S, Tamba E, Sheriff A, Conteh S, Smith T, Tobin S, Brooks T, Houlihan C, Cummings R, Fletcher T (2015) Clinical presentation, biochemical, and haematological parameters and their association with outcome in patients with Ebola virus disease: an observational cohort study. Lancet Infect Dis 15(11):1292–1299. doi:10.1016/S1473-3099(15)00144-9

Huttner A, Dayer JA, Yerly S, Combescure C, Auderset F, Desmeules J, Kaya G et al (2015) The effect of dose on the safety and immunogenicity of the VSV Ebola candidate vaccine: a randomised double-blind, placebo-controlled phase 1/2 trial. Lancet Infect Dis 15(10):1156–1166. doi:10.1016/S1473-3099(15)00154-1

Jacobs M, Aarons E, Bhagani S, Buchanan R, Cropley I, Hopkins S, Lester R, Martin D, Marshall N, Mepham S, Warren S, Rodger A (2015) Post-exposure prophylaxis against Ebola virus disease with experimental antiviral agents: a case-series of health-care workers. Lancet Infect Dis 15(11):1300–1304. doi:10.1016/s1473-3099(15)00228-5

Jahrling PB, Frame JD, Rhoderick JB, Monson MH (1985) Endemic Lassa fever in Liberia. IV. Selection of optimally effective plasma for treatment by passive immunization. Trans R Soc Trop Med Hyg 79(3):380–384

Jahrling PB, Geisbert J, Swearengen JR, Jaax GP, Lewis T, Huggins JW, Schmidt JJ, LeDuc JW, Peters CJ (1996) Passive immunization of Ebola virus-infected cynomolgus monkeys with immunoglobulin from hyperimmune horses. Arch Virol Suppl 11:135–140

Jahrling PB, Peters CJ (1984) Passive antibody therapy of Lassa fever in cynomolgus monkeys: importance of neutralizing antibody and Lassa virus strain. Infect Immun 44(2):528–533

Jahrling PB, Geisbert JB, Swearengen JR, Larsen T, Geisbert TW (2007) Ebola hemorrhagic fever: evaluation of passive immunotherapy in nonhuman primates. J Infect Dis 196(2), S400–3. doi:10.1086/520587

Jin J, Simmons G (2016) Inhibitory antibodies targeting emerging viruses: advancements and mechanisms. Clin Vaccine Immunol 23(7), 535–539. doi:10.1128/CVI.00136-16

Johansen LM, Brannan JM, Delos SE, Shoemaker CJ, Stossel A, Lear C, Hoffstrom BG, Dewald LE, Schornberg KL, Scully C, Lehar J, Hensley LE, White JM, Olinger GG (2013) FDA-approved selective estrogen receptor modulators inhibit Ebola virus infection. Sci Transl Med 5 (190):190ra179. doi:10.1126/scitranslmed.3005471

Johansen LM, DeWald LE, Shoemaker CJ, Hoffstrom BG, Lear-Rooney CM, Stossel A, Nelson E, Delos SE, Simmons JA, Grenier JM, Pierce LT, Pajouhesh H, Lehar J, Hensley LE, Glass PJ, White JM, Olinger GG (2015) A screen of approved drugs and molecular probes identifies therapeutics with anti-Ebola virus activity. Sci Transl Med 7 (290):290ra289. doi:10.1126/scitranslmed.aaa5597

Johnson JC, Martinez O, Honko AN, Hensley LE, Olinger GG, Basler CF (2014) Pyridinyl imidazole inhibitors of p38 MAP kinase impair viral entry and reduce cytokine induction by Zaire ebolavirus in human dendritic cells. Antiviral Res 107:102–109. doi:10.1016/j.antiviral.2014.04.014

Krasnianskii BP, Mikhailov VV, Borisevich IV, Gradoboev VN, Evseev AA, Pshenichnov VA (1995) Preparation of hyperimmune horse serum against Ebola virus. Vopr Virusol 40(3):138–140

Krasnianskii VP, Mikhailov VV, Borisevich IV, Gradoboev VN, Evseev AA, Pshenichnov VA (1994) The isolation of hyperimmune horse serum to the Ebola virus. Vopr Virusol 39(2):91–92

Kudoyarova-Zubavichene NM, Sergeyev NN, Chepurnov AA, Netesov SV (1999) Preparation and use of hyperimmune serum for prophylaxis and therapy of Ebola virus infections. J Infect Dis 179(Suppl 1):218–223. doi:10.1086/514294

Kugelman JR, Kugelman-Tonos J, Ladner JT, Pettit J, Keeton CM, Nagle ER, Garcia KY, Froude JW, Kuehne AI, Kuhn JH, Bavari S, Zeitlin L, Dye JM, Olinger GG, Sanchez-Lockhart M, Palacios GF (2015) Emergence of Ebola Virus Escape Variants in Infected Nonhuman Primates Treated with the MB-003 Antibody Cocktail. Cell reports 12(12):2111–2120. doi:10.1016/j.celrep.2015.08.038

Larsen T, Stevens EL, Davis KJ, Geisbert JB, Daddario-DiCaprio KM, Jahrling PB, Hensley LE, Geisbert TW (2007) Pathologic findings associated with delayed death in nonhuman primates experimentally infected with Zaire Ebola virus. J Infect Dis 196(Suppl 2):323–328. doi:10.1086/520589

Leligdowicz A, Fischer WA, 2nd Uyeki TM, Fletcher TE, Adhikari NK, Portella G, Lamontagne F, Clement C, Jacob ST, Rubinson L, Vanderschuren A, Hajek J, Murthy S, Ferri M, Crozier I, Ibrahima E, Lamah MC, Schieffelin JS, Brett-Major D, Bausch DG, Shindo N, Chan AK, O'Dempsey T, Mishra S, Jacobs M, Dickson S, Lyon GM 3rd, Fowler RA (2016) Ebola virus disease and critical illness. Crit Care 20(1), 217. doi:10.1186/s13054-016-1325-2

Liddell AM, Davey RT Jr, Mehta AK, Varkey JB, Kraft CS, Tseggay GK, Badidi O, Faust AC, Brown KV, Suffredini AF, Barrett K, Wolcott MJ, Marconi VC, Lyon GM 3rd, Weinstein GL, Weinmeister K, Sutton S, Hazbun M, Albarino CG, Reed Z, Cannon D, Stroher U, Feldman M, Ribner BS, Lane HC, Fauci AS, Uyeki TM (2015) Characteristics and Clinical Management of a Cluster of 3 Patients With Ebola Virus Disease, Including the First Domestically Acquired Cases in the United States. Ann Intern Med 163(2):81–90. doi:10.7326/M15-0530

Madelain V, Nguyen TH, Olivo A, de Lamballerie X, Guedj J, Taburet AM, Mentre F (2016) Ebola virus infection: review of the Pharmacokinetic and Pharmacodynamic properties of drugs considered for testing in human efficacy trials. Clin Pharmacokinet 55(8):907–923. doi:10.1007/s40262-015-0364-1

Markin VA, Mikhailov VV, Krasnianskii VP, Borisevich IV, Firsova IV (1997) Developing principles for emergency prevention and treatment of Ebola fever. Vopr Virusol 42(1):31–34

Mattia JG, Vandy MJ, Chang JC, Platt DE, Dierberg K, Bausch DG, Brooks T, Conteh S, Crozier I, Fowler RA, Kamara AP, Kang C, Mahadevan S, Mansaray Y, Marcell L, McKay G, O'Dempsey T, Parris V, Pinto R, Rangel A, Salam AP, Shantha J, Wolfman V, Yeh S, Chan AK, Mishra S (2016) Early clinical sequelae of Ebola virus disease in Sierra Leone: a cross-sectional study. Lancet Infect Dis 16(3), 331–338. doi:10.1016/S1473-3099(15)00489-2

McCarthy SD, Majchrzak-Kita B, Racine T, Kozlowski HN, Baker DP, Hoenen T, Kobinger GP, Fish EN, Branch DR (2016) A Rapid Screening Assay Identifies Monotherapy with Interferon-ss and Combination Therapies with Nucleoside Analogs as Effective Inhibitors of Ebola Virus. PLoS Negl Trop Dis 10(1):e0004364. doi:10.1371/journal.pntd.0004364

McMullan LK, Flint M, Dyall J, Albarino C, Olinger GG, Foster S, Sethna P, Hensley LE, Nichol ST, Lanier ER, Spiropoulou CF (2016) The lipid moiety of brincidofovir is required for in vitro antiviral activity against Ebola virus. Antiviral Res 125:71–78. doi:10.1016/j.antiviral.2015.10.010

Mikhailov VV, Borisevich IV, Chernikova NK, Potryvaeva NV, Krasnianskii VP (1994) The evaluation in hamadryas baboons of the possibility for the specific prevention of Ebola fever. Vopr Virusol 39(2):82–84

Mire CE, Geisbert JB, Agans KN, Thi EP, Lee AC, Fenton KA, Geisbert TW (2016) Passive Immunotherapy: assessment of Convalescent Serum Against Ebola Virus Makona Infection in Nonhuman Primates. J Infect Dis 214(suppl 3):S367–S374. doi:10.1093/infdis/jiw333

Misasi J, Chandran K, Yang JY, Considine B, Filone CM, Cote M, Sullivan N, Fabozzi G, Hensley L, Cunningham J (2012) Filoviruses require endosomal cysteine proteases for entry but exhibit distinct protease preferences. J Virol 86(6):3284–3292. doi:10.1128/jvi.06346-11

Moekotte AL, Huson MA, van der Ende AJ, Agnandji ST, Huizenga E, Goorhuis A, Grobusch MP (2016) Monoclonal antibodies for the treatment of Ebola virus disease. Expert Opin Investig Drugs 25(11), 1325-1335. doi:10.1080/13543784.2016.1240785

Mupapa K, Massamba M, Kibadi K, Kuvula K, Bwaka A, Kipasa M, Colebunders R, Muyembe-Tamfum JJ (1999) Treatment of Ebola hemorrhagic fever with blood transfusions from convalescent patients. International Scientific and Technical Committee. J Infect Dis 179 (Suppl 1):S18–23. doi:10.1086/514298

Oestereich L, Ludtke A, Wurr S, Rieger T, Munoz-Fontela C, Gunther S (2014) Successful treatment of advanced Ebola virus infection with T-705 (favipiravir) in a small animal model. Antiviral Res 105:17–21. doi:10.1016/j.antiviral.2014.02.014

Olinger GG Jr, Pettitt J, Kim D, Working C, Bohorov O, Bratcher B, Hiatt E, Hume SD, Johnson AK, Morton J, Pauly M, Whaley KJ, Lear CM, Biggins JE, Scully C, Hensley L, Zeitlin L (2012) Delayed treatment of Ebola virus infection with plant-derived monoclonal antibodies provides protection in rhesus macaques. Proc Natl Acad Sci USA 109(44):18030–18035. doi:10.1073/pnas.1213709109

Parren PW, Geisbert TW, Maruyama T, Jahrling PB, Burton DR (2002) Pre- and postexposure prophylaxis of Ebola virus infection in an animal model by passive transfer of a neutralizing human antibody. J Virol 76(12):6408–6412

Pattabhi S, Wilkins CR, Dong R, Knoll ML, Posakony J, Kaiser S, Mire CE, Wang ML, Ireton RC, Geisbert TW, Bedard KM, Iadonato SP, Loo YM, Gale M Jr (2016) Targeting Innate Immunity for Antiviral Therapy through Small Molecule Agonists of the RLR Pathway. J Virol 90(5):2372–2387. doi:10.1128/jvi.02202-15

Pecheur EI, Borisevich V, Halfmann P, Morrey JD, Smee DF, Prichard M, Mire CE, Kawaoka Y, Geisbert TW, Polyak SJ (2016) The Synthetic Antiviral Drug Arbidol Inhibits Globally Prevalent Pathogenic Viruses. J Virol 90(6):3086–3092. doi:10.1128/jvi.02077-15

Pettitt J, Zeitlin L, Kim DH, Working C, Johnson JC, Bohorov O, Bratcher B, Hiatt E, Hume SD, Johnson AK, Morton J, Pauly MH, Whaley KJ, Ingram MF, Zovanyi A, Heinrich M, Piper A, Zelko J, Olinger GG (2013) Therapeutic intervention of Ebola virus infection in rhesus macaques with the MB-003 monoclonal antibody cocktail. Science translational medicine 5 (199):199ra113. doi:10.1126/scitranslmed.3006608

Qiu X, Audet J, Wong G, Pillet S, Bello A, Cabral T, Strong JE, Plummer F, Corbett CR, Alimonti JB, Kobinger GP (2012a) Successful treatment of ebola virus-infected cynomolgus macaques with monoclonal antibodies. Science translational medicine 4 (138):138ra181. doi:10.1126/scitranslmed.3003876

Qiu X, Fernando L, Melito PL, Audet J, Feldmann H, Kobinger G, Alimonti JB, Jones SM (2012b) Ebola GP-specific monoclonal antibodies protect mice and guinea pigs from lethal Ebola virus infection. PLoS neglected tropical diseases 6(3):e1575. doi:10.1371/journal.pntd.0001575

Qiu X, Wong G, Audet J, Bello A, Fernando L, Alimonti JB, Fausther-Bovendo H, Wei H, Aviles J, Hiatt E, Johnson A, Morton J, Swope K, Bohorov O, Bohorova N, Goodman C, Kim D, Pauly MH, Velasco J, Pettitt J, Olinger GG, Whaley K, Xu B, Strong JE, Zeitlin L, Kobinger GP (2014) Reversion of advanced Ebola virus disease in nonhuman primates with ZMapp. Nature 514(7520):47–53. doi:10.1038/nature13777

Rasmussen AL, Okumura A, Ferris MT, Green R, Feldmann F, Kelly SM, Scott DP, Safronetz D, Haddock E, LaCasse R, Thomas MJ, Sova P, Carter VS, Weiss JM, Miller DR, Shaw GD, Korth MJ, Heise MT, Baric RS, de Villena FP, Feldmann H, Katze MG (2014) Host genetic diversity enables Ebola hemorrhagic fever pathogenesis and resistance. Sci 346(6212), 987–991. doi:10.1126/science.1259595

Rhein BA, Powers LS, Rogers K, Anantpadma M, Singh BK, Sakurai Y, Bair T, Miller-Hunt C, Sinn P, Davey RA, Monick MM, Maury W (2015) Interferon-gamma Inhibits Ebola Virus Infection. PLoS Pathog 11(11):e1005263. doi:10.1371/journal.ppat.1005263

Richardson T, Johnston AM, Draper H (2017) A systematic review of Ebola treatment trials to assess the extent to which they adhere to ethical guidelines. PLoS One 12(1), e0168975. doi:10.1371/journal.pone.0168975

Russek-Cohen E, Rubin D, Price D, Sun W, Cox E, Borio L (2016) A US Food and Drug Administration perspective on evaluating medical products for Ebola. Clinical trials 13(1):105–109. doi:10.1177/1740774515620613

Saeed MF, Kolokoltsov AA, Freiberg AN, Holbrook MR, Davey RA (2008) Phosphoinositide-3 kinase-Akt pathway controls cellular entry of Ebola virus. PLoS Pathog 4(8):e1000141. doi:10.1371/journal.ppat.1000141

Sandritter T (1999) Palivizumab for respiratory syncytial virus prophylaxis. J Pediatr Health Care 13 (4):191–195; quiz 196–197. doi:10.1016/S0891-5245(99)90039-1

Sissoko D, Keita M, Diallo B, Aliabadi N, Fitter DL, Dahl BA, Bore JA, Koundouno FR, Singethan K, Meisel S, Enkirch T, Mazzarelli A, Amburgey V, Faye O, Sall AA, Magassouba N, Carroll MW, Anglaret X, Malvy D, Formenty P, Aylward RB, Keita S, Djingarey MH, Loman NJ, Gunther S, Duraffour S (2016a) Ebola virus persistence in breast milk after no reported illness: a likely source of virus transmission from mother to child. Clin Infect Dis Official Publication Infect Dis Soc Am. doi:10.1093/cid/ciw793

Sissoko D, Laouenan C, Folkesson E, M'lebing AB, Beavogui AH, Laouenan S, Carazo S et al (2016b) Experimental treatment with favipiravir for Ebola virus disease (the JIKI Trial): a historically controlled, single-arm proof-of-concept trial in Guinea. PLoS Med 13(3): e1001967. doi:10.1371/journal.pmed.1001967

Smith DR, McCarthy S, Chrovian A, Olinger G, Stossel A, Geisbert TW, Hensley LE, Connor JH (2010) Inhibition of heat-shock protein 90 reduces Ebola virus replication. Antiviral Res 87 (2):187–194. doi:10.1016/j.antiviral.2010.04.015

Smith LM, Hensley LE, Geisbert TW, Johnson J, Stossel A, Honko A, Yen JY, Geisbert J, Paragas J, Fritz E, Olinger G, Young HA, Rubins KH, Karp CL (2013) Interferon-beta therapy prolongs survival in rhesus macaque models of Ebola and Marburg hemorrhagic fever. J Infect Dis 208(2), 310–318. doi:10.1093/infdis/jis921

Smither SJ, Eastaugh LS, Steward JA, Nelson M, Lenk RP, Lever MS (2014) Post-exposure efficacy of oral T-705 (Favipiravir) against inhalational Ebola virus infection in a mouse model. Antiviral Res 104:153–155. doi:10.1016/j.antiviral.2014.01.012

Soares MM, King SW, Thorpe PE (2008) Targeting inside-out phosphatidylserine as a therapeutic strategy for viral diseases. Nat Med 14(12):1357–1362. doi:10.1038/nm.1885

Sobarzo A, Ochayon DE, Lutwama JJ, Balinandi S, Guttman O, Marks RS, Kuehne AI, Dye JM, Yavelsky V, Lewis EC, Lobel L (2013) Persistent immune responses after Ebola virus infection. The New England journal of medicine 369(5):492–493. doi:10.1056/NEJMc1300266

Sueblinvong V, Johnson DW, Weinstein GL, Connor MJ Jr, Crozier I, Liddell AM, Franch HA, Wall BR, Kalil AC, Feldman M, Lisco SJ, Sevransky JE (2015) Critical care for multiple organ failure secondary to Ebola virus disease in the United States. Crit Care Med 43(10):2066–2075. doi:10.1097/CCM.0000000000001197

Swenson DL, Warfield KL, Warren TK, Lovejoy C, Hassinger JN, Ruthel G, Blouch RE, Moulton HM, Weller DD, Iversen PL, Bavari S (2009) Chemical modifications of antisense morpholino oligomers enhance their efficacy against Ebola virus infection. Antimicrob Agents Chemother 53(5):2089–2099. doi:10.1128/aac.00936-08

Thi EP, Mire CE, Lee AC, Geisbert JB, Zhou JZ, Agans KN, Snead NM, Deer DJ, Barnard TR, Fenton KA, MacLachlan I, Geisbert TW (2015) Lipid Nanoparticle siRNA Treatment Ebola-Virus-Makona-Infected Nonhuman Primates. Nature 521(7552):362–365. doi:10.1038/nature14442

Thorpe PE (2010) Targeting anionic phospholipids on tumor blood vessels and tumor cells. Thromb Res 125(Suppl 2):134–137. doi:10.1016/S0049-3848(10)70031-1

van Griensven J, Edwards T, Baize S, Ebola-Tx C (2016a) Efficacy of Convalescent Plasma in Relation to Dose of Ebola Virus Antibodies. New Engl J Med 375(23):2307–2309. doi:10.1056/NEJMc1609116

van Griensven J, Edwards T, de Lamballerie X, Semple MG, Gallian P, Baize S, Horby PW, Raoul H, Magassouba N, Antierens A, Lomas C, Faye O, Sall AA, Fransen K, Buyze J, Ravinetto R, Tiberghien P, Claeys Y, De Crop M, Lynen L, Bah EI, Smith PG, Delamou A, De Weggheleire A, Haba N, Ebola-Tx C (2016b) Evaluation of Convalescent Plasma for Ebola Virus Disease in Guinea. New Engl J Med 374(1):33–42. doi:10.1056/NEJMoa1511812

van Griensven J, Edwards T, Gallian P, Ebola-Tx C (2016c) Convalescent Plasma for Ebola Virus Disease. New Engl J Med 374(25):2500. doi:10.1056/NEJMc1602284

Warfield KL, Swenson DL, Olinger GG, Nichols DK, Pratt WD, Blouch R, Stein DA, Aman MJ, Iversen PL, Bavari S (2006) Gene-specific countermeasures against Ebola virus based on antisense phosphorodiamidate morpholino oligomers. PLoS Pathog 2(1):e1. doi:10.1371/journal.ppat.0020001

Warren TK, Jordan R, Lo MK, Ray AS, Mackman RL, Soloveva V, Siegel D, Perron M, Bannister R, Hui HC, Larson N, Strickley R, Wells J, Stuthman KS, Van Tongeren SA, Garza NL, Donnelly G, Shurtleff AC, Retterer CJ, Gharaibeh D, Zamani R, Kenny T, Eaton BP, Grimes E, Welch LS, Gomba L, Wilhelmsen CL, Nichols DK, Nuss JE, Nagle ER, Kugelman JR, Palacios G, Doerffler E, Neville S, Carra E, Clarke MO, Zhang L, Lew W, Ross B, Wang Q, Chun K, Wolfe L, Babusis D, Park Y, Stray KM, Trancheva I, Feng JY, Barauskas O, Xu Y, Wong P, Braun MR, Flint M, McMullan LK, Chen SS, Fearns R, Swaminathan S, Mayers DL, Spiropoulou CF, Lee WA, Nichol ST, Cihlar T, Bavari S (2016) Therapeutic efficacy of the small molecule GS-5734 against Ebola virus in rhesus monkeys. Nature 531(7594):381–385. doi:10.1038/nature17180

Warren TK, Warfield KL, Wells J, Swenson DL, Donner KS, Van Tongeren SA, Garza NL, Dong L, Mourich DV, Crumley S, Nichols DK, Iversen PL, Bavari S (2010) Advanced antisense therapies for postexposure protection against lethal filovirus infections. Nat Med 16(9):991–994. doi:10.1038/nm.2202

Warren TK, Wells J, Panchal RG, Stuthman KS, Garza NL, Van Tongeren SA, Dong L, Retterer CJ, Eaton BP, Pegoraro G, Honnold S, Bantia S, Kotian P, Chen X, Taubenheim BR, Welch LS, Minning DM, Babu YS, Sheridan WP, Bavari S (2014) Protection against filovirus diseases by a novel broad-spectrum nucleoside analogue BCX4430. Nature 508(7496):402–405. doi:10.1038/nature13027

Wec AZ, Nyakatura EK, Herbert AS, Howell KA, Holtsberg FW, Bakken RR, Mittler E, Christin JR, Shulenin S, Jangra RK, Bharrhan S, Kuehne AI, Bornholdt ZA, Flyak AI, Saphire EO, Crowe JE Jr, Aman MJ, Dye JM, Lai JR, Chandran K (2016) A "Trojan horse" bispecific-antibody strategy for broad protection against ebolaviruses. Science 354(6310): 350–354. doi:10.1126/science.aag3267

Wong G, Qiu X, de La Vega MA, Fernando L, Wei H, Bello A, Fausther-Bovendo H, Audet J, Kroeker A, Kozak R, Tran K, He S, Tierney K, Soule G, Moffat E, Gunther S, Gao GF, Strong J, Embury-Hyatt C, Kobinger G (2016) Pathogenicity Comparison Between the Kikwit and Makona Ebola Virus Variants in Rhesus Macaques. J Infect Dis 214(suppl 3):S281–S289. doi:10.1093/infdis/jiw267

Xu Z, Jin B, Teng G, Rong Y, Sun L, Zhang J, Du N, Liu L, Su H, Yuan Y, Chen H (2016) Epidemiologic characteristics, clinical manifestations, and risk factors of 139 patients with Ebola virus disease in western Sierra Leone. Am J Infect Control 44(11), 1285–1290. doi:10.1016/j.ajic.2016.04.216

Yen HL (2016) Current and novel antiviral strategies for influenza infection. Curr Opin Virol 18:126–134. doi:10.1016/j.coviro.2016.05.004

Zhao Y, Ren J, Harlos K, Jones DM, Zeltina A, Bowden TA, Padilla-Parra S, Fry EE, Stuart DI (2016) Toremifene interacts with and destabilizes the Ebola virus glycoprotein. Nature 535 (7610):169–172. doi:10.1038/nature18615

Zhou Y, Vedantham P, Lu K, Agudelo J, Carrion R Jr, Nunneley JW, Barnard D, Pohlmann S, McKerrow JH, Renslo AR, Simmons G (2015) Protease inhibitors targeting coronavirus and filovirus entry. Antiviral Res 116:76–84. doi:10.1016/j.antiviral.2015.01.011

Part III
Replication: at Cellular Level and Below

Filovirus Strategies to Escape Antiviral Responses

Judith Olejnik, Adam J. Hume, Daisy W. Leung, Gaya K. Amarasinghe, Christopher F. Basler and Elke Mühlberger

Abstract This chapter describes the various strategies filoviruses use to escape host immune responses with a focus on innate immune and cell death pathways. Since filovirus replication can be efficiently blocked by interferon (IFN), filoviruses have evolved mechanisms to counteract both type I IFN induction and IFN response signaling pathways. Intriguingly, marburg- and ebolaviruses use different strategies to inhibit IFN signaling. This chapter also summarizes what is known about the role of IFN-stimulated genes (ISGs) in filovirus infection. These fall into three categories: those that restrict filovirus replication, those whose activation is inhibited by filoviruses, and those that have no measurable effect on viral replication. In addition to innate immunity, mammalian cells have evolved strategies to counter viral infections, including the induction of cell death and stress response pathways, and we summarize our current knowledge of how filoviruses interact with these pathways. Finally, this chapter delves into the interaction of EBOV with myeloid dendritic cells and macrophages and the associated inflammatory response,

---

J. Olejnik · A.J. Hume · E. Mühlberger (✉)
Department of Microbiology and National Emerging Infectious Diseases Laboratories, Boston University School of Medicine, 620 Albany Street, Boston, MA 02118 USA
e-mail: muehlber@bu.edu

J. Olejnik
e-mail: jolejnik@bu.edu

A.J. Hume
e-mail: hume@bu.edu

D.W. Leung · G.K. Amarasinghe
Department of Pathology and Immunology, Washington University School of Medicine in St. Louis, St. Louis, MO 63110 USA
e-mail: dwleung@wustl.edu

G.K. Amarasinghe
e-mail: gamarasinghe@wustl.edu

C.F. Basler
Microbial Pathogenesis, Georgia State University, Institute for Biomedical Sciences, Atlanta, GA 30303 USA
e-mail: Cbasler@gsu.edu

Current Topics in Microbiology and Immunology (2017) 411:293–322
DOI 10.1007/82_2017_13

Published Online: 08 July 2017

which differs dramatically between these cell types when they are infected with EBOV. In summary, we highlight the multifaceted nature of the host-viral interactions during filoviral infections.

Contents

1 Introduction

The filovirus family includes three genera, *Ebolavirus*, *Marburgvirus*, and *Cuevavirus*. The genus *Marburgvirus* has two members (Marburg and Ravn viruses), and Lloviu virus is the only member of the genus *Cuevavirus*. Five ebolaviruses (Bundibugyo, Ebola, Reston, Sudan, and Taï Forest virus) have been assigned to the genus *Ebolavirus* (Fig. 1; see chapter *Guide to the Correct Use of Filoviral Nomenclature* in this book for a closer analysis of filovirus taxonomy). Although the members of the filovirus family differ in their virulence (Mahanty and Bray 2004), they take similar but not identical approaches to the circumvention or inhibition of host defense pathways.

Filoviruses encode at least four proteins that counteract host antiviral defense strategies: glycoprotein (GP), viral protein (VP) 24, VP35, and VP40. These proteins have additional roles in viral attachment, transcription, replication, or virion formation (see chapters *Inside the Cell* and *Filovirus Structural Biology: The Molecules in the Machine* in this book for in-depth descriptions of these latter functions).

Much work has been done to elucidate how filoviruses interact with the interferon (IFN) pathway, which is one of the best-studied antiviral host defense systems. Activation of the type I IFN signaling pathway leads to the expression of IFN-stimulated genes (ISGs), the mediators of IFN-induced inhibition to pathogens

Filoviridae

Ebolavirus	***Marburgvirus***	***Cuevavirus***
Ebola virus (EBOV)	Marburg virus (MARV)	Lloviu virus (LLOV)
Sudan virus (SUDV)	Ravn virus (RAVV)	
Bundibugyo virus (BDBV)		
Taï Forest virus (TAFV)		
Reston virus (RESTV)		

Fig. 1 The virus family *Filoviridae* includes three genera, *Ebolavirus*, *Marburgvirus*, and *Cuevavirus*. The genus *Ebolavirus* has five members: Ebola virus (EBOV), Sudan virus (SUDV), Bundibugyo virus (BDBV), Tai Forest virus (TAFV), and Reston virus (RESTV). The genus *Marburgvirus* has two members: Marburg virus (MARV) and Ravn virus (RAVV). The genus *Cuevavirus* has only one member, Lloviu virus (LLOV). Viruses in red are pathogenic to humans, those in green appear to be non-pathogenic to humans, and those in blue are not known to have been in contact with humans

(Hoffmann et al. 2015). Pretreatment of cells with type I IFNs (IFNα and IFNβ) or type II IFN (IFNγ) efficiently blocks filovirus replication, indicating that at least some of the ISGs are potent inhibitors of filovirus replication (Pinto et al. 2015; Rhein et al. 2015). This chapter will highlight the molecular mechanisms that filoviruses use to inhibit IFN induction and signaling. It will also summarize what is known about the role of ISGs in filovirus infection, which falls into three categories: those that restrict filovirus replication, those that the virus blocks from restricting replication, and others that have no effect on viral replication.

The IFN system is not the only strategy used by cells to control viral infection. Various forms of cell death are used as a crude, last-ditch effort to limit viral replication; by destroying the host cell, virus replication is, at least temporarily, halted. While some viruses actively inhibit cell death signaling, other viruses have co-opted this host defense strategy to facilitate virus spread. Here, we will describe the role that cell death plays during filovirus infection.

Finally, this chapter will review how filovirus infection impacts the functions of macrophages and dendritic cells (DCs). These cells are key players in innate virus control and play central roles in linking innate and adaptive immunity. Importantly, they are also early target cells during filovirus infection.

2 Immune Evasion by Filovirus Proteins

2.1 *Inhibition of IFN Induction*

The innate immune system is an evolutionarily conserved branch of the host response that serves as the first line of defense against invading pathogens. Detection of viral infection is mediated by pattern recognition receptors (PRRs), which recognize pathogen associated molecular patterns (PAMPs) absent in the host. Two major types of PRRs are RIG-I-like receptors (RLRs) and Toll-like receptors (TLRs) (Akira et al. 2006; Gerlier 2011). RLRs, including retinoic acid inducible gene I (RIG-I) and melanoma associated differentiation factor 5 (MDA5), are cytoplasmic proteins containing N-terminal caspase activation and recruitment domains (CARDs), central DExD/H box helicase domains, and C-terminal RNA binding domains that recognize PAMPs from viral genomic material, including those generated due to viral replication within the host cells (Leung et al. 2012; Takeuchi and Akira 2008). RIG-I is activated by short double stranded RNA (dsRNA) with 5′-ppp or 5′-OH groups, whereas MDA5 preferentially binds to longer lengths of dsRNA (Leung et al. 2012; Takeuchi and Akira 2008; Schlee et al. 2009; Yoneyama et al. 2005; Schmidt et al. 2009; Li et al. 2009; Akira and Takeda 2004). Signaling through CARD-CARD interaction between RLRs and the mitochondrial activator of viral signaling (MAVS, also known as IPS-1, VISA, or Cardif) located on mitochondria (Sun et al. 2006; Berke and Modis 2012) (Fig. 2) results in the activation of TANK binding kinase (TBK1) and inhibitor of nuclear factor κ-B kinase subunit ε (IKKε) through association with TANK (TRAF family member-associated NFκB activator) and TRAF3 (TNF receptor-associated factor 3). TBK1/IKKε kinases phosphorylate the transcription factors IFN regulatory factors 3 and 7 (IRF3 and IRF7), which homodimerize and translocate into the nucleus to activate the expression of type I IFNs (Akira et al. 2006).

The filoviral VP35 proteins are multifunctional. In addition to their functions as viral polymerase cofactors (see chapters *Inside the Cell* and *Filovirus Structural Biology: The Molecules in the Machine* in this book for more information), they antagonize the antiviral response. Ebola virus (EBOV) and Marburg virus (MARV) VP35 proteins bind viral dsRNA to prevent their recognition by RIG-I and MDA5 (Prins et al. 2010a; Ramanan et al. 2012; Leung et al. 2009, 2010), representing some of the better studied systems of viral antagonism of IFN production through PAMP sequestration. Structural studies of the C-terminal IFN inhibitory domain (IID) of EBOV VP35 in complex with an 8-bp in vitro transcribed RNA show that a series of conserved basic residues facilitate binding of EBOV VP35 to the phosphodiester backbone of dsRNA (Fig. 2) (Leung et al. 2010). Mutational analysis of these residues, particularly Arg312, shows that these basic residues are important for dsRNA binding and IFN inhibition (Prins et al. 2010a). In fact, substitution of Lys319 and Arg322 with alanine residues renders a guinea pig-adapted EBOV avirulent (Prins et al. 2010a). EBOV VP35 also caps the blunt ends of dsRNA through hydrophobic residues Phe235 and Phe239 (Leung et al. 2010). This dual interaction provides a mechanism that allows EBOV VP35 to efficiently sequester dsRNA from detection by and activation of RIG-I and MDA5. Similarly, MARV VP35 coats the dsRNA backbone. Although MARV VP35 is structurally homologous to EBOV VP35, the backbones of their crystal structures have a root mean square deviation of less than 1.0 Ångstrom (Ramanan et al. 2012), MARV VP35 is unable to endcap dsRNA and preferentially binds longer dsRNA that are targeted by MDA5 (Berke and Modis 2012; Ramanan et al. 2012; Peisley et al. 2011; Edwards et al. 2016) (Fig. 2). These differences in MARV VP35 recognition of

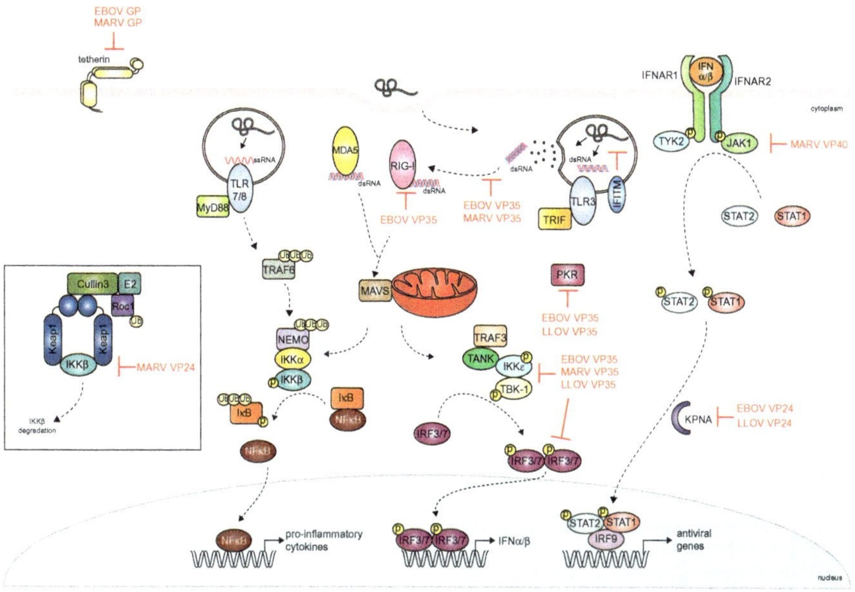

Fig. 2 Filoviruses inhibit the type I IFN response at multiple steps. This figure shows a simplified schematic of the type I IFN signal pathway. Viral PAMPs (e.g., dsRNA, endosomal ssRNA) are detected by host PRRs, such as RLRs (e.g., MDA5, RIG-I) and TLRs, which lead to the production of type I IFNs (IFNα/β) and pro-inflammatory cytokines. Binding of IFNα/β to the receptor complex IFNAR1/2 activates the JAK/STAT pathway leading to the expression of IFN-stimulated genes (ISGs). Filovirus proteins (in red) target different steps of these pathways

dsRNA motifs manifests in less-efficient inhibition of RIG-I signaling compared to EBOV VP35 (Edwards et al. 2016). EBOV VP35 also interacts with and sequesters individual components of stress granules, cytoplasmic structures which translationally silence RNAs in response to stress (Nelson et al. 2016; Le Sage et al. 2016). The correspondence between in vitro studies that evaluate RNA sequestration with in vivo studies of corresponding mutant viral infections support the relevance of these mechanisms (Prins et al. 2010a). The inhibition of type I IFN induction by VP35 appears to be a pan-filovirus characteristic, as it has been shown for the VP35 proteins of all five ebolaviruses, both marburgviruses, and Lloviu virus (LLOV) (Edwards et al. 2016; Feagins and Basler 2015; Guito et al. 2016). However, there are species-specific differences. Thus, marburgvirus VP35 proteins might be less efficient in counteracting IFN induction compared to their ebolavirus counterparts (Edwards et al. 2016; Guito et al. 2016).

2.1.1 Suppression of PACT Induced RIG-I Activation

RIG-I can be activated by the cellular protein kinase R (PKR) activator (PACT) (Kok et al. 2011; Iwamura et al. 2001; Luthra et al. 2013), although the exact molecular mechanism is poorly understood. Some viral proteins, such as EBOV and MARV VP35, target PACT to inhibit RIG-I signaling (Edwards et al. 2016; Luthra et al. 2013). PACT induces potent activation of RIG-I-dependent IFNβ promoter activity, which is correlated with an increase in RIG-I ATPase activity. Expression of the EBOV VP35 C-terminal domain suppresses RIG-I ATPase activity, as well as IFNβ promoter activity (Edwards et al. 2016; Luthra et al. 2013). Coimmunoprecipitation studies show that EBOV VP35 binds PACT and disrupts the interaction between PACT and RIG-I in a RNA-independent manner (Luthra et al. 2013). Moreover, EBOV VP35 residues critical for dsRNA binding, including Arg312, Arg322, and Phe239, are required for PACT binding. Interestingly, dsRNA binding does not appear to mediate the interaction between EBOV VP35 and PACT. Further studies are needed to elucidate upon the molecular mechanism of how PACT binding to VP35 regulates RIG-I activity, as well as the role of PACT binding to VP35 on viral polymerase activity as VP35 functions as a cofactor for the filoviral replication complex (Luthra et al. 2013; Prins et al. 2010b; Becker et al. 1998). The function of PACT in promoting translational inhibition through PKR activation requires additional studies to define the cellular role of its impact.

2.1.2 Inhibition of IRF3 and IRF7 Activation

In addition to facilitating RLR signaling, IRF3 and IRF7 are part of the signaling cascade for a number of other PRRs including TLRs and cytoplasmic DNA sensors [reviewed in (Hiscott 2007)]. Whether it is for the purpose of redundant inhibition of the above-mentioned RLR pathway or for the purpose of inhibiting other PRRs, EBOV VP35 also inhibits IRF3- and IRF7-mediated signaling, in part by binding to and inhibiting the function of the upstream kinases TBK1 and IKKε (reviewed in Basler and Amarasinghe 2009; Prins et al. 2009, Fig. 2). Similar to EBOV, MARV VP35 also inhibits IRF3 phosphorylation and IRF3 reporter gene activity, even in the presence of over-expressed TBK1 and IKKε (Ramanan et al. 2012). In a comparative analysis of ebola- and marburgvirus VP35 proteins, marburgvirus VP35 proteins were slightly less efficient at inhibiting IRF3 reporter gene activation than ebolavirus VP35 proteins (Guito et al. 2016). LLOV VP35 is also capable of inhibiting IRF3 activation (Feagins and Basler 2015).

2.2 *Inhibition of IFN Signaling by Filoviral Proteins*

Type I IFNs, including IFNα and IFNβ are master regulators of antiviral responses (Rawlings et al. 2004; O'Shea et al. 2015). IFNα is predominantly produced by hematopoietic cells, including plasmacytoid dendritic cells, while IFNβ is more broadly expressed (Ivashkiv and Donlin 2014; Chow and Gale 2015). Type I IFNs can act in an autocrine or paracrine fashion and bind to IFNα/β receptor (IFNAR) to activate Janus kinase 1 (JAK1) and tyrosine kinase 2 (TYK2), which phosphorylate Signal transducer and activator of transcription 1 (STAT1) and STAT2. Phosphorylated STAT1 is recognized by a subset of the karyopherin α (KPNA) family of nuclear transport proteins, the NPI-1 subfamily, which translocate STAT1-containing complexes to the nucleus (McBride et al. 2002; Sekimoto et al. 1997). The phosphorylated STAT1/STAT2 heterodimer forms a ternary complex along with IRF9 in the nucleus that induces transcription of ISGs through the IFN-stimulated gene response elements (ISREs). ISGs can inhibit different stages of viral infection including entry, replication, transcription, translation, assembly, and egress (Schneider et al. 2014; Sadler and Williams 2008). As a result of the actions of ISG expression, an overall antiviral state is achieved within the infected and neighboring cells. In addition to inhibition of IFN induction, filoviruses also encode proteins that inhibit the IFN signaling pathway. Below we describe how filoviral components have developed ways to inhibit JAK/STAT activity, interfere with the nuclear transport of transcription factors STAT1 and STAT2, as well as inhibit the activity of ISGs.

2.2.1 Inhibition of Phosphorylation of JAK/STAT Pathway Proteins

IFNα/β binding to IFNAR leads to the phosphorylation of JAK1 and TYK2. Although MARV and EBOV have a similar genome organization, only MARV VP40 inhibits JAK1-dependent signaling pathways (Valmas et al. 2010) (Fig. 2). Expression of MARV VP40 inhibits the tyrosine phosphorylation of JAK1, TYK2, STAT1, and STAT2 in response to IFNγ- and IL6-mediated phosphorylation of STAT1 and STAT3 (Guito et al. 2016; Valmas et al. 2010). This process is similar to a JAK1-deficient phenotype, suggesting that MARV VP40 targets JAK1, and not TYK2. Furthermore, MARV VP40 residues Ala57 and Ala165 appear to be important for inhibition of IFN signaling as mutation of these residues results in loss of JAK1 inhibition (Valmas et al. 2010).

Recent studies analyzing ebola- and marburgvirus proteins for their abilities to counteract the antiviral response suggest that the VP40 proteins of MARV, RAVV, and to a lesser extent TAFV and SUDV, inhibit IFN- and Sendai virus (SeV)-induced ISG production (Guito et al. 2016). Initial studies analyzing LLOV show that LLOV VP40 is unable to inhibit IFN signaling, similar to EBOV VP40 (Feagins and Basler 2015).

2.2.2 Inhibition of Nuclear Transport of STATs

In contrast to MARV VP40, which inhibits JAK1 phosphorylation, ebolavirus VP24 proteins block the nuclear translocation of phosphorylated STAT1 (pY-STAT1) complexes by targeting the STAT1 transporter KPNA (Guito et al. 2016; Reid et al. 2006, 2007; Xu et al. 2014). Recent biochemical and structural studies have elucidated how EBOV VP24 affects STAT1 nuclear accumulation while maintaining other KPNA-mediated cargo delivery (Xu et al. 2014).

All KPNA recognize cargo containing a classical nuclear localization signal (CNLS) through a major site on armadillo repeats (ARM) 2–4 and a minor site on ARM 6–8 (Conti et al. 1998; Conti and Kuriyan 2000; Chook and Blobel 2001; Conti and Izaurralde 2001). However, pY-STAT1 is transported by the nucleoprotein-interacting protein 1 (NPI-1) subfamily, including KPNA1, KPNA5, and KPNA6, which can recognize a relatively uncharacterized nonclassical NLS (ncNLS) (Sekimoto et al. 1997). EBOV VP24 binds to KPNA with a significantly higher affinity than pY-STAT1, suggesting that EBOV VP24 competes with pY-STAT1 for binding to KPNA (Xu et al. 2014). The crystal structure of EBOV VP24 in complex with the minimal binding region of KPNA5 has a large surface area of interaction with a hydrophobic core and high shape complementarity (Xu et al. 2014). The binding surface is formed by residues in KPNA that are conserved only among the NPI-1 subfamily and residues of EBOV VP24 that vary in the closely related MARV VP24 (Xu et al. 2014). Use of the ncNLS allows STAT1 transport to occur independently of other nucleocytoplasmic trafficking of cargoes containing cNLSs. This may be important in EBOV pathogenesis by maintaining certain cellular functions that can facilitate viral replication.

VP24 from all ebolaviruses, but not MARV VP24, inhibit IFN signaling, although BDBV and RESTV VP24 are less efficient at inhibition. This may in part be due to decreased KPNA binding (Guito et al. 2016; Schwarz 2016). LLOV VP24 inhibits the IFN pathway in a manner similar to EBOV VP24 by inhibiting IFN-induced STAT1 nuclear translocation and ISG induction presumably due to its ability to bind KPNA (Feagins and Basler 2015).

2.3 ISGs in Filovirus Infection

As mentioned above, pretreatment of cells with IFNα/β or IFNγ drastically reduces filovirus infectivity, suggesting an inhibitory role for ISGs (Pinto et al. 2015; Rhein et al. 2015; Bjorndal et al. 2003). Some well-studied ISGs include the antiviral proteins myxovirus resistance 1 (Mx1), IFN-inducible dsRNA-dependent protein kinase R (PKR), 2′-5′-oligoadenylate synthetase (OAS), IFN-induced proteins with tetratricopeptide repeats (IFITs), apolipoprotein B mRNA-editing enzyme catalytic polypeptide (APOBEC1), tripartite motif-containing proteins (TRIM) molecules, and tetherin (Schneider et al. 2014; Sadler and Williams 2008; Yan and Chen 2012; Haller et al. 2015; Garcia et al. 2006; Rebouillat and Hovanessian 1999; Diamond and Farzan 2013; Vladimer et al. 2014; Ozato et al. 2008; Kuhl et al. 2011). Here we review the literature regarding the ability of ISGs to restrict filovirus infection.

2.3.1 PKR

Protein kinase R (PKR) is a PRR that uses an N-terminal dsRNA binding domain to sense viral infection [reviewed in (Garcia et al. 2007)]. Kinase activity of PKR is activated upon dsRNA binding, leading to the phosphorylation of targets, including eukaryotic translation initiation factor 2α (eIF2α), which results in the shutdown of translation and viral protein synthesis (Garcia et al. 2007; Nanduri et al. 1998; Dauber and Wolff 2009). In addition to inhibiting the previously mentioned dsRNA-sensing proteins RIG-I and MDA5, EBOV VP35 can prevent dsRNA-dependent PKR activation (Feng et al. 2007; Schümann et al. 2009). Three residues within the C-terminus of EBOV VP35, R305, K309, and R312, are critical for inhibition of PKR as mutation of all three of these amino acids to alanine greatly reduces the ability of VP35 to inhibit PKR activation (Schümann et al. 2009). Intriguingly, inhibition of PKR activation by EBOV VP35 does not appear to be dependent upon the ability to bind dsRNA, since the R312A mutant of VP35, which lacks dsRNA binding ability, can still prevent PKR activation (Schümann et al. 2009). More recently, LLOV VP35 was shown to inhibit SeV-induced PKR activation, indicating that inhibition of PKR activation by VP35 may be a pan-filovirus function (Feagins and Basler 2015).

2.3.2 PML Protein

Promyelocytic leukemia (PML) protein is a component of antiviral ISG forming nuclear bodies, often referred to as PML nuclear bodies (PML-NB) or Nuclear Domain 10 (ND10), which aggregate the replication and transcriptional machinery of many viruses (Everett and Chelbi-Alix 2007; Geoffroy and Chelbi-Alix 2011). While present in non-activated cells, PML expression is greatly enhanced upon stimulation of cells with type I or type II IFNs (Regad and Chelbi-Alix 2001; Chelbi-Alix et al. 1995; Lavau et al. 1995). The antiviral capacity of PML applies to certain RNA viruses as PML overexpression results in the restriction of both vesicular stomatitis Indiana virus and influenza A virus, but not encephalomyocarditis virus (Chelbi-Alix et al. 1998; Jin et al. 2014).

As of today, there is only one study looking at the role of PML in filovirus infection. PML expression is upregulated in EBOV-infected IFN-competent MCF7 cells (human breast cancer cell line), whereas only a minor change of PML expression occurs in infected Vero cells (african green monkey kidney epithelial cell line), which cannot produce IFN. Interestingly, PML expression does not change in surrounding, noninfected cells, indicating an IFN-independent mechanism of increased PML expression directly induced by EBOV infection (Bjorndal et al. 2003). Viral replication occurs readily in PML-expressing cells, suggesting that PML has no major inhibitory effect on EBOV replication, which was not analyzed in this study (Bjorndal et al. 2003).

2.3.3 IFIT Proteins

The IFN-induced protein with tetratricopeptide repeats (IFIT) family of proteins contains five members in humans, IFIT1, 1B, 1C, 2, 3, and 5. With the exception of

IFIT1B, the expression of IFIT proteins are strongly induced by IFN and possess antiviral activity against specific RNA viruses (Pinto et al. 2015; Vladimer et al. 2014; Fensterl and Sen 2015; Hyde and Diamond 2015; Fensterl and Sen 2011; Young et al. 2016). There are two antiviral mechanisms that have been attributed to members of the IFIT family. IFIT1 binds to and inhibits the translation of viral mRNAs lacking 2′-*O* cap methylation. It also sequesters RNA of uncapped viral genomes and anti-genomes containing 5′ triphosphates and prevents their translation with the help of IFIT2 and IFIT3 (Vladimer et al. 2014; Hyde and Diamond 2015; Pichlmair et al. 2011; Kumar et al. 2014; Daffis et al. 2010). The antiviral activity of IFIT1 is lacking or is minimal against wild-type flaviviruses but is greater against mutant forms of these viruses lacking 2′-*O* methyltransferases (2′OMTases) (Daffis et al. 2010; Szretter et al. 2012; Li et al. 2013; Kimura et al. 2013). These data indicate that the antiviral activity of IFIT1 against some viruses is overcome by viral 2′OMTase activity. The EBOV polymerase is predicted to have 2′OMTase activity (Zhao et al. 2016; Bujnicki and Rychlewski 2002; Ferron et al. 2002). Consistent with this prediction, EBOV replication does not seem to be affected by IFIT1. EBOV titers are similar in mouse macrophages lacking IFIT1 compared to wild-type (wt) cells (Pinto et al. 2015). Whether other IFITs have an antiviral function against EBOV or whether MARV is influenced by the IFIT family has not yet been reported.

2.3.4 IFITM Proteins

The IFN-induced transmembrane (IFITM) family includes IFITM1, 2, 3, and 5 in humans. A broad range of enveloped RNA viruses are restricted by at least one member of the IFITM family. IFITM proteins act at late stages of viral entry and restrict fusion from late endosomes (Huang et al. 2011; Alber and Staeheli 1996; Brass et al. 2009; Mudhasani et al. 2013; Lu et al. 2011; Everitt et al. 2013; Anafu et al. 2013; Bailey et al. 2014). The various members of the IFITM protein family are able to inhibit filoviral GP-mediated entry, as shown for retroviruses (murine leukemia virus) pseudotyped with distinct filoviral GPs, including those of all ebolavirus species, MARV, and LLOV to varying degrees (Huang et al. 2011; Wrensch et al. 2015) (Fig. 2). Importantly, the IFITM proteins also inhibit infection with EBOV and MARV (Huang et al. 2011). The inhibition of filovirus GP-mediated entry likely occurs in the late endosome (Huang et al. 2011). The exact mechanism remains unclear, but it appears that modulation of cathepsin activity is not involved (Huang et al. 2011).

2.3.5 Tetherin

Tetherin is a cell surface-localized transmembrane protein which acts as an antiviral ISG by inhibiting the budding of various viruses (Sakuma et al. 2009; Neil et al. 2008). Budding of filoviral particles is mediated by the matrix protein VP40, and tetherin blocks the release of virus-like particles (VLPs) from cells expressing

EBOV, MARV, or RAVV VP40 (Sakuma et al. 2009; Kaletsky et al. 2009; Radoshitzky et al. 2010; Feagins and Basler 2014). However, tetherin-mediated restriction of VLP release can be overcome by co-expression of GP (Fig. 2). While the mechanism of the antagonizing function of GP has not been determined, EBOV GP and tetherin interact directly, suggesting a direct mechanism of action (Kaletsky et al. 2009; Lopez et al. 2010). While the mucin-like domain of EBOV GP was initially hypothesized to play a role in overcoming tetherin-mediated restriction, it does not seem to play a critical role in this process (Kaletsky et al. 2009; Radoshitzky et al. 2010). Interestingly, mutations in the VP40 gene in mouse-adapted RAVV increase the sensitivity of the virus to restriction by human, but not mouse, tetherin (Feagins and Basler 2014).

2.3.6 ISG15

ISG15 is a small, ubiquitin-like protein whose main antiviral function appears to be ISGylation, the covalent attachment to cellular and viral proteins in a manner similar to ubiquitinyl or SUMOyl conjugation. While the targets and functions of ISGylation remain relatively poorly characterized, ISG15 is generally characterized as harboring antiviral activity (Zhang and Zhang 2011). Similar to tetherin, ISG15 inhibits budding of VLPs containing EBOV VP40. Overexpression of a dominant-negative form of the ubiquitin ligase Nedd4 abolishes this inhibition, indicating a role for Nedd4 in ISG15-mediated restriction of EBOV (Malakhova and Zhang 2008; Okumura et al. 2008). Inhibition of VLPs containing EBOV VP40 is observed in the absence of other viral proteins. It is not known if ISG15 is able to inhibit budding of other filoviruses besides EBOV.

2.3.7 ZAP

Zinc finger antiviral protein (ZAP) is a zinc finger-containing ISG which has antiviral activity against a number of RNA viruses, although its antiviral activity is not universal (Gao et al. 2002; Bick et al. 2003). While the exact mechanism of ZAP antiviral function remains to be determined, initial studies show that ZAP post-transcriptionally reduces viral RNAs in a zinc finger-dependent manner (Mao et al. 2013). Additional studies indicate that ZAP binds to and destabilizes target viral RNAs by facilitating the removal of the mRNA caps and poly A tails as well as recruiting the 3′–5′ exosome to facilitate transcript degradation of RNAs (Guo et al. 2007; Zhu et al. 2011). EBOV and SUDV and, to a lesser extent MARV, replicate to lower titers in Rat2 and 293T cells expressing rat ZAP (Muller et al. 2007). Expression of ZAP leads to a reduction in the mRNA amounts of mainly EBOV and MARV L mRNAs, suggesting that ZAP interferes with L mRNA synthesis or stability (Muller et al. 2007).

2.4 Inhibition of RNA Interference Pathways

The related miRNA and RNAi pathways serve as cellular antiviral defense systems aside from their roles in normal cellular functions (Umbach and Cullen 2009; Haasnoot and Berkhout 2011; Vasselon et al. 2013). Three EBOV proteins, VP30, VP35, and VP40, each inhibit RNA interference (RNAi) in co-transfection experiments (Haasnoot et al. 2007; Fabozzi et al. 2011). EBOV VP35, the most effective of these EBOV proteins at inhibiting the RNAi pathway, does so in a dsRNA binding-dependent manner (Haasnoot et al. 2007). Inhibition the RNAi pathway by VP30 and VP35 may in part be due to their ability to interact with individual components of the RNAi pathway; VP30 interacts with Dicer and TRBP, while VP35 interacts with Dicer, TRBP, and PACT, independent of its dsRNA binding domain (Fabozzi et al. 2011). The VP35 proteins from both EBOV and MARV are able to inhibit Dicer-dependent production of virus-derived siRNAs (vsiRNAs) during infection of cells with influenza A virus lacking NS1, suggesting that VP35 may perform the same function during filovirus infections (Li et al. 2016). Interestingly, a lipid nanoparticle-delivered siRNA cocktail targeting VP35 and L is efficacious in rhesus monkeys up to 3 days post exposure to an otherwise lethal EBOV dose (Thi et al. 2015).

3 Cell Death in Filovirus Infection

Viruses rely on a functional cellular machinery to replicate. Cell death is an efficient way to disable this cellular machinery and therefore, is considered as an antiviral strategy. Some forms of cell death trigger a pro-inflammatory response, leading to additional cell damage (Labbe and Saleh 2008). Cell death caused by filovirus infection occurs in both animal (in vivo) and cell culture (in vitro) models.

3.1 Filovirus-Induced Cell Death in Animal Models and Patients

3.1.1 Cell Death of Permissive Cells in Vivo

Both in vivo and in vitro infection models suggest that filovirus-infected cells undergo non-apoptotic cell death, including necrosis (Baskerville et al. 1978; Connolly et al. 1999; Geisbert et al. 2000, 2003a; Ryabchikova et al. 1996a, 1999; Ryabchikova 2004; Murphy et al. 1971; Warfield et al. 2007; Groseth et al. 2012; Warfield et al. 2009; Qiu et al. 2014; Bird et al. 2016; Cross et al. 2015; Herbert et al. 2015; Ludtke et al. 2015; Zumbrun et al. 2012; Ebihara et al. 2013; Lever et al. 2012; Warren et al. 2010; Gibb et al. 2001; Bray et al. 2001; Olejnik et al. 2011; Schmidt et al. 2011).

Although expression of apoptotic markers increases during EBOV infection in humans and in animal models (Geisbert et al. 2003a; Baize et al. 1999; Hensley et al. 2002; Leroy et al. 2001; Rubins et al. 2007), apoptosis is not observed in infected cells directly (Baskerville et al. 1978; Connolly et al. 1999; Geisbert et al. 2000, 2003a; Ryabchikova et al. 1996b, 1999; Ryabchikova 2004; Murphy et al. 1971; Gibb et al. 2001; Olejnik et al. 2011, 2013). A hallmark of filovirus infection is liver damage, which is associated with tissue necrosis and hepatocyte death (Baskerville et al. 1978; Ryabchikova et al. 1999; Ryabchikova 2004; Murphy et al. 1971; Warfield et al. 2007; Groseth et al. 2012; Warfield et al. 2009; Qiu et al. 2014; Bird et al. 2016; Cross et al. 2015; Herbert et al. 2015; Zumbrun et al. 2012; Ebihara et al. 2013; Lever et al. 2012; Ryabchikova et al. 1996b; Ellis et al. 1978; Zaki et al. 1999; Gedigk et al. 1968; Rippey et al. 1984). Hepatocytes and Kupffer cells in various rodent models of EBOV disease undergo apoptosis (Groseth et al. 2012; Ebihara et al. 2013; Bradfute et al. 2010). However, it is not clear whether apoptosis is induced in infected cells or in noninfected bystander cells or in both (Groseth et al. 2012; Ebihara et al. 2013; Bradfute et al. 2010). Inhibition of apoptosis delays liver dysfunction in mice, suggesting that the extent of hepatic cell death plays a crucial role in EBOV pathogenesis (Groseth et al. 2012; Ebihara et al. 2013; Bradfute et al. 2010).

3.1.2 Cell Death of Non-Permissive Cells in Vivo

Although lymphocytes are not permissive to filovirus infection, they are depleted in infected patients and in infected animals, presumably through apoptotic cell death (Geisbert et al. 2000, 2003a; Warfield et al. 2007, 2009; Cross et al. 2015; Ebihara et al. 2013; Baize et al. 1999, 2002; Leroy et al. 2000; Bradfute et al. 2007, 2008, 2010; Gupta et al. 2007; Reed et al. 2004; Brannan et al. 2015). Apoptotic lymphocytes are detected by various methods, including histology staining, electron microscopy, TUNEL assay, and flow cytometry in the tissues or peripheral blood mononuclear cells (PBMCs) of filovirus-infected animals (Geisbert et al. 2000, 2003a; Warfield et al. 2007, 2009; Cross et al. 2015; Ebihara et al. 2013; Bradfute et al. 2008, 2010; Reed et al. 2004; Brannan et al. 2015). Various apoptotic markers, including cleavage of the 41/7 nuclear matrix protein (NMP), DNA fragmentation, and increased Fas/CD95 expression are detected in lymphocytes from fatally EBOV-infected patients and macaques (Baize et al. 1999; Reed et al. 2004; Wauquier et al. 2010). In a mouse model of EBOV infection, lymphocyte apoptosis was shown to be dependent on both death receptor and intrinsic apoptosis signaling (Bradfute et al. 2010). Surprisingly, blocking lymphocyte apoptosis in an EBOV mouse model does not increase survival (Bradfute et al. 2010). A recent study analyzing the immune signature in patients infected with EBOV did not observe significant differences in the number of circulating white blood cells between fatal cases and survivors of infection. However, fatal cases showed a high percentage of $CD4^+$ and $CD8^+$ T cells expressing CTLA-4 and PD-1, both markers of T cell exhaustion (Ruibal et al. 2016). An increase in the number of white blood cells followed by only moderate lymphopenia was observed in nonhuman primates

(NHPs) fatally infected with EBOV (Marzi et al. 2015). Initial increases in white blood cell numbers are also observed in survivors of EBOV infection (Kreuels et al. 2014; Wolf et al. 2015), suggesting that temporal changes in the T cell compartment might play a role in EBOV pathogenesis.

3.2 *In Vitro Analysis of Filovirus-Induced Cell Death*

In cell culture models, filovirus infection generally leads to a visible cytopathic effect (CPE) including cell blebbing, cell rounding, vacuolization, and detachment (Schmidt et al. 2011; Olejnik et al. 2013; Alazard-Dany et al. 2006; Barrientos and Rollin 2007; Hoenen et al. 2013; Geisbert et al. 2003b; Boehmann et al. 2005; Gupta et al. 2010). The degree of CPE caused by infection varies with different ebolaviruses (Boehmann et al. 2005). EBOV-infected cells, including human monocyte-derived macrophages, undergo necrotic cell death rather than apoptosis (Olejnik et al. 2013). There are conflicting results about the fate of EBOV-infected monocytes and macrophages in a human PBMC infection model. Although no signs of apoptosis were detected in EBOV-infected macrophages during PBMC infection in one study (Geisbert et al. 2000), apoptotic markers were found on monocytes and macrophages in EBOV- and BDBV-infected PBMCs by another group (Gupta et al. 2007, 2010). Whether differences in the experimental protocol account for the conflicting results remains to be determined.

EBOV-infected cells remain sensitive to the induction of apoptosis through both the intrinsic and extrinsic pathways (Olejnik et al. 2013). Interestingly, stimulation of dsRNA-dependent apoptosis in EBOV-infected cells inhibits EBOV replication (Olejnik et al. 2013). These data indicate that apoptotic pathways can be successfully induced to inhibit EBOV propagation. Intriguingly and in contrast to EBOV, MARV infection leads to the activation of cytoprotective responses in infected cells to prolong cell viability (as discussed below) (Page et al. 2014; Edwards and Basler 2015; Zhang et al. 2014; Johnson et al. 2016).

3.2.1 Filovirus Cytotoxic Proteins

Signs of cytotoxicity, including cell rounding and detachment have been associated with EBOV GP and VP40 (Alazard-Dany et al. 2006; Takada et al. 2000; Yang et al. 2000; Chan et al. 2000; Volchkov et al. 2001; Simmons et al. 2002; Ray et al. 2004; Sullivan et al. 2005; Han et al. 2007; Francica et al. 2009; Hacke et al. 2015; Melito et al. 2008). The mechanism of cell death in VP40-expressing cells remains undetermined but seems to be related to transient overexpression of the protein (Alazard-Dany et al. 2006; Melito et al. 2008). Since VP40 mediates budding (Hartlieb and Weissenhorn 2006), it is conceivable that the observed cytopathic effects are caused by massive membrane loss.

The mechanisms of GP-induced cytotoxicity are not entirely understood. Reduced activation of the ERK2 kinase and reduced integrin cell surface levels might play a role (Sullivan et al. 2005; Francica et al. 2009; Zampieri et al. 2007). While some studies report the induction of cell death in GP-expressing cells (Yang et al. 2000; Ray et al. 2004; Zampieri et al. 2007), other reports did not observe cell disruption (Chan et al. 2000; Simmons et al. 2002). A possible explanation for these conflicting results might be differences in the expression levels of GP in the various studies. Notably, low levels of EBOV GP expression comparable to GP levels in EBOV-infected cells do not lead to the induction of cytotoxic effects (Alazard-Dany et al. 2006). GP is processed from a precursor protein (preGP) into two subunits, GP_1 and GP_2, which both have cytotoxic effects (Yang et al. 2000; Han et al. 2007; Francica et al. 2009; Hacke et al. 2015; Zampieri et al. 2007). For GP_1, the heavily glycosylated mucin-like domain is sufficient to mediate these effects (Yang et al. 2000; Francica et al. 2009; Zampieri et al. 2007). Increased expression of membrane-bound GP from a recombinant EBOV resulted in elevated cytotoxicity (Volchkov et al. 2001). Interestingly, this recombinant virus was less virulent in a guinea pig model, suggesting that early cell death of infected cells limits viral spread (Volchkova et al. 2015). A decrease in tumor necrosis factor α–converting enzyme (TACE)-mediated shedding of GP leads to increased cytotoxic effects and enhanced viral growth and infectivity (Dolnik et al. 2004, 2015). This suggests that GP-mediated cytotoxicity in EBOV infection is controlled by regulating both the levels of expression and shedding of GP. Intriguingly, no cytotoxic effects have been associated with overexpression of MARV GP (Chan et al. 2000). Cell surface expression of EBOV, MARV, or LLOV GP interferes with Fas/CD95 binding by steric shielding, thereby preventing the subsequent induction of apoptosis (Noyori et al. 2013). Whether this effect plays a role in the prevention of apoptosis in infected cells has yet to be determined.

3.2.2 Regulation of the Oxidative Stress Response by MARV VP24

MARV activates expression of genes with promoters that contain antioxidant response elements (AREs). ARE genes encode proteins that facilitate cellular survival under conditions of stress, including oxidative stress. MARV VP24 interacts with KELCH-like ECH-associated protein 1 (Keap1) (Page et al. 2014; Edwards and Basler 2015; Pichlmair et al. 2012). Keap1 interacts with proteins such as the transcription factor nuclear factor (erythroid-derived 2)-like 2 (Nrf2), and directs their Cul3 ubiquitin ligase-dependent poly-ubiquitinylation, thereby activating expression of ARE-containing genes. However, cell stresses trigger alteration in Keap1–Nrf2 interaction, leading to Nrf2 stabilization and activation of ARE gene expression (Copple et al. 2008). MARV VP24 interacts with Keap1 via one of the C-terminal six-bladed beta propeller Kelch domains. Interaction occurs through a loop (K-loop), which projects out from the MARV VP24 structure, and contains a GE amino acid motif that is preceded by acidic residues. This domain is similar to interaction motifs found in other Keap1 targets, including Nrf2. Biophysical studies indicate that MARV VP24 and Nrf2 interact with the same region of the KELCH domain (Edwards and Basler 2015; Zhang et al. 2014; Johnson et al. 2016). MARV VP24 binding to Keap1 disrupts Keap1-Nrf2 binding leading to ARE gene expression (Page et al. 2014; Edwards and Basler 2015). ARE gene expression is upregulated after MARV infection, but not after EBOV infection, which is consistent with the observation that MARV VP24 interacts with Keap1 but EBOV VP24 does not (Edwards and Basler 2015). Further, Nrf2-deficient mice are resistant to disease caused by mouse-adapted MARV (Page et al. 2014). These data suggest that MARV VP24 activates Nrf2 to facilitate viral replication, perhaps by activating the cytoprotective ARE response; infected cells survive longer and produce more virus. Alternatively, the ARE response may protect viral products from oxidative damage to facilitate replication. Keap1 also modulates stability of the kinase IKKβ, which regulates NF-κB expression, and MARV VP24 relieves the suppressive activity of Keap1 on NF-κB activation. As NF-κB can also exert cytoprotective effects, MARV VP24–Keap1 interaction may represent a general strategy of the virus to sustain the viability of infected cells (Edwards and Basler 2015).

4 Interaction of Filoviruses with Host Immune Cells

Antigen-presenting cells, including monocytes, macrophages, and myeloid dendritic cells, are considered the primary target cells of filoviruses and play crucial roles in filovirus pathogenesis (Bray and Geisbert 2005; Martinez et al. 2012). Infection of these cells by filoviruses severely impacts their function in mediating an appropriate innate immune response and might play a role in the excessive pro-inflammatory response observed during filovirus disease (Geisbert et al. 2003a, c; Hensley et al. 2002; Leroy et al. 2001; Rubins et al. 2007; Baize et al. 2002; Wauquier et al. 2010; Ruibal et al. 2016; Rougeron et al. 2015; Feldmann and Geisbert 2011; Villinger et al. 1999; McElroy et al. 2016; Gupta et al. 2001; McElroy et al. 2014a, b; Sanchez et al. 2004; Hutchinson and Rollin 2007; Ignatiev et al. 2000; Ebihara et al. 2011; Marzi et al. 2013; Martins et al. 2015; van Paassen et al. 2012; Fritz et al. 2008; Hensley et al. 2011; Alves et al. 2010; Geisbert et al. 2007; Lin et al. 2015; Connor et al. 2015; Fernando et al. 2015; Ignat'ev et al. 1995; Caballero et al. 2014, 2016; Marzi et al. 2016). In addition, comparative data on the cytokine and chemokine responses in EBOV-infected patients and nonhuman primates suggest that a broadly dysregulated inflammatory response is associated with severe or fatal EBOV disease (Baize et al. 1999, 2002; McElroy et al. 2014a; Martins et al. 2015).

4.1 Monocytes and Macrophages

Substantial data from studies of human infections and experimental animal infections demonstrate the replication of filoviruses in macrophage infection in vivo [reviewed in (Martinez et al. 2012)]. Macrophages are among the earliest cell types infected and remain targets of infection throughout the course of disease (Geisbert et al. 2003a). Human monocyte-derived macrophages are permissive for filovirus infection in vitro (Feldmann et al. 1996; Stroher et al. 2001; Martinez et al. 2013). Monocytes obtained from human blood also appear to be permissive for filovirus infection in vitro (Stroher et al. 2001). This observation seems to contradict studies that found monocytes to be resistant to viral entry by viruses pseudotyped with EBOV GP (Yonezawa et al. 2005; Dube et al. 2008). An explanation that may resolve this apparent contradiction is that as monocytes differentiate towards a macrophage or DC phenotype, they acquire permissiveness for GP-mediated entry (Martinez et al. 2013). This hypothesis suggests that monocytes that are infected by EBOV are in the process of differentiating, a process that may be facilitated by interaction of the cells with GP.

In vitro studies suggest that monocytes and macrophages may be significant sources of the cytokines that are characteristic of severe filovirus disease. Substantial cytokine and chemokine expression is induced early in filovirus infection of monocytes or macrophages (Hensley et al. 2002; Gupta et al. 2001; Stroher et al. 2001; Martinez et al. 2013; Wahl-Jensen et al. 2011; Ayithan et al. 2014; Olejnik et al. 2017). Similar responses are elicited by inactivated EBOV and by Ebola VLPs generated by co-expressing EBOV VP40 and GP (Hensley et al. 2002; Stroher et al. 2001; Wahl-Jensen et al. 2011; Olejnik et al. 2017; Wahl-Jensen et al. 2005). This shows that viral genome replication and transcription within the infected cells are not required for the induction of the observed inflammatory response. VLP studies revealed that EBOV GP is sufficient to stimulate cytokine and chemokine expression. Intriguingly, EBOV GP activates Toll-like receptor 4 (TLR4) and induces downstream inflammatory responses (Okumura et al. 2010; Escudero-Perez et al. 2014). TLR4 activation is mediated by both GP inserted on the surface of viral particles and shed GP, a truncated form of membrane-bound GP lacking the transmembrane domain (Olejnik et al. 2017; Escudero-Perez et al. 2014). In contrast, soluble GP, the carboxy terminal part of which differs from transmembrane GP and shed GP, cannot activate TLR4-mediated inflammatory responses (Wahl-Jensen et al. 2005; Escudero-Perez et al. 2014). There are conflicting results regarding the activation of human macrophages by RESTV infection. In one study, RESTV infection led to a robust activation of human monocytes and macrophages similar to EBOV, whereas RESTV-infected macrophages remained remarkably silent in another study (Stroher et al. 2001; Olejnik et al. 2017). It is conceivable that different procedures to generate virus stocks might account for this discrepancy (Hartman et al. 2008). VLPs containing RESTV GP fail to stimulate a pro-inflammatory response in primary human macrophages, indicating that RESTV GP does not trigger TLR4 signaling (Olejnik et al. 2017). The inability of RESTV GP to activate human macrophages through TLR4 might contribute to lower pathogenicity by preventing the cytokine storm observed in EBOV infection. Interestingly, EBOV GP-mediated TLR4 activation of macrophages can be inhibited by using TLR4 antagonists or anti-TLR4 antibodies which opens up potential treatment options (Olejnik et al. 2017; Escudero-Perez et al. 2014). Further studies are needed, however, to determine to what extent GP drives monocyte and macrophage pro-inflammatory responses, whether TLR4 signaling is the dominant pathway inducing this response and if TLR4 activation is correlated with pathogenicity. Although it seems counterproductive for a virus to induce a strong inflammatory response in the infected cells, GP-mediated TLR4 activation might be beneficial for the virus. Among the upregulated host proteins detected in GP-activated macrophages are suppressor of cytokine signaling 1 and 3 (SOCS1 and SOCS3). Intriguingly, SOCS3 enhances EBOV particle budding, supporting the hypothesis that the induction of a pro-inflammatory response promotes EBOV infection (Olejnik et al. 2017; Okumura et al. 2010, 2015).

4.2 Dendritic Cells

DCs play a critical role in linking the innate and adaptive immunity. They respond to pathogens by undergoing a maturation process that facilitates presentation of antigen to and stimulation of T cell responses (Lanzavecchia 1999). Like macrophages, DCs are productively infected by filoviruses (Bosio et al. 2003; Mahanty et al. 2003). However, whereas EBOV infection of monocytes and macrophages leads to substantial cytokine production, as discussed above, a much more muted cytokine response is produced by infected human monocyte-derived DCs. In each case, a minimal response and an absence of IFN responses, as assessed by cytokine production or gene expression changes, occurs following infection (Hensley et al. 2002; Bosio et al. 2003; Mahanty et al. 2003; Lubaki et al. 2013). Further, other measures of DC maturation are also suppressed, including the upregulation of cell surface markers and activation of T cell responses (Bosio et al. 2003; Mahanty et al. 2003; Lubaki et al. 2013, 2016).

Stimulation of human DCs with VLPs exposing EBOV GP or recombinant shed EBOV GP leads to activation of cells and secretion of cytokines and chemokines (Escudero-Perez et al. 2014; Bosio et al. 2004; Martinez et al. 2007; Ye et al. 2006). Inflammatory responses after stimulation with recombinant shed EBOV GP are inhibited by anti-TLR4 antibodies, suggesting GP-TLR4 mediated activation comparable to macrophages as discussed above (Escudero-Perez et al. 2014). In addition, EBOV GP delivered by VLPs interacts with C-type lectin domain family 4 member G (CLEC4G/LSECtin) on DCs, leading to the induction of an inflammatory response (Zhao et al. 2016).

The major inhibitor of DC maturation in EBOV infection appears to be the VP35 protein. VP35, when delivered to DCs with alphavirus replicons, from herpes simplex virus (HSV), or from lentiviral vectors, is effective in suppressing DC responses (Bosio et al. 2003; Yen and Basler 2016; Yen et al. 2014; Jin et al. 2010). Delivery by lentivirus allowed the comparison of wild-type and mutant VP35s and their capacity to counteract different DC maturation stimuli and signaling pathways (Yen et al. 2014). Wild-type EBOV VP35 effectively suppresses DC maturation induced by stimuli that signal via either the RIG-I or MDA5 pathways. The suppression is manifested by impairment of *IFNA* or *IFNB* gene expression, a decrease in inflammatory cytokine production, suppressed upregulation of cell surface markers of maturation, and impaired capacity to stimulate T cell responses (Yen et al. 2014). VP35 mutants previously described to lose dsRNA binding and RIG-I inhibitory activities lost the capacity to suppress each of these measures of DC maturation and function (Yen et al. 2014). Delivery of MARV VP35 has the same effect as EBOV VP35 (Yen and Basler 2016). Delivery of the other defined innate immune antagonists of either EBOV or MARV, including EBOV VP24, MARV VP40, and MARV VP24, has only limited impact on DC maturation (Yen and Basler 2016). Both EBOV VP24 and MARV VP40 effectively suppress ISG upregulation following either infection with the RIG-I activating SeV or following treatment of cells with IFN. Nonetheless, these effects of EBOV VP24 or MARV VP40 are insufficient to prevent upregulation of cell surface markers of maturation or to impair T cell activation. Similarly, although MARV VP24 can modulate expression of genes connected to the antioxidant response, it had no impact on other measures of DC maturation (Yen and Basler 2016).

The findings obtained by studying individual filovirus proteins are consistent with the results obtained with recombinant EBOVs possessing mutations in VP35 or VP24 proteins important for inhibition of IFN induction and signaling. In contrast to EBOV encoding wild-type IFN antagonists, recombinant EBOV containing mutant VP35 triggers robust IFN and cytokine responses, upregulates cell surface markers, and becomes able to stimulate T cell responses (Lubaki et al. 2013). Mutation of EBOV VP24 had modest impact on DC maturation (Lubaki et al. 2013). These findings therefore suggest that VP35 plays a critical role in suppressing DC maturation in the context of EBOV infection. Studies of lentivirus-transduced DCs indicate that VP35 potently inhibits RIG-I and MDA5 signaling but has only a very modest effect on DC responses after LPS induced signaling, mediated by TLR4 (Yen et al. 2011). These data therefore suggest that RLR signaling plays the determining role in DC maturation induced by the VP35 mutant. Why GP signaling via TLR4 is not sufficient to induce DC maturation remains to be clarified.

Plasmacytoid DCs (pDCs) represent a special class of DCs that produce copious amounts of IFNα in response to viral stimulation. In pDCs, response to RNA viruses such as EBOV is largely TLR7-dependent. When the EBOV VP35 protein is delivered to pDCs via a recombinant Newcastle disease virus (NDV), it is ineffective at countering pDC IFNα production (Leung et al. 2011). Consistent with an inability to block IFNα production in this system, VP35 is unable to block signaling by TLR7 when TLR7 is expressed by transfection in 293T cells. Interestingly, in in vitro infections EBOV fails to trigger IFNα production by pDCs and EBOV particles fail to enter pDCs (Leung et al. 2011). IFNα–producing pDCs have been detected in the spleen of MARV-infected macaques, but it was not determined if these cells were infected with MARV or were noninfected cells (Fritz et al. 2008). It remains to be determined whether pDCs are activated during EBOV infection in vivo and if EBOV and MARV differ in their ability to activate IFN responses in this cell type.

References

Akira S, Takeda K (2004) Toll-like receptor signalling. Nat Rev Immunol 4(7):499–511

Akira S, Uematsu S, Takeuchi O (2006) Pathogen recognition and innate immunity. Cell 124 (4):783–801

Alazard-Dany N et al (2006) Ebola virus glycoprotein GP is not cytotoxic when expressed constitutively at a moderate level. J Gen Virol 87(Pt 5):1247–1257

Alber D, Staeheli P (1996) Partial inhibition of vesicular stomatitis virus by the interferon-induced human 9-27 protein. J Interferon Cytokine Res 16(5):375–380

Alves DA et al (2010) Aerosol exposure to the Angola strain of Marburg virus causes lethal viral hemorrhagic fever in cynomolgus macaques. Vet Pathol 47(5):831–851

Anafu AA et al (2013) Interferon-inducible transmembrane protein 3 (IFITM3) restricts reovirus cell entry. J Biol Chem 288(24):17261–17271

Ayithan N et al (2014) Ebola virus-like particles stimulate type I interferons and proinflammatory cytokine expression through the toll-like receptor and interferon signaling pathways. J Interferon Cytokine Res 34(2):79–89

Bailey CC et al (2014) IFITM-family proteins: the cell's first line of antiviral defense. Annu Rev Virol 1:261–283

Baize S et al (1999) Defective humoral responses and extensive intravascular apoptosis are associated with fatal outcome in Ebola virus-infected patients. Nat Med 5(4):423–426

Baize S et al (2002) Inflammatory responses in Ebola virus-infected patients. Clin Exp Immunol 128(1):163–168

Barrientos LG, Rollin PE (2007) Release of cellular proteases into the acidic extracellular milieu exacerbates Ebola virus-induced cell damage. Virology 358(1):1–9

Baskerville A et al (1978) The pathology of experimental Ebola virus infection in monkeys. J Pathol 125(3):131–138

Basler CF, Amarasinghe GK (2009) Evasion of interferon responses by Ebola and Marburg viruses. J Interferon Cytokine Res 29(9):511–520

Becker S et al (1998) Interactions of Marburg virus nucleocapsid proteins. Virol 249(2):406–417

Berke IC, Modis Y (2012) MDA5 cooperatively forms dimers and ATP-sensitive filaments upon binding double-stranded RNA. EMBO J 31(7):1714–1726

Bick MJ et al (2003) Expression of the zinc-finger antiviral protein inhibits alphavirus replication. J Virol 77(21):11555–11562

Bird BH et al (2016) Humanized mouse model of Ebola virus disease mimics the immune responses in human disease. J Infect Dis 213(5):703–711

Bjorndal AS, Szekely L, Elgh F (2003) Ebola virus infection inversely correlates with the overall expression levels of promyelocytic leukaemia (PML) protein in cultured cells. BMC Microbiol 3:6

Boehmann Y et al (2005) A reconstituted replication and transcription system for Ebola virus Reston and comparison with Ebola virus Zaire. Virology 332(1):406–417

Bosio CM et al (2003) Ebola and Marburg viruses replicate in monocyte-derived dendritic cells without inducing the production of cytokines and full maturation. J Infect Dis 188(11): 1630–1638

Bosio CM et al (2004) Ebola and Marburg virus-like particles activate human myeloid dendritic cells. Virology 326(2):280–287
Bradfute SB et al (2007) Lymphocyte death in a mouse model of Ebola virus infection. J Infect Dis 196(Suppl 2):S296–S304
Bradfute SB, Warfield KL, Bavari S (2008) Functional CD8 + T cell responses in lethal Ebola virus infection. J Immunol 180(6):4058–4066
Bradfute SB et al (2010) Mechanisms and consequences of ebolavirus-induced lymphocyte apoptosis. J Immunol 184(1):327–335
Brannan JM et al (2015) Interferon alpha/beta receptor-deficient mice as a model for ebola virus disease. J Infect Dis 212(Suppl 2):S282–S294
Brass AL et al (2009) The IFITM proteins mediate cellular resistance to influenza A H1N1 virus, West Nile virus, and dengue virus. Cell 139(7):1243–1254
Bray M, Geisbert TW (2005) Ebola virus: the role of macrophages and dendritic cells in the pathogenesis of Ebola hemorrhagic fever. Int J Biochem Cell Biol 37(8):1560–1566
Bray M et al (2001) Haematological, biochemical and coagulation changes in mice, guinea-pigs and monkeys infected with a mouse-adapted variant of Ebola Zaire virus. J Comp Pathol 125 (4):243–253
Bujnicki JM, Rychlewski L (2002) In silico identification, structure prediction and phylogenetic analysis of the 2′-O-ribose (cap 1) methyltransferase domain in the large structural protein of ssRNA negative-strand viruses. Protein Eng 15(2):101–108
Caballero IS et al (2014) Lassa and Marburg viruses elicit distinct host transcriptional responses early after infection. BMC Genom 15:960
Caballero IS et al (2016) In vivo Ebola virus infection leads to a strong innate response in circulating immune cells. BMC Genom 17:707
Chan SY, Ma MC, Goldsmith MA (2000) Differential induction of cellular detachment by envelope glycoproteins of Marburg and Ebola (Zaire) viruses. J Gen Virol 81(Pt 9):2155–2159
Chelbi-Alix MK et al (1995) Induction of the PML protein by interferons in normal and APL cells. Leukemia 9(12):2027–2033
Chelbi-Alix MK et al (1998) Resistance to virus infection conferred by the interferon-induced promyelocytic leukemia protein. J Virol 72(2):1043–1051
Chook YM, Blobel G (2001) Karyopherins and nuclear import. Curr Opin Struct Biol 11(6):703–715
Chow KT, Gale M Jr (2015) Snapshot: interferon signaling. Cell 163:1808
Connolly BM et al (1999) Pathogenesis of experimental Ebola virus infection in guinea pigs. J Infect Dis 179(Suppl 1):S203–S217
Connor JH et al (2015) Transcriptional profiling of the immune response to Marburg virus infection. J Virol 89(19):9865–9874
Conti E, Izaurralde E (2001) Nucleocytoplasmic transport enters the atomic age. Curr Opin Cell Biol 13(3):310–319
Conti E, Kuriyan J (2000) Crystallographic analysis of the specific yet versatile recognition of distinct nuclear localization signals by karyopherin alpha. Struct 8(3):329–338
Conti E et al (1998) Crystallographic analysis of the recognition of a nuclear localization signal by the nuclear import factor karyopherin alpha. Cell 94(2):193–204
Copple IM et al (2008) The Nrf2-Keap1 defence pathway: role in protection against drug-induced toxicity. Toxicology 246(1):24–33
Cross RW et al (2015) Modeling the disease course of Zaire ebolavirus Infection in the outbred Guinea Pig. J Infect Dis 212(Suppl 2):S305–S315
Daffis S et al (2010) 2′-O methylation of the viral mRNA cap evades host restriction by IFIT family members. Nature 468(7322):452–456
Dauber B, Wolff T (2009) Activation of the antiviral kinase PKR and viral countermeasures. Viruses 1(3):523–544
Diamond MS, Farzan M (2013) The broad-spectrum antiviral functions of IFIT and IFITM proteins. Nat Rev Immunol 13(1):46–57

Dolnik O et al (2004) Ectodomain shedding of the glycoprotein GP of Ebola virus. EMBO J 23 (10):2175–2184
Dolnik O et al (2015) Shedding of Ebola virus surface glycoprotein is a mechanism of self-regulation of cellular cytotoxicity and has a direct effect on virus infectivity. J Infect Dis 212(Suppl 2):S322–S328
Dube D et al (2008) Cell adhesion promotes Ebola virus envelope glycoprotein-mediated binding and infection. J Virol 82(14):7238–7242
Ebihara H et al (2011) Host response dynamics following lethal infection of rhesus macaques with Zaire ebolavirus. J Infect Dis 204(Suppl 3):S991–S999
Ebihara H et al (2013) A Syrian golden hamster model recapitulating ebola hemorrhagic fever. J Infect Dis 207(2):306–318
Edwards MR, Basler CF (2015) Marburg virus VP24 protein relieves suppression of the NF-kappaB pathway through interaction with kelch-like ECH-associated protein 1. J Infect Dis 212(Suppl 2):S154–S159
Edwards MR et al (2016) Differential regulation of interferon responses by Ebola and Marburg virus VP35 proteins. Cell Rep 14(7):1632–1640
Ellis DS et al (1978) Ultrastructure of Ebola virus particles in human liver. J Clin Pathol 31 (3):201–208
Escudero-Perez B et al (2014) Shed GP of Ebola virus triggers immune activation and increased vascular permeability. PLoS Pathog 10(11):e1004509
Everett RD, Chelbi-Alix MK (2007) PML and PML nuclear bodies: implications in antiviral defence. Biochimie 89(6–7):819–830
Everitt AR et al (2013) Defining the range of pathogens susceptible to Ifitm3 restriction using a knockout mouse model. PLoS ONE 8(11):e80723
Fabozzi G et al (2011) Ebolavirus proteins suppress the effects of small interfering RNA by direct interaction with the mammalian RNA interference pathway. J Virol 85(6):2512–2523
Feagins AR, Basler CF (2014) The VP40 protein of Marburg virus exhibits impaired budding and increased sensitivity to human tetherin following mouse adaptation. J Virol 88(24):14440–14450
Feagins AR, Basler CF (2015) Lloviu virus VP24 and VP35 proteins function as innate immune antagonists in human and bat cells. Virol 485:145–152
Feldmann H, Geisbert TW (2011) Ebola haemorrhagic fever. Lancet 377(9768):849–862
Feldmann H et al (1996) Filovirus-induced endothelial leakage triggered by infected monocytes/macrophages. J Virol 70(4):2208–2214
Feng Z et al (2007) The VP35 protein of Ebola virus inhibits the antiviral effect mediated by double-stranded RNA-dependent protein kinase PKR. J Virol 81(1):182–192
Fensterl V, Sen GC (2011) The ISG56/IFIT1 gene family. J Interferon Cytokine Res 31(1):71–78
Fensterl V, Sen GC (2015) Interferon-induced Ifit proteins: their role in viral pathogenesis. J Virol 89(5):2462–2468
Fernando L et al (2015) Immune response to Marburg virus angola infection in nonhuman primates. J Infect Dis 212(Suppl 2):S234–S241
Ferron F et al (2002) Viral RNA-polymerases—a predicted 2′-O-ribose methyltransferase domain shared by all Mononegavirales. Trends Biochem Sci 27(5):222–224
Francica JR, Matukonis MK, Bates P (2009) Requirements for cell rounding and surface protein down-regulation by Ebola virus glycoprotein. Virology 383(2):237–247
Fritz EA et al (2008) Cellular immune response to Marburg virus infection in cynomolgus macaques. Viral Immunol 21(3):355–363
Gao G, Guo X, Goff SP (2002) Inhibition of retroviral RNA production by ZAP, a CCCH-type zinc finger protein. Science 297(5587):1703–1706
Garcia MA et al (2006) Impact of protein kinase PKR in cell biology: from antiviral to antiproliferative action. Microbiol Mol Biol Rev 70(4):1032–1060
Garcia MA, Meurs EF, Esteban M (2007) The dsRNA protein kinase PKR: virus and cell control. Biochimie 89(6–7):799–811

Gedigk P, Bechtelsheimer H, Korb G (1968) Pathological anatomy of the "Marburg virus" disease (the so-called "Marburg monkey disease"). Dtsch Med Wochenschr 93(12):590–601

Geisbert TW et al (2000) Apoptosis induced in vitro and in vivo during infection by Ebola and Marburg viruses. Lab Invest 80(2):171–186

Geisbert TW et al (2003a) Pathogenesis of Ebola hemorrhagic fever in cynomolgus macaques: evidence that dendritic cells are early and sustained targets of infection. Am J Pathol 163 (6):2347–2370

Geisbert TW et al (2003b) Pathogenesis of Ebola hemorrhagic fever in primate models: evidence that hemorrhage is not a direct effect of virus-induced cytolysis of endothelial cells. Am J Pathol 163(6):2371–2382

Geisbert TW et al (2003c) Treatment of Ebola virus infection with a recombinant inhibitor of factor VIIa/tissue factor: a study in rhesus monkeys. Lancet 362(9400):1953–1958

Geisbert TW et al (2007) Marburg virus Angola infection of rhesus macaques: pathogenesis and treatment with recombinant nematode anticoagulant protein c2. J Infect Dis 196(Suppl 2): S372–S381

Geoffroy MC, Chelbi-Alix MK (2011) Role of promyelocytic leukemia protein in host antiviral defense. J Interferon Cytokine Res 31(1):145–158

Gerlier D, Lyles DS (2011) Interplay between innate immunity and negative-strand RNA viruses: towards a rational model. Microbiol Mol Biol Rev 75(3):468–490, (second page of table of contents)

Gibb TR et al (2001) Pathogenesis of experimental Ebola Zaire virus infection in BALB/c mice. J Comp Pathol 125(4):233–242

Groseth A et al (2012) The Ebola virus glycoprotein contributes to but is not sufficient for virulence in vivo. PLoS Pathog 8(8):e1002847

Guito JC et al (2016) Novel activities by ebolavirus and marburgvirus interferon antagonists revealed using a standardized in vitro reporter system. Virol 501:147–165

Guo X et al (2007) The zinc-finger antiviral protein recruits the RNA processing exosome to degrade the target mRNA. Proc Natl Acad Sci U S A 104(1):151–156

Gupta M et al (2001) Monocyte-derived human macrophages and peripheral blood mononuclear cells infected with ebola virus secrete MIP-1alpha and TNF-alpha and inhibit poly-IC-induced IFN-alpha in vitro. Virology 284(1):20–25

Gupta M, Spiropoulou C, Rollin PE (2007) Ebola virus infection of human PBMCs causes massive death of macrophages, CD4 and CD8 T cell sub-populations in vitro. Virology 364 (1):45–54

Gupta M et al (2010) Reduced virus replication, proinflammatory cytokine production, and delayed macrophage cell death in human PBMCs infected with the newly discovered Bundibugyo ebolavirus relative to Zaire ebolavirus. Virology 402(1):203–208

Haasnoot J, Berkhout B (2011) RNAi and cellular miRNAs in infections by mammalian viruses. Methods Mol Biol 721:23–41

Haasnoot J et al (2007) The Ebola virus VP35 protein is a suppressor of RNA silencing. PLoS Pathog 3(6):e86

Hacke M et al (2015) Inhibition of Ebola virus glycoprotein-mediated cytotoxicity by targeting its transmembrane domain and cholesterol. Nat Commun 6:7688

Haller O et al (2015) Mx GTPases: dynamin-like antiviral machines of innate immunity. Trends Microbiol 23(3):154–163

Han Z et al (2007) Permeabilization of the plasma membrane by Ebola virus GP2. Virus Genes 34 (3):273–281

Hartlieb B, Weissenhorn W (2006) Filovirus assembly and budding. Virology 344(1):64–70

Hartman AL et al (2008) Whole-genome expression profiling reveals that inhibition of host innate immune response pathways by Ebola virus can be reversed by a single amino acid change in the VP35 protein. J Virol 82(11):5348–5358

Hensley LE et al (2002) Proinflammatory response during Ebola virus infection of primate models: possible involvement of the tumor necrosis factor receptor superfamily. Immunol Lett 80 (3):169–179

Hensley LE et al (2011) Pathogenesis of Marburg hemorrhagic fever in cynomolgus macaques. J Infect Dis 204(Suppl 3):S1021–S1031
Herbert AS et al (2015) Niemann-pick C1 is essential for ebolavirus replication and pathogenesis in vivo. MBio 6(3):e00565-15
Hiscott J (2007) Convergence of the NF-kappaB and IRF pathways in the regulation of the innate antiviral response. Cytokine Growth Factor Rev 18(5–6):483–490
Hoenen T et al (2013) A novel Ebola virus expressing luciferase allows for rapid and quantitative testing of antivirals. Antiviral Res 99(3):207–213
Hoffmann HH, Schneider WM, Rice CM (2015) Interferons and viruses: an evolutionary arms race of molecular interactions. Trends Immunol 36(3):124–138
Huang IC et al (2011) Distinct patterns of IFITM-mediated restriction of filoviruses, SARS coronavirus, and influenza A virus. PLoS Pathog 7(1):e1001258
Hutchinson KL, Rollin PE (2007) Cytokine and chemokine expression in humans infected with Sudan Ebola virus. J Infect Dis 196(Suppl 2):S357–S363
Hyde JL, Diamond MS (2015) Innate immune restriction and antagonism of viral RNA lacking 2-O methylation. Virology 479–480:66–74
Ignat'ev GM et al (1995) Mechanisms of protective immune response in models of Marburg fever in monkeys. Vopr Virusol 40(3):109–113
Ignatiev GM et al (2000) Immune and pathophysiological processes in baboons experimentally infected with Ebola virus adapted to guinea pigs. Immunol Lett 71(2):131–140
Ivashkiv LB, Donlin LT (2014) Regulation of type I interferon responses. Nat Rev Immunol 14 (1):36–49
Iwamura T et al (2001) PACT, a double-stranded RNA binding protein acts as a positive regulator for type I interferon gene induced by Newcastle disease virus. Biochem Biophys Res Commun 282(2):515–523
Jin H et al (2010) The VP35 protein of Ebola virus impairs dendritic cell maturation induced by virus and lipopolysaccharide. J Gen Virol 91(Pt 2):352–361
Jin G, Gao Y, Lin HK (2014) Cytoplasmic PML: from molecular regulation to biological functions. J Cell Biochem 115(5):812–818
Johnson B et al (2016) Dimerization controls Marburg virus VP24-dependent modulation of host antioxidative stress responses. J Mol Biol 428(17):3483–3494
Kaletsky RL et al (2009) Tetherin-mediated restriction of filovirus budding is antagonized by the Ebola glycoprotein. Proc Natl Acad Sci U S A 106(8):2886–2891
Kimura T et al (2013) Ifit1 inhibits Japanese encephalitis virus replication through binding to 5′ capped 2′-O unmethylated RNA. J Virol 87(18):9997–10003
Kok KH et al (2011) The double-stranded RNA-binding protein PACT functions as a cellular activator of RIG-I to facilitate innate antiviral response. Cell Host Microbe 9(4):299–309
Kreuels B et al (2014) A case of severe Ebola virus infection complicated by gram-negative septicemia. N Engl J Med 371(25):2394–2401
Kuhl A et al (2011) The Ebola virus glycoprotein and HIV-1 Vpu employ different strategies to counteract the antiviral factor tetherin. J Infect Dis 204(Suppl 3):S850–S860
Kumar P et al (2014) Inhibition of translation by IFIT family members is determined by their ability to interact selectively with the 5′-terminal regions of cap0-, cap1- and 5′ppp- mRNAs. Nucleic Acids Res 42(5):3228–3245
Labbe K, Saleh M (2008) Cell death in the host response to infection. Cell Death Differ 15 (9):1339–1349
Lanzavecchia A (1999) Dendritic cell maturation and generation of immune responses. Haematologica 84(Suppl EHA-4):23–25
Lavau C et al (1995) The acute promyelocytic leukaemia-associated PML gene is induced by interferon. Oncogene 11(5):871–876
Le Sage V et al (2016) Ebola virus VP35 blocks stress granule assembly. Virol 502:73–83
Leroy EM et al (2000) Human asymptomatic Ebola infection and strong inflammatory response. Lancet 355(9222):2210–2215

Leroy EM et al (2001) Early immune responses accompanying human asymptomatic Ebola infections. Clin Exp Immunol 124(3):453–460
Leung DW et al (2009) Structure of the Ebola VP35 interferon inhibitory domain. Proc Natl Acad Sci U S A 106(2):411–416
Leung DW et al (2010) Structural basis for dsRNA recognition and interferon antagonism by Ebola VP35. Nat Struct Mol Biol 17(2):165–172
Leung LW et al (2011a) Ebolavirus VP35 suppresses IFN production from conventional but not plasmacytoid dendritic cells. Immunol Cell Biol 89(7):792–802
Leung LW et al (2011b) Ebola virus failure to stimulate plasmacytoid dendritic cell interferon responses correlates with impaired cellular entry. J Infect Dis 204(Suppl 3):S973–S977
Leung DW, Basler CF, Amarasinghe GK (2012) Molecular mechanisms of viral inhibitors of RIG-I-like receptors. Trends Microbiol 20(3):139–146
Lever MS et al (2012) Lethality and pathogenesis of airborne infection with filoviruses in A129 alpha/beta-/-interferon receptor-deficient mice. J Med Microbiol 61(Pt 1):8–15
Li X et al (2009) Structural basis of double-stranded RNA recognition by the RIG-I like receptor MDA5. Arch Biochem Biophys 488(1):23–33
Li SH et al (2013) Rational design of a flavivirus vaccine by abolishing viral RNA 2′-O methylation. J Virol 87(10):5812–5819
Li Y et al (2016) Induction and suppression of antiviral RNA interference by influenza A virus in mammalian cells. Nat Microbiol 2:16250
Lin KL et al (2015) Temporal characterization of Marburg virus Angola infection following aerosol challenge in Rhesus Macaques. J Virol 89(19):9875–9885
Lopez LA et al (2010) Ebola virus glycoprotein counteracts BST-2/Tetherin restriction in a sequence-independent manner that does not require tetherin surface removal. J Virol 84 (14):7243–7255
Lu J et al (2011) The IFITM proteins inhibit HIV-1 infection. J Virol 85(5):2126–2137
Lubaki NM et al (2013) The lack of maturation of Ebola virus-infected dendritic cells results from the cooperative effect of at least two viral domains. J Virol 87(13):7471–7485
Lubaki NM et al (2016) The Ebola interferon inhibiting domains attenuate and dysregulate cell-mediated immune responses. PLoS Pathog 12(12):e1006031
Ludtke A et al (2015) Ebola virus disease in mice with transplanted human hematopoietic stem cells. J Virol 89(8):4700–4704
Luthra P et al (2013) Mutual antagonism between the Ebola virus VP35 protein and the RIG-I activator PACT determines infection outcome. Cell Host Microbe 14(1):74–84
Mahanty S, Bray M (2004) Pathogenesis of filoviral haemorrhagic fevers. Lancet Infect Dis 4 (8):487–498
Mahanty S et al (2003) Cutting edge: impairment of dendritic cells and adaptive immunity by Ebola and Lassa viruses. J Immunol 170(6):2797–2801
Malakhova OA, Zhang DE (2008) ISG15 inhibits Nedd4 ubiquitin E3 activity and enhances the innate antiviral response. J Biol Chem 283(14):8783–8787
Mao R et al (2013) Inhibition of hepatitis B virus replication by the host zinc finger antiviral protein. PLoS Pathog 9(7):e1003494
Martinez O, Valmas C, Basler CF (2007) Ebola virus-like particle-induced activation of NF-kappaB and Erk signaling in human dendritic cells requires the glycoprotein mucin domain. Virology 364(2):342–354
Martinez O, Leung LW, Basler CF (2012) The role of antigen-presenting cells in filoviral hemorrhagic fever: gaps in current knowledge. Antiviral Res 93(3):416–428
Martinez O et al (2013) Ebola virus exploits a monocyte differentiation program to promote its entry. J Virol 87(7):3801–3814
Martins K et al (2015) Characterization of clinical and immunological parameters during Ebola virus infection of rhesus macaques. Viral Immunol 28(1):32–41
Marzi A et al (2013) Antibodies are necessary for rVSV/ZEBOV-GP-mediated protection against lethal Ebola virus challenge in nonhuman primates. Proc Natl Acad Sci U S A 110(5): 1893–1898

Marzi A et al (2015) Delayed disease progression in Cynomolgus macaques infected with Ebola virus Makona strain. Emerg Infect Dis 21(10):1777–1783

Marzi A et al (2016) A hamster model for Marburg virus infection accurately recapitulates Marburg hemorrhagic fever. Sci Rep 6:39214

McBride KM et al (2002) Regulated nuclear import of the STAT1 transcription factor by direct binding of importin-alpha. EMBO J 21(7):1754–1763

McElroy AK et al (2014a) Ebola hemorrhagic fever: novel biomarker correlates of clinical outcome. J Infect Dis 210(4):558–566

McElroy AK et al (2014b) Biomarker correlates of survival in pediatric patients with Ebola virus disease. Emerg Infect Dis 20(10):1683–1690

McElroy AK et al (2016) Kinetic analysis of biomarkers in a cohort of US patients with Ebola virus disease. Clin Infect Dis 63(4):460–467

Melito PL et al (2008) The creation of stable cell lines expressing Ebola virus glycoproteins and the matrix protein VP40 and generating Ebola virus-like particles utilizing an ecdysone inducible mammalian expression system. J Virol Methods 148(1–2):237–243

Mudhasani R et al (2013) IFITM-2 and IFITM-3 but not IFITM-1 restrict Rift Valley fever virus. J Virol 87(15):8451–8464

Muller S et al (2007) Inhibition of filovirus replication by the zinc finger antiviral protein. J Virol 81(5):2391–2400

Murphy FA et al (1971) Marburg virus infection in monkeys. Ultrastructural studies. Lab Invest 24 (4):279–291

Nanduri S et al (1998) Structure of the double-stranded RNA-binding domain of the protein kinase PKR reveals the molecular basis of its dsRNA-mediated activation. EMBO J 17(18): 5458–5465

Neil SJ, Zang T, Bieniasz PD (2008) Tetherin inhibits retrovirus release and is antagonized by HIV-1 Vpu. Nature 451(7177):425–430

Nelson EV et al (2016) Ebola virus does not induce stress granule formation during infection and sequesters stress granule proteins within viral inclusions. J Virol 90(16):7268–7284

Noyori O et al (2013) Suppression of Fas-mediated apoptosis via steric shielding by filovirus glycoproteins. Biochem Biophys Res Commun 441(4):994–998

Okumura A, Pitha PM, Harty RN (2008) ISG15 inhibits Ebola VP40 VLP budding in an L-domain-dependent manner by blocking Nedd4 ligase activity. Proc Natl Acad Sci U S A 105 (10):3974–3979

Okumura A et al (2010) Interaction between Ebola virus glycoprotein and host toll-like receptor 4 leads to induction of proinflammatory cytokines and SOCS1. J Virol 84(1):27–33

Okumura A et al (2015) Suppressor of cytokine signaling 3 is an inducible host factor that regulates virus egress during Ebola virus infection. J Virol 89(20):10399–10406

Olejnik J et al (2011) Intracellular events and cell fate in filovirus infection. Viruses 3(8):1501–1531

Olejnik J et al (2013) Ebola virus does not block apoptotic signaling pathways. J Virol 87 (10):5384–5396

Olejnik J et al (2017) Ebolaviruses associated with differential pathogenicity induce distinct host responses in human macrophages. J Virol, (in press)

O'Shea JJ et al (2015) The JAK-STAT pathway: impact on human disease and therapeutic intervention. Annu Rev Med 66:311–328

Ozato K et al (2008) TRIM family proteins and their emerging roles in innate immunity. Nat Rev Immunol 8(11):849–860

Page A et al (2014) Marburgvirus hijacks nrf2-dependent pathway by targeting nrf2-negative regulator keap1. Cell Rep 6(6):1026–1036

Peisley A et al (2011) Cooperative assembly and dynamic disassembly of MDA5 filaments for viral dsRNA recognition. Proc Natl Acad Sci U S A 108(52):21010–21015

Pichlmair A et al (2011) IFIT1 is an antiviral protein that recognizes 5′-triphosphate RNA. Nat Immunol 12(7):624–630

Pichlmair A et al (2012) Viral immune modulators perturb the human molecular network by common and unique strategies. Nature 487(7408):486–490
Pinto AK et al (2015) Human and Murine IFIT1 proteins do not restrict infection of negative-sense rna viruses of the orthomyxoviridae, bunyaviridae, and filoviridae families. J Virol 89 (18):9465–9476
Prins KC, Cardenas WB, Basler CF (2009) Ebola virus protein VP35 impairs the function of interferon regulatory factor-activating kinases IKKepsilon and TBK-1. J Virol 83(7): 3069–3077
Prins KC et al (2010a) Mutations abrogating VP35 interaction with double-stranded RNA render Ebola virus avirulent in guinea pigs. J Virol 84(6):3004–3015
Prins KC et al (2010b) Basic residues within the ebolavirus VP35 protein are required for its viral polymerase cofactor function. J Virol 84(20):10581–10591
Qiu X et al (2014) Establishment and characterization of a lethal mouse model for the Angola strain of Marburg virus. J Virol 88(21):12703–12714
Radoshitzky SR et al (2010) Infectious Lassa virus, but not filoviruses, is restricted by BST-2/tetherin. J Virol 84(20):10569–10580
Ramanan P et al (2012) Structural basis for Marburg virus VP35-mediated immune evasion mechanisms. Proc Natl Acad Sci U S A 109(50):20661–20666
Rawlings JS, Rosler KM, Harrison DA (2004) The JAK/STAT signaling pathway. J Cell Sci 117 (Pt 8):1281–1283
Ray RB et al (2004) Ebola virus glycoprotein-mediated anoikis of primary human cardiac microvascular endothelial cells. Virology 321(2):181–188
Rebouillat D, Hovanessian AG (1999) The human 2′,5′-oligoadenylate synthetase family: interferon-induced proteins with unique enzymatic properties. J Interferon Cytokine Res 19 (4):295–308
Reed DS et al (2004) Depletion of peripheral blood T lymphocytes and NK cells during the course of ebola hemorrhagic fever in cynomolgus macaques. Viral Immunol 17(3):390–400
Regad T, Chelbi-Alix MK (2001) Role and fate of PML nuclear bodies in response to interferon and viral infections. Oncogene 20(49):7274–7286
Reid SP et al (2006) Ebola virus VP24 binds karyopherin alpha1 and blocks STAT1 nuclear accumulation. J Virol 80(11):5156–5167
Reid SP et al (2007) Ebola virus VP24 proteins inhibit the interaction of NPI-1 subfamily karyopherin alpha proteins with activated STAT1. J Virol 81(24):13469–13477
Rhein BA et al (2015) Interferon-gamma inhibits Ebola virus infection. PLoS Pathog 11(11): e1005263
Rippey JJ, Schepers NJ, Gear JH (1984) The pathology of Marburg virus disease. S Afr Med J 66 (2):50–54
Rougeron V et al (2015) Ebola and Marburg haemorrhagic fever. J Clin Virol 64:111–119
Rubins KH et al (2007) The temporal program of peripheral blood gene expression in the response of nonhuman primates to Ebola hemorrhagic fever. Genome Biol 8(8):R174
Ruibal P et al (2016) Unique human immune signature of Ebola virus disease in Guinea. Nature 533(7601):100–104
Ryabchikova E, Price BBS (2004) Ebola and Marburg viruses: a view of infection using electron microscopy. Columbus, Ohio, USA: Battelle Press
Ryabchikova E et al (1996a) Ebola virus infection in guinea pigs: presumable role of granulomatous inflammation in pathogenesis. Arch Virol 141(5):909–921
Ryabchikova E et al (1996b) Respiratory Marburg virus infection in guinea pigs. Arch Virol 141 (11):2177–2190
Ryabchikova EI, Kolesnikova LV, Luchko SV (1999) An analysis of features of pathogenesis in two animal models of Ebola virus infection. J Infect Dis 179(Suppl 1):S199–S202
Sadler AJ, Williams BR (2008) Interferon-inducible antiviral effectors. Nat Rev Immunol 8 (7):559–568

Sakuma T et al (2009) Inhibition of Lassa and Marburg virus production by tetherin. J Virol 83 (5):2382–2385
Sanchez A et al (2004) Analysis of human peripheral blood samples from fatal and nonfatal cases of Ebola (Sudan) hemorrhagic fever: cellular responses, virus load, and nitric oxide levels. J Virol 78(19):10370–10377
Schlee M et al (2009) Recognition of 5′ triphosphate by RIG-I helicase requires short blunt double-stranded RNA as contained in panhandle of negative-strand virus. Immun 31(1):25–34
Schmidt A et al (2009) 5′-triphosphate RNA requires base-paired structures to activate antiviral signaling via RIG-I. Proc Natl Acad Sci U S A 106(29):12067–12072
Schmidt KM et al (2011) Recombinant Marburg virus expressing EGFP allows rapid screening of virus growth and real-time visualization of virus spread. J Infect Dis 204(Suppl 3):S861–S870
Schneider WM, Chevillotte MD, Rice CM (2014) Interferon-stimulated genes: a complex web of host defenses. Annu Rev Immunol 32:513–545
Schümann M, Gantke T, Mühlberger E (2009) Ebola virus VP35 antagonizes PKR activity through its C-terminal interferon inhibitory domain. J Virol 83(17):8993–8997
Schwarz TM, et al (2016) VP24-karyopherin alpha binding affinities differ between Ebolavirus species influencing interferon inhibition and VP24 stability. J Virol
Sekimoto T et al (1997) Extracellular signal-dependent nuclear import of Stat1 is mediated by nuclear pore-targeting complex formation with NPI-1, but not Rch1. EMBO J 16(23):7067–7077
Simmons G et al (2002) Ebola virus glycoproteins induce global surface protein down-modulation and loss of cell adherence. J Virol 76(5):2518–2528
Stroher U et al (2001) Infection and activation of monocytes by Marburg and Ebola viruses. J Virol 75(22):11025–11033
Sullivan NJ et al (2005) Ebola virus glycoprotein toxicity is mediated by a dynamin-dependent protein-trafficking pathway. J Virol 79(1):547–553
Sun Q et al (2006) The specific and essential role of MAVS in antiviral innate immune responses. Immun 24(5):633–642
Szretter KJ et al (2012) 2′-O methylation of the viral mRNA cap by West Nile virus evades ifit1-dependent and -independent mechanisms of host restriction in vivo. PLoS Pathog 8(5): e1002698
Takada A et al (2000) Downregulation of beta1 integrins by Ebola virus glycoprotein: implication for virus entry. Virology 278(1):20–26
Takeuchi O, Akira S (2008) MDA5/RIG-I and virus recognition. Curr Opin Immunol 20(1):17–22
Thi EP et al (2015) Lipid nanoparticle siRNA treatment of Ebola-virus-Makona-infected nonhuman primates. Nature 521(7552):362–365
Umbach JL, Cullen BR (2009) The role of RNAi and microRNAs in animal virus replication and antiviral immunity. Genes Dev 23(10):1151–1164
Valmas C et al (2010) Marburg virus evades interferon responses by a mechanism distinct from ebola virus. PLoS Pathog 6(1):e1000721
van Paassen J et al (2012) Acute liver failure, multiorgan failure, cerebral oedema, and activation of proangiogenic and antiangiogenic factors in a case of Marburg haemorrhagic fever. Lancet Infect Dis 12(8):635–642
Vasselon T et al (2013) RNAi and retroviruses: are they in RISC? Biomol Concepts 4(1):43–52
Villinger F et al (1999) Markedly elevated levels of interferon (IFN)-gamma, IFN-alpha, interleukin (IL)-2, IL-10, and tumor necrosis factor-alpha associated with fatal Ebola virus infection. J Infect Dis 179(Suppl 1):S188–S191
Vladimer GI, Gorna MW, Superti-Furga G (2014) IFITs: emerging roles as key anti-viral proteins. Front Immunol 5:94
Volchkov VE et al (2001) Recovery of infectious Ebola virus from complementary DNA: RNA editing of the GP gene and viral cytotoxicity. Science 291(5510):1965–1969

Volchkova VA et al (2015) RNA editing of the GP gene of Ebola virus is an important pathogenicity factor. J Infect Dis 212(Suppl 2):S226–S233
Wahl-Jensen V et al (2005) Role of Ebola virus secreted glycoproteins and virus-like particles in activation of human macrophages. J Virol 79(4):2413–2419
Wahl-Jensen V et al (2011) Ebola virion attachment and entry into human macrophages profoundly effects early cellular gene expression. PLoS Negl Trop Dis 5(10):e1359
Warfield KL et al (2007) Development of a model for marburgvirus based on severe-combined immunodeficiency mice. Virol J 4:108
Warfield KL et al (2009) Development and characterization of a mouse model for Marburg hemorrhagic fever. J Virol 83(13):6404–6415
Warren TK et al (2010) Antiviral activity of a small-molecule inhibitor of filovirus infection. Antimicrob Agents Chemother 54(5):2152–2159
Wauquier N et al (2010) Human fatal Zaire Ebola virus infection is associated with an aberrant innate immunity and with massive lymphocyte apoptosis. PLoS Negl Trop Dis 4(10)
Wolf T et al (2015) Severe Ebola virus disease with vascular leakage and multiorgan failure: treatment of a patient in intensive care. Lancet 385(9976):1428–1435
Wrensch F et al (2015) Interferon-induced transmembrane protein-mediated inhibition of host cell entry of Ebolaviruses. J Infect Dis 212(Suppl 2):S210–S218
Xu W et al (2014) Ebola virus VP24 targets a unique NLS binding site on karyopherin alpha 5 to selectively compete with nuclear import of phosphorylated STAT1. Cell Host Microbe 16 (2):187–200
Yan N, Chen ZJ (2012) Intrinsic antiviral immunity. Nat Immunol 13(3):214–222
Yang ZY et al (2000) Identification of the Ebola virus glycoprotein as the main viral determinant of vascular cell cytotoxicity and injury. Nat Med 6(8):886–889
Ye L et al (2006) Ebola virus-like particles produced in insect cells exhibit dendritic cell stimulating activity and induce neutralizing antibodies. Virology 351(2):260–270
Yen BC, Basler CF (2016) Effects of filovirus interferon antagonists on responses of human monocyte-derived dendritic cells to RNA virus infection. J Virol 90(10):5108–5118
Yen JY et al (2011) Therapeutics of Ebola hemorrhagic fever: whole-genome transcriptional analysis of successful disease mitigation. J Infect Dis 204(Suppl 3):S1043–S1052
Yen B et al (2014) Molecular basis for ebolavirus VP35 suppression of human dendritic cell maturation. J Virol 88(21):12500–12510
Yoneyama M et al (2005) Shared and unique functions of the DExD/H-box helicases RIG-I, MDA5, and LGP2 in antiviral innate immunity. J Immunol 175(5):2851–2858
Yonezawa A, Cavrois M, Greene WC (2005) Studies of ebola virus glycoprotein-mediated entry and fusion by using pseudotyped human immunodeficiency virus type 1 virions: involvement of cytoskeletal proteins and enhancement by tumor necrosis factor alpha. J Virol 79(2): 918–926
Young DF et al (2016) Human IFIT1 inhibits mRNA translation of rubulaviruses but not other members of the Paramyxoviridae family. J Virol
Zaki SR et al (1999) A novel immunohistochemical assay for the detection of Ebola virus in skin: implications for diagnosis, spread, and surveillance of Ebola hemorrhagic fever. Commission de Lutte contre les Epidemies a Kikwit. J Infect Dis 179(Suppl 1):S36–47
Zhang AP et al (2014) Crystal structure of Marburg virus VP24. J Virol
Zampieri CA et al (2007) The ERK mitogen-activated protein kinase pathway contributes to Ebola virus glycoprotein-induced cytotoxicity. J Virol 81(3):1230–1240
Zhang D, Zhang DE (2011) Interferon-stimulated gene 15 and the protein ISGylation system. J Interferon Cytokine Res 31(1):119–130
Zhao Z et al (2016a) Drug repurposing to target Ebola virus replication and virulence using structural systems pharmacology. BMC Bioinform 17:90

Zhao D et al (2016b) The Myeloid LSECtin is a DAP12-coupled receptor that is crucial for inflammatory response induced by Ebola virus glycoprotein. PLoS Pathog 12(3):e1005487

Zhu Y et al (2011) Zinc-finger antiviral protein inhibits HIV-1 infection by selectively targeting multiply spliced viral mRNAs for degradation. Proc Natl Acad Sci U S A 108(38):15834–15839

Zumbrun EE et al (2012) Development of a murine model for aerosolized ebolavirus infection using a panel of recombinant inbred mice. Viruses 4(12):3468–3493

Mechanisms of Filovirus Entry

R.A. Davey, O. Shtanko, M. Anantpadma, Y. Sakurai, K. Chandran and W. Maury

Abstract Filovirus entry into cells is complex, perhaps as complex as any viral entry mechanism identified to date. However, over the past 10 years, the important events required for filoviruses to enter into the endosomal compartment and fuse with vesicular membranes have been elucidated (Fig. 1). Here, we highlight the important steps that are required for productive entry of filoviruses into mammalian cells.

Contents

R.A. Davey · O. Shtanko · M. Anantpadma · Y. Sakurai
Department of Virology and Immunology, Texas Biomedical Research Institute, San Antonio, TX, USA
e-mail: rdavey@txbiomed.org

O. Shtanko
e-mail: oshtanko@txbiomed.org

M. Anantpadma
e-mail: manantpadma@txbiomed.org

Y. Sakurai
e-mail: ysakurai@txbiomed.org

K. Chandran
Department of Microbiology and Immunology, Albert Einstein College of Medicine, Bronx, NY, USA
e-mail: kartik.chandran@einstein.yu.edu

W. Maury (✉)
Department of Microbiology, The University of Iowa, Iowa City, IA, USA
e-mail: wendy-maury@uiowa.edu

Current Topics in Microbiology and Immunology (2017) 411:323–352
DOI 10.1007/82_2017_14

Published Online: 11 June 2017

1 Filovirus Morphology

Filovirions are structurally pleomorphic, but are characterized by a commonly observed long filamentous structure. These helically symmetrical virions that contain a negative strand RNA genome are ~80 nm in width and as long as 1400 nm (Feldmann et al. 2013). Ebolavirus particles are thought to be longer than Marburg virus (MARV) (Geisbert and Jahrling 1995). Filovirus particles are enveloped, acquiring the membrane during viral budding from the host cell plasma membrane. The membrane is studded with the virally encoded glycoprotein (GP) that can be seen to protrude from the membrane in electron micrographs. VP40, which lines the inner leaflet of the envelope, is the main protein controlling morphology, although nucleoprotein (NP) and several minor structural proteins including VP24 define the virion cross-section diameter (Harty et al. 2000; Noda et al. 2002; Makino et al. 2011). Within the virion, the RNA genome is surrounded and protected by the nucleocapsid complex composed of NP, VP30, and VP35 (Makino et al. 2011) with the matrix proteins VP40 and VP24 also contributing to the nucleocapsid structure (Hoenen et al. 2010; Bharat et al. 2012). The viral RNA-dependent RNA polymerase (RdRp, L) is thought to be recruited to this complex by interactions with VP35 and VP30 (Becker et al. 1998; Groseth et al. 2009).

2 The Biology of the Filovirus Glycoprotein

The biology of filovirus glycoproteins has been extensively studied [see reviews (Miller and Chandran 2012; Simmons 2013; Rhein and Maury 2015; Martin et al. 2016)]. Filovirus GP is a class I viral membrane fusion glycoprotein that is similar to other well-studied glycoproteins of this class such as HIV-1 Env and influenza A virus hemagglutinin [recently reviewed (Harrison 2015)]. The major protein product transcribed from the ebolavirus GP gene is a secreted, soluble GP (sGP) (Sanchez et al. 1998). The function of sGP is still unclear, but it may be important in viral immune evasion [recently reviewed (Audet and Kobinger 2015)]. Full-length, membrane-associated ebolaviral GP is produced by a cotranscriptional frameshift that results in the insertion of a non-templated adenosine residue during transcription of the GP gene (Volchkov et al. 1995; Sanchez et al. 1996). In contrast, the marburgvirus GP gene directly encodes membrane-associated GP (Feldmann et al. 1999). The mature filoviral GP is generated by posttranslational furin cleavage of the pro-protein, producing a disulfide-linked heterodimeric protein composed of GP1 and GP2 (Volchkov et al. 1998). The GP1 subunit is required for receptor

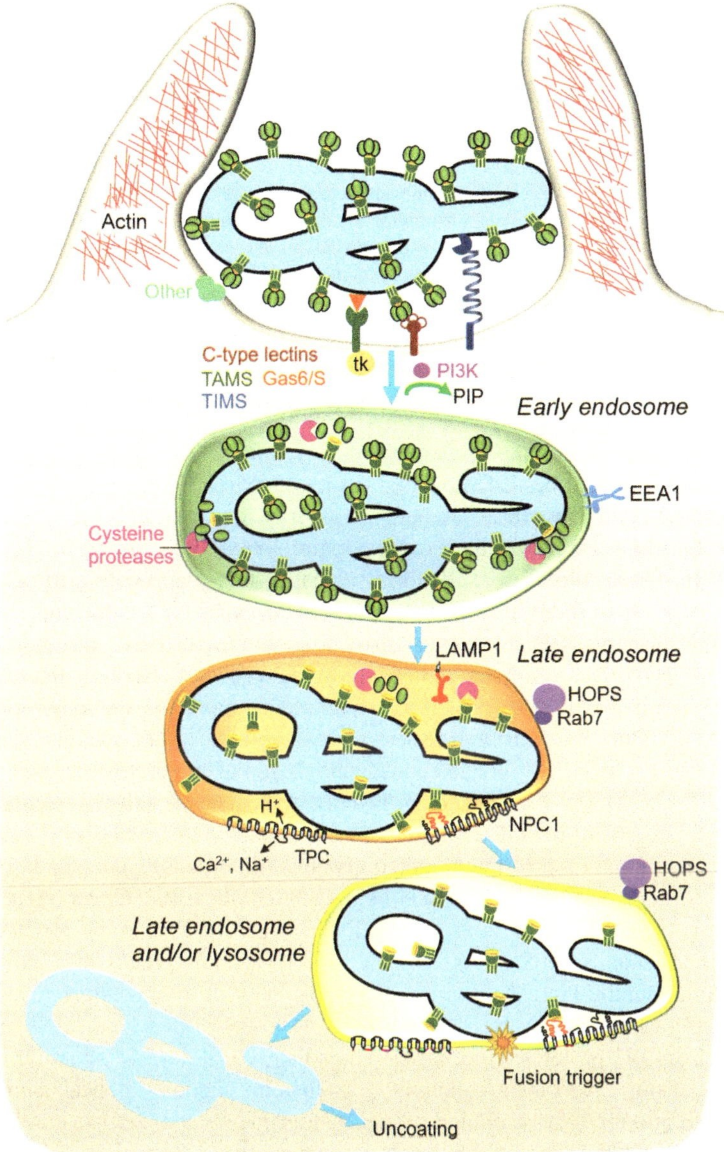
Actin
Other
C-type lectins
TAMS
Gas6/S
TIMS
tk
PI3K
PIP
Early endosome
EEA1
Cysteine
proteases
LAMP1
Late endosome
HOPS
Rab7
H+
NPC1
Ca2+, Na+
TPC
HOPS
Rab7
Late endosome
and/or lysosome
Fusion trigger
Uncoating

◂**Fig. 1** Schematic of cell entry pathway of filoviruses showing major host cell factors important for virus infection. Heavily glycosylated trimeric glycoproteins are present on the surface of filovirions. Interactions with phosphatidylserine receptors, such as the TIMs and TAMS (through Gas6 or Protein S), and C-type lectins mediate virion internalization into the early endosomal compartment by macropinocytosis and requires signaling events such as activation of the PI3K pathway and tyrosine kinase activity (tk). These signaling proteins may help trigger actin polymerization and aid in formation of the macropinosome. Once inside, the endosome processing by endosomal proteases, in particular cysteine proteases, cleave the glycoprotein to a smaller form, exposing residues within the receptor-binding domain. Trafficking of the virus particle in the endosome requires the HOPS complex, (includes RILP) and other associated trafficking factors including TPC ion channels. Upon arrival in a $NPC1^+/TPC^+$ late endosome/lysosome, the processed glycoprotein binds to NPC1. The glycoprotein/NPC1 interaction and additional uncharacterized events stimulate virion/cellular membrane fusion, allowing the release of the viral nucleocapsid into the cell cytosol

interactions and membrane-associated GP2 is required for membrane fusion. Like other class I viral membrane fusion glycoproteins, filovirus GPs are found on virions as trimers. Crystal structures of both ebolavirus and marburgvirus trimeric GP ectodomains have been solved (Lee and Saphire 2009; Dias et al. 2011; Hashiguchi et al. 2015). In all filoviruses, GP forms a chalice-like shape with a trimer of heterodimers of GP1/GP2. GP2 serves as the base, whereas GP1 is the cup.

Mature filovirus GP1 has four distinct subdomains: base, receptor-binding domain (RBD), glycan cap, and mucin-like domain (MLD). The base interacts with GP2 ectodomain residues, providing structural support for the other domains. Residues within the RBD interact with an intracellular cellular receptor, Niemann-Pick C1 (NPC1) in late endosomal/lysosomal compartments, as discussed below. The MLD and glycan cap are heavily glycosylated with *N*-linked glycans. These glycans shield the GP ectodomain from neutralizing antibodies (Lennemann et al. 2014). In addition to the *N*-linked glycans, filovirus GPs contain as many as 80 *O*-linked glycans on the MLD (Feldmann et al. 1991; Geyer et al. 1992; Feldmann et al. 1994; Jeffers et al. 2002). The MLD can be deleted from EBOV GP, and its removal is thought to enhance GP processing and surface expression, thereby increasing virion entry efficiency in tissue culture (Yang et al. 2000; Jeffers et al. 2002).

Filovirus GP2 proteins have similar functionality and domain structure as other class I fusion proteins, containing a fusion peptide, helical repeats, a membrane-proximal extracellular region, a transmembrane domain, and a short cytoplasmic tail. The fusion peptide is located within a loop near the N-terminal sequence of GP2 rather than directly at the N terminus as occurring in many class I viral glycoproteins. This internal fusion loop (IFL) is delineated by a disulfide bond and contains hydrophobic residues at the tip of the loop (Ito et al. 1999; Jeffers et al. 2002; White et al. 2008; Gregory et al. 2011). Premature triggering of fusion events is prevented by interactions of the hydrophobic loop sequences with GP1 residues on adjacent GP subunits (Lee and Saphire 2009).

3 Filovirus Interactions with the Surface of Cells

Despite efforts to identify filoviral GP interactions with cognate cellular protein receptors on the surface of cells, no one cell protein is sufficient to define the broad tropism of the virus. Instead, filoviruses appear to bind to cells through at least two types of alternative surface receptors interactions: filovirus GP binding to carbohydrate-binding receptors and virion lipid binding to phosphatidylserine receptors.

Carbohydrate-binding receptors. *N*- and *O*-linked glycans on EBOV GP bind to two structurally different groups of carbohydrate-binding receptors: C-type lectins and glycosaminoglycans. A series of structurally related C-type lectin receptors (CLRs) have been shown to enhance filovirus infection by binding to specific sugars on the GP glycans (Becker et al. 1995; Alvarez et al. 2002; Takada et al. 2004; Powlesland et al. 2008; Lennemann et al. 2014; Dahlmann et al. 2015). CLRs that enhance filovirus entry include dendritic cell-specific ICAM-3-grabbing non-integrin (DC-SIGN) (Alvarez et al. 2002; Baribaud et al. 2002b; Colmenares et al. 2002; Geijtenbeek and van Kooyk 2003; Lasala et al. 2003; Lin et al. 2003; Simmons et al. 2003a; Marzi et al. 2004), liver/lymph node-specific ICAM-3 grabbing non-integrin (L-SIGN) (Bashirova et al. 2001; Alvarez et al. 2002; Marzi et al. 2004; Marzi et al. 2007), lymph node sinusoidal endothelial cell C-type lectin (LSECTin) (Gramberg et al. 2005; Powlesland et al. 2008; Pipirou et al. 2011), asialoglycoprotein receptor 1 (ASGPRI) (Becker et al. 1995), and human macrophage galactose- and acetylgalactosamine-specific C-type lectin (hMGL) (Takada et al. 2004). These five CLRs are but a small subset of all C-type lectin receptors and it is likely that additional CLRs also bind to filovirus GP glycans and enhance infection.

CLRs are found on a variety of primary cells and each has a specific carbohydrate interaction. For instance, DC-SIGN, found on myeloid cells, and L-SIGN (also called DC-SIGNR), found on some endothelial cells, bind to high mannose *N*-linked glycans (Alvarez et al. 2002; Lin et al. 2003; Simmons et al. 2003; Marzi et al. 2006, 2007; Lennemann et al. 2014). Early work demonstrated that EBOV and MARV GPs contain abundant high mannose *N*-linked glycans (Feldmann et al. 1991; Geyer et al. 1992; Feldmann et al. 1994). Therefore, not surprisingly, loss of *N*-linked glycans on either the glycan cap or MLD of GP1 abrogates EBOV entry facilitated by these CLRs (Lennemann et al. 2014). LSECTin on liver and lymph node sinusoidal endothelial cells binds to *N*-acetylglucosamine-β1, 2-mannose (Powlesland et al. 2008). ASGPRI and hMGL are expressed on hepatocytes and monocyte-derived immature dendritic cells or macrophages, respectively, and both bind to galactose and *N*-acetylgalactosamine (Becker et al. 1995; Takada et al. 2004; Lennemann et al. 2014).

The role of EBOV GP *N*-linked glycans on CLR-dependent entry has been evaluated by a number of different approaches. Early work with glycosidase treatments and lectin binding analysis defined the extensive presence of *N*- and *O*-linked glycans on filovirus GPs (Feldmann et al. 1994). The interaction of the GP-associated glycans with CLRs was initially determined by overexpression studies and CLR antibody and/or inhibitor studies (Alvarez et al. 2002; Baribaud et al. 2002a, b; Takada et al. 2004; Gramberg et al. 2005; Powlesland et al. 2008; Pipirou et al. 2011). This work established the importance of CLRs for filovirus entry into a variety of cell types. A more recent study that genetically eliminated *N*-linked glycans extended these findings by identifying which *N*-linked glycans affect CLR binding (Lennemann et al. 2014). Loss of the eight *N*-linked glycans on the mucin-like domain or the seven *N*-linked glycans on the glycan cap of GP1 profoundly decreased virus entry mediated by DC-SIGN or L-SIGN. Since removal of glycans from either GP1 domain was equally effective at inhibiting mannose-dependent interactions with DC-SIGN or L-SIGN, this finding suggests that reducing the total number of *N*-linked glycans on GP1 alters glycan processing in producer cells. This presumably results in more of the *N*-linked glycans being processed to more complex structures. The impact of the loss of EBOV GP1 *N*-glycans was more subdomain-specific for interactions with LSECtin, ASGPR1, and hMGL. The removal of *N*-glycans from the core of GP1 had little to no effect on entry via these CLRs, whereas loss of *N*-glycans on the MLD more significantly decreased entry. It should be noted that some of these CLRs likely interact with *O*-glycans on the EBOV GP1 MLD. The effect of partial or complete *O*-glycan removal from the full-length EBOV GP1 has yet to be investigated.

Soluble C-type lectins such as mannose-binding lectin (MBL) and ficolin-1 also bind to filovirus GPs (Ji et al. 2005; Brudner et al. 2013; Favier et al. 2016). Soluble lectins serve as serum pattern recognition receptors with MBL binding to *N*-acetyl glucosamine (GlcNAc), mannose, and fucose, whereas ficolin-1 binds to GlcNAc and sialic acid on host cells and pathogens (Hansen and Holmskov 1998; Aoyagi et al. 2005; Gout et al. 2010). Binding of these soluble lectins to EBOV enhances virus infection of Vero E6 cells or primary human macrophages under low complement conditions (Brudner et al. 2013; Favier et al. 2016). Consistent with a role of MBL in enhancing virus entry, a number of genetic polymorphisms have been described in human population that reduce the levels of the soluble lectin (Turner 2003; Brudner et al. 2013), suggesting these genes are actively under positive selection. However, for viruses, interactions with soluble CLECs can also be detrimental. Soluble CLECs interact with the serine proteases, MASPs, to initiate the complement cascade and thereby clear virus from circulation [for a review, see: (Mason and Tarr 2015)].

A final group of glycan binding moieties, glycosaminoglycans (GAGs), has recently been shown to bind to EBOV GP (O'Hearn et al. 2015). Addition of a series of exogenous GAGs, including heparin, heparin sulfate, chondroitin and chondroitin sulfate, to permissive cells blocked EBOV transduction at very early times during infection (Cheng et al. 2015; O'Hearn et al. 2015). Further, knock down of exostosin 1, which is critical for the production of heparin sulfate, reduces EBOV and MARV infection in tissue culture (O'Hearn et al. 2015).

It is likely that the interactions of C-type lectins and GAGs with filovirions play an important role facilitating virus infection in vivo, but this has yet to be directly studied. No studies have explored the role of these receptors during filovirus infection of animals. Of special interest would be the impact of CLRs on the early and important cellular targets of filoviruses: dendritic cells and macrophages. Interestingly, the murine homolog of human DC-SIGN (SIGNR1) does not enhance transduction of pseudovirions bearing EBOV GP (Caminschi et al. 2006; Gramberg et al. 2006). However, at least four other DC-SIGN-related genes encode similar proteins in mice (Park et al. 2001; Powlesland et al. 2006) and one or more of these proteins may serve to enhance filovirus entry into cells of these animals.

Phosphatidylserine receptors. Although filovirus GP glycan interactions with the host factors enhance virus entry into a variety of cells, it is not clear that these carbohydrate interactions directly lead to virus internalization, and cells lacking expression of these receptors remain permissive for EBOV infection. Thus, additional cell surface receptors must interact with filoviruses and mediate viral entry.

Shimojima et al. identified that the tyrosine kinase receptors, Tyro3, Axl, and Mer (TAM receptors) enhance filovirus transduction (Shimojima et al. 2006). The involvement of Axl for virus entry was later confirmed by Kondratowicz et al., using a high-throughput screening approach. The same study also identified T cell, immunoglobulin, mucin domain receptor-1 (TIM-1) as being important for EBOV cell entry (Kondratowicz et al. 2011). A direct role in EBOV entry was confirmed by Jemielity et al., and by Moller-Tank et al., who showed that TIM directly binds PS on the outer leaflet of the filovirus envelope membrane. This interaction is independent of the presence of the viral glycoprotein (Jemielity et al. 2013; Moller-Tank et al. 2013). Formal demonstration that TAMs also interact with filovirus lipids has yet to occur, but TAM interactions occur with PS present on other enveloped viruses such as dengue virus 2 (Meertens et al. 2012). Both the TIM and TAM families of cell surface receptors have previously been described to bind to PS on the surface of apoptotic bodies, mediating their clearance (Linger et al. 2008; DeKruyff et al. 2010; Freeman et al. 2010; Ravichandran 2010, 2011). The concept of PS present on the surface of virions mediating virus binding of and internalization into cells was termed apoptotic mimicry by Mercer and Helenius (Mercer and Helenius 2008). This uptake mechanism is not only important for entry of filoviruses, but also for a number of other viruses and viral vectors that are either enveloped or associated with cellular membranes (Morizono et al. 2011; Meertens et al. 2012; Feng et al. 2013; Jemielity et al. 2013; Moller-Tank et al. 2013; Morizono and Chen 2014). These include poxviruses, flaviviruses, alphaviruses, and hepatitis A virus as well as viral vectors such as lentiviruses pseudotyped with the alphaviral glycoprotein.

TIM receptors. Genes for three TIM family members (TIM-1, TIM-3 and TIM-4) are present in the human genome, whereas at least four TIMs are expressed in mice (TIM-1 through TIM-4) (Freeman et al. 2010). TIMs are type I, cell surface glycoproteins that share a common structure. Their amino terminal immunoglobulin variable-like domain (IgV) contains a binding pocket for PS (Santiago et al. 2007a, b). This pocket is conserved across the entire family and contains invariant asparagine and aspartic acid residues that coordinate a metal ion and are required for lipid interaction (Santiago et al. 2007a; Meertens et al. 2012; Moller-Tank et al. 2013; Rhein et al. 2016). In addition to the PS pocket residues, other murine and human TIM-4 IgV residues are important for filovirus binding and internalization (Rhein et al. 2016), through direct interaction or stabilization of PS binding (Tietjen et al. 2014). The TIM IgV domain is adjacent to an *O*-glycosylated mucin-like domain and is anchored to the cell membrane by a transmembrane domain followed by a cytoplasmic tail. Expression of exogenous TIM-1 in T cells signals through phosphorylation of cytoplasmic tail tyrosine (de Souza et al. 2005; Binne et al. 2007; de Souza et al. 2008). However, none of the TIM family members requires their cytoplasmic domain for virus binding and internalization, suggesting that this signaling mechanism is not required for enveloped virus uptake (Meertens et al. 2012; Moller-Tank et al. 2014).

While all three human TIM family members bind PS and enhance the uptake of apoptotic bodies (Kobayashi et al. 2007; DeKruyff et al. 2010), TIM-3 does not enhance filovirus entry. Mapping studies indicated that the TIM-3 mucin-like domain is ~2.5-times shorter than that of TIM-1 and -4, suggesting that the length of the mucin-like domain is important for function as a filovirus receptor (Moller-Tank et al. 2014). Consistent with this observation, reducing the TIM-1 mucin-like domain resulted in reduced filovirus receptor efficiency (Moller-Tank et al. 2014). Thus, both the amino acid residues within IgV binding pocket and length of the mucin-like domain determine the efficacy of the TIM molecule to serve as a filovirus receptor.

TAM receptors. Tyro3, Axl, and Mer (TAM receptors) are highly related cell surface receptors that contain two N-terminal immunoglobulin-like domains, two fibronectin type III domains, a single transmembrane domain, and a cytoplasmic protein tyrosine kinase (PTK) domain. TAMs do not bind directly to lipids. Instead, the N-termini of TAMs bind to the serum protein growth arrest specific factor 6 (Gas6), which in turn binds to apoptotic body or virion lipids. Tyro3 and Mer, but not Axl, also bind to a second PS-binding serum protein, Protein S (Stitt et al. 1995; Prasad et al. 2006). Both Gas6 and Protein S consist of an N-terminal domain rich in γ-carboxyglutamic acid residues that bind to PS, a loop region, four epidermal growth factor-like repeats, and two C-terminal laminin G-like domains forming a globulin-like structure that binds to the Ig-like domains of the TAM receptors (Lemke and Rothlin 2008; Linger et al. 2008). Dimerization of TAM receptors occurs after binding of their ligands, resulting in a complex (Sasaki et al. 2006). Formation of this complex causes TAM receptor signaling and autophosphorylation of tyrosines within the PTK domain (Stitt et al. 1995) and PI3K activation (Zhong et al. 2010). While TAM signaling is not required for enhancing enveloped virus infection (Bhattacharyya et al. 2013), mapping of regions of Axl required for optimal viral transduction demonstrated that motifs within both the ectodomain and known signaling motifs in the cytoplasmic tail were required (Shimojima et al. 2007). As TAM cytoplasmic tail signaling results in reduced type 1 interferon responses (Bhattacharyya et al. 2013), it appears that TAM signaling does not directly enhance virus entry, but instead dampens early innate immune responses. By dampening these responses, virus replication is enhanced. Thus, two different mechanisms are at play that stimulate TAM-dependent virus replication: direct virus binding and internalization and inhibition of type 1 interferon responses.

Although TIMs and TAMs have been demonstrated to enhance filovirus attachment and infection, the specific mechanism(s) by which these factors might induce internalization of virus particles remains unknown and the role of these receptors in vivo has yet to be explored. Thus, similar to the carbohydrate-binding receptors discussed above, it is unclear whether the loss of one of these proteins impacts filovirus infection and pathogenesis or if these proteins are sufficiently redundant in their function that loss of them individually would have little consequence to virus entry. If the loss of specific receptors in vivo proves to be important for determining virus loads or aberrant cytokine profiles that result from filovirus infection, development of antibodies or small molecules to inhibit filovirus interactions with these receptors may prove to be important antivirals.

Other PS receptors. Although a number of other PS receptors and PS-binding proteins have been described, only two additional PS receptors have been identified to enhance entry of enveloped viruses or enveloped viral vectors. A complex composed of integrin αV/β3 or β5 and the serum protein, MFG-E8, facilitates uptake of lentiviral particles bearing a modified Sindbis virus GP in a PS-dependent manner (Morizono and Chen 2014). In addition, the PS receptor CD300a mediates uptake of dengue virus 2 into cells (Carnec et al. 2015). Whether either of these receptor/receptor complexes is important for filovirus entry into cells has yet to be evaluated.

4 Filovirus Internalization and Endocytic Trafficking

Lipid signaling requirements. After binding to the cell surface, the virus particle is internalized into the endosomal network, from which the capsid escapes into the cell cytoplasm (Empig and Goldsmith 2002; Simmons et al. 2003b; Sanchez 2007). Endocytic uptake involves invagination of the cell surface lipid membrane, encapsulation of the virus particle, and severing of the vesicle membrane from the cell surface after which the vesicle and its contents are trafficked through the cell cytoplasm. The alteration of membrane structure that leads to endocytosis involves multiple cell proteins, specific lipid microdomains, consumption of energy and is often triggered through signals generated from the lipids themselves. Using virion-like particles (VLPs), pseudotyped virus and wild-type viruses, it was shown that the EBOV GP leads to activation of phosphoinositide-3-kinase (PI3K), an enzyme that phosphorylates the lipid, phosphatidylinositol to generate a variety of secondary PIP messengers. These in turn activate other signaling proteins, with phosphorylation of Akt-1 known to occur after interaction of GP with the cell. One role of Akt-1 is to activate Rac1, which can trigger actin polymerization through a series of intermediate proteins including Pak1 (Saeed et al. 2008; Quinn et al. 2009). Another lipid signaling pathway that at least in some cells is important for EBOV infection involves phospholipase C. Phospholipase C inhibitors inhibit Axl-dependent entry of EBOV GP pseudovirions (Hunt et al. 2011). Other lipids are also involved in productive EBOV infection of cells. Acid sphingomyelinase

(ASM) activity and presence of its lipid substrate, sphingomyelin, which is a component of cholesterol-rich lipid microdomains, termed lipid rafts, is needed for EBOV infection (Miller et al. 2012b). ASM-generated ceramide works in concert with PI3K to modulate PI3K-mediated signaling and membrane dynamics (Gao et al. 2011). These findings suggest that tightly regulated lipid-mediated signaling is involved in the earliest steps of EBOV infection but the exact mechanism remains to be defined.

Routes of filovirus uptake. Cellular receptors that undergo recycling and many viruses share common internalization mechanisms such as clathrin-coated pits, caveolae, or other routes termed non-clathrin/non-caveolin-mediated endocytosis. In case of filoviruses, after binding to the cell surface, the virus particles traffic through the endosomal network to low pH compartment(s) (Empig and Goldsmith 2002; Simmons et al. 2003b; Sanchez 2007). Filovirus movement through the endosomal compartment allows the virus to enter the cell cytoplasm deep within the cell rather than at the periphery. This provides a way for the virus to avoid barriers imposed by the cytoskeleton and interaction with other cytoplasmic proteins that may degrade the virus (Franco and Shuman 2012). Early electron microscopy showed EBOV particles protruding out of highly structured membrane invaginations on the cell surface, that were structurally similar to clathrin-coated pits (Geisbert and Jahrling 1995), although it is unclear if these structures result in virus internalization and productive infection. Both clathrin-coated vesicles and caveolae rely heavily on free membrane cholesterol to promote the high curvature of the vesicle membrane. Since disruption of membrane cholesterol abrogates EBOV infection (Empig and Goldsmith 2002; Brindley and Maury 2008), it was predicted that one of these pathways would be involved in uptake of this virus family. Later work showed a fraction of virus particles were associated with clathrin, and depletion of clathrin heavy chain (CHC) by siRNA resulted in a partial drop in infectivity (Aleksandrowicz et al. 2011). In addition, chlorpromazine, a small molecule that disrupts the formation of clathrin-coated pits, showed a partial block to both EBOV pseudotype and wild-type virus infection (Bhattacharyya et al. 2010; Hunt et al. 2011). However, chlorpromazine is known to also interfere with biogenesis of other vesicle types, including the closely related uptake mechanisms of phagosomes and macropinosomes (Ivanov 2008) which are also cholesterol-dependent. Furthermore, CHC plays multiple roles in general membrane biogenesis including phagocytosis (O'Halloran and Anderson 1992; Chen et al. 2013).

Other work indicated that clathrin-coated pits were not involved in productive infection. In one study, the majority of virus particles colocalized with caveolin-1, a marker of caveolae, on the cell surface, but not transferrin, a marker of the clathrin-mediated endocytosis (Empig and Goldsmith 2002), suggesting that filoviruses entered cells through caveolae and not clathrin-coated pits. However, it was later shown that several cell lines lacking caveolae still supported transduction of lentivirions pseudotyped with EBOV GP (Simmons et al. 2003b). Together, these findings provide evidence that EBOV uptake routes may differ from cell to cell, possibly being driven by the type of receptor being used, and suggest that clathrin-coated pits and caveolae may not be primary uptake pathways for filoviruses.

More recently, macropinocytosis has been shown to play a clear and predominant role in EBOV infection mechanism. Macropinocytosis involves large-scale rearrangement of the plasma membrane to engulf extracellular fluid and particles. It is initiated with plasma membrane ruffling at the cell periphery (Amstutz et al. 2008; Mercer and Helenius 2009; Pernet et al. 2009; Raghu et al. 2009; Eierhoff et al. 2010; Mercer et al. 2010; Schelhaas et al. 2012). Actin-dependent membrane protrusions extend from the cell and then collapse back onto the plasma membrane creating uncoated, fluid-filled vacuoles termed macropinosomes (Swanson 2008). These vacuoles allow cellular uptake of large cargo. In contrast to clathrin-coated pits and caveolae generated vacuoles that have diameters of around 200 nm, the diameter of macropinosome can range up to microns in size and can easily accommodate filovirus particles. Consistent with using macropinocytosis of uptake, EBOV infection is heavily dependent upon actin function. Inhibition of actin polymerization by cytochalasin D (May et al. 1998) blocks uptake of EBOV particles into cells (Nanbo et al. 2010; Saeed et al. 2010). Interestingly, the EBOV GP is capable of triggering aggregation of the actin-related protein 2, Arp2, an actin nucleation factor (Saeed et al. 2010) suggesting the capability to induce macropinocytosis. Both membrane ruffling and macropinosome biogenesis depend on cellular kinases, including PI3K, protein kinase C (PKC), and phospholipases Cγ (PLCγ) and A_2 (PLC-A_2), all of which are implicated as important for EBOV infection. A key correlate came when EBOV particles were seen colocalized with high molecular weight dextran, a fluid phase marker that allows macropinosomes to be visualized. Disruption of Pak1, CtBP/BARS, PKC, Cdc42, or AMPK function in cells, which all control actin polymerization and are all needed for macropinosome formation (Ridley et al. 1992; Dharmawardhane et al. 2000; West et al. 2000; Norbury 2006; Liberali et al. 2008; Koivusalo et al. 2010; Kondratowicz et al. 2013) also disrupted EBOV infection, internalization of virus particles and dextran uptake (Nanbo et al. 2010; Saeed et al. 2010; Kondratowicz et al. 2013). The pharmacological inhibitor 5-(*N*-ethyl-*N*-isopropyl)-amiloride (EIPA), an inhibitor of a Na^+ channel needed for macropinosome formation also blocked EBOV infection (Fretz et al. 2006; Hunt et al. 2011). A final line of evidence supporting the role of macropinocytosis in filovirus uptake was that internalized virions localize to structures positive for sorting-nexin 5 (SNX5) (Nanbo et al. 2010), a component of macropinosomes required for vesicular maturation (Kerr et al. 2006; Lim et al. 2008). An important caveat is that the majority of work has been performed using immortalized cell lines. Given the diverse receptor usage of filoviruses, specific cell types may express and display different receptors and thereby utilize alternative uptake routes. However, macropinocytosis is seen in most cell types, and so is likely to play a significant role in infection.

Viral endocytic trafficking to membrane fusion sites. Once internalized, filoviruses are trafficked through the endocytic network and require endosomal acidification. Pharmacological agents that inhibit endosomal acidification, including bafilomycin A1, monensin, chloroquine, and ammonium chloride, block filovirus infection as well as transduction with virus pseudotypes bearing the EBOV or MARV GP (Takada et al. 1997; Wool-Lewis and Bates 1998). The requirement for acidification is not uncommon, with viruses from more than half of all established virus families utilizing low pH as a trigger for membrane fusion (Kielian 2014).

Further details of the uptake process were revealed using specific markers of endosomal compartments. EEA1 and LAMP1 are proteins that reside in early and late endosomes, respectively, and have been well characterized (Luzio et al. 2007b). Both wild-type virus and VLPs were seen associated with each of these markers in cells, indicating that virions access both early and late endosomal compartments (Saeed et al. 2010; Favier et al. 2016). Rab5 and Rab7 are small GTPases that regulate vesicular trafficking from the early to late endosomes and from the late endosomes to lysosomes, respectively (Numrich and Ungermann 2014). Both dominant-negative and constitutively active forms disrupt normal endocytosis and efficiently block EBOV infection (Saeed et al. 2010). Furthermore, recent live cell imaging studies of filovirus uptake showed that filovirus membrane fusion is activated in compartments containing Rab7. Both lines of evidence indicate that virus requires access to at least the late endosome or endolysosomes for productive infection (Simmons et al. 2015; Spence et al. 2016). The precise site of nucleocapsid release into the cytoplasm remains to be determined.

Other cellular host factors involved in endocytic trafficking have been implicated in filovirus entry. A haploid genetic screen identified the HOPS (homotypic fusion and protein sorting) complex as an entry host factor (Carette et al. 2011). Additional studies also indicated a role for the HOPS-interacting factors RILP and ORP1 K (van der Kant et al. 2013). Although more work is needed, current evidence suggests that (i) endosomal fusion events mediated by HOPS deliver virions to one or more compartments that can support viral membrane fusion and/or (ii) deliver fusion cofactors to these compartments.

Calcium is an important mediator in many cell functions including vesicle transport (Hay 2007; Luzio et al. 2007a) and consequently, for infection by many viruses types (Cheshenko et al. 2007; Bozym et al. 2010). Calcium enters the cytoplasm through calcium channels found on the plasma membrane, on intracellular vesicles such as endosomes and the endoplasmic reticulum. The change in cytoplasmic calcium concentration is detected by sensing proteins that modify cell behavior including endosome trafficking and cytoskeletal remodeling (Abe and Puertollano 2011; Shen et al. 2011; Gandini and Felix 2015). Using siRNA screening, host factors important for cell entry of pseudotypes carrying EBOV GP, calcium sensing proteins including calmodulin and Ca^{2+}/calmodulin-dependent protein kinase II (CaMKII) were identified (Kolokoltsov et al. 2009). Separately, screening FDA-approved drugs for repurposing against EBOV demonstrated that multiple L-type calcium channel blocking agents inhibited both pseudotypes as well as EBOV and MARV (Madrid et al. 2013; Gehring et al. 2014; Johansen et al. 2015). The finding that L-type channel blocking compounds also inhibit two-pore channels (TPCs), found in endosomes/lysosomes (Sakurai et al. 2015), suggested involvement in virus uptake. TPCs are endosomal ion channels that transport calcium as well as sodium ions and can be activated by nicotinic acid adenine dinucleotide phosphate (NAADP) and phosphatidylinositol-3,5-bisphosphate (PI(3,5) P2), a product of PI3 K, PIKfyve, and polyphosphoinositide phosphate activity

(Patel 2015). The latter 3 proteins are known to be important for EBOV infection (Saeed et al. 2008; Carette et al. 2011). TPCs regulate vesicle trafficking in endosomal/lysosomal systems and other cellular compartments (Davis et al. 2012; Grimm et al. 2014; Ambrosio et al. 2015, 2016). They physically interact with the regulators of endosome maturation, Rab5 and Rab7 (Grimm et al. 2014; Lin-Moshier et al. 2014). Consistent with this role, TPCs affect EBOV movement through the endocytic pathway, and disruption of their function through pharmacological inhibitors stops virus uptake prior to membrane fusion (Spence et al. 2016). In addition to NAADP antagonist Ned19, tetrandrine, an alkaloid originally identified from an Asian herb, inhibits TPC-mediated calcium flux and blocks filovirus entry in multiple cell types and prevented disease in the mouse model of EBOV infection (Rathbun et al. 2015; Sakurai et al. 2015; Hoffmann et al. 2016).

5 Proteolytic Processing of Filovirus GP in Endosomes

Like other class I virus glycoproteins, filovirus GP requires proteolytic cleavage during maturation (Weissenhorn et al. 1998; Malashkevich et al. 1999; White et al. 2008). The preGP precursor polypeptide is cleaved at a polybasic site in the Golgi by the subtilin/kexin-like endopeptidase furin, to yield GP1 and GP2 subunits, and liberate a hydrophobic internal fusion loop (IFL) near the N–terminus of GP2 (Volchkov et al. 1998; Wool-Lewis and Bates 1999). Unexpectedly, however, furin cleavage is dispensable for infection in vitro and in vivo (Wool-Lewis and Bates 1999; Neumann et al. 2002). This conundrum was resolved by the discovery that filovirus GPs undergo further proteolytic priming by host cysteine proteases (cysteine cathepsins) in endosomes of target cells (Chandran et al. 2005). Biochemical and structural studies provided evidence that cathepsin B (CatB) and cathepsin L combine to initiate cleavage of EBOV GP1 within the β13–β14 loop. Consequently, the GP1 mucin-like and glycan cap subdomains are released from the trimeric GP spike, and/or degraded (Chandran et al. 2005; Schornberg et al. 2006; Dube et al. 2009; Hood et al. 2010; Bale et al. 2011). Although this proteolytic processing step to generate GP_{CL} does not itself appear to induce conformational rearrangements within the pre-fusion GP (Bale et al. 2011; Bornholdt et al. 2015; Wang et al. 2016), it is necessary for viral entry and membrane fusion (Chandran et al. 2005; Schornberg et al. 2006; Wong et al. 2010; Simmons et al. 2015; Spence et al. 2016). Subsequent work determined at least two reasons for this requirement: first, the extensive proteolytic remodeling of GP unmasks the RBS, thereby allowing GP_{CL} to recognize NPC1 (see below) (Miller et al. 2012a; Bornholdt et al. 2015; Hashiguchi et al. 2015; Wang et al. 2016) and second, it appears to prepare GP to undergo subsequent fusion-relevant conformational changes. This effect is likely mediated in part by proteolytic removal of the β13-β14 loop in GP1, which overlies the GP2 IFL (Lee et al. 2008; Zhao et al. 2016). Furthermore, a structure of the GP pre-fusion trimer revealed that residues in the β13–β14 loop are "tucked" into a solvent-accessible pocket at the base of GP trimer, and can be displaced by a small molecule, toremifene, that destabilizes GP (Zhao et al. 2016). Therefore, proteolytic removal of these residues during entry might favor, through multiple mechanisms, a GP conformation susceptible to fusion triggering. Providing further evidence for this hypothesis, stepwise cleavage of GP by CatL and CatB (or by surrogate proteases mediating similar steps) leads to the stepwise destabilization of the pre-fusion conformation of GP, and serial viral passage in the presence of a CatB inhibitor selects destabilizing mutations in GP (Wong et al. 2010; Brecher et al. 2012). Finally, although CatB cleavage of GP in vitro affords bypass of the CatB requirement within cells, entry by GP_{CL}-bearing viruses can still be blocked by the cysteine protease inhibitor E-64, suggesting an additional cathepsin-dependent step downstream of the biochemically defined GP $\rightarrow$ GP_{CL} priming events (see Fusion triggering, below).

Cysteine cathepsin activation of GP is shared by all known filoviruses, but has also painted a more complex picture of the dependence of filoviruses on individual proteases (Misasi et al. 2012; Ng et al. 2014). Specifically, EBOV and TAFV GP-mediated entry is strongly CatB-dependent, whereas SUDV, RESTV, MARV, and LLOV GP-mediated entry has a diminished requirement for CatB (SUDV) or no requirement at all (RESTV, MARV, LLOV). CatL is dispensable in both grivet (Vero) cells and human peripheral blood mononuclear cell (PBMC)-derived dendritic cells, but acts as a CatB-accessory or redundant factor for filovirus entry (Chandran et al. 2005; Schornberg et al. 2006; Martinez et al. 2010). The molecular basis of these naturally occurring virus-dependent differences in protease dependence remains to be identified. They are apparently distinct from the amino acid changes near the N–terminus of GP1 and at GP1–GP2 subunit interface associated with CatB-independent GP mutants obtained under inhibitor selection (Wong et al. 2010).

In contrast to these in vitro studies, $Ctsb^{-/-}$ and $CtsL^{-/-}$ mice lacking CatB or CatL, respectively, are susceptible to EBOV infection and develop disease, suggesting that viral entry can proceed in the absence of either enzyme in vivo (Marzi et al. 2012). While more work is needed, one possible explanation for this apparent discrepancy is that multiple murine cysteine cathepsins play redundant roles in priming EBOV GP. Given the broad substrate preferences of cysteine cathepsins and their known capacity for redundancy, experiments in animal models with broad-spectrum (but class-specific) cysteine cathepsin inhibitors like E-64 are warranted, but their interpretation may be complicated by the murine requirements for CatL and CatS in MHC II invariant chain degradation and antigen presentation in thymus and antigen-presenting cells, respectively (Riese et al. 1998; Sevenich et al. 2010).

6 NPC1 Is a Critical Intracellular Receptor for Filoviruses

Niemann-Pick C1 (NPC1) is a highly conserved, ubiquitously expressed, polytopic membrane protein localized to the limiting membrane of late endosomes (Carstea et al. 1997; Higgins et al. 1999; Davies and Ioannou 2000; Naureckiene et al. 2000). Working in concert with a small cholesterol-binding protein, NPC2, NPC1 regulates efflux of cholesterol from late endosomes and lysosomes to the endoplasmic reticulum and other cellular sites, and loss of either gene results in the development of a progressive neurodegenerative disease, Niemann-Pick type C disease (Carstea et al. 1997; Naureckiene et al. 2000). Using a genome-wide loss-of-function genetic screen in human haploid cells, Carette and co-workers identified NPC1 to be essential for filovirus entry and infection (Carette et al. 2011). Côté el al. also found NPC1 to be a critical filovirus entry factor in a small-molecule inhibitor screen (Côté et al. 2011). Importantly, studies in $Npc1^{-/-}$ mice revealed that NPC1 is indispensable for filovirus replication and in vivo pathogenesis (Herbert et al. 2015). Unexpectedly, even heterozygous $Npc1^{+/-}$ mice, which do not suffer from Niemann-Pick type C disease, were substantially protected from EBOV and MARV infection—they developed plasma viral titers similar to WT mice, but were able to clear infection by 7 days post-challenge (Carette et al. 2011; Herbert et al. 2015). The molecular basis of this phenotype is currently unknown, but may reflect the requirement for high levels of NPC1 in some cell types critical for viral pathogenesis. The requirement for NPC1 by all known filoviruses implicates this protein as a universal filovirus entry factor (Miller et al. 2012a; Ng et al. 2014).

Mechanistic studies showed that NPC1's role in endosomal cholesterol trafficking could be decoupled from its function as a filovirus entry factor. Mutations in NPC1 that abrogate its cholesterol trafficking function had little effect on filovirus entry, and NPC2, which is required for NPC1-dependent cholesterol trafficking, is dispensable for filovirus infection both in vitro and in vivo (Côté et al. 2011; Miller et al. 2012a; Herbert et al. 2015). Moreover, proteolytically primed GP_{CL}, but not intact uncleaved GP, binds directly to the second luminal domain of NPC1, domain C, in a manner that is required for filovirus entry and infection (Krishnan et al. 2012; Miller et al. 2012a). These and other studies have delineated the previously proposed GP1 RBS as the site of NPC1 domain C (NPC1–C) binding, a conclusion corroborated by the recent structural elucidation of a GP_{CL}:NPC1–C complex (Wang et al. 2016). This structure further revealed that two loops in NPC1–C mediate NPC1's interaction with GP_{CL}. Intriguingly, these loops also appear to be required for transient NPC2:NPC1–C binding during the handoff of cholesterol from NPC2 to NPC1 (Li et al. 2016), raising the possibility that GP_{CL}:NPC1–C recognition evolved as a form of molecular mimicry. Finally, a cryo-electron microscopy-based structure of full-length NPC1 in detergent micelles in complex with EBOV GP_{CL} suggests that GP_{CL} contacts NPC1–C but not NPC1's other luminal domains, A and I (Gong et al. 2016). However, roles for the latter in proper domain C presentation and/or for downstream events in entry cannot yet be fully discounted.

Together, these and other findings indicate that NPC1 is a novel viral receptor, engaging its viral glycoprotein partner only in an intracellular compartment, and not at the cell surface (Miller et al. 2012a; White and Schornberg 2012; Jae and Brummelkamp 2015). Two mechanisms appear to preclude GP-NPC1 interaction at the cell surface. First, endosomal cleavage of GP by host cysteine cathepsins is necessary for the interaction, because it unmasks the NPC1-binding site in GP1. Second, essentially no NPC1 is present at the cell surface (Scott et al. 2004; Berger et al. 2007; Miller et al. 2012a). Consistent with the obligate intracellular role of NPC1, cell biological studies provided evidence that viral particles accumulate in LAMP1-positive late endosomal/lysosomal compartments in NPC1-deficient cells, indicating that this receptor is dispensable for viral attachment, internalization, and delivery to late endosomes (Carette et al. 2011; Krishnan et al. 2012; Miller et al. 2012a). These findings indicate instead that NPC1 acts at a late step at or near the site of membrane fusion. Live cell imaging studies support this hypothesis by showing lipid mixing between viral and endosomal membranes, an initial step in viral membrane fusion, which is initiated in compartments that contain NPC1 as well as TPC (Simmons et al. 2015; Aman 2016; Spence et al. 2016). Moreover, both loss of cellular NPC1 and mutation of viral GP to abrogate GP_{CL}:NPC1 binding drastically inhibited viral lipid mixing (Spence et al. 2016), strongly suggesting that NPC1 interaction is a prerequisite for triggering viral membrane fusion.

The mechanism for requirement of GP_{CL}:NPC1 binding in late endosomes for filovirus membrane fusion remains unclear but several hypotheses have been proposed (Brecher et al. 2012; Hunt et al. 2012; Miller and Chandran 2012; Miller et al. 2012a; Wang et al. 2016; White and Whittaker 2016). First, NPC1-dependent tethering of viral particles to the limiting endosomal membrane may play a critical membrane-localizing role in fusion and/or may protect GP from proteolytic destruction in the hostile environment of the late endosome or endolysosome. Second, binding may induce conformational changes required for fusion (see Fusion triggering below).

Recent work also indicates that NPC1 influences filovirus host tropism and viral interspecies transmission. Variation at a single amino acid residue in "loop 2" of NPC1–C accounts for the species-specific susceptibility of cells derived from the African fruit bat *Eidolon helvum* (Ng et al. 2015). A second, highly conservative, amino acid change, also in loop 2, rendered cells from a snake, Russell's viper (*Daboia russellii*), completely resistant to filovirus entry and infection (Ndungo et al. 2016). In both cases, *NPC1* sequence variation affected GP-receptor recognition but not NPC1-dependent cholesterol trafficking, suggesting loop 2 can be varied without impacting its housekeeping function. Indeed, *NPC1* is evolving under positive selection in both mammals at large, and in bats, with codons in NPC1–C and at GP contact sites overrepresented (Ng et al. 2015; Pontremoli et al. 2016). These findings prompted the hypothesis that *NPC1* in mammals has been shaped by a long-term coevolutionary arms race between NPC1-dependent filovirus-like agents and their hosts.

The unusual obligate intracellular property of the filovirus receptor interaction also has potential implications for the capacity of the adaptive immune response to control filovirus infections. Specifically, it may serve as a means for viral particles to evade neutralizing antibodies. This is because, as discussed above, GP:NPC1 recognition necessitates GP $\rightarrow$ GP_{CL} cleavage with concomitant removal of the glycan cap and mucin sequences that comprise the viral "glycan shield." As noted above, loss of the glycan shield through site directed mutagenesis enhances anti-EBOV antisera neutralization of virion particles (Lennemann et al. 2014). Conversely, conventional antibodies directed against the highly conserved GP_{CL}

RBS are largely unable to access it within endosomes (Bornholdt et al. 2015; Flyak et al. 2015; Hashiguchi et al. 2015; Wec et al. 2016). Taking advantage of the endosomal exposure of the RBS, Wec et al. (2016) engineered bispecific antibodies that were delivered to NPC1 containing vesicles on the virus itself. The antibodies then bound NPC1–C or the GP_{CL} RBS and disrupted their interaction. The engineered bispecific antibodies protected mice against both EBOV and SUDV in postexposure challenge studies and represent promising broadly active anti-filovirus immunotherapeutics.

7 Membrane Fusion

By analogy to other class I viral fusion proteins, the fully primed, NPC1-bound GP intermediate generated within endosomes is proposed to undergo a series of conformational changes that culminate in fusion of viral and cellular membranes, and cytoplasmic escape of the viral ribonucleocapsid (Hunt et al. 2012; Miller and Chandran 2012; White and Whittaker 2016). These changes include: (i) partial or complete release of GP1-GP2 contacts, which permits GP2 to undergo fusion-related rearrangements (Lee et al. 2008; Lee and Saphire 2009; Dias et al. 2011); (ii) rearrangement of the three N-terminal heptad repeats (NHR) in GP2 to form a protruding coiled coil, a so-called "extended intermediate", which projects the N-terminal fusion loop into the endosomal membrane (White et al. 2008; Gregory et al. 2011); and (iii) folding back of the three C-terminal heptad repeats (CHR) into grooves along the NHR core trimer to create a stable post-fusion six-helix bundle (6HB) conformation (Weissenhorn et al. 1998; Malashkevich et al. 1999). These rearrangements are proposed to drive close apposition and merger of viral and endosomal lipid bilayers.

Fusion triggering. The final molecular trigger that initiates this conformational cascade remains incompletely defined. Current evidence indicates that GP $\rightarrow$ GP_{CL} cleavage by CatB/CatL is insufficient to trigger filovirus membrane fusion, at either neutral or acid pH (Chandran et al. 2005; Schornberg et al. 2006; Bale et al. 2011; Miller and Chandran 2012; Miller et al. 2012a). Further, although GP_{CL}:NPC1 binding is critically required for fusion triggering, its precise role in this process is unclear. This step itself does not appear to trigger fusion-related conformational changes at either neutral or acid pH, at least in vitro (Miller et al. 2012a), although this might just represent a limitation inherent in the soluble NPC1–C proteins used for these studies. Interestingly, however, comparison of recent apo and NPC1-bound GP structures raises the possibility that NPC1–C binding does indeed induce conformational changes in sequences at the GP1–GP2 interface and in the GP2 IFL. More work is needed to uncover their role, if any, in viral membrane fusion.

The finding that virions bearing cleaved GP still maintain their sensitivity to the broad cysteine protease inhibitor E-64, suggests additional proteolytic cleavage steps of NPC1-bound GP could also play a role in triggering membrane fusion

(Schornberg et al. 2006; Wong et al. 2010; Miller et al. 2012a). Further, in vitro treatment of fully primed GP with mildly reducing reagents at physiological temperatures induced fusion loop-mediated GP binding to liposomes, suggesting that reduction of the GP1–GP2 intersubunit disulfide bond by an unknown mechanism may be involved in fusion (Brecher et al. 2012).

Fusion execution. The structural similarity of the coiled-coil region of GP2 to other class I fusion proteins led to successful development of C-peptides that inhibited EBOV and MARV infection. Consistent with activation of membrane fusion in the endosome, the C-peptides derived from EBOV GP2 were shown to be effective only when targeted to the endosome (Watanabe et al. 2000; Miller et al. 2011; Higgins et al. 2013). Further work emphasized the role of endosomal acidification in filovirus membrane fusion (in addition to roles in activating endosomal cysteine proteases). First, acid pH induces a conformational change in the GP2 fusion loop that reorients and compacts a hydrophobic patch at its apex; this may destabilize the lipid bilayer and promote fusion of viral and cellular membranes (Gregory et al. 2011, 2014; Lee et al. 2016). Second, acid pH stabilizes the post-fusion conformation of both EBOV and MARV GP2, at least in part by protonating a belt of anionic residues on the surface of the six-helix bundle (the "anionic stripe"). This destabilizes the six-helix bundle through electrostatic repulsion (Harrison et al. 2011, 2012; Koellhoffer et al. 2012; Harrison et al. 2013). Other acid-dependent class I viral fusion proteins may also be regulated through this type of mechanism (Aydin et al. 2014; Koellhoffer et al. 2014).

A full accounting of the molecular mechanism of filovirus membrane fusion has been complicated by the extensive involvement of endosomal host factors. This last feature at least partly explains observations that EBOV GP-expressing cells do not undergo classical cell–cell fusion leading to easily detectable syncytia formation, as seen with many other class I membrane fusion proteins. Using a highly sensitive, reporter-based cell–cell content-mixing assay to study EBOV GP-mediated fusion, Bar et al. found that exposure of the EBOV GP-expressing cells to acid pH could induce cell–cell fusion (2006). More recently, Markosyan et al. (2016) reported an assay for EBOV GP-dependent content mixing between cells loaded with aqueous fluorescent dyes, and showed that this process required known filovirus entry host factors. However, the efficiency of fusion in both of these assays was relatively low in comparison to that obtained with other class I fusion glycoproteins, including influenza A virus hemagglutinin and Semliki Forest virus E1/E2, likely because only small amounts of the necessary host factors, such as NPC1, are available at the cell surface. Development of a robust fusion assay that recapitulates the known requirements for host factors (e.g., cysteine cathepsins, NPC1, acid pH) and that could be used to dissect additional steps and molecular players remains a high priority.

Overall, work on filovirus cell entry has revealed a complex interaction with the host. While initial interaction with the host cell is through cell factors found on diverse sets of cells, and through interaction of host PS receptors and CLRs with the virus envelope and viral glycoprotein glycans, the virus makes up for specificity by interacting with specific endosome proteins such as NPC1. A schematic of filovirus entry events is shown in (Fig. 1). While the field has advanced significantly over the past 10 years, much remains to be learnt. Outstanding questions to be answered are: How does the expression profile of receptor proteins impact virus trafficking and infection outcome in specific cell types? Do virus protein interactions with known surface host factors directly result in virion internalization by macropinocytosis or are other unknown factors involved? What additional endosomal proteases are needed to process the virus glycoprotein for productive infection and where does each act? What is the direct trigger of membrane fusion and what are the sites for productive release of the filovirus particle into the cell cytoplasm? Answers to each of these questions will aid in understanding the complex biology of filoviruses and in development of new therapy strategies.

Acknowledgements We thank Ms. M. Messenger for design of the figure.

References

Abe K, Puertollano R (2011) Role of TRP channels in the regulation of the endosomal pathway. Physiology (Bethesda) 26(1):14–22

Aleksandrowicz P, Marzi A, Biedenkopf N, Beimforde N, Becker S, Hoenen T, Feldmann H, Schnittler HJ (2011) Ebola virus enters host cells by macropinocytosis and clathrin-mediated endocytosis. J Infect Dis 204(Suppl 3):S957–S967

Alvarez CP, Lasala F, Carrillo J, Muniz O, Corbi AL, Delgado R (2002) C-type lectins DC-SIGN and L-SIGN mediate cellular entry by Ebola virus in cis and in trans. J Virol 76(13):6841–6844

Aman MJ (2016) Chasing Ebola through the Endosomal Labyrinth. MBio 7(2)

Ambrosio AL, Boyle JA, Aradi AE, Christian KA, Di Pietro SM (2016) TPC2 controls pigmentation by regulating melanosome pH and size. Proc Natl Acad Sci U S A 113(20): 5622–5627

Ambrosio AL, Boyle JA, Di Pietro SM (2015) TPC2 mediates new mechanisms of platelet dense granule membrane dynamics through regulation of Ca2+ release. Mol Biol Cell 26(18): 3263–3274

Amstutz B, Gastaldelli M, Kalin S, Imelli N, Boucke K, Wandeler E, Mercer J, Hemmi S, Greber UF (2008) Subversion of CtBP1-controlled macropinocytosis by human adenovirus serotype 3. EMBO J 27(7):956–969

Aoyagi Y, Adderson EE, Min JG, Matsushita M, Fujita T, Takahashi S, Okuwaki Y, Bohnsack JF (2005) Role of L-ficolin/mannose-binding lectin-associated serine protease complexes in the opsonophagocytosis of type III group B streptococci. J Immunol 174(1):418–425

Audet J, Kobinger GP (2015) Immune evasion in ebolavirus infections. Viral Immunol 28(1): 10–18

Aydin H, Al-Khooly D, Lee JE (2014) Influence of hydrophobic and electrostatic residues on SARS-coronavirus S2 protein stability: insights into mechanisms of general viral fusion and inhibitor design. Protein Sci 23(5):603–617

Bale S, Liu T, Li S, Wang Y, Abelson D, Fusco M, Woods VL, Saphire EO (2011) Ebola virus glycoprotein needs an additional trigger, beyond proteolytic priming for membrane fusion. PLoS Neglected Trop Dis 5(11):e1395

Bar S, Takada A, Kawaoka Y, Alizon M (2006) Detection of cell-cell fusion mediated by Ebola virus glycoproteins. J Virol 80(6):2815–2822

Baribaud F, Doms RW, Pohlmann S (2002a) The role of DC-SIGN and DC-SIGNR in HIV and Ebola virus infection: can potential therapeutics block virus transmission and dissemination? Expert Opin Ther Targets 6(4):423–431

Baribaud F, Pohlmann S, Leslie G, Mortari F, Doms RW (2002b) Quantitative expression and virus transmission analysis of DC-SIGN on monocyte-derived dendritic cells. J Virol 76(18): 9135–9142

Bashirova AA, Geijtenbeek TB, van Duijnhoven GC, van Vliet SJ, Eilering JB, Martin MP, Wu L, Martin TD, Viebig N, Knolle PA, KewalRamani VN, van Kooyk Y, Carrington M (2001) A dendritic cell-specific intercellular adhesion molecule 3-grabbing nonintegrin (DC-SIGN)-related protein is highly expressed on human liver sinusoidal endothelial cells and promotes HIV-1 infection. J Exp Med 193(6):671–678

Becker S, Rinne C, Hofsass U, Klenk HD, Mühlberger E (1998) Interactions of Marburg virus nucleocapsid proteins. Virology 249(2):406–417

Becker S, Spiess M, Klenk HD (1995) The asialoglycoprotein receptor is a potential liver-specific receptor for Marburg virus. J Gen Virol 76(Pt 2):393–399

Berger AC, Salazar G, Styers ML, Newell-Litwa KA, Werner E, Maue RA, Corbett AH, Faundez V (2007) The subcellular localization of the Niemann-Pick Type C proteins depends on the adaptor complex AP-3. J Cell Sci 120(Pt 20):3640–3652

Bharat TA, Noda T, Riches JD, Kraehling V, Kolesnikova L, Becker S, Kawaoka Y, Briggs JA (2012) Structural dissection of Ebola virus and its assembly determinants using cryo-electron tomography. Proc Natl Acad Sci U S A 109(11):4275–4280

Bhattacharyya S, Warfield KL, Ruthel G, Bavari S, Aman MJ, Hope TJ (2010) Ebola virus uses clathrin-mediated endocytosis as an entry pathway. Virology 401(1):18–28

Bhattacharyya S, Zagorska A, Lew ED, Shrestha B, Rothlin CV, Naughton J, Diamond MS, Lemke G, Young JA (2013) Enveloped viruses disable innate immune responses in dendritic cells by direct activation of TAM receptors. Cell Host Microbe 14(2):136–147

Binne LL, Scott ML, Rennert PD (2007) Human TIM-1 associates with the TCR complex and up-regulates T cell activation signals. J Immunol 178(7):4342–4350

Bornholdt ZA, Ndungo E, Fusco ML, Bale S, Flyak AI, Crowe JE Jr, Chandran K, Saphire EO (2015) Host-primed Ebola virus GP exposes a hydrophobic NPC1 receptor-binding pocket, revealing a target for broadly neutralizing antibodies. MBio 7(1)

Bozym RA, Morosky SA, Kim KS, Cherry S, Coyne CB (2010) Release of intracellular calcium stores facilitates coxsackievirus entry into polarized endothelial cells. PLoS Pathog 6(10): e1001135

Brecher M, Schornberg KL, Delos SE, Fusco ML, Saphire EO, White JM (2012) Cathepsin cleavage potentiates the Ebola virus glycoprotein to undergo a subsequent fusion-relevant conformational change. J Virol 86(1):364–372

Brindley MA, Maury W (2008) Equine infectious anemia virus entry occurs through clathrin-mediated endocytosis. J Virol 82(4):1628–1637

Brudner M, Karpel M, Lear C, Chen L, Yantosca LM, Scully C, Sarraju A, Sokolovska A, Zariffard MR, Eisen DP, Mungall BA, Kotton DN, Omari A, Huang IC, Farzan M, Takahashi K, Stuart L, Stahl GL, Ezekowitz AB, Spear GT, Olinger GG, Schmidt EV, Michelow IC (2013) Lectin-dependent enhancement of Ebola virus infection via soluble and transmembrane C-type lectin receptors. PLoS ONE 8(4):e60838

Caminschi I, Corbett AJ, Zahra C, Lahoud M, Lucas KM, Sofi M, Vremec D, Gramberg T, Pohlmann S, Curtis J, Handman E, van Dommelen SL, Fleming P, Degli-Esposti MA, Shortman K, Wright MD (2006) Functional comparison of mouse CIRE/mouse DC-SIGN and human DC-SIGN. Int Immunol 18(5):741–753

Carette JE, Raaben M, Wong AC, Herbert AS, Obernosterer G, Mulherkar N, Kuehne AI, Kranzusch PJ, Griffin AM, Ruthel G, Dal Cin P, Dye JM, Whelan SP, Chandran K, Brummelkamp TR (2011) Ebola virus entry requires the cholesterol transporter Niemann-Pick C1. Nature 477(7364):340–343

Carnec X, Meertens L, Dejarnac O, Perera-Lecoin M, Lamine Hafirassou M, Kitaura J, Ramdasi R, Schwartz O, Amara A (2015) The phosphatidylserine and phosphatidylethanolamine receptor CD300a binds dengue virus and enhances infection. J Virol 90(1):92–102

Carstea ED, Morris JA, Coleman KG, Loftus SK, Zhang D, Cummings C, Gu J, Rosenfeld MA, Pavan WJ, Krizman DB, Nagle J, Polymeropoulos MH, Sturley SL, Ioannou YA, Higgins ME, Comly M, Cooney A, Brown A, Kaneski CR, Blanchette-Mackie EJ, Dwyer NK, Neufeld EB, Chang TY, Liscum L, Strauss JF, Ohno K, Zeigler M, Carmi R, Sokol J, Markie D, O'Neill RR, van Diggelen OP, Elleder M, Patterson MC, Brady RO, Vanier MT, Pentchev PG, Tagle DA (1997) Niemann-Pick C1 disease gene: homology to mediators of cholesterol homeostasis. Science (New York, N.Y.) 277(5323):228–231

Chandran K, Sullivan NJ, Felbor U, Whelan SP, Cunningham JM (2005) Endosomal proteolysis of the Ebola virus glycoprotein is necessary for infection. Science (New York, N.Y.) 308(5728):1643–1645

Chen D, Jian Y, Liu X, Zhang Y, Liang J, Qi X, Du H, Zou W, Chen L, Chai Y, Ou G, Miao L, Wang Y, Yang C (2013) Clathrin and AP2 are required for phagocytic receptor-mediated apoptotic cell clearance in Caenorhabditis elegans. PLoS Genet 9(5):e1003517

Cheng H, Lear-Rooney CM, Johansen L, Varhegyi E, Chen ZW, Olinger GG, Rong L (2015) Inhibition of Ebola and Marburg virus entry by G protein-coupled receptor antagonists. J Virol 89(19):9932–9938

Cheshenko N, Liu W, Satlin LM, Herold BC (2007) Multiple receptor interactions trigger release of membrane and intracellular calcium stores critical for herpes simplex virus entry. Mol Biol Cell 18(8):3119–3130

Colmenares M, Puig-Kroger A, Pello OM, Corbi AL, Rivas L (2002) Dendritic cell (DC)-specific intercellular adhesion molecule 3 (ICAM-3)-grabbing nonintegrin (DC-SIGN, CD209), a C-type surface lectin in human DCs, is a receptor for Leishmania amastigotes. J Biol Chem 277(39):36766–36769

Côté M, Misasi J, Ren T, Bruchez A, Lee K, Filone CM, Hensley L, Li Q, Ory D, Chandran K, Cunningham J (2011) Small molecule inhibitors reveal Niemann-Pick C1 is essential for Ebola virus infection. Nature 477(7364):344–348

Dahlmann F, Biedenkopf N, Babler A, Jahnen-Dechent W, Karsten CB, Gnirss K, Schneider H, Wrensch F, O'Callaghan CA, Bertram S, Herrler G, Becker S, Pohlmann S, Hofmann-Winkler H (2015) Analysis of Ebola Virus entry Into macrophages. J Infect Dis 212(Suppl 2): S247–S257

Davies JP, Ioannou YA (2000) Topological analysis of Niemann-Pick C1 protein reveals that the membrane orientation of the putative sterol-sensing domain is identical to those of 3-hydroxy-3-methylglutaryl-CoA reductase and sterol regulatory element binding protein cleavage-activating protein. J Biol Chem 275(32):24367–24374

Davis LC, Morgan AJ, Chen JL, Snead CM, Bloor-Young D, Shenderov E, Stanton-Humphreys MN, Conway SJ, Churchill GC, Parrington J, Cerundolo V, Galione A (2012) NAADP activates two-pore channels on T cell cytolytic granules to stimulate exocytosis and killing. Curr Biol 22(24):2331–2337

de Souza AJ, Oak JS, Jordanhazy R, DeKruyff RH, Fruman DA, Kane LP (2008) T cell Ig and mucin domain-1-mediated T cell activation requires recruitment and activation of phosphoinositide 3-kinase. J Immunol 180(10):6518–6526

de Souza AJ, Oriss TB, O'Malley KJ, Ray A, Kane LP (2005) T cell Ig and mucin 1 (TIM-1) is expressed on in vivo-activated T cells and provides a costimulatory signal for T cell activation. Proc Natl Acad Sci U S A 102(47):17113–17118

DeKruyff RH, Bu X, Ballesteros A, Santiago C, Chim YL, Lee HH, Karisola P, Pichavant M, Kaplan GG, Umetsu DT, Freeman GJ, Casasnovas JM (2010) T cell/transmembrane, Ig, and mucin-3 allelic variants differentially recognize phosphatidylserine and mediate phagocytosis of apoptotic cells. J Immunol 184(4):1918–1930

Dharmawardhane S, Schurmann A, Sells MA, Chernoff J, Schmid SL, Bokoch GM (2000) Regulation of macropinocytosis by p21-activated kinase-1. Mol Biol Cell 11(10):3341–3352

Dias JM, Kuehne AI, Abelson DM, Bale S, Wong AC, Halfmann P, Muhammad MA, Fusco ML, Zak SE, Kang E, Kawaoka Y, Chandran K, Dye JM, Saphire EO (2011) A shared structural solution for neutralizing ebolaviruses. Nat Struct Mol Biol 18(12):1424–1427

Dube D, Brecher MB, Delos SE, Rose SC, Park EW, Schornberg KL, Kuhn JH, White JM (2009) The primed ebolavirus glycoprotein (19-kilodalton GP1,2): sequence and residues critical for host cell binding. J Virol 83(7):2883–2891

Eierhoff T, Hrincius ER, Rescher U, Ludwig S, Ehrhardt C (2010) The epidermal growth factor receptor (EGFR) promotes uptake of influenza A viruses (IAV) into host cells. PLoS Pathog 6(9):e1001099

Empig CJ, Goldsmith MA (2002) Association of the caveola vesicular system with cellular entry by filoviruses. J Virol 76(10):5266–5270

Favier AL, Gout E, Reynard O, Ferraris O, Kleman JP, Volchkov V, Peyrefitte C, Thielens NM (2016) Enhancement of Ebola Virus Infection via Ficolin-1 Interaction with the Mucin Domain of GP Glycoprotein. J Virol 90(11):5256–5269

Feldmann H, Nichol ST, Klenk HD, Peters CJ, Sanchez A (1994) Characterization of filoviruses based on differences in structure and antigenicity of the virion glycoprotein. Virology 199(2): 469–473

Feldmann H, Sanchez A, Geisbert TW (2013) Filoviridae: Marburg and Ebola viruses. In: Cohen JI, Griffin DE, Lamb RA, Martin MA, Racaniello VR, Boizman B (eds) Fields Virology. 6th edn. Philadelphia, Wolers Kluwer. pp 923–956

Feldmann H, Volchkov VE, Volchkova VA, Klenk HD (1999) The glycoproteins of Marburg and Ebola virus and their potential roles in pathogenesis. Arch Virol Suppl 15:159–169

Feldmann H, Will C, Schikore M, Slenczka W, Klenk HD (1991) Glycosylation and oligomerization of the spike protein of Marburg virus. Virology 182(1):353–356

Feng Z, Hensley L, McKnight KL, Hu F, Madden V, Ping L, Jeong SH, Walker C, Lanford RE, Lemon SM (2013) A pathogenic picornavirus acquires an envelope by hijacking cellular membranes. Nature 496(7445):367–371

Flyak AI, Ilinykh PA, Murin CD, Garron T, Shen X, Fusco ML, Hashiguchi T, Bornholdt ZA, Slaughter JC, Sapparapu G, Klages C, Ksiazek TG, Ward AB, Saphire EO, Bukreyev A, Crowe JE Jr (2015) Mechanism of human antibody-mediated neutralization of Marburg virus. Cell 160(5):893–903

Franco IS, Shuman HA (2012) A pathogen's journey in the host cell: bridges between actin and traffic. Bioarchitecture 2(2):38–42

Freeman GJ, Casasnovas JM, Umetsu DT, DeKruyff RH (2010) TIM genes: a family of cell surface phosphatidylserine receptors that regulate innate and adaptive immunity. Immunol Rev 235(1):172–189

Fretz M, Jin J, Conibere R, Penning NA, Al-Taei S, Storm G, Futaki S, Takeuchi T, Nakase I, Jones AT (2006) Effects of Na+/H+ exchanger inhibitors on subcellular localisation of endocytic organelles and intracellular dynamics of protein transduction domains HIV-TAT peptide and octaarginine. J Control Release 116(2):247–254

Gandini MA, Felix R (2015) Molecular and functional interplay of voltage-gated Ca(2)(+) channels with the cytoskeleton. Curr Mol Pharmacol 8(1):69–80

Gao X, Lowry PR, Zhou X, Depry C, Wei Z, Wong GW, Zhang J (2011) PI3K/Akt signaling requires spatial compartmentalization in plasma membrane microdomains. Proc Natl Acad Sci U S A 108(35):14509–14514

Gehring G, Rohrmann K, Atenchong N, Mittler E, Becker S, Dahlmann F, Pohlmann S, Vondran FW, David S, Manns MP, Ciesek S, von Hahn T (2014) The clinically approved drugs amiodarone, dronedarone and verapamil inhibit filovirus cell entry. J Antimicrob Chemother 69(8):2123–2131

Geijtenbeek TB, van Kooyk Y (2003) Pathogens target DC-SIGN to influence their fate DC-SIGN functions as a pathogen receptor with broad specificity. Apmis 111(7–8):698–714

Geisbert TW, Jahrling PB (1995) Differentiation of filoviruses by electron microscopy. Virus Res 39(2–3):129–150

Geyer H, Will C, Feldmann H, Klenk HD, Geyer R (1992) Carbohydrate structure of Marburg virus glycoprotein. Glycobiology 2(4):299–312

Gong X, Qian H, Zhou X, Wu J, Wan T, Cao P, Huang W, Zhao X, Wang X, Wang P, Shi Y, Gao GF, Zhou Q, Yan N (2016) Structural Insights into the Niemann-Pick C1 (NPC1)-mediated cholesterol transfer and Ebola infection. Cell 165(6):1467–1478

Gout E, Garlatti V, Smith DF, Lacroix M, Dumestre-Perard C, Lunardi T, Martin L, Cesbron JY, Arlaud GJ, Gaboriaud C, Thielens NM (2010) Carbohydrate recognition properties of human ficolins: glycan array screening reveals the sialic acid binding specificity of M-ficolin. J Biol Chem 285(9):6612–6622

Gramberg T, Caminschi I, Wegele A, Hofmann H, Pohlmann S (2006) Evidence that multiple defects in murine DC-SIGN inhibit a functional interaction with pathogens. Virology 345(2): 482–491

Gramberg T, Hofmann H, Moller P, Lalor PF, Marzi A, Geier M, Krumbiegel M, Winkler T, Kirchhoff F, Adams DH, Becker S, Munch J, Pohlmann S (2005) LSECtin interacts with filovirus glycoproteins and the spike protein of SARS coronavirus. Virology 340(2):224–236

Gregory SM, Harada E, Liang B, Delos SE, White JM, Tamm LK (2011) Structure and function of the complete internal fusion loop from Ebolavirus glycoprotein 2. Proc Natl Acad Sci U S A 108(27):11211–11216

Gregory SM, Larsson P, Nelson EA, Kasson PM, White JM, Tamm LK (2014) Ebolavirus entry requires a compact hydrophobic fist at the tip of the fusion loop. J Virol 88(12):6636–6649

Grimm C, Holdt LM, Chen CC, Hassan S, Muller C, Jors S, Cuny H, Kissing S, Schroder B, Butz E, Northoff B, Castonguay J, Luber CA, Moser M, Spahn S, Lullmann-Rauch R, Fendel C, Klugbauer N, Griesbeck O, Haas A, Mann M, Bracher F, Teupser D, Saftig P, Biel M, Wahl-Schott C (2014) High susceptibility to fatty liver disease in two-pore channel 2-deficient mice. Nat Commun 5:4699

Groseth A, Charton JE, Sauerborn M, Feldmann F, Jones SM, Hoenen T, Feldmann H (2009) The Ebola virus ribonucleoprotein complex: a novel VP30-L interaction identified. Virus Res 140(1–2):8–14

Hansen S, Holmskov U (1998) Structural aspects of collectins and receptors for collectins. Immunobiology 199(2):165–189

Harrison JS, Higgins CD, Chandran K, Lai JR (2011) Designed protein mimics of the Ebola virus glycoprotein GP2 α-helical bundle: stability and pH effects. Mol Cell 44(2):1587–1596

Harrison JS, Higgins CD, O'Meara MJ, Koellhoffer JF, Kuhlman BA, Lai JR (2013) Role of electrostatic repulsion in controlling pH-dependent conformational changes of viral fusion proteins. Structure (London, England: 1993) 21(7): 1085–1096

Harrison JS, Koellhoffer JF, Chandran K, Lai JR (2012) Marburg virus glycoprotein GP2: pH-dependent stability of the ectodomain α-helical bundle. Biochemistry 51(12):2515–2525

Harrison SC (2015) Viral membrane fusion. Virology 479–480:498–507

Harty RN, Brown ME, Wang G, Huibregtse J, Hayes FP (2000) A PPxY motif within the VP40 protein of Ebola virus interacts physically and functionally with a ubiquitin ligase: implications for filovirus budding. Proc Natl Acad Sci U S A 97(25):13871–13876

Hashiguchi T, Fusco ML, Bornholdt ZA, Lee JE, Flyak AI, Matsuoka R, Kohda D, Yanagi Y, Hammel M, Crowe JE Jr, Saphire EO (2015) Structural basis for Marburg virus neutralization by a cross-reactive human antibody. Cell 160(5):904–912

Hay JC (2007) Calcium: a fundamental regulator of intracellular membrane fusion? EMBO Rep 8(3):236–240

Herbert AS, Davidson C, Kuehne AI, Bakken R, Braigen SZ, Gunn KE, Whelan SP, Brummelkamp TR, Twenhafel NA, Chandran K, Walkley SU, Dye JM (2015) Niemann-pick C1 is essential for ebolavirus replication and pathogenesis in vivo. MBio 6(3): e00565–e00515

Higgins CD, Koellhoffer JF, Chandran K, Lai JR (2013) C-peptide inhibitors of Ebola virus glycoprotein-mediated cell entry: effects of conjugation to cholesterol and side chain-side chain crosslinking. Bioorg Med Chem Lett 23(19):5356–5360

Higgins ME, Davies JP, Chen FW, Ioannou YA (1999) Niemann-Pick C1 is a late endosome-resident protein that transiently associates with lysosomes and the trans-Golgi network. Mol Genet Metab 68(1):1–13

Hoenen T, Jung S, Herwig A, Groseth A, Becker S (2010) Both matrix proteins of Ebola virus contribute to the regulation of viral genome replication and transcription. Virology 403(1): 56–66

Hoffmann M, Gonzalez Hernandez M, Berger E, Marzi A, Pohlmann S (2016) The Glycoproteins of all filovirus species use the same host factors for entry into bat and human cells but entry efficiency is species dependent. PLoS ONE 11(2):e0149651

Hood CL, Abraham J, Boyington JC, Leung K, Kwong PD, Nabel GJ (2010) Biochemical and structural characterization of cathepsin L-processed Ebola virus glycoprotein: implications for viral entry and immunogenicity. J Virol 84(6):2972–2982

Hunt CL, Kolokoltsov AA, Davey RA, Maury W (2011) The tyro3 receptor kinase axl enhances macropinocytosis of Zaire Ebola virus. J Virol 85(1):334–347

Hunt CL, Lennemann NJ, Maury W (2012) Filovirus entry: a novelty in the viral fusion world. Viruses 4(2):258–275

Ito H, Watanabe S, Sanchez A, Whitt MA, Kawaoka Y (1999) Mutational analysis of the putative fusion domain of Ebola virus glycoprotein. J Virol 73(10):8907–8912

Ivanov AI (2008) Pharmacological inhibition of endocytic pathways: is it specific enough to be useful? Methods Mol Biol 440:15–33

Jae LT, Brummelkamp TR (2015) Emerging intracellular receptors for hemorrhagic fever viruses. Trends Microbiol 23(7):392–400

Jeffers SA, Sanders DA, Sanchez A (2002) Covalent modifications of the ebola virus glycoprotein. J Virol 76(24):12463–12472

Jemielity S, Wang JJ, Chan YK, Ahmed AA, Li W, Monahan S, Bu X, Farzan M, Freeman GJ, Umetsu DT, Dekruyff RH, Choe H (2013) TIM-family proteins promote infection of multiple enveloped viruses through virion-associated phosphatidylserine. PLoS Pathog 9(3):e1003232

Ji X, Olinger GG, Aris S, Chen Y, Gewurz H, Spear GT (2005) Mannose-binding lectin binds to Ebola and Marburg envelope glycoproteins, resulting in blocking of virus interaction with DC-SIGN and complement-mediated virus neutralization. J Gen Virol 86(Pt 9):2535–2542

Johansen LM, DeWald LE, Shoemaker CJ, Hoffstrom BG, Lear-Rooney CM, Stossel A, Nelson E, Delos SE, Simmons JA, Grenier JM, Pierce LT, Pajouhesh H, Lehar J, Hensley LE, Glass PJ, White JM, Olinger GG (2015) A screen of approved drugs and molecular probes identifies therapeutics with anti-Ebola virus activity. Sci Transl Med 7(290): 290ra289

Kerr MC, Lindsay MR, Luetterforst R, Hamilton N, Simpson F, Parton RG, Gleeson PA, Teasdale RD (2006) Visualisation of macropinosome maturation by the recruitment of sorting nexins. J Cell Sci 119(Pt 19):3967–3980

Kielian M (2014) Mechanisms of virus membrane fusion proteins. Annu Rev Virol 1(1):171–189

Kobayashi N, Karisola P, Pena-Cruz V, Dorfman DM, Jinushi M, Umetsu SE, Butte MJ, Nagumo H, Chernova I, Zhu B, Sharpe AH, Ito S, Dranoff G, Kaplan GG, Casasnovas JM, Umetsu DT, Dekruyff RH, Freeman GJ (2007) TIM-1 and TIM-4 glycoproteins bind phosphatidylserine and mediate uptake of apoptotic cells. Immunity 27(6):927–940

Koellhoffer JF, Dai Z, Malashkevich VN, Stenglein MD, Liu Y, Toro R, Harrison JS, Chandran K, Derisi JL, Almo SC, Lai JR (2014) Structural characterization of the glycoprotein GP2 core domain from the CAS virus, a novel Arenavirus-like species. J Mol Biol 426(7):1452–1468

Koellhoffer JF, Malashkevich VN, Harrison JS, Toro R, Bhosle RC, Chandran K, Almo SC, Lai JR (2012) Crystal structure of the Marburg virus GP2 core domain in its postfusion conformation. Biochemistry 51(39):7665–7675

Koivusalo M, Welch C, Hayashi H, Scott CC, Kim M, Alexander T, Touret N, Hahn KM, Grinstein S (2010) Amiloride inhibits macropinocytosis by lowering submembranous pH and preventing Rac1 and Cdc42 signaling. J Cell Biol 188(4):547–563

Kolokoltsov AA, Saeed MF, Freiberg AN, Holbrook MR, Davey RA (2009) Identification of novel cellular targets for therapeutic intervention against Ebola virus infection by siRNA screening. Drug Dev Res 70(4):255–265

Kondratowicz AS, Hunt CL, Davey RA, Cherry S, Maury WJ (2013) AMP-activated protein kinase is required for the macropinocytic internalization of ebolavirus. J Virol 87(2):746–755
Kondratowicz AS, Lennemann NJ, Sinn PL, Davey RA, Hunt CL, Moller-Tank S, Meyerholz DK, Rennert P, Mullins RF, Brindley M, Sandersfeld LM, Quinn K, Weller M, McCray PB Jr, Chiorini J, Maury W (2011) T-cell immunoglobulin and mucin domain 1 (TIM-1) is a receptor for Zaire Ebola virus and Lake Victoria Marburg virus. Proc Natl Acad Sci U S A 108(20): 8426–8431
Krishnan A, Miller EH, Herbert AS, Ng M, Ndungo E, Whelan SP, Dye JM, Chandran K (2012) Niemann-Pick C1 (NPC1)/NPC1-like1 chimeras define sequences critical for NPC1's function as a flovirus entry receptor. Viruses 4(11):2471–2484
Lasala F, Arce E, Otero JR, Rojo J, Delgado R (2003) Mannosyl glycodendritic structure inhibits DC-SIGN-mediated Ebola virus infection in cis and in trans. Antimicrob Agents Chemother 47(12):3970–3972
Lee J, Gregory SM, Nelson EA, White JM, Tamm LK (2016) The roles of histidines and charged residues as potential triggers of a conformational change in the fusion loop of Ebola virus glycoprotein. PLoS ONE 11(3):e0152527
Lee JE, Fusco ML, Hessell AJ, Oswald WB, Burton DR, Saphire EO (2008) Structure of the Ebola virus glycoprotein bound to an antibody from a human survivor. Nature 454(7201):177–182
Lee JE, Saphire EO (2009) Ebolavirus glycoprotein structure and mechanism of entry. Future Virol 4(6):621–635
Lemke G, Rothlin CV (2008) Immunobiology of the TAM receptors. Nat Rev Immunol 8(5): 327–336
Lennemann NJ, Rhein BA, Ndungo E, Chandran K, Qiu X, Maury W (2014) Comprehensive functional analysis of N-linked glycans on Ebola virus GP1. MBio 5(1):e00862–e00813
Li X, Saha P, Li J, Blobel G, Pfeffer SR (2016) Clues to the mechanism of cholesterol transfer from the structure of NPC1 middle lumenal domain bound to NPC2. Proc Natl Acad Sci U S A 113(36):10079–10084
Liberali P, Kakkonen E, Turacchio G, Valente C, Spaar A, Perinetti G, Bockmann RA, Corda D, Colanzi A, Marjomaki V, Luini A (2008) The closure of Pak1-dependent macropinosomes requires the phosphorylation of CtBP1/BARS. EMBO J 27(7):970–981
Lim JP, Wang JT, Kerr MC, Teasdale RD, Gleeson PA (2008) A role for SNX5 in the regulation of macropinocytosis. BMC Cell Biol 9:58
Lin-Moshier Y, Keebler MV, Hooper R, Boulware MJ, Liu X, Churamani D, Abood ME, Walseth TF, Brailoiu E, Patel S, Marchant JS (2014) The two-pore channel (TPC) interactome unmasks isoform-specific roles for TPCs in endolysosomal morphology and cell pigmentation. Proc Natl Acad Sci U S A 111(36):13087–13092
Lin G, Simmons G, Pohlmann S, Baribaud F, Ni H, Leslie GJ, Haggarty BS, Bates P, Weissman D, Hoxie JA, Doms RW (2003) Differential N-linked glycosylation of human immunodeficiency virus and Ebola virus envelope glycoproteins modulates interactions with DC-SIGN and DC-SIGNR. J Virol 77(2):1337–1346
Linger RM, Keating AK, Earp HS, Graham DK (2008) TAM receptor tyrosine kinases: biologic functions, signaling, and potential therapeutic targeting in human cancer. Adv Cancer Res 100:35–83
Luzio JP, Bright NA, Pryor PR (2007a) The role of calcium and other ions in sorting and delivery in the late endocytic pathway. Biochem Soc Trans 35(Pt 5):1088–1091
Luzio JP, Pryor PR, Bright NA (2007b) Lysosomes: fusion and function. Nat Rev Mol Cell Biol 8(8):622–632
Madrid PB, Chopra S, Manger ID, Gilfillan L, Keepers TR, Shurtleff AC, Green CE, Iyer LV, Dilks HH, Davey RA, Kolokoltsov AA, Carrion R Jr, Patterson JL, Bavari S, Panchal RG, Warren TK, Wells JB, Moos WH, Burke RL, Tanga MJ (2013) A systematic screen of FDA-approved drugs for inhibitors of biological threat agents. PLoS ONE 8(4):e60579
Makino A, Yamayoshi S, Shinya K, Noda T, Kawaoka Y (2011) Identification of amino acids in Marburg virus VP40 that are important for virus-like particle budding. J Infect Dis 204(Suppl 3):S871–S877

Malashkevich VN, Schneider BJ, McNally ML, Milhollen MA, Pang JX, Kim PS (1999) Core structure of the envelope glycoprotein GP2 from Ebola virus at 1.9-A resolution. Proc Natl Aca of Sci U S A 96(6):2662–2667

Markosyan RM, Miao C, Zheng YM, Melikyan GB, Liu SL, Cohen FS (2016) Induction of cell-cell fusion by Ebola virus glycoprotein: low pH Is not a trigger. PLoS Pathog 12(1): e1005373

Martin B, Hoenen T, Canard B, Decroly E (2016) Filovirus proteins for antiviral drug discovery: a structure/function analysis of surface glycoproteins and virus entry. Antiviral Res 135:1–14

Martinez O, Johnson J, Manicassamy B, Rong L, Olinger GG, Hensley LE, Basler CF (2010) Zaire Ebola virus entry into human dendritic cells is insensitive to cathepsin L inhibition. Cell Microbiol 12(2):148–157

Marzi A, Akhavan A, Simmons G, Gramberg T, Hofmann H, Bates P, Lingappa VR, Pohlmann S (2006) The signal peptide of the ebolavirus glycoprotein influences interaction with the cellular lectins DC-SIGN and DC-SIGNR. J Virol 80(13):6305–6317

Marzi A, Gramberg T, Simmons G, Moller P, Rennekamp AJ, Krumbiegel M, Geier M, Eisemann J, Turza N, Saunier B, Steinkasserer A, Becker S, Bates P, Hofmann H, Pohlmann S (2004) DC-SIGN and DC-SIGNR interact with the glycoprotein of Marburg virus and the S protein of severe acute respiratory syndrome coronavirus. J Virol 78(21):12090–12095

Marzi A, Moller P, Hanna SL, Harrer T, Eisemann J, Steinkasserer A, Becker S, Baribaud F, Pohlmann S (2007) Analysis of the interaction of Ebola virus glycoprotein with DC-SIGN (dendritic cell-specific intercellular adhesion molecule 3-grabbing nonintegrin) and its homologue DC-SIGNR. J Infect Dis 196(Suppl 2):S237–S246

Marzi A, Reinheckel T, Feldmann H (2012) Cathepsin B & L are not required for Ebola virus replication. PLoS Neglected Trop Dis 6(12):e1923

Mason CP, Tarr AW (2015) Human lectins and their roles in viral infections. Molecules 20(2): 2229–2271

May JA, Ratan H, Glenn JR, Losche W, Spangenberg P, Heptinstall S (1998) GPIIb-IIIa antagonists cause rapid disaggregation of platelets pre-treated with cytochalasin D. Evidence that the stability of platelet aggregates depends on normal cytoskeletal assembly. Platelets 9(3–4):227–232

Meertens L, Carnec X, Lecoin MP, Ramdasi R, Guivel-Benhassine F, Lew E, Lemke G, Schwartz O, Amara A (2012) The TIM and TAM families of phosphatidylserine receptors mediate dengue virus entry. Cell Host Microbe 12(4):544–557

Mercer J, Helenius A (2008) Vaccinia virus uses macropinocytosis and apoptotic mimicry to enter host cells. Science 320(5875):531–535

Mercer J, Helenius A (2009) Virus entry by macropinocytosis. Nat Cell Biol 11(5):510–520

Mercer J, Knebel S, Schmidt FI, Crouse J, Burkard C, Helenius A (2010) Vaccinia virus strains use distinct forms of macropinocytosis for host-cell entry. Proc Natl Acad Sci U S A 107(20): 9346–9351

Miller EH, Chandran K (2012) Filovirus entry into cells—new insights. Curr Opin Virol 2(2): 206–214

Miller EH, Harrison JS, Radoshitzky SR, Higgins CD, Chi X, Dong L, Kuhn JH, bavari S, Lai JR, Chandran K (2011) Inhibition of Ebola virus entry by a C-peptide targeted to endosomes. J Biol Chem 286(18):15854–15861

Miller EH, Obernosterer G, Raaben M, Herbert AS, Deffieu MS, Krishnan A, Ndungo E, Sandesara RG, Carette JE, Kuehne AI, Ruthel G, Pfeffer SR, Dye JM, Whelan SP, Brummelkamp TR, Chandran K (2012a) Ebola virus entry requires the host-programmed recognition of an intracellular receptor. EMBO J 31(10):1947–1960

Miller ME, Adhikary S, Kolokoltsov AA, Davey RA (2012b) Ebolavirus requires acid sphingomyelinase activity and plasma membrane sphingomyelin for infection. J Virol 86(14): 7473–7483

Misasi J, Chandran K, Yang J-Y, Considine B, Filone CM, Côté M, Sullivan N, Fabozzi G, Hensley L, Cunningham J (2012) Filoviruses require endosomal cysteine proteases for entry but exhibit distinct protease preferences. J Virol 86(6):3284–3292

Moller-Tank S, Albritton LM, Rennert PD, Maury W (2014) Characterizing functional domains for TIM-mediated enveloped virus entry. J Virol 88(12):6702–6713
Moller-Tank S, Kondratowicz AS, Davey RA, Rennert PD, Maury W (2013) Role of the phosphatidylserine receptor TIM-1 in enveloped-virus entry. J Virol 87(15):8327–8341
Morizono K, Chen IS (2014) Role of phosphatidylserine receptors in enveloped virus infection. J Virol 88(8):4275–4290
Morizono K, Xie Y, Olafsen T, Lee B, Dasgupta A, Wu AM, Chen IS (2011) The soluble serum protein Gas6 bridges virion envelope phosphatidylserine to the TAM receptor tyrosine kinase Axl to mediate viral entry. Cell Host Microbe 9(4):286–298
Nanbo A, Imai M, Watanabe S, Noda T, Takahashi K, Neumann G, Halfmann P, Kawaoka Y (2010) Ebolavirus is internalized into host cells via macropinocytosis in a viral glycoprotein-dependent manner. PLoS Pathog 6(9):e1001121
Naureckiene S, Sleat DE, Lackland H, Fensom A, Vanier MT, Wattiaux R, Jadot M, Lobel P (2000) Identification of HE1 as the second gene of Niemann-Pick C disease. Science 290 (5500):2298–2301
Ndungo E, Herbert AS, Raaben M, Obernosterer G, Biswas R, Miller EH, Wirchnianski AS, Carette JE, Brummelkamp TR, Whelan SP, Dye JM, Chandran K (2016) A single residue in Ebola virus receptor NPC1 influences cellular host range in reptiles. mSphere 1(2): 7–16
Neumann G, Feldmann H, Watanabe S, Lukashevich I, Kawaoka Y (2002) Reverse genetics demonstrates that proteolytic processing of the Ebola virus glycoprotein is not essential for replication in cell culture. J Virol 76(1):406–410
Ng M, Ndungo E, Jangra RK, Cai Y, Postnikova E, Radoshitzky SR, Dye JM, Ramirez de Arellano E, Negredo A, Palacios G, Kuhn JH, Chandran K (2014) Cell entry by a novel European filovirus requires host endosomal cysteine proteases and Niemann-Pick C1. Virology 468–470:637–646
Ng M, Ndungo E, Kaczmarek ME, Herbert AS, Binger T, Kuehne AI, Jangra RK, Hawkins JA, Gifford RJ, Biswas R, Demogines A, James RM, Yu M, Brummelkamp TR, Drosten C, Wang LF, Kuhn JH, Muller MA, Dye JM, Sawyer SL, Chandran K (2015) Filovirus receptor NPC1 contributes to species-specific patterns of ebolavirus susceptibility in bats. Elife 4: e11785
Noda T, Sagara H, Suzuki E, Takada A, Kida H, Kawaoka Y (2002) Ebola virus VP40 drives the formation of virus-like filamentous particles along with GP. J Virol 76(10):4855–4865
Norbury CC (2006) Drinking a lot is good for dendritic cells. Immunology 117(4):443–451
Numrich J, Ungermann C (2014) Endocytic Rabs in membrane trafficking and signaling. Biol Chem 395(3):327–333
O'Halloran TJ, Anderson RG (1992) Clathrin heavy chain is required for pinocytosis, the presence of large vacuoles, and development in Dictyostelium. J Cell Biol 118(6):1371–1377
O'Hearn A, Wang M, Cheng H, Lear-Rooney CM, Koning K, Rumschlag-Booms E, Varhegyi E, Olinger G, Rong L (2015) Role of EXT1 and glycosaminoglycans in the early stage of filovirus entry. J Virol 89(10):5441–5449
Park CG, Takahara K, Umemoto E, Yashima Y, Matsubara K, Matsuda Y, Clausen BE, Inaba K, Steinman RM (2001) Five mouse homologues of the human dendritic cell C-type lectin, DC-SIGN. Int Immunol 13(10):1283–1290
Patel S (2015) Function and dysfunction of two-pore channels. Sci Signal 8(384): re7
Pernet O, Pohl C, Ainouze M, Kweder H, Buckland R (2009) Nipah virus entry can occur by macropinocytosis. Virology 395(2):298–311
Pipirou Z, Powlesland AS, Steffen I, Pohlmann S, Taylor ME, Drickamer K (2011) Mouse LSECtin as a model for a human Ebola virus receptor. Glycobiology 21(6):806–812
Pontremoli C, Forni D, Cagliani R, Filippi G, De Gioia L, Pozzoli U, Clerici M, Sironi M (2016) Positive selection drives evolution at the host-filovirus interaction surface. Mol Biol Evol 33(11):2836–2847
Powlesland AS, Fisch T, Taylor ME, Smith DF, Tissot B, Dell A, Pohlmann S, Drickamer K (2008) A novel mechanism for LSECtin binding to Ebola virus surface glycoprotein through truncated glycans. J Biol Chem 283(1):593–602

Powlesland AS, Ward EM, Sadhu SK, Guo Y, Taylor ME, Drickamer K (2006) Widely divergent biochemical properties of the complete set of mouse DC-SIGN-related proteins. J Biol Chem 281(29):20440–20449

Prasad D, Rothlin CV, Burrola P, Burstyn-Cohen T, Lu Q, Garcia de Frutos P, Lemke G (2006) TAM receptor function in the retinal pigment epithelium. Mol Cell Neurosci 33(1):96–108

Quinn K, Brindley MA, Weller ML, Kaludov N, Kondratowicz A, Hunt CL, Sinn PL, McCray PB Jr, Stein CS, Davidson BL, Flick R, Mandell R, Staplin W, Maury W, Chiorini JA (2009) Rho GTPases modulate entry of Ebola virus and vesicular stomatitis virus pseudotyped vectors. J Virol 83(19):10176–10186

Raghu H, Sharma-Walia N, Veettil MV, Sadagopan S, Chandran B (2009) Kaposi's sarcoma-associated herpesvirus utilizes an actin polymerization-dependent macropinocytic pathway to enter human dermal microvascular endothelial and human umbilical vein endothelial cells. J Virol 83(10):4895–4911

Rathbun JY, Droniou ME, Damoiseaux R, Haworth KG, Henley JE, Exline CM, Choe H, Cannon PM (2015) Novel arenavirus entry inhibitors discovered by using a minigenome rescue system for high-throughput drug screening. J Virol 89(16):8428–8443

Ravichandran KS (2010) Find-me and eat-me signals in apoptotic cell clearance: progress and conundrums. J Exp Med 207(9):1807–1817

Ravichandran KS (2011) Beginnings of a good apoptotic meal: the find-me and eat-me signaling pathways. Immunity 35(4):445–455

Rhein BA, Brouillette RB, Schaack GA, Chiorini JA, Maury W (2016) Characterization of human and murine T-cell immunoglobulin mucin domain 4 (TIM-4) IgV domain residues critical for Ebola virus entry. J Virol 90(13):6097–6111

Rhein BA, Maury WJ (2015) Ebola virus entry into host cells: identifying therapeutic strategies. Curr Clin Microbiol Rep 2(3):115–124

Ridley AJ, Paterson HF, Johnston CL, Diekmann D, Hall A (1992) The small GTP-binding protein rac regulates growth factor-induced membrane ruffling. Cell 70(3):401–410

Riese RJ, Mitchell RN, Villadangos JA, Shi GP, Palmer JT, Karp ER, De Sanctis GT, Ploegh HL, Chapman HA (1998) Cathepsin S activity regulates antigen presentation and immunity. J Clin Invest 101(11):2351–2363

Saeed MF, Kolokoltsov AA, Albrecht T, Davey RA (2010) Cellular entry of Ebola virus involves uptake by a macropinocytosis-like mechanism and subsequent trafficking through early and late endosomes. PLoS Pathog 6(9):e1001110

Saeed MF, Kolokoltsov AA, Freiberg AN, Holbrook MR, Davey RA (2008) Phosphoinositide-3 kinase-Akt pathway controls cellular entry of Ebola virus. PLoS Pathog 4(8):e1000141

Sakurai Y, Kolokoltsov AA, Chen CC, Tidwell MW, Bauta WE, Klugbauer N, Grimm C, Wahl-Schott C, Biel M, Davey RA (2015) Ebola virus. Two-pore channels control Ebola virus host cell entry and are drug targets for disease treatment. Science 347(6225):995–998

Sanchez A (2007) Analysis of filovirus entry into Vero e6 cells, using inhibitors of endocytosis, endosomal acidification, structural integrity, and cathepsin (B and L) activity. J Infect Dis 196 (Suppl 2):S251–S258

Sanchez A, Trappier SG, Mahy BW, Peters CJ, Nichol ST (1996) The virion glycoproteins of Ebola viruses are encoded in two reading frames and are expressed through transcriptional editing. Proc Natl Acad Sci U S A 93(8):3602–3607

Sanchez A, Yang ZY, Xu L, Nabel GJ, Crews T, Peters CJ (1998) Biochemical analysis of the secreted and virion glycoproteins of Ebola virus. J Virol 72(8):6442–6447

Santiago C, Ballesteros A, Martinez-Munoz L, Mellado M, Kaplan GG, Freeman GJ, Casasnovas JM (2007a) Structures of T cell immunoglobulin mucin protein 4 show a metal-Ion-dependent ligand binding site where phosphatidylserine binds. Immunity 27(6): 941–951

Santiago C, Ballesteros A, Tami C, Martinez-Munoz L, Kaplan GG, Casasnovas JM (2007b) Structures of T cell immunoglobulin mucin receptors 1 and 2 reveal mechanisms for regulation of immune responses by the TIM receptor family. Immunity 26(3):299–310

Sasaki T, Knyazev PG, Clout NJ, Cheburkin Y, Gohring W, Ullrich A, Timpl R, Hohenester E (2006) Structural basis for Gas6-Axl signalling. EMBO J 25(1):80–87

Schelhaas M, Shah B, Holzer M, Blattmann P, Kuhling L, Day PM, Schiller JT, Helenius A (2012) Entry of human papillomavirus type 16 by actin-dependent, clathrin- and lipid raft-independent endocytosis. PLoS Pathog 8(4):e1002657

Schornberg K, Matsuyama S, Kabsch K, Delos S, Bouton A, White J (2006) Role of endosomal cathepsins in entry mediated by the Ebola virus glycoprotein. J Virol 80(8):4174–4178

Scott C, Higgins ME, Davies JP, Ioannou YA (2004) Targeting of NPC1 to late endosomes involves multiple signals, including one residing within the putative sterol-sensing domain. J Biol Chem 279(46):48214–48223

Sevenich L, Hagemann S, Stoeckle C, Tolosa E, Peters C, Reinheckel T (2010) Expression of human cathepsin L or human cathepsin V in mouse thymus mediates positive selection of T helper cells in cathepsin L knock-out mice. Biochimie 92(11):1674–1680

Shen D, Wang X, Xu H (2011) Pairing phosphoinositides with calcium ions in endolysosomal dynamics: phosphoinositides control the direction and specificity of membrane trafficking by regulating the activity of calcium channels in the endolysosomes. BioEssays 33(6):448–457

Shimojima M, Ikeda Y, Kawaoka Y (2007) The mechanism of Axl-mediated Ebola virus infection. J Infect Dis 196(Suppl 2):S259–S263

Shimojima M, Takada A, Ebihara H, Neumann G, Fujioka K, Irimura T, Jones S, Feldmann H, Kawaoka Y (2006) Tyro3 family-mediated cell entry of Ebola and Marburg viruses. J Virol 80 (20):10109–10116

Simmons G (2013) Filovirus entry. Adv Exp Med Biol 790:83–94

Simmons G, Reeves JD, Grogan CC, Vandenberghe LH, Baribaud F, Whitbeck JC, Burke E, Buchmeier MJ, Soilleux EJ, Riley JL, Doms RW, Bates P, Pohlmann S (2003a) DC-SIGN and DC-SIGNR bind Ebola glycoproteins and enhance infection of macrophages and endothelial cells. Virology 305(1):115–123

Simmons G, Rennekamp AJ, Chai N, Vandenberghe LH, Riley JL, Bates P (2003b) Folate receptor alpha and caveolae are not required for Ebola virus glycoprotein-mediated viral infection. J Virol 77(24):13433–13438

Simmons JA, D'Souza RS, Ruas M, Galione A, Casanova JE, White JM (2015) Ebolavirus glycoprotein directs fusion through NPC1+ endolysosomes. J Virol 90(1):605–610

Spence JS, Krause TB, Mittler E, Jangra RK, Chandran K (2016) Direct visualization of Ebola virus fusion triggering in the endocytic pathway. MBio 7(1):e01857–e01815

Stitt TN, Conn G, Gore M, Lai C, Bruno J, Radziejewski C, Mattsson K, Fisher J, Gies DR, Jones PF et al (1995) The anticoagulation factor protein S and its relative, Gas6, are ligands for the Tyro 3/Axl family of receptor tyrosine kinases. Cell 80(4):661–670

Swanson JA (2008) Shaping cups into phagosomes and macropinosomes. Nat Rev Mol Cell Biol 9 (8):639–649

Takada A, Fujioka K, Tsuiji M, Morikawa A, Higashi N, Ebihara H, Kobasa D, Feldmann H, Irimura T, Kawaoka Y (2004) Human macrophage C-type lectin specific for galactose and N-acetylgalactosamine promotes filovirus entry. J Virol 78(6):2943–2947

Takada A, Robison C, Goto H, Sanchez A, Murti KG, Whitt MA, Kawaoka Y (1997) A system for functional analysis of Ebola virus glycoprotein. Proc Natl Acad Sci U S A 94(26):14764–14769

Tietjen GT, Gong Z, Chen CH, Vargas E, Crooks JE, Cao KD, Heffern CT, Henderson JM, Meron M, Lin B, Roux B, Schlossman ML, Steck TL, Lee KY, Adams EJ (2014) Molecular mechanism for differential recognition of membrane phosphatidylserine by the immune regulatory receptor Tim4. Proc Natl Acad Sci U S A 111(15):E1463–E1472

Turner MW (2003) The role of mannose-binding lectin in health and disease. Mol Immunol 40(7):423–429

van der Kant R, Fish A, Janssen L, Janssen H, Krom S, Ho N, Brummelkamp T, Carette J, Rocha N, Neefjes J (2013) Late endosomal transport and tethering are coupled processes controlled by RILP and the cholesterol sensor ORP1L. J Cell Sci 126(Pt 15):3462–3474

Volchkov VE, Becker S, Volchkova VA, Ternovoj VA, Kotov AN, Netesov SV, Klenk HD (1995) GP mRNA of Ebola virus is edited by the Ebola virus polymerase and by T7 and vaccinia virus polymerases. Virology 214(2):421–430

Volchkov VE, Feldmann H, Volchkova VA, Klenk HD (1998) Processing of the Ebola virus glycoprotein by the proprotein convertase furin. Proc Natl Acad Sci U S A 95(10):5762–5767

Wang H, Shi Y, Song J, Qi J, Lu G, Yan J, Gao GF (2016) Ebola viral glycoprotein bound to its endosomal receptor Niemann-Pick C1. Cell 164(1–2):258–268

Watanabe SS, Takada AA, Watanabe TT, Ito HH, Kida HH, Kawaoka YY (2000) Functional importance of the coiled-coil of the Ebola virus glycoprotein. J Virol 74(21):10194–10201

Wec AZ, Nyakatura EK, Herbert AS, Howell KA, Holtsberg FW, Bakken RR, Mittler E, Christin JR, Shulenin S, Jangra RK, Bharrhan S, Kuehne AI, Bornholdt ZA, Flyak AI, Saphire EO, Crowe JE Jr, Aman MJ, Dye JM, Lai JR, Chandran K (2016) A "Trojan horse" bispecific antibody strategy for broad protection against ebolaviruses. Science 354(6310): 350–354

Weissenhorn W, Carfí A, Lee KH, Skehel JJ, Wiley DC (1998) Crystal structure of the Ebola virus membrane fusion subunit, GP2, from the envelope glycoprotein ectodomain. Mol Cell 2(5): 605–616

West MA, Prescott AR, Eskelinen EL, Ridley AJ, Watts C (2000) Rac is required for constitutive macropinocytosis by dendritic cells but does not control its downregulation. Curr Biol 10(14): 839–848

White JM, Delos SE, Brecher M, Schornberg K (2008) Structures and mechanisms of viral membrane fusion proteins: multiple variations on a common theme. Crit Rev Biochem Mol Biol 43(3):189–219

White JM, Schornberg KL (2012) A new player in the puzzle of filovirus entry. Nat Rev Microbiol 10(5):317–322

White JM, Whittaker GR (2016) Fusion of enveloped viruses in endosomes. Traffic

Wong AC, Sandesara RG, Mulherkar N, Whelan SP, Chandran K (2010) A forward genetic strategy reveals destabilizing mutations in the Ebolavirus glycoprotein that alter its protease dependence during cell entry. J Virol 84(1):163–175

Wool-Lewis RJ, Bates P (1998) Characterization of Ebola virus entry by using pseudotyped viruses: identification of receptor-deficient cell lines. J Virol 72(4):3155–3160

Wool-Lewis RJ, Bates P (1999) Endoproteolytic processing of the Ebola virus envelope glycoprotein: cleavage is not required for function. J Virol 73(2):1419–1426

Yang ZY, Duckers HJ, Sullivan NJ, Sanchez A, Nabel EG, Nabel GJ (2000) Identification of the Ebola virus glycoprotein as the main viral determinant of vascular cell cytotoxicity and injury. Nat Med 6(8):886–889

Zhao Y, Ren J, Harlos K, Jones DM, Zeltina A, Bowden TA, Padilla-Parra S, Fry EE, Stuart DI (2016) Toremifene interats with and destabilizes the Ebola virus glycoprotein. Nature 535 (7610):169–172

Zhong Z, Wang Y, Guo H, Sagare A, Fernandez JA, Bell RD, Barrett TM, Griffin JH, Freeman RS, Zlokovic BV (2010) Protein S protects neurons from excitotoxic injury by activating the TAM receptor Tyro3-phosphatidylinositol 3-kinase-Akt pathway through its sex hormone-binding globulin-like region. J Neurosci 30(46):15521–15534

Inside the Cell: Assembly of Filoviruses

Larissa Kolesnikova, Asuka Nanbo, Stephan Becker and Yoshihiro Kawaoka

Abstract This chapter reviews our current knowledge about the spatiotemporal assembly of filoviral particles. We will follow particles from nucleocapsid entry into the cytoplasm until the nucleocapsids are enveloped at the plasma membrane. We will also highlight the currently open scientific questions surrounding filovirus assembly.

L. Kolesnikova (✉) · S. Becker (✉)
Institute of Virology, Philipps University of Marburg, Hans-Meerwein-Str. 2, 35043 Marburg, Germany
e-mail: kolesnik@staff.uni-marburg.de

S. Becker
e-mail: becker@staff.uni-marburg.de

A. Nanbo
Department of Cell Physiology, Hokkaido University Graduate School of Medicine, Kita-Ku, Kita 15 Nishi 7, Sapporo 060-8638, Japan
e-mail: nanboa@med.hokudai.ac.jp

S. Becker
German Center for Infection Research (DZIF), Institute of Virology, Philipps University of Marburg, Hans-Meerwein-Str. 2, 35043 Marburg, Germany

Y. Kawaoka
Division of Virology, Department of Microbiology and Immunology, Institute of Medical Science, University of Tokyo, Minato-Ku, Tokyo, Japan
e-mail: yoshihiro.kawaoka@wisc.edu

Y. Kawaoka
Department of Special Pathogens, International Research Center for Infectious Diseases, Institute of Medical Science, University of Tokyo, Minato-Ku,Tokyo, Japan

Y. Kawaoka
Department of Pathobiological Sciences, School of Veterinary Medicine, University of Wisconsin—Madison, Madison, WI 53711, USA

Current Topics in Microbiology and Immunology (2017) 411:353–380
DOI 10.1007/82_2017_15

Published Online: 11 June 2017

Contents

1 Introduction

Studies on filovirus assembly began almost simultaneously with the discovery of Ebola (EBOV) and Marburg virus (MARV), and were initially based on the ultrastructural analysis of virus-infected cells (Peters et al. 1971; Ellis et al. 1979; Murphy et al. 1978). These studies have already defined morphological signs of filoviral infection and the main steps of assembly, such as the formation of perinuclear viral inclusions consisting of tubule-like filamentous nucleocapsids, the redistribution of these filamentous nucleocapsids to the cell periphery, and finally, their envelopment at the plasma membrane. Further biochemical studies identified seven filoviral structural proteins, NP, VP35, VP40, GP, VP30, VP24, and L, and characterized the single-stranded negative-sense RNA genome, reviewed in (Beer et al. 1999; Feldmann and Kiley 1999). The development of viral life cycle modeling systems (e.g., minigenome, delta-VP30 system and virus-like particle systems) allowed scientists to simulate different stages of virus assembly and to further advance our understanding of viral protein functions. For example, the essential roles of nucleocapsid proteins (NP, VP35, VP30, and L) in the transcription and replication of filoviral negative-strand RNA have been clarified by use of a minigenome system, reviewed in (Mühlberger 2007). Similarly, the key role of VP40 in the attraction of viral surface protein GP and the nucleocapsid to the site of filovirus budding, and in budding per se, has been dissected by using an infectious VLP system, reviewed in (Jasenosky and Kawaoka 2004; Aman et al. 2003; Hartlieb and Weissenhorn 2006; Ascenzi et al. 2008; Dolnik et al. 2008; Olejnik et al. 2011). Molecular biological and structural studies have shown that filoviral proteins form homo-oligomers, and heterologous protein–protein complexes. Remarkably, the formation of NP and VP40 higher order homo-oligomers takes place in specific cellular compartments. For example, the ectopic expression of filoviral NP alone leads to left-handed helical structures in the perinuclear region, whereas the ectopic expression of VP40 alone results in the accumulation of regular protein arrays beneath the plasma membrane, reviewed in (Hartlieb and Weissenhorn 2006). However, during filoviral infection, the formation of “empty” VP40-enriched membrane particles (or NP helices without other nucleocapsid proteins and viral RNA) is not observed. Therefore, mechanisms must exist that coordinate the onset and completion of the different steps of filovirus assembly ultimately leading to the formation of infectious particles. In this review, we used the currently available data to reconstruct the different steps of filovirus assembly with an emphasis on the probable mechanisms involved in coordinating the assembly processes.

2 Uncoating

The mechanisms of filovirus uncoating are not fully understood. Filovirus uncoating likely comprises multiple steps that involve viral protein modifications (e.g., selective proteolysis) and are triggered by virus–cell interactions, as has been shown for other viruses, reviewed by (Greber et al. 1994; Haywood 2010). In the following paragraphs, we address the likely events that occur after infection and emphasize aspects that still need to be analyzed.

As described in the "*Mechanisms of Filovirus Entry*" chapter in this book, fusion of the viral and cellular membranes during filoviral entry leads to the release of the nucleocapsids into the cytoplasm, while the viral envelope remains attached to the endosomal/lysosomal membrane (Fig. 1a). Thus, two main viral components gain direct access to the cytosol, the nucleocapsid and the inner layer of the viral envelope (Fig. 1b). The inner layer of the filoviral envelope is composed of a regular protein lattice of VP40, which is attached to the inner side of the virus membrane, and the four (EBOV)- or eight (MARV)-amino-acid long cytoplasmic domains of virus surface glycoproteins, GPs, which are the type I transmembrane proteins, reviewed in (Mittler et al. 2013; Hartlieb and Weissenhorn 2006; Schibli and Weissenhorn 2004). The fate of the VP40 lattice and the cytoplasmic domain of GP, which after fusion gain access to the cytosol of the target cell, is not known. Likewise, we do not know what happens to the endosome-associated viral envelope. Studies are needed to determine whether it is delivered to the lysosomes and degraded, recycles to the plasma membrane, or plays a signaling role at the initial stage of host–virus interactions.

The incoming nucleocapsid contains the single-stranded, non-segmented RNA genome of negative polarity and five viral proteins: the nucleoprotein NP, the RNA-dependent RNA polymerase L, the polymerase cofactor VP35, the transcription factor VP30, and VP24. Four nucleocapsid proteins (NP, VP35, L, and VP30) together with viral RNA are necessary and sufficient to support filovirus replication and transcription, as demonstrated using minireplicon systems (Mühlberger et al. 1998, 1999; Weik et al. 2002; Boehmann et al. 2005; Groseth et al. 2005).

The fate and function of the fifth protein (VP24) in the incoming nucleocapsid remain unclear. VP24, often called the minor matrix protein, displays both membrane-binding properties and is implicated in nucleocapsid assembly (Han et al. 2003; Bamberg et al. 2005; Huang et al. 2002; Watanabe et al. 2006).

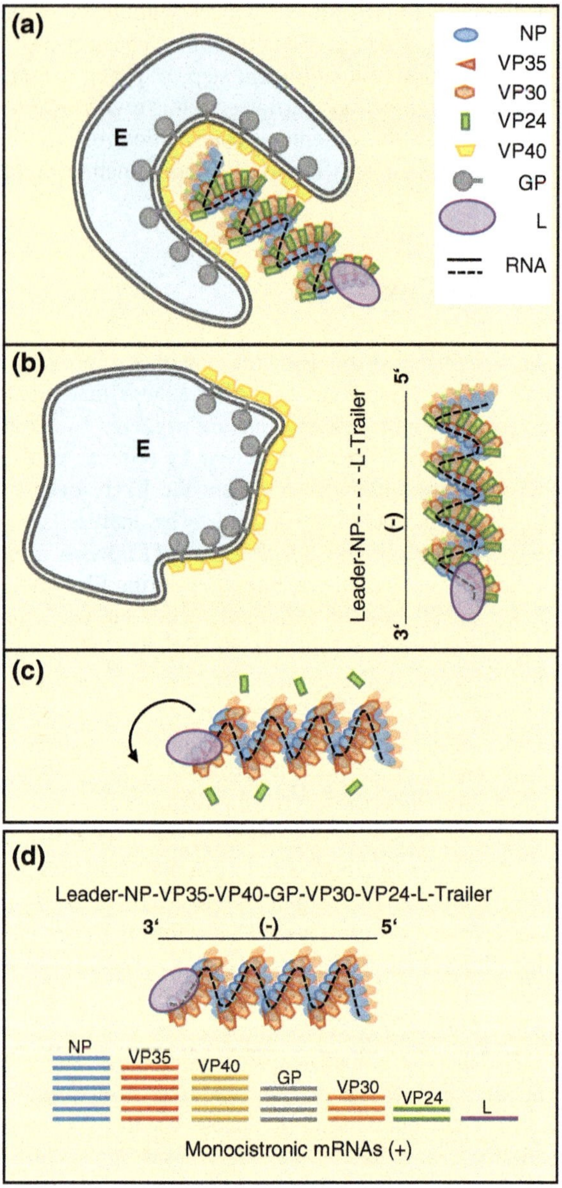

Fig. 1 Hypothetical model of filovirus uncoating and primary transcription. (**a**, **b**) Envelope of incoming viral particle is fused with endosomal (E) membrane and nucleocapsid enters the cytoplasm. (**c**) Dissociation of VP24 and activation of viral polymerase for primary transcription (*round arrow*). (**d**) Primary transcription. The mRNAs for viral proteins are sequentially transcribed from the 3' end of the RNA genome, and transcription terminates at conserved transcription stop signals. It is then reinitiated at nearby conserved transcription start signals. A gradient of mRNAs is produced because reinitiation does not occur in all cases

Analysis of the radial positions of the proteins in the MARV virions detected VP24 in close proximity to NP and VP35 (Bharat et al. 2011), suggesting that VP24 has a closer connection to the nucleocapsid than to the viral envelope, and stays associated with the nucleocapsid after uncoating. Although VP24 is important for nucleocapsid formation (Huang et al. 2002; Noda et al. 2007a; Mateo et al. 2011; Bamberg et al. 2005), it has an inhibitory effect on viral transcription and/or replication in the EBOV minireplicon system (Watanabe et al. 2007; Hoenen et al. 2010b). Therefore, it has been suggested that the dissociation of VP24 from the surface of the incoming nucleocapsid (Fig. 1c) might serve as a signal for the onset of primary transcription (Sanchez et al. 2001; Watt et al. 2014).

3 Primary Transcription

Time course RT-PCR analysis of mRNAs in EBOV-infected cells has determined that the onset of primary transcription occurs between 2 and 4 h post-infection (Hoenen et al. 2012).

During primary transcription, the viral RNA in the incoming nucleocapsid serves as a template, and the RNA-encapsidating proteins NP, VP35, L, and VP30 collaboratively synthesize the novel viral mRNA. During transcription, the seven viral genes are sequentially transcribed into monocistronic mRNAs, which are capped and polyadenylated (Weik et al. 2002; Mühlberger et al. 1996; Sanchez and Kiley 1987). The mRNAs are not encapsidated. As with all negative-sense single-strand RNA viruses, the filoviral polymerase is believed to gain access to the viral genes through a single polymerase binding site at the 3' end of the genome. Once bound, the polymerase complex proceeds along the RNA template by stopping and reinitiating at each gene junction, thereby transcribing the individual genes sequentially in their 3' to 5' order (Whelan et al. 2004). The transcriptional stop signal seems to slow down the polymerase complex, allowing a poly A tail to be added to the nascent mRNA. During this process, the polymerase falls off the template with a certain probability; therefore, not all polymerase complexes reinitiate transcription of the next gene located downstream. Since this process takes place at every gene boundary, it has been suggested that the first gene, NP, should be transcribed at the highest levels, and the last gene, L, should be transcribed at the lowest levels (Fig. 1d) (Mühlberger et al. 1996; Mühlberger 2007). Indeed, recent studies have shown that although mRNA levels for different viral genes increase during the course of the infection, the ratio of NP mRNAs to L mRNAs was almost constantly 30 to 1 at 6, 12, and 24 h post-infection (Shabman et al. 2013, 2014).

Transcription of EBOV is dependent on VP30 and, in particular, on the phosphorylation and oligomerization state of VP30. Non- or weakly phosphorylated VP30 activates transcription of EBOV minigenomes, whereas fully phosphorylated VP30 is inactive in transcription (Modrof et al. 2002, 2003; Mühlberger et al. 1999; Biedenkopf et al. 2013). The use of a reverse genetics system showed that mutations in VP30 phosphorylation domains alter EBOV fitness (Martinez et al. 2011), and non-phosphorylated VP30 is a requisite for transcription reinitiation of downstream genes (Martínez et al. 2008). Recent study showed that phosphorylation and dephosphorylation events are necessary to fully support EBOV transcription (Biedenkopf et al. 2016). EBOV VP30 homo-oligomerizes into hexamers, which can be detected in viral particles (Hartlieb et al. 2007). Oligomerization-deficient VP30 mutants are unable to support transcription (Hartlieb et al. 2003).

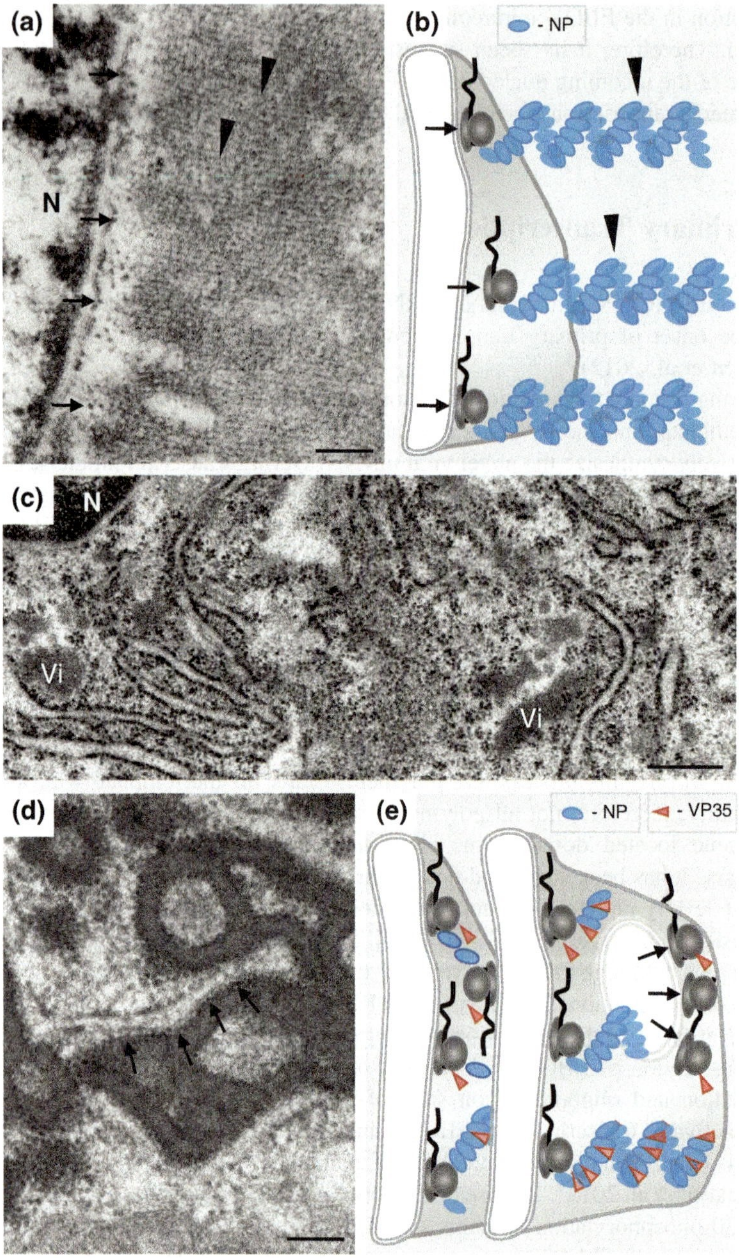
(a)
N
(b)
- NP
(c)
N
Vi
Vi
(d)
(e)
- NP
- VP35

◂**Fig. 2** Spatial organization of NP (or NP and VP35) translation, and perinuclearly located viral inclusion. Morphological evidence of the connection between the synthesis of nucleocapsid proteins and the ER-bound ribosomes. (**a**) Electron micrograph of cells expressing MARV NP. NP helices sectioned along longitudinal axes appear to be located in parallel tubes (*arrowheads*). The NP helices are located near to the outer membrane of the nuclear (N) envelope covered with ribosomes (*arrows*). (**b**) Scheme shows probable association between ER-bound ribosomes (arrows) and synthesis of NP helices (*arrowheads*). (**c**) Electron micrograph of a human macrophage infected with MARV, 24 h post-infection. Viral inclusions (Vi) are surrounded by circular rosettes of polysomes and ER-bound ribosomes, located near nucleus (N). (**d**) Electron micrograph of cells expressing MARV NP and VP35. Electron dense complexes of viral proteins are lined with ER-bound ribosomes (*arrows*). (**e**) Schematic presentation of synthesis of NP and VP35 complexes on the surface of rough ER. Arrows show ribosomes on oblique section of membrane. Bars, (**a**) 200 nm, (**c**) 500 nm, (**d**) 200 nm

In contrast to EBOV VP30, MARV VP30 is not essential for the transcription of MARV minigenomes (Mühlberger et al. 1996). However, the phosphorylation state of MARV VP30 influences the interaction of VP30 with NP-induced inclusions; only phosphorylated MARV VP30 accumulates in NP inclusions (Modrof et al. 2001). Moreover, MARV VP30 was necessary to rescue recombinant MARV in an infectious clone system, suggesting that MARV VP30 is essential for the primary transcription of a full-length genome (Enterlein et al. 2006).

It is unknown whether the primary transcripts are randomly distributed in the cytoplasm, accumulate near incoming nucleocapsids, or are targeted to specific subcellular compartments to produce the viral proteins where they are needed.

4 Spatial Organization of Viral Protein Translation

The earliest possible detection of filoviral protein expression comes from immunofluorescence analyses of NP in infected cells. The first signs of filoviral NP expression were detected at around 6–10 h post-infection (Nanbo et al. 2013; Hoenen et al. 2012; Becker et al. 1998), which is 2–6 h later than the onset of viral mRNA transcription. Remarkably, there is no diffuse distribution of NP at earlier time points of infection. NP signals always appear as small bright spots in the perinuclear region of the infected cells. This observation suggests that viral protein synthesis occurs at distinct sites of the infected cells. Interestingly, of all of the nucleocapsid proteins, only recombinant NP accumulates at distinct perinuclear sites when expressed alone (Noda et al. 2011; Groseth et al. 2009; Dolnik et al. 2010; Becker et al. 1998). Ectopic expression of all of the other nucleocapsid proteins (VP35, VP30, L, and VP24) results in their diffuse intracellular distribution (Becker et al. 1998; Reid et al. 2006; Basler et al. 2003; Mateo et al. 2010; Bamberg et al. 2005; Modrof et al. 2002; Groseth et al. 2009). These findings indicate that an inherent function of NP is to ensure that its synthesis occurs at the perinuclear region. A closer look at NP synthesis by ultrastructural analysis showed that upon single expression, filoviral NP molecules homo-oligomerize to form loose helices (Mavrakis et al. 2002; Noda et al. 2011; Noda et al. 2010; DiCarlo et al. 2007). In ultrathin sections, these NP helices resemble parallel thin-walled tubes when the plane of the section coincides with the longitudinal axis of the helices (Fig. 2a, b). While one end of these tubes is located near to the ER-bound ribosomes, the other end protrudes into the cytoplasm (Noda et al. 2011; Watanabe et al. 2006; Kolesnikova et al. 2000). This suggests that the formation of several NP helices might take place simultaneously from one surface (e.g., the surface of the ER sheet), where newly synthesized NP molecules on the ER-bound ribosomes can bind to the previously synthesized ones. Although we currently have no direct evidence that NP mRNAs are recruited to the ER-bound ribosomes, the distinctiveness of the intracellular localization of NP-induced inclusions as well as viral inclusions (Fig. 2a–c) supports this hypothesis despite NP being a cytosolic protein. A recent study that investigated the ribosome loading of ER-associated mRNAs on a global scale found that the ER is a major site for the translation initiation of various mRNAs, not only those coding for proteins destined for the ER but also for cytosolic proteins (Reid and Nicchitta 2012; Hermesh and Jansen 2013; Weis et al. 2013).

Where does translation of the other nucleocapsid proteins (VP35, L, VP30, and VP24) take place? Upon viral infection, all nucleocapsid proteins are concentrated in the perinuclear region where they are colocalized with NP (Becker et al. 1998; Nanbo et al. 2013). The absence of diffusely distributed VP35, VP30, L, and VP24 during the course of a filoviral infection points to mechanisms of specific localization and timing of translation of the nucleocapsid proteins, which would minimize protein transport expenditure. Such mechanisms are absent when individual nucleocapsid proteins except for NP are ectopically overexpressed. Remarkably, co-expression of recombinant EBOV or MARV NP and VP35 is accompanied by recruitment of ribosomes to the NP- and VP35-induced aggregates whose surfaces are literally covered by ribosomes (Noda et al. 2011; Dolnik et al. 2008). This phenomenon, which is not observed upon expression of NP alone, suggests that either VP35 mRNA contains information for its partitioning to the ER-bound ribosomes in concert with NP mRNA, or that VP35 mRNA, or a complex of NP and VP35 mRNAs, induces recruitment of the cellular translation machinery more efficiently than NP alone. Thus, targeting of NP translation to the ER-bound ribosomes might occur independently of other viral proteins; however, the robustness of this process is increased upon co-translation of NP and VP35. The association of ribosomes with membranes of cells expressing NP and VP35 is not always visible in ultrathin sections most likely for technical reasons (e.g., oblique sections of membranes, Fig. 2d, e).

5 Protein Prerequisites for the Synthesis of Anti-Genomic and Genomic RNA

During replication, the encapsidated viral RNA genome is transcribed into full-length positive-sense replicative intermediates, antigenomes. The antigenomes serve as templates for the production of progeny genomes, reviewed in (Mühlberger

2007). Both the genomes and the antigenomes are encapsidated by the nucleocapsid proteins. A minimum set of nucleocapsid proteins required to support MARV or EBOV replication in a minireplicon system includes NP, VP35, and L (Mühlberger et al. 1998, 1999).

How is genome replication organized in time and space? The onset of viral RNA replication, namely, the appearance of negative-sense genomes, has been detected at 6–12 h post-infection (Nanbo et al. 2013; Hoenen et al. 2012). This is coincident with the onset of filoviral NP expression as described above (Hoenen et al. 2012; Nanbo et al. 2013). This simultaneity of viral protein synthesis and replication is to be expected because the formation of antigenomic and genomic RNA is dependent on the cooperative function of the nucleocapsid proteins (NP, VP35, and L). NP is necessary for RNA encapsidation, VP35 acts as a cofactor for the RNA-dependent RNA polymerase, and L contains the enzymatic functions of the RNA-dependent RNA polymerase (Elliott et al. 1985; Mühlberger et al. 1998, 1999). These three viral proteins have a complicated mode of interaction. NP, VP35, and EBOV L can form homo-oligomers. As stated previously, EBOV and MARV NP, when expressed alone, assembles into long loose helices that associated with non-specific cellular RNA (Mavrakis et al. 2002; Noda et al. 2010). EBOV and MARV VP35 proteins also form homo-oligomers (Moller et al. 2005; Reid et al. 2005) and EBOV L was also shown recently to homo-oligomerize (Trunschke et al. 2013). In contrast, L and NP do not interact directly with each other, but use VP35 as a bridging molecule (Becker et al. 1998; Groseth et al. 2009; Reid et al. 2005). Homo-oligomerization of MARV VP35 is a prerequisite for L binding, and, hence, for a functional polymerase complex (Moller et al. 2005). To provide the specificity of viral RNA transcription/replication, the interaction between NP, VP35, and L must be organized in such a way as to allow attachment of each new component to the forming encapsidated RNA only after conformational changes occur in the preformed protein complex. The existence of some ordered mode of nucleocapsid assembly has been suggested based on the results of a cryo-tomographic study of the MARV nucleocapsid. This study showed that alternate NPs within the helical nucleocapsid are not equivalent; instead, a boomerang-like protrusion was found between every other NP (Bharat et al. 2011). Thus, it has been suggested that this nonequivalence can be introduced at one end of the nucleocapsid concomitant with the synthesis and assembly of the nucleocapsid, and propagated along the helix. This mode of interaction would prevent the homo-oligomerization of NP and the binding of nonspecific RNA nucleotides (which has been observed during the expression of NP alone), and would result in the successive construction of antigenomic or genomic RNA in the 5′–3′ direction enwrapped in NP coupled with VP35 (Fig. 3a). These RNA-NP-VP35 (ribonucleoproteins, RNP) complexes represent the structural and functional basis for the replicative activity of filoviral polymerase L. During the synthesis of encapsidated antigenomic or genomic RNAs, the newly formed RNP complexes need to be close to the template, at distance reachable by the L polymerase. This mode of assembly results in the accumulation of morphologically similar RNP complexes (thin-walled tubes) located close to each other (Fig. 3b, c).

As mentioned above, the site for the onset of viral RNA replication coincides with the site of NP synthesis in the perinuclearly located small spots (Nanbo et al. 2013; Hoenen et al. 2012). The presence of an incoming nucleocapsid, as a

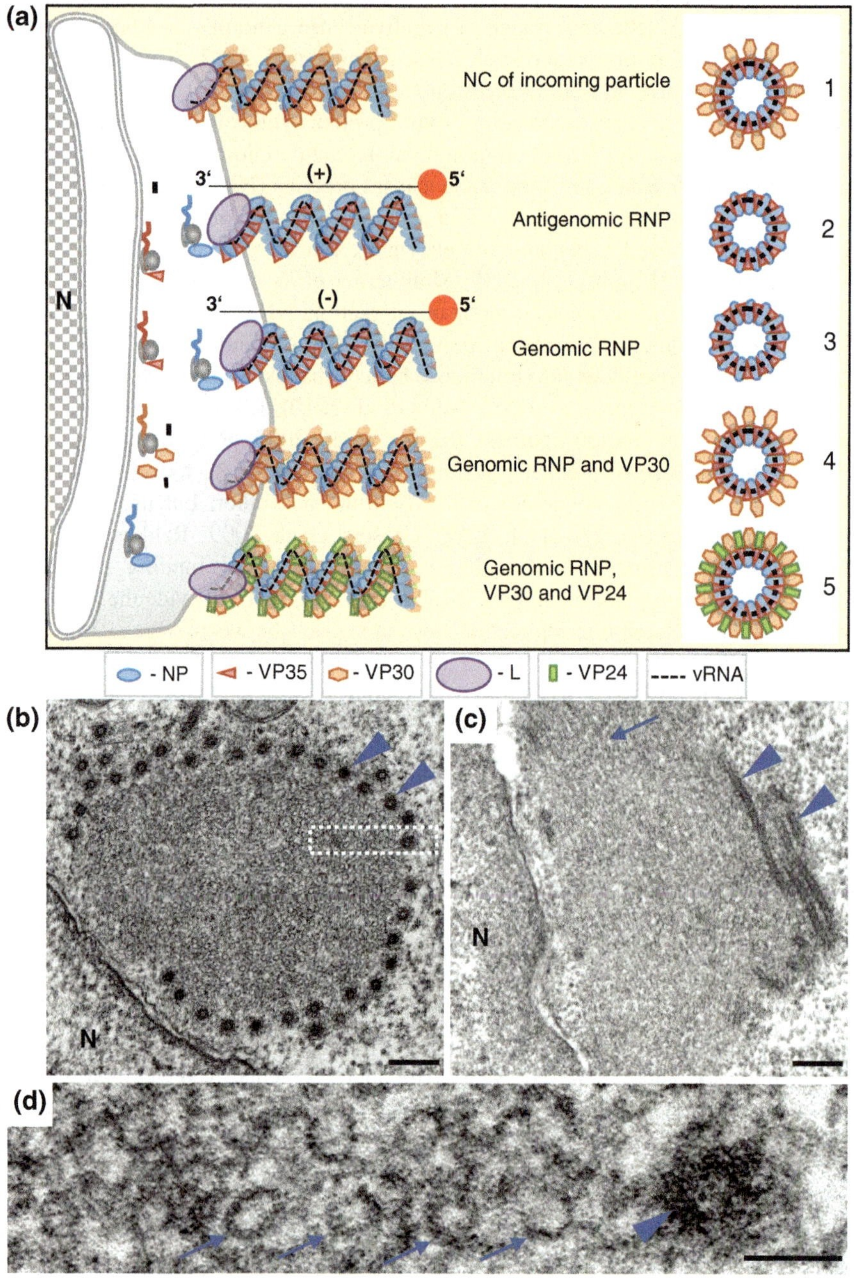

◂**Fig. 3** Replication of the viral genome and formation of viral inclusions. (**a**) Schematic presentation of viral antigenomic and genomic RNP synthesis, and of nucleocapsid maturation. The encapsidated viral RNA in the incoming particle (*1*) is copied to produce a full-length positive-sense replicative intermediate, the antigenomic RNP (*2*). The antigenomic RNP is used as template for the production of the progeny genome (*3*). *Red circles* indicate the site of initiation of RNA synthesis from *5′* to the *3′* end. Both the antigenomic and the genomic RNAs are encapsidated by NP and VP35 (*2* and *3*). Binding of VP30 and VP24 to the nucleocapsid (*4* and *5*) makes wall of nucleocapsid thicker. (**b** and **c**) Electron micrographs of MARV viral inclusions with cross-sectioned (**b**) or longitudinally sectioned nucleocapsids (**c**). *Arrowheads* indicate thick-walled tubes (mature nucleocapsids), *blue arrows* indicate thin-walled tubes (immature nucleocapsids). (**d**) panel shows the boxed area of (**b**) at higher magnification. Bars, (**b** and **c**) 200 nm; (**d**) 50 nm

template, is absolutely necessary for the onset of replication, which means that the incoming nucleocapsid must be targeted to the perinuclear region, close to the rough ER membrane, and anchored there. The mechanisms that provide such targeting and anchoring remain unknown. The incoming nucleocapsid might be delivered to the perinuclear region within the endosome, reviewed in (Miller and Chandran 2012). Interestingly, EBOV VP24 colocalizes with Sec61a, a component of the heterotrimeric Sec61 complex that promotes protein translocation at the ER membranes (Iwasa et al. 2011). Iwasa and coauthors found that depletion of Sec61a reduces the polymerase activity even in the absence of VP24, suggesting that VP24 might provide a connection for the incoming nucleocapsid to the ER membrane and modulate the viral polymerase activity by dissociating from the incoming nucleocapsid.

6 Assembly of Mature Nucleocapsids

The term “mature” has been used for completely formed and released filovirus progeny and for viral inclusions (Geisbert and Jahrling 1995); however, the morphological changes of filoviral nucleocapsids that occur during the course of infection prompted us to also use this term for nucleocapsids (Noda et al. 2006; Kolesnikova et al. 2000). In ultrathin sections, the structure of cross-sectioned newly synthesized genomic RNA encapsidated with the NP-VP35-L complex has the appearance of a tube with an electron translucent central core (around 20 nm in diameter), and an electron dense thin wall, approximately 5-nm thick (Fig. 3b, c). These thin-walled nucleocapsids were termed immature. In contrast, nucleocapsid inside a filoviral particle has the appearance of a tube (50–55 nm in diameter) with a wall of high electron density, or, in another words, a thick wall (Fig. 3b–d). Morphologically similar nucleocapsids were detected in the cytoplasm near the plasma membrane, or during budding in filoviral-infected cells. These thick-walled nucleocapsids were named “mature” in contrast to the thin-walled nucleocapsids, many of which were present in the perinuclear region during the earlier stages of infection (Kolesnikova et al. 2000).

According to cryotomographic data, nucleocapsid within filoviral particles has a left-handed helical structure with polarity induced by characteristic boomerang-like protrusions located at an angle to the helical axis, so-called "barbed" and "pointed" ends by analogy with actin filaments (Bharat et al. 2011, 2012; Noda et al. 2005). Left-handed helical nucleocapsid, together with polarity, is a common feature of Mononegavirales (Egelman et al. 1989; Ge et al. 2010; Schoehn et al. 2004; Bakker et al. 2013; Desfosses et al. 2011), although a right-handed nucleocapsid has been reported for EBOV also (Booth et al. 2013; Beniac et al. 2012). Filoviral NPs, in particular, the N-termini of NPs can form left-handed helices, and therefore NP is considered to be a main determinant of the helical structure of the nucleocapsid (Bharat et al. 2011, 2012; Noda et al. 2005). A comparison of a MARV nucleocapsid 3D reconstruction with immunogold labeling data on the radial positions of the viral proteins suggests that the innermost density of the MARV nucleocapsid corresponds to NP, and the boomerang-like protrusions correspond to VP24 and VP35 (Bharat et al. 2011). The positions of VP30 and L remain unclear for both EBOV and MARV mature nucleocapsids.

The molecular mechanisms that accompany the maturation of nucleocapsids, or the morphologically observed changes in the thickness of the nucleocapsid wall, include the binding of VP30 and VP24 to the immature nucleocapsid and their oligomerization. Several studies have shown that the set of expressed nucleocapsid proteins influence the wall thickness of the nucleocapsid-like structures. For example, singly expressed NP formed thin-walled tubes, whereas co-expression of NP with other nucleocapsid proteins (VP35, VP24, VP30) resulted in thick-walled tubes that looked similar to the nucleocapsids detected in virus-infected cells in the pre-budding stage (Watanabe et al. 2006; Huang et al. 2002; Dolnik et al. 2008). Remarkably, thick-walled tubes were found at the edge of the thin-walled NP-induced tubes, suggesting that binding of VP24 molecules to the preformed NP helices occurs in succession from one end of helix (Noda et al. 2006, 2011). These observations suggest that directionality of nucleocapsid assembly might represent a common mechanism that is maintained not only during encapsidation of the viral RNA, but also during later stages of nucleocapsid maturation.

The sequence, or spatiotemporal steps of VP30 and VP24 binding to the immature nucleocapsid remain unclear. EBOV VP30 may interact with NP by its N- and C-terminal domains (Hartlieb et al. 2007), and, as mentioned earlier, is necessary for the transcription and replication of the viral genome (Biedenkopf et al. 2013; Martínez et al. 2008). VP24 is important for nucleocapsid formation, although it associates loosely with RNPs (Kiley et al. 1988; Huang et al. 2002; Bamberg et al. 2005; Noda et al. 2007a; Han et al. 2003; Mateo et al. 2011). As mentioned previously, VP24 has an inhibitory effect on viral transcription and/or replication in the EBOV minireplicon system (Watanabe et al. 2007; Hoenen et al. 2010b). Based on these observations, the binding of VP24 to the immature nucleocapsid appears to occur after the binding of VP30.

Thus, a hypothetical reconstruction of nucleocapsid maturation might include: First (i) the formation of an immature nucleocapsid that represents the viral genomic RNA encapsidated with the NP-VP35-L protein complex with a left-handed helical structure (Fig. 3a). Then, (ii) binding of VP30 to the immature nucleocapsid, which might make thin-walled tubes thicker, regulating the function of the RNP complex either as a transcriptase or replicase depending on the VP30 phosphorylation state (Biedenkopf et al. 2013). Last, (iii) binding of VP24, which stabilizes the nucleocapsid complex and probably locks transcription and replication (Watanabe et al. 2007; Hoenen et al. 2006, 2010b), and hence prepares the nucleocapsid for transport to the budding site.

7 Formation of Perinuclear Viral Inclusions

As noted above, filoviral inclusions appear at 6–12 h post-infection as small NP-positive spots located in the perinuclear region. Time course studies, based on immunofluorescence protein staining, viral RNA detection in fixed cells, and live cell imaging analyses of infected cells, demonstrated that these small filoviral inclusions represent the site of virus replication (Becker et al. 1998; Nanbo et al. 2013; Hoenen et al. 2012). Inclusions could increase in size and/or fuse with one another to form larger inclusions with a pleomorphic shape. At the same time, many new small spots appear distal from the nucleus as the infection progresses (Nanbo et al. 2013; Hoenen et al. 2012).

At the ultrastructural level, filoviral inclusions appear as aggregates of thin- and thick-walled nucleocapsids located close to the nucleus or rough ER (Geisbert and Jahrling 1995; Kolesnikova et al. 2000), see also (Figs. 2c and 3b, c). Depending on the plane of the sections, nucleocapsids can appear to be located in parallel tubes (Fig. 3c), or as hexagonally arranged circles (Fig. 3b, d). Sometimes viral inclusions look like amorphous electron dense material (Fig. 2c). Mature nucleocapsids might be located either at the periphery of viral inclusions or inside (Fig. 3b, c), and the number of mature nucleocapsids inside the viral inclusions increases during the course of the infection (Geisbert and Jahrling 1995).

All inclusions remain in the perinuclear region and display a sway-like movement, which is not influenced by depolymerization of actin filaments, but can be inhibited by depolymerization of microtubules (Hoenen et al. 2012). In highly motile macrophages, the shape of filoviral inclusions often changes in line with the cellular shape; in round cells, viral inclusions are predominantly round and in elongated cells they are drawn out along the long axis (Fig. 4) (Kolesnikova et al. 2002). During mitosis, EBOV viral inclusions are transformed from a condensed to a dispersed pattern that allows the equal distribution of the encapsidated viral RNAs into the daughter cells, in which viral inclusions are again transformed to perinuclear condensed inclusions (Hoenen et al. 2012). This synchronization of viral inclusion transformation with cytokinesis or cell movement suggests that a delicate connection exists between viral inclusions and perinuclearly located cellular structures that provides the viral inclusions with maneuverability yet does not limit their growth. One of the possible mechanisms, which still need to be tested, might be the anchoring of viral protein synthesis to the ER-bound ribosomes. Rough ER is

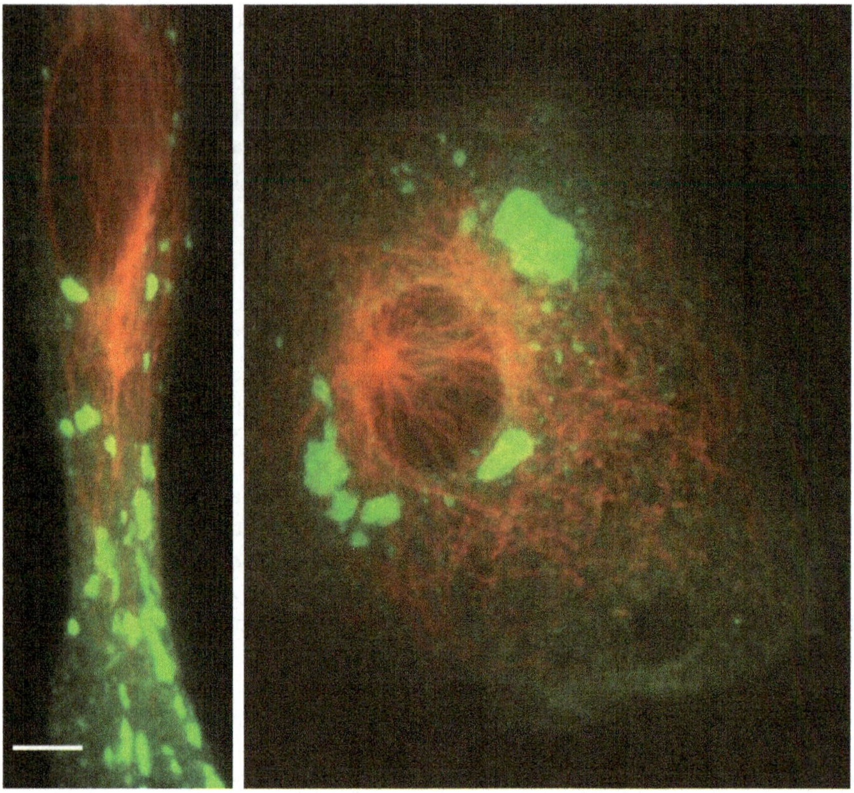

Fig. 4 Viral inclusions. An elongated and a round human macrophage infected with MARV, 2 days post-infection. Viral inclusions (*green*), microtubules (*red*). Bar, 10 μm

composed of the nuclear envelope and perinuclearly located sheet-like cisternae with relatively flat areas where the membrane extends for many microns with little membrane curvature (Shibata et al. 2006). ER distribution and the sheet/tubule balance are influenced by the cytoskeleton, mainly the microtubules (Gurel et al. 2014), and ER reorganization during mitosis ensures the equal distribution of this single interconnected network into the daughter cells (Puhka et al. 2007, 2012).

Further correlative light-electron studies of the location of viral RNAs and nucleocapsid proteins are needed to better understand the mechanisms of formation and transformation of viral inclusions. Many questions remain open. For example, what is the fate of the antigenomic RNP complexes? Degradation, accumulation, or sorting away from the viral inclusions? Is the hexagonal order of nucleocapsids inside the viral inclusions induced by dense packing (Starostin 2006), or by a special location order of the newly formed encapsidated viral RNA? If this location order exists, does it induce gradual nucleocapsid redistribution, for example, to the periphery of the viral inclusions during the course of their maturation?

8 Transport of Nucleocapsids to the Sites of Budding

Viral inclusions remain in the perinuclear region during the course of infection; therefore, mature nucleocapsids must cross the cytoplasm to reach the budding sites at the plasma membrane. It has been shown that only when filoviral nucleocapsid proteins are co-expressed with the viral matrix protein VP40, can they be found beneath the plasma membrane and inside filamentous VLPs released into the supernatant (Hoenen et al. 2006, 2011; Spiegelberg et al. 2011; Wenigenrath et al. 2010; Watanabe et al. 2004; Dolnik et al. 2008; Noda et al. 2006). On the basis of these observations, it is believed that the interaction of nucleocapsids with VP40 is required to make them competent for transport.

The ability of filoviral VP40 to reach the plasma membrane and induce the formation of filamentous VLPs is well known (Timmins et al. 2001; Noda et al. 2002; Kolesnikova et al. 2004a; Swenson et al. 2004; Hartlieb and Weissenhorn 2006); however, the molecular mechanisms responsible for VP40 transport to the plasma membrane remain enigmatic. EBOV and MARV VP40s are peripheral membrane proteins (Jasenosky et al. 2001; Kolesnikova et al. 2002). In cells expressing VP40, approximately 80% of the VP40 molecules became membrane associated under steady state conditions, and were found on the cytosolic side of the plasma membrane, of small vesicles, and of the multivesicular bodies (MVBs) (Kolesnikova et al. 2004a). In filovirus-infected cells, VP40 was detected in multiple cellular compartments, but more often VP40 displayed a weak diffuse distribution pattern in the cytoplasm with a more intense signal in the perinuclear structures and bright distinct signals at the cell periphery. VP40 can be seen sometimes inside the nucleus (Bjorndal et al. 2003; Nanbo et al. 2013; Kolesnikova et al. 2002) as reported for many paramyxovirus matrix proteins, reviewed in Harrison et al. (2010). Essentially, bright distinct VP40 signals at the cell periphery represent higher order oligomers of VP40 beneath patches or subdomains of plasma membrane but not beneath the whole cell surface. Targeting to these subdomains was clearly apparent in studies of polarized cells where MARV VP40 was targeted to the patches of the basolateral domain (Kolesnikova et al. 2007b). Analyses of ectopically expressed filoviral VP40s indicate that they might have a connection to various components of the cellular transport systems (Fig. 5a), including actin (Han and Harty 2005; Adu-Gyamfi et al. 2012; Kolesnikova et al. 2007a), IQGAP1 (Lu et al. 2013), Sec24C (Yamayoshi et al. 2008), and microtubules (Ruthel et al. 2005; Noda et al. 2006). The list of components involved in VP40 trafficking might be even longer if one considers that VP40 has been shown to colocalize with the cholesterol-enriched plasma membrane domain (Panchal et al. 2003; Koehler et al. 2015) suggesting a potential association between VP40 transport and cholesterol trafficking, reviewed in (King and Sharom 2012; Du and Yang 2013).

How is the co-transport of VP40 and mature nucleocapsids regulated? What mechanisms provide the specificity and timing of the VP40 interactions with the viral proteins, and with different intracellular membranes and proteins? The answers to the questions remain obscure. Structural and functional studies have shown that

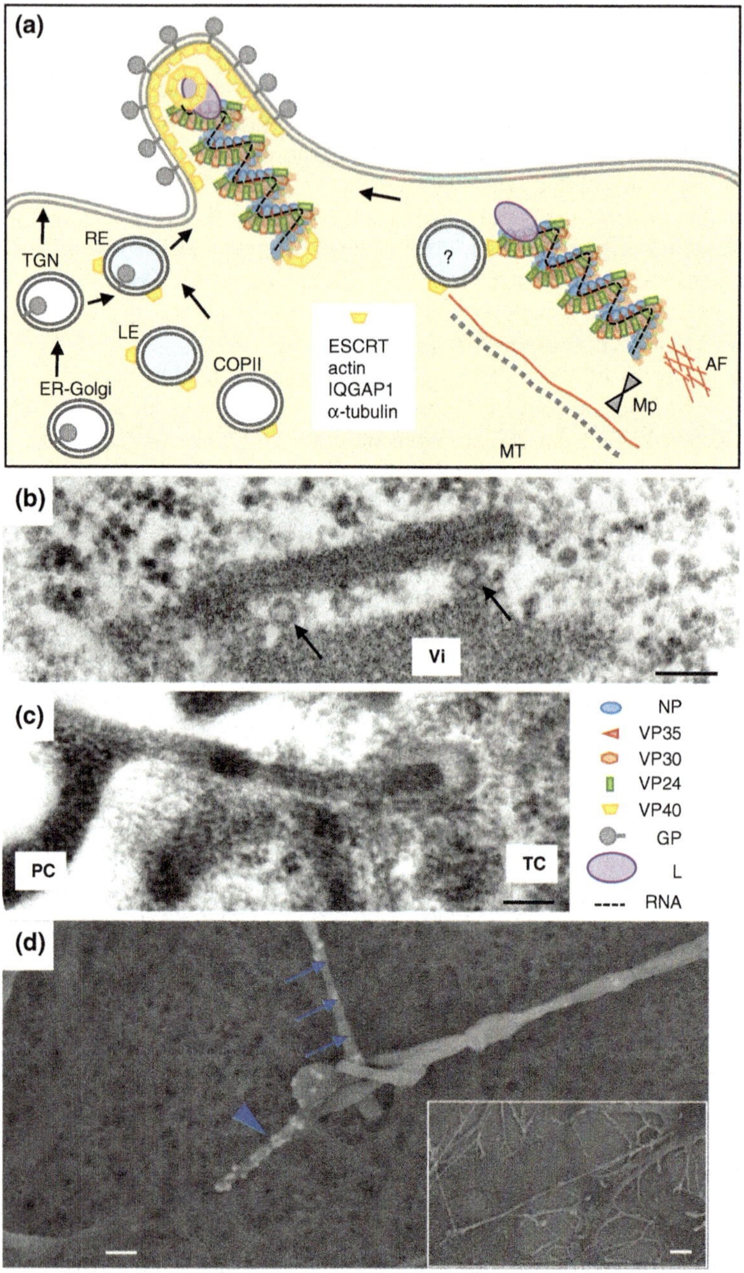
(a)
RE
TGN
ER-Golgi
LE
COPII
?
ESCRT
actin
IQGAP1
α-tubulin
AF
Mp
MT
(b)
Vi
(c)
PC
TC
NP
VP35
VP30
VP24
VP40
GP
L
RNA
(d)

◂**Fig. 5** Transport of filovirus nucleocapsid, VP40 and GP to the site of budding, and mode of interaction between infected and target cells. (**a**) Schematic presentation of the transport routes and cellular interacting partners. GP is transported along the secretory pathway (ER, Golgi, TGN, trans-Golgi), VP40 is transported along the retrograde late endosomal (LE) pathway, or in association with COPII vesicles. GP and VP40 might be sorted into intracellular recycling endosomal (RE) compartment before delivery to the plasma membrane. In addition, VP40 might interact with the cellular proteins listed in the white box. Nucleocapsid is transported with the "pointed" end first either along cytoskeletal filaments (AF, actin filaments, MT, microtubules) with the help of motor proteins (Mp), or propulsed by actin polymerization at the rear end of the nucleocapsid (comet-like actin tail). The association of the nucleocapsid with the small vesicles (**b**) might influence the efficiency of nucleocapsid transport. (**b**) Two small vesicles (*arrows*) located between the mature nucleocapsid and the viral inclusions (Vi). Ultrathin section of MARV-infected Vero cells, 48 h post-infection. (**c**) Contact between a target cell (TC) and a MARV virion located at the tip of a filopodium and still connected to the producer cell (PC). Ultrathin section of MARV infected Huh-7 cells, 24 h post-infection. (**d**) Contact between MARV-infected and noninfected Huh-7 cells imaged by scanning electron microscopy and immunogold labeling of MARV GP. The infected cell and virus particle are recognized by gold particles (*white spots*), whereas the noninfected cells do not have gold-labeled GP on the surface. Six-shaped virion (*arrowhead*) located at the tip of the filopodium (*arrows*) is lassoed by the long filopodium of the uninfected cell. Inset shows this contact at a lower magnification. Bars, (b and c) 100 nm; (d) 200 nm, inset, 1000 nm

VP40 adopts different conformations, reviewed in (Hartlieb and Weissenhorn 2006; Stahelin 2014). In addition, VP40s can undergo tyrosine phosphorylation (Kolesnikova et al. 2012; García et al. 2012), and ubiquitination (Okumura et al. 2008), and can be sorted to cholesterol-enriched membranes (Bavari et al. 2002; Panchal et al. 2003; Koehler et al. 2015). Therefore, the regulation of VP40 co-transport with mature nucleocapsids might be based on the different VP40 conformations and posttranslational modifications, each of which occurs after the interplay of VP40 with a specific protein, lipid, or RNA, and each of which allows VP40 to perform a distinct function. For example, the interaction of an EBOV VP40 construct (31–326 aa) with liposomes in vitro induced a conformational change in VP40 that resulted in hexamerization (Scianimanico et al. 2000).

Although filoviral VP40s have been detected inside viral inclusions (Geisbert and Jahrling 1995; Kolesnikova et al. 2000; Schudt et al. 2013), only tiny amounts of MARV VP40 have been found in association with nucleocapsids during transport by using immunogold labeling (Kolesnikova et al. 2002). This amount of VP40 most likely was under the detection level for live cell imaging analysis, which showed that nucleocapsids leaving viral inclusions do not contain VP40, and only nucleocapsids located close to the plasma membrane colocalized with VP40 (Schudt et al. 2013). Remarkably, EBOV VP40 octamers can bind RNA (Gomis-Ruth et al. 2003; Hoenen et al. 2005), and the RNA-binding activity of VP40 octamers is essential for the negative regulation of viral transcription in the minireplicon system (Hoenen et al. 2010a; Bornholdt et al. 2013). Are VP40 octamers present in transport-competent nucleocapsids in virus-infected cells? And if so, are they located at the tips of the nucleocapsid (Fig. 5a)? These questions need to be answered.

Interestingly, filoviral filamentous nucleocapsids are always transported along their long axis; lateral transport has not been detected (Schudt et al. 2013; Bharat et al. 2011). This mode of transport suggests that the site of connection between the nucleocapsids and the cellular transport machinery is not located along the lateral surface, but is close to the nucleocapsid tips. Small vesicles located near the nucleocapsid tips were detected on ultrathin sections of infected cells (Fig. 5b); however, the protein composition of these vesicles and their role, if any, in nucleocapsid transport remain unknown (Fig. 5a).

Nucleocapsid movement in and release of particles from virus-infected cells has been shown to be predominantly actin-based with different transport modes and velocities (Kolesnikova et al. 2007a; Schudt et al. 2013). From viral inclusions to the cell periphery, nucleocapsids are transported either over long distances along actin filaments (speed approx. 200 nm/s), or over short distances (switching from one actin filament to another with a speed of approx. 400 nm/s) (Schudt et al. 2013, 2015). A comet-like actin tail has sometimes been observed at the rear end of nucleocapsids, suggesting that nucleocapsids can be propulsed by actin polymerization. In one study, at the cell periphery, nucleocapsids colocalized with a bright VP40 signal, were recruited to filopodia, and then co-transported with host motor protein myosin-10 toward the budding sites at the tip or side of the filopodia with a low velocity of approx. 100 nm/s (Schudt et al. 2013).

Further live cell imaging studies are needed to clarify the transport mechanisms used by filoviral nucleocapsids.

9 Budding

Envelope formation. Filoviral nucleocapsids that come close to the plasma membrane are enveloped by a lipid bilayer that contains a regular lattice of VP40 on the cytosolic side and inserted viral surface GP, reviewed in (Mittler et al. 2013; Hartlieb and Weissenhorn 2006; Schibli and Weissenhorn 2004). Remarkably, the formation of the filoviral envelope can occur upon co-expression of VP40 and GP, and is accompanied by the redistribution of GP to the VP40-enriched patches of the plasma membrane (Kolesnikova et al. 2004b; Noda et al. 2002). Evidence of GP sorting into the VP40-enriched membrane (but not vice versa) has been obtained in polarized cells, where singly expressed GP accumulated at the apical domain, and singly expressed VP40 formed VP40-enriched membrane patches at the basolateral domain. Upon co-expression of both proteins, approx. 20% of the GP was redistributed to the VP40-enriched patches at the basolateral domain where MARV budding takes place (Sanger et al. 2001; Kolesnikova et al. 2007b). A study on the targeting of chimeric GP constructs containing different combinations of ecto-, transmembrane, and cytoplasmic domains of MARV GP and Lassavirus surface protein GPC to the VP40-enriched membrane patches showed that the transmembrane domain is essential and sufficient for the recruitment of GP into the VP40-enriched membrane (Mittler et al. 2007). Interestingly, EBOV and MARV viral surface glycoproteins can be targeted to the heterologous VP40-enriched membrane, and form hybrid VLPs (Warfield et al. 2005). VLPs containing homologous or heterologous VP40 and GP have been shown to provide protection from filovirus infection upon VLP immunization (Warfield et al. 2003, 2004; Swenson et al. 2005). The mechanisms responsible for GP sorting into the VP40-enriched membrane remain a mystery. The characteristics of the transmembrane domain of GP, such as its acylation, length, and hydrophobicity (Sharpe et al. 2010; Shao and Hegde 2011), as well as the likelihood that the VP40 and GP transport routes intersect, can be considered important factors. Surprisingly, the budding of filoviruses has been detected not only at the plasma membrane, but also into the intracellular membrane with a multivesicular appearance (Feldmann et al. 1996; Kolesnikova et al. 2004b), which might represent a point of intersection of the VP40 and GP transport routes.

Both conventional ultrastructural studies and more advanced cryoelectron tomography analyses show an "empty" space between the nucleocapsids and the filoviral envelope, lacking proteins (Geisbert and Jahrling 1995; Beniac et al. 2012; Bharat et al. 2011). Moreover, the ordered VP40 matrix layer does not follow the helical symmetry of the nucleocapsid, suggesting that the association of the viral envelope with the nucleocapsid takes place through flexible interactions (Geisbert and Jahrling 1995; Beniac et al. 2012; Bharat et al. 2011). These interactions restrict the radial position of the VP40 layer and the associated membrane, but do not precisely define the lateral position of VP40 (Beniac et al. 2012; Bharat et al. 2011). Such flexible interactions can mediate envelopment of a preformed nucleocapsid by a preformed VP40 lattice through a Velcro-like interaction. The VP40 layer may grab the nucleocapsid via the C-terminus of NP, which has been shown to be important for the incorporation of EBOV NP into VP40-derived VLPs (Licata et al. 2004; Noda et al. 2007b). Remarkably, a comparison of the VP40 lattice in VLPs and MARV by using cryotomography showed that the VP40 layers are arranged differently, and only within the virion does the VP40 layer contain features that repeat regularly around the circumference of the virion (Bharat et al. 2011).

Envelopment of nucleocapsids and directionality. The coating of filovirus nucleocapsids with VP40- and GP-enriched membrane occurs predominantly at the plasma membrane (Geisbert and Jahrling 1995). MARV budding has been detected at the tip or side of long thin filopodia-like extensions (Welsch et al. 2010; Kolesnikova et al. 2007a). A cryoelectron tomographic study of MARV-infected cells revealed a clear directionality of nucleocapsid transport inside filopodia and during envelopment (Bharat et al. 2011). The study found that the beginning of the envelopment of MARV nucleocapsids was specifically associated with the pointed end of the helical nucleocapsid, and nucleocapsids located inside the filopodia had a preferential orientation with the pointed end facing away from the cell (Bharat et al. 2011). These data suggest that "factors" responsible for the transport of the nucleocapsid and for the initiation of envelopment are associated with the pointed end and not with the whole surface of the nucleocapsid. What factors determine the directionality of MARV budding remains an open question. It is essential to mention that the recruitment of filovirus NP or RNP complexes into filamentous

VLPs is genus-specific (Spiegelberg et al. 2011), and impaired when the RNP complexes or VP40 are derived from different filovirus genera.

The directionality of MARV nucleocapsid transport and budding has been detected at 22 h post-infection, corresponding to the time required for one replication cycle (Bharat et al. 2011). Whether this directionality is preserved at 2 or 3 days after infection needs to be elucidated. It may be that at later stages, when infected cells are overcrowded with viral proteins, another mode of envelopment takes place. For example, a horizontal mode of budding has been suggested for EBOV nucleocapsids based on transmission and scanning electron microscopy examination of EBOV-infected cells at 2 days post-infection (Noda et al. 2006).

Fission and/or cell-to-cell spread. Mechanisms of filoviral particle fission and the probable involvement of the endosomal sorting complex required for transport (ESCRT)-associated proteins has been extensively studied. ESCRT complexes are essential to induce inward budding and pinching off of vesicles into MVBs, reviewed in (Henne et al. 2011). Inward budding of internal MVBs is topologically similar to virus budding, and several virus systems have shown that the interaction between ESCRT-associated proteins and so-called "late domains" in viral proteins can induce or even enable the budding of viral particles (Bieniasz 2006; Calistri et al. 2009). Filoviral VP40s contain late domain motifs: 16-PPPY-19 in MARV VP40, and two overlapping late domains 7-PTAPPEY-13 in EBOV VP40. These domains have been shown to mediate interactions with the ESCRT-associated proteins Nedd4 and Tsg101 (Timmins et al. 2003; Martin-Serrano et al. 2001; Harty et al. 2000; Urata et al. 2007), and inhibition of ESCRT function has been shown to affect VP40 VLP formation (Licata et al. 2003; Yasuda et al. 2003; Urata and Yasuda 2010). However, mutations in the VP40 late domains or inhibition of the ESCRT machinery do not completely inhibit the release of filoviruses into the extracellular space. For example, recombinant EBOVs containing mutations in one or both of the VP40 late domains were attenuated by one log unit (Neumann et al. 2005), and the budding of MARV in cells expressing a dominant-negative vacuolar protein sorting-associated protein 4 (VPS4) was also only partially inhibited (Kolesnikova et al. 2009). These data suggest that the vacuolar protein sorting pathway is only one of several so far unknown mechanisms that drive filoviral budding.

We must keep in mind that filovirus transmission from cell to cell might occur not only via free extracellular particles, but alternatively, the virus may be passed on at contacts established between infected and noninfected target cells (Fig. 5c, d). A recent study of recombinant MARV containing a mutation in the NP late domain demonstrated that the level of the attenuation of these two modes of virus transmission might be different (Dolnik et al. 2014). It has been shown for other human pathogenic viruses that cell-to-cell spread facilitates rapid viral dissemination, promotes immune evasion, and influences disease outcomes, reviewed in (Sattentau 2008). The potential impact of these two routes of transmission should be taken into consideration in future studies on filovirus pathogenesis.

This review summarizes our current understanding of the relationship between the cellular compartments and filovirus assembly. While some details of this relationship have recently emerged, the regulation of the filovirus assembly process remains unclear; the viral and cellular signals that link one step of virus assembly to the next have yet to be identified. Further development of systems that model the different steps of filovirus assembly, together with new imaging techniques, will allow us to tackle questions regarding how posttranslational modifications of viral proteins and variations in their composition influence their specific interactions with host cell proteins and provide spatiotemporal coordination of virus assembly. On the basis of current knowledge, we present hypotheses that might stimulate the field and direct future research.

References

Adu-Gyamfi E, Digman MA, Gratton E, Stahelin RV (2012) Single-particle tracking demonstrates that actin coordinates the movement of the Ebola virus matrix protein. Biophys J 103(9): L41–43. doi:10.1016/j.bpj.2012.09.026

Aman MJ, Bosio CM, Panchal RG, Burnett JC, Schmaljohn A, Bavari S (2003) Molecular mechanisms of filovirus cellular trafficking. Microbes Infect 5(7):639–649

Ascenzi P, Bocedi A, Heptonstall J, Capobianchi MR, Di Caro A, Mastrangelo E, Bolognesi M, Ippolito G (2008) Ebolavirus and Marburgvirus: insight the Filoviridae family. Mol Aspects Med 29(3):151–185. doi:10.1016/j.mam.2007.09.005

Bakker SE, Duquerroy S, Galloux M, Loney C, Conner E, Eléouët JF, Rey FA, Bhella D (2013) The respiratory syncytial virus nucleoprotein-RNA complex forms a left-handed helical nucleocapsid. J Gen Virol 94(Pt 8):1734–1738. doi:10.1099/vir.0.053025-0

Bamberg S, Kolesnikova L, Moller P, Klenk HD, Becker S (2005) VP24 of Marburg virus influences formation of infectious particles. J Virol 79(21):13421–13433

Basler CF, Mikulasova A, Martinez-Sobrido L, Paragas J, Mühlberger E, Bray M, Klenk HD, Palese P, García-Sastre A (2003) The Ebola virus VP35 protein inhibits activation of interferon regulatory factor 3. J Virol 77(14):7945–7956

Bavari S, Bosio CM, Wiegand E, Ruthel G, Will AB, Geisbert TW, Hevey M, Schmaljohn C, Schmaljohn A, Aman MJ (2002) Lipid raft microdomains: a gateway for compartmentalized trafficking of Ebola and Marburg viruses. J Exp Med 195(5):593–602

Becker S, Rinne C, Hofsass U, Klenk HD, Mühlberger E (1998) Interactions of Marburg virus nucleocapsid proteins. Virology 249(2):406–417

Beer B, Kurth R, Bukreyev A (1999) Characteristics of Filoviridae: Marburg and Ebola viruses. Naturwissenschaften 86(1):8–17

Beniac DR, Melito PL, Devarennes SL, Hiebert SL, Rabb MJ, Lamboo LL, Jones SM, Booth TF (2012) The organisation of Ebola virus reveals a capacity for extensive, modular polyploidy. PLoS ONE 7(1):e29608. doi:10.1371/journal.pone.0029608

Bharat TA, Riches JD, Kolesnikova L, Welsch S, Krähling V, Davey N, Parsy ML, Becker S, Briggs JA (2011) Cryo-electron tomography of Marburg virus particles and their morphogenesis within infected cells. PLoS Biol 9(11):e1001196. doi:10.1371/journal.pbio.1001196

Bharat TA, Noda T, Riches JD, Kraehling V, Kolesnikova L, Becker S, Kawaoka Y, Briggs JA (2012) Structural dissection of Ebola virus and its assembly determinants using cryo-electron tomography. Proc Natl Acad Sci USA 109(11):4275–4280. doi:10.1073/pnas.1120453109

Biedenkopf N, Hartlieb B, Hoenen T, Becker S (2013) Phosphorylation of Ebola virus VP30 influences the composition of the viral nucleocapsid complex: impact on viral transcription and replication. J Biol Chem 288(16):11165–11174. doi:10.1074/jbc.M113.461285

Biedenkopf N, Lier C, Becker S (2016) Dynamic phosphorylation of VP30 Is essential for Ebola Virus life cycle. J Virol 90(10):4914–4925. doi:10.1128/jvi.03257-15

Bieniasz PD (2006) Late budding domains and host proteins in enveloped virus release. Virology 344(1):55–63

Bjorndal AS, Szekely L, Elgh F (2003) Ebola virus infection inversely correlates with the overall expression levels of promyelocytic leukaemia (PML) protein in cultured cells. BMC Microbiol 3(1):6

Boehmann Y, Enterlein S, Randolf A, Mühlberger E (2005) A reconstituted replication and transcription system for Ebola virus Reston and comparison with Ebola virus Zaire. Virology 332(1):406–417. doi:10.1016/j.virol.2004.11.018

Booth TF, Rabb MJ, Beniac DR (2013) How do filovirus filaments bend without breaking? Trends Microbiol 21(11):583–593. doi:10.1016/j.tim.2013.08.001

Bornholdt ZA, Noda T, Abelson DM, Halfmann P, Wood MR, Kawaoka Y, Saphire EO (2013) Structural rearrangement of Ebola Virus VP40 begets multiple functions in the virus life cycle. Cell 154(4):763–774. doi:10.1016/j.cell.2013.07.015

Calistri A, Salata C, Parolin C, Palù G (2009) Role of multivesicular bodies and their components in the egress of enveloped RNA viruses. Rev Med Virol 19(1):31–45. doi:10.1002/rmv.588

Desfosses A, Goret G, Farias Estrozi L, Ruigrok RW, Gutsche I (2011) Nucleoprotein-RNA orientation in the measles virus nucleocapsid by three-dimensional electron microscopy. J Virol 85(3):1391–1395. doi:10.1128/jvi.01459-10

DiCarlo A, Möller P, Lander A, Kolesnikova L, Becker S (2007) Nucleocapsid formation and RNA synthesis of Marburg virus is dependent on two coiled coil motifs in the nucleoprotein. Virol J 4:105. doi:10.1186/1743-422x-4-105

Dolnik O, Kolesnikova L, Becker S (2008) Filoviruses: interactions with the host cell. Cell Mol Life Sci 65(5):756–776. doi:10.1007/s00018-007-7406-2

Dolnik O, Kolesnikova L, Stevermann L, Becker S (2010) Tsg101 is recruited by a late domain of the nucleocapsid protein to support budding of Marburg virus-like particles. J Virol 84(15): 7847–7856. doi:10.1128/jvi.00476-10

Dolnik O, Kolesnikova L, Welsch S, Strecker T, Schudt G, Becker S (2014) Interaction with tsg101 is necessary for the efficient transport and release of nucleocapsids in marburg virus-infected cells. PLoS Pathog 10(10):e1004463. doi:10.1371/journal.ppat.1004463

Du X, Yang H (2013) Endosomal cholesterol trafficking: protein factors at a glance. Acta Biochim Biophys Sin (Shanghai) 45(1):11–17. doi:10.1093/abbs/gms095

Egelman EH, Wu SS, Amrein M, Portner A, Murti G (1989) The Sendai virus nucleocapsid exists in at least four different helical states. J Virol 63(5):2233–2243

Elliott LH, Kiley MP, McCormick JB (1985) Descriptive analysis of Ebola virus proteins. Virology 147(1):169–176

Ellis DS, Stamford S, Lloyd G, Bowen ET, Platt GS, Way H, Simpson DI (1979) Ebola and Marburg viruses: I. Some ultrastructural differences between strains when grown in Vero cells. J Med Virol 4(3):201–211

Enterlein S, Volchkov V, Weik M, Kolesnikova L, Volchkova V, Klenk HD, Mühlberger E (2006) Rescue of recombinant Marburg virus from cDNA is dependent on nucleocapsid protein VP30. J Virol 80(2):1038–1043. doi:10.1128/jvi.80.2.1038-1043.2006

Feldmann H, Kiley MP (1999) Classification, structure, and replication of filoviruses. Curr Top Microbiol Immunol 235:1–21

Feldmann H, Bugany H, Mahner F, Klenk HD, Drenckhahn D, Schnittler HJ (1996) Filovirus-induced endothelial leakage triggered by infected monocytes/macrophages. J Virol 70(4):2208–2214

García M, Cooper A, Shi W, Bornmann W, Carrion R, Kalman D, Nabel GJ (2012) Productive replication of Ebola virus is regulated by the c-Abl1 tyrosine kinase. Sci Transl Med 4(123): 123ra124. doi:10.1126/scitranslmed.3003500

Ge P, Tsao J, Schein S, Green TJ, Luo M, Zhou ZH (2010) Cryo-EM model of the bullet-shaped vesicular stomatitis virus. Science 327(5966):689–693. doi:10.1126/science.1181766

Geisbert TW, Jahrling PB (1995) Differentiation of filoviruses by electron microscopy. Virus Res 39(2–3):129–150

Gomis-Ruth FX, Dessen A, Timmins J, Bracher A, Kolesnikowa L, Becker S, Klenk HD, Weissenhorn W (2003) The matrix protein VP40 from Ebola virus octamerizes into pore-like structures with specific RNA binding properties. Structure (Camb) 11(4):423–433

Greber UF, Singh I, Helenius A (1994) Mechanisms of virus uncoating. Trends Microbiol 2(2): 52–56

Groseth A, Feldmann H, Theriault S, Mehmetoglu G, Flick R (2005) RNA polymerase I-driven minigenome system for Ebola viruses. J Virol 79(7):4425–4433. doi:10.1128/jvi.79.7.4425-4433.2005

Groseth A, Charton JE, Sauerborn M, Feldmann F, Jones SM, Hoenen T, Feldmann H (2009) The Ebola virus ribonucleoprotein complex: a novel VP30-L interaction identified. Virus Res 140(1–2):8–14. doi:10.1016/j.virusres.2008.10.017

Gurel PS, Hatch AL, Higgs HN (2014) Connecting the Cytoskeleton to the Endoplasmic Reticulum and Golgi. Curr Biol 24(14):R660–R672. doi:10.1016/j.cub.2014.05.033

Han Z, Harty RN (2005) Packaging of actin into Ebola virus VLPs. Virol J 2(1):92

Han Z, Boshra H, Sunyer JO, Zwiers SH, Paragas J, Harty RN (2003) Biochemical and functional characterization of the Ebola virus VP24 protein: implications for a role in virus assembly and budding. J Virol 77(3):1793–1800

Harrison MS, Sakaguchi T, Schmitt AP (2010) Paramyxovirus assembly and budding: building particles that transmit infections. Int J Biochem Cell Biol 42(9):1416–1429. doi:10.1016/j.biocel.2010.04.005

Hartlieb B, Weissenhorn W (2006) Filovirus assembly and budding. Virology 344(1):64–70. doi:10.1016/j.virol.2005.09.018

Hartlieb B, Modrof J, Mühlberger E, Klenk H-D, Becker S (2003) Oligomerization of Ebola virus VP30 is essential for viral transcription and can be inhibited by a synthetic peptide. J Biol Chem 278(43):41830–41836

Hartlieb B, Muziol T, Weissenhorn W, Becker S (2007) Crystal structure of the C-terminal domain of Ebola virus VP30 reveals a role in transcription and nucleocapsid association. Proc Natl Acad Sci USA 104(2):624–629. doi:10.1073/pnas.0606730104

Harty RN, Brown ME, Wang G, Huibregtse J, Hayes FP (2000) A PPxY motif within the VP40 protein of Ebola virus interacts physically and functionally with a ubiquitin ligase: implications for filovirus budding. Proc Natl Acad Sci USA 97(25):13871–13876

Haywood AM (2010) Membrane uncoating of intact enveloped viruses. J Virol 84(21):10946–10955. doi:10.1128/jvi.00229-10

Henne WM, Buchkovich NJ, Emr SD (2011) The ESCRT pathway. Dev Cell 21(1):77–91. doi:10.1016/j.devcel.2011.05.015

Hermesh O, Jansen RP (2013) Take the (RN) A-train: localization of mRNA to the endoplasmic reticulum. Biochim Biophys Acta 1833(11):2519–2525. doi:10.1016/j.bbamcr.2013.01.013

Hoenen T, Volchkov V, Kolesnikova L, Mittler E, Timmins J, Ottmann M, Reynard O, Becker S, Weissenhorn W (2005) VP40 Octamers are essential for ebola virus replication. J Virol 79(3): 1898–1905

Hoenen T, Groseth A, Kolesnikova L, Theriault S, Ebihara H, Hartlieb B, Bamberg S, Feldmann H, Ströher U, Becker S (2006) Infection of naive target cells with virus-like particles: implications for the function of ebola virus VP24. J Virol 80(14):7260–7264. doi:10.1128/jvi.00051-06

Hoenen T, Biedenkopf N, Zielecki F, Jung S, Groseth A, Feldmann H, Becker S (2010a) Oligomerization of Ebola virus VP40 is essential for particle morphogenesis and regulation of viral transcription. J Virol 84(14):7053–7063. doi:10.1128/jvi.00737-10

Hoenen T, Jung S, Herwig A, Groseth A, Becker S (2010b) Both matrix proteins of Ebola virus contribute to the regulation of viral genome replication and transcription. Virology 403(1):56–66. doi:10.1016/j.virol.2010.04.002

Hoenen T, Groseth A, de Kok-Mercado F, Kuhn JH, Wahl-Jensen V (2011) Minigenomes, transcription and replication competent virus-like particles and beyond: reverse genetics systems for filoviruses and other negative stranded hemorrhagic fever viruses. Antiviral Res 91(2):195–208. doi:10.1016/j.antiviral.2011.06.003

Hoenen T, Shabman RS, Groseth A, Herwig A, Weber M, Schudt G, Dolnik O, Basler CF, Becker S, Feldmann H (2012) Inclusion bodies are a site of ebolavirus replication. J Virol 86(21):11779–11788. doi:10.1128/jvi.01525-12

Huang Y, Xu L, Sun Y, Nabel GJ (2002) The assembly of Ebola virus nucleocapsid requires virion-associated proteins 35 and 24 and posttranslational modification of nucleoprotein. Mol Cell 10(2):307–316

Iwasa A, Halfmann P, Noda T, Oyama M, Kozuka-Hata H, Watanabe S, Shimojima M, Watanabe T, Kawaoka Y (2011) Contribution of Sec61α to the life cycle of Ebola virus. J Infect Dis 204(Suppl 3):S919–926. doi:10.1093/infdis/jir324

Jasenosky LD, Kawaoka Y (2004) Filovirus budding. Virus Res 106(2):181–188

Jasenosky LD, Neumann G, Lukashevich I, Kawaoka Y (2001) Ebola virus VP40-induced particle formation and association with the lipid bilayer. J Virol 75(11):5205–5214

Kiley MP, Cox NJ, Elliott LH, Sanchez A, DeFries R, Buchmeier MJ, Richman DD, McCormick JB (1988) Physicochemical properties of Marburg virus: evidence for three distinct virus strains and their relationship to Ebola virus. J Gen Virol 69(Pt 8):1957–1967

King G, Sharom FJ (2012) Proteins that bind and move lipids: MsbA and NPC1. Crit Rev Biochem Mol Biol 47(1):75–95. doi:10.3109/10409238.2011.636505

Koehler A, Kolesnikova L, Welzel U, Schudt G, Herwig A, Becker S (2015) A single amino acid change in the Marburg virus matrix protein VP40 provides a replicative advantage in a species-specific manner. J Virol. doi:10.1128/jvi.02670-15

Kolesnikova L, Muhlberger E, Ryabchikova E, Becker S (2000) Ultrastructural organization of recombinant Marburg virus nucleoprotein: comparison with Marburg virus inclusions. J Virol 74(8):3899–3904

Kolesnikova L, Bugany H, Klenk HD, Becker S (2002) VP40, the matrix protein of Marburg virus, is associated with membranes of the late endosomal compartment. J Virol 76(4): 1825–1838

Kolesnikova L, Bamberg S, Berghofer B, Becker S (2004a) The matrix protein of Marburg virus is transported to the plasma membrane along cellular membranes: exploiting the retrograde late endosomal pathway. J Virol 78(5):2382–2393

Kolesnikova L, Berghofer B, Bamberg S, Becker S (2004b) Multivesicular bodies as a platform for formation of the Marburg virus envelope. J Virol 78(22):12277–12287

Kolesnikova L, Bohil AB, Cheney RE, Becker S (2007a) Budding of Marburgvirus is associated with filopodia. Cell Microbiol 9(4):939–951

Kolesnikova L, Ryabchikova E, Shestopalov A, Becker S (2007b) Basolateral budding of Marburg virus: VP40 retargets viral glycoprotein GP to the basolateral surface. J Infect Dis 196(Suppl 2):S232–236. doi:10.1086/520584

Kolesnikova L, Strecker T, Morita E, Zielecki F, Mittler E, Crump C, Becker S (2009) Vacuolar protein sorting pathway contributes to the release of Marburg virus. J Virol 83(5):2327–2337. doi:10.1128/jvi.02184-08

Kolesnikova L, Mittler E, Schudt G, Shams-Eldin H, Becker S (2012) Phosphorylation of Marburg virus matrix protein VP40 triggers assembly of nucleocapsids with the viral envelope at the plasma membrane. Cell Microbiol 14(2):182–197. doi:10.1111/j.1462-5822.2011.01709.x

Licata JM, Simpson-Holley M, Wright NT, Han Z, Paragas J, Harty RN (2003) Overlapping motifs (PTAP and PPEY) within the Ebola virus VP40 protein function independently as late budding domains: involvement of host proteins TSG101 and VPS-4. J Virol 77(3):1812–1819

Licata JM, Johnson RF, Han Z, Harty RN (2004) Contribution of ebola virus glycoprotein, nucleoprotein, and VP24 to budding of VP40 virus-like particles. J Virol 78(14):7344–7351

Lu J, Qu Y, Liu Y, Jambusaria R, Han Z, Ruthel G, Freedman BD, Harty RN (2013) Host IQGAP1 and Ebola virus VP40 interactions facilitate virus-like particle egress. J Virol 87(13): 7777–7780. doi:10.1128/jvi.00470-13

Martinez MJ, Volchkova VA, Raoul H, Alazard-Dany N, Reynard O, Volchkov VE (2011) Role of VP30 phosphorylation in the Ebola virus replication cycle. J Infect Dis 204(Suppl 3):S934–940. doi:10.1093/infdis/jir320

Martínez MJ, Biedenkopf N, Volchkova V, Hartlieb B, Alazard-Dany N, Reynard O, Becker S, Volchkov V (2008) Role of Ebola virus VP30 in transcription reinitiation. J Virol 82(24): 12569–12573. doi:10.1128/jvi.01395-08

Martin-Serrano J, Zang T, Bieniasz PD (2001) HIV-1 and Ebola virus encode small peptide motifs that recruit Tsg101 to sites of particle assembly to facilitate egress. Nat Med 7(12):1313–1319

Mateo M, Reid SP, Leung LW, Basler CF, Volchkov VE (2010) Ebolavirus VP24 binding to karyopherins is required for inhibition of interferon signaling. J Virol 84(2):1169–1175. doi:10.1128/jvi.01372-09

Mateo M, Carbonnelle C, Martinez MJ, Reynard O, Page A, Volchkova VA, Volchkov VE (2011) Knockdown of Ebola virus VP24 impairs viral nucleocapsid assembly and prevents virus replication. J Infect Dis 204(Suppl 3):S892–896. doi:10.1093/infdis/jir311

Mavrakis M, Kolesnikova L, Schoehn G, Becker S, Ruigrok RW (2002) Morphology of Marburg virus NP-RNA. Virology 296(2):300–307

Miller EH, Chandran K (2012) Filovirus entry into cells - new insights. Curr Opin Virol 2(2):206–214. doi:10.1016/j.coviro.2012.02.015

Mittler E, Kolesnikova L, Strecker T, Garten W, Becker S (2007) Role of the transmembrane domain of marburg virus surface protein GP in assembly of the viral envelope. J Virol 81(8): 3942–3948

Mittler E, Kolesnikova L, Herwig A, Dolnik O, Becker S (2013) Assembly of the Marburg virus envelope. Cell Microbiol 15(2):270–284. doi:10.1111/cmi.12076

Modrof J, Möritz C, Kolesnikova L, Konakova T, Hartlieb B, Randolf A, Mühlberger E, Becker S (2001) Phosphorylation of Marburg virus VP30 at serines 40 and 42 is critical for its interaction with NP inclusions. Virology 287(1):171–182. doi:10.1006/viro.2001.1027

Modrof J, Muhlberger E, Klenk HD, Becker S (2002) Phosphorylation of VP30 impairs ebola virus transcription. J Biol Chem 277(36):33099–33104

Modrof J, Becker S, Muehlberger E (2003) Ebola virus transcription activator VP30 is a zinc-binding protein. J Virol 77(5):3334–3338

Moller P, Pariente N, Klenk HD, Becker S (2005) Homo-oligomerization of Marburgvirus VP35 is essential for its function in replication and transcription. J Virol 79(23):14876–14886. doi:10.1128/JVI.79.23.14876-14886.2005

Mühlberger E (2007) Filovirus replication and transcription. Future Virol 2(2):205–215. doi:10.2217/17460794.2.2.205

Mühlberger E, Trommer S, Funke C, Volchkov V, Klenk HD, Becker S (1996) Termini of all mRNA species of Marburg virus: sequence and secondary structure. Virology 223(2):376–380

Mühlberger E, Lotfering B, Klenk HD, Becker S (1998) Three of the four nucleocapsid proteins of Marburg virus, NP, VP35, and L, are sufficient to mediate replication and transcription of Marburg virus-specific monocistronic minigenomes. J Virol 72(11):8756–8764

Mühlberger E, Weik M, Volchkov VE, Klenk HD, Becker S (1999) Comparison of the transcription and replication strategies of marburg virus and Ebola virus by using artificial replication systems. J Virol 73(3):2333–2342

Murphy FA, van der Groen G, Whiteflied SG, Lange JV (1978) Ebola and Marburg virus morphology and taxonomy. In: S. P (ed) Ebola virus haemorrhagic fever. Elsevier/North Holland, Amsterdam, pp 61–84

Nanbo A, Watanabe S, Halfmann P, Kawaoka Y (2013) The spatio-temporal distribution dynamics of Ebola virus proteins and RNA in infected cells. Sci Rep 3:1206. doi:10.1038/srep01206

Neumann G, Ebihara H, Takada A, Noda T, Kobasa D, Jasenosky LD, Watanabe S, Kim JH, Feldmann H, Kawaoka Y (2005) Ebola virus VP40 late domains are not essential for viral replication in cell culture. J Virol 79(16):10300–10307

Noda T, Sagara H, Suzuki E, Takada A, Kida H, Kawaoka Y (2002) Ebola virus VP40 drives the formation of virus-like filamentous particles along with GP. J Virol 76(10):4855–4865

Noda T, Aoyama K, Sagara H, Kida H, Kawaoka Y (2005) Nucleocapsid-like structures of Ebola virus reconstructed using electron tomography. J Vet Med Sci 67(3):325–328

Noda T, Ebihara H, Muramoto Y, Fujii K, Takada A, Sagara H, Kim JH, Kida H, Feldmann H, Kawaoka Y (2006) Assembly and budding of Ebolavirus. PLoS Pathog 2(9):e99. doi:10.1371/journal.ppat.0020099

Noda T, Halfmann P, Sagara H, Kawaoka Y (2007a) Regions in Ebola virus VP24 that are important for nucleocapsid formation. J Infect Dis 196(Suppl 2):S247–250. doi:10.1086/520596

Noda T, Watanabe S, Sagara H, Kawaoka Y (2007b) Mapping of the VP40-binding regions of the nucleoprotein of Ebola virus. J Virol 81(7):3554–3562. doi:10.1128/jvi.02183-06

Noda T, Hagiwara K, Sagara H, Kawaoka Y (2010) Characterization of the Ebola virus nucleoprotein-RNA complex. J Gen Virol 91(Pt 6):1478–1483. doi:10.1099/vir.0.019794-0

Noda T, Kolesnikova L, Becker S, Kawaoka Y (2011) The importance of the NP: VP35 ratio in Ebola virus nucleocapsid formation. J Infect Dis 204(Suppl 3):S878–883. doi:10.1093/infdis/jir310

Okumura A, Pitha PM, Harty RN (2008) ISG15 inhibits Ebola VP40 VLP budding in an L-domain-dependent manner by blocking Nedd4 ligase activity. Proc Natl Acad Sci USA 105 (10):3974–3979. doi:10.1073/pnas.0710629105

Olejnik J, Ryabchikova E, Corley RB, Mühlberger E (2011) Intracellular events and cell fate in filovirus infection. Viruses 3(8):1501–1531. doi:10.3390/v3081501

Panchal RG, Ruthel G, Kenny TA, Kallstrom GH, Lane D, Badie SS, Li L, Bavari S, Aman MJ (2003) In vivo oligomerization and raft localization of Ebola virus protein VP40 during vesicular budding. Proc Natl Acad Sci USA 100(26):15936–15941

Peters D, Müller G, Slenczka WG (1971) Morphology, development, and classification of the Marburg virus. Marburg virus disease. Springer, Berlin

Puhka M, Vihinen H, Joensuu M, Jokitalo E (2007) Endoplasmic reticulum remains continuous and undergoes sheet-to-tubule transformation during cell division in mammalian cells. J Cell Biol 179(5):895–909. doi:10.1083/jcb.200705112

Puhka M, Joensuu M, Vihinen H, Belevich I, Jokitalo E (2012) Progressive sheet-to-tubule transformation is a general mechanism for endoplasmic reticulum partitioning in dividing mammalian cells. Mol Biol Cell 23(13):2424–2432. doi:10.1091/mbc.E10-12-0950

Reid DW, Nicchitta CV (2012) Primary role for endoplasmic reticulum-bound ribosomes in cellular translation identified by ribosome profiling. J Biol Chem 287(8):5518–5527. doi:10.1074/jbc.M111.312280

Reid SP, Cárdenas WB, Basler CF (2005) Homo-oligomerization facilitates the interferon-antagonist activity of the ebolavirus VP35 protein. Virology 341(2):179–189. doi:10.1016/j.virol.2005.06.044

Reid SP, Leung LW, Hartman AL, Martinez O, Shaw ML, Carbonnelle C, Volchkov VE, Nichol ST, Basler CF (2006) Ebola virus VP24 binds karyopherin alpha1 and blocks STAT1 nuclear accumulation. J Virol 80(11):5156–5167. doi:10.1128/jvi.02349-05

Ruthel G, Demmin GL, Kallstrom G, Javid MP, Badie SS, Will AB, Nelle T, Schokman R, Nguyen TL, Carra JH, Bavari S, Aman MJ (2005) Association of ebola virus matrix protein VP40 with microtubules. J Virol 79(8):4709–4719. doi:10.1128/jvi.79.8.4709-4719.2005

Sanchez A, Kiley MP (1987) Identification and analysis of Ebola virus messenger RNA. Virology 157(2):414–420

Sanchez A, Khan AS, Zaki SR, Nabel GJ, Ksiazek TG, Peters CJ (2001) Filoviridae: Marburg and Ebola viruses. In: Knipe DM, Howley PM (eds) Fields virology (4th edn). Lippincott, Williams & Wilkins, Philadelphia

Sanger C, Muhlberger E, Ryabchikova E, Kolesnikova L, Klenk HD, Becker S (2001) Sorting of Marburg virus surface protein and virus release take place at opposite surfaces of infected polarized epithelial cells. J Virol 75(3):1274–1283

Sattentau Q (2008) Avoiding the void: cell-to-cell spread of human viruses. Nat Rev Microbiol 6(11):815–826. doi:10.1038/nrmicro1972

Schibli DJ, Weissenhorn W (2004) Class I and class II viral fusion protein structures reveal similar principles in membrane fusion. Mol Membr Biol 21(6):361–371. doi:10.1080/09687860400017784

Schoehn G, Mavrakis M, Albertini A, Wade R, Hoenger A, Ruigrok RW (2004) The 12 A structure of trypsin-treated measles virus N-RNA. J Mol Biol 339(2):301–312. doi:10.1016/j.jmb.2004.03.073

Schudt G, Kolesnikova L, Dolnik O, Sodeik B, Becker S (2013) Live-cell imaging of Marburg virus-infected cells uncovers actin-dependent transport of nucleocapsids over long distances. Proc Natl Acad Sci USA 110(35):14402–14407. doi:10.1073/pnas.1307681110

Schudt G, Dolnik O, Kolesnikova L, Biedenkopf N, Herwig A, Becker S (2015) Transport of Ebolavirus nucleocapsids is dependent on actin polymerization: live-cell imaging analysis of Ebolavirus-infected cells. J Infect Dis 212(Suppl 2):S160–166. doi:10.1093/infdis/jiv083

Scianimanico S, Schoehn G, Timmins J, Ruigrok RH, Klenk HD, Weissenhorn W (2000) Membrane association induces a conformational change in the Ebola virus matrix protein. EMBO J 19(24):6732–6741

Shabman RS, Hoenen T, Groseth A, Jabado O, Binning JM, Amarasinghe GK, Feldmann H, Basler CF (2013) An upstream open reading frame modulates ebola virus polymerase translation and virus replication. PLoS Pathog 9(1):e1003147. doi:10.1371/journal.ppat.1003147

Shabman RS, Jabado OJ, Mire CE, Stockwell TB, Edwards M, Mahajan M, Geisbert TW, Basler CF (2014) Deep sequencing identifies noncanonical editing of Ebola and Marburg Virus RNAs in infected cells. MBio 5(6). doi:10.1128/mBio.02011-14

Shao S, Hegde RS (2011) Membrane protein insertion at the endoplasmic reticulum. Annu Rev Cell Dev Biol 27:25–56. doi:10.1146/annurev-cellbio-092910-154125

Sharpe HJ, Stevens TJ, Munro S (2010) A comprehensive comparison of transmembrane domains reveals organelle-specific properties. Cell 142(1):158–169. doi:10.1016/j.cell.2010.05.037

Shibata Y, Voeltz GK, Rapoport TA (2006) Rough sheets and smooth tubules. Cell 126(3):435–439. doi:10.1016/j.cell.2006.07.019

Spiegelberg L, Wahl-Jensen V, Kolesnikova L, Feldmann H, Becker S, Hoenen T (2011) Genus-specific recruitment of filovirus ribonucleoprotein complexes into budding particles. J Gen Virol 92(Pt 12):2900–2905. doi:10.1099/vir.0.036863-0

Stahelin RV (2014) Membrane binding and bending in Ebola VP40 assembly and egress. Front Microbiol 5:300. doi:10.3389/fmicb.2014.00300

Starostin EL (2006) On the perfect hexagonal packing of rods. J Phys: Condens Matter 18(14): S187–S204

Swenson DL, Warfield KL, Kuehl K, Larsen T, Hevey MC, Schmaljohn A, Bavari S, Aman MJ (2004) Generation of Marburg virus-like particles by co-expression of glycoprotein and matrix protein. FEMS Immunol Med Microbiol 40(1):27–31

Swenson DL, Warfield KL, Negley DL, Schmaljohn A, Aman MJ, Bavari S (2005) Virus-like particles exhibit potential as a pan-filovirus vaccine for both Ebola and Marburg viral infections. Vaccine 23(23):3033–3042. doi:10.1016/j.vaccine.2004.11.070

Timmins J, Scianimanico S, Schoehn G, Weissenhorn W (2001) Vesicular release of ebola virus matrix protein VP40. Virology 283(1):1–6

Timmins J, Schoehn G, Ricard-Blum S, Scianimanico S, Vernet T, Ruigrok RW, Weissenhorn W (2003) Ebola virus matrix protein VP40 interaction with human cellular factors Tsg101 and Nedd4. J Mol Biol 326(2):493–502

Trunschke M, Conrad D, Enterlein S, Olejnik J, Brauburger K, Mühlberger E (2013) The L-VP35 and L-L interaction domains reside in the amino terminus of the Ebola virus L protein and are potential targets for antivirals. Virology 441(2):135–145. doi:10.1016/j.virol.2013.03.013

Urata S, Yasuda J (2010) Regulation of Marburg virus (MARV) budding by Nedd4.1: a different WW domain of Nedd4.1 is critical for binding to MARV and Ebola virus VP40. J Gen Virol 91 (Pt 1):228–234. doi:10.1099/vir.0.015495-0

Urata S, Noda T, Kawaoka Y, Morikawa S, Yokosawa H, Yasuda J (2007) Interaction of Tsg101 with Marburg Virus VP40 depends on the PPPY Motif, but not the PT/SAP motif as in the case of Ebola Virus, and Tsg101 plays a critical role in the budding of Marburg virus-like particles induced by VP40, NP, and GP. J Virol 81(9):4895–4899

Warfield KL, Bosio CM, Welcher BC, Deal EM, Mohamadzadeh M, Schmaljohn A, Aman MJ, Bavari S (2003) Ebola virus-like particles protect from lethal Ebola virus infection. Proc Natl Acad Sci USA 100(26):15889–15894

Warfield KL, Swenson DL, Negley DL, Schmaljohn AL, Aman MJ, Bavari S (2004) Marburg virus-like particles protect guinea pigs from lethal Marburg virus infection. Vaccine 22(25–26): 3495–3502. doi:10.1016/j.vaccine.2004.01.063

Warfield KL, Swenson DL, Demmin G, Bavari S (2005) Filovirus-like particles as vaccines and discovery tools. Expert Rev Vaccines 4(3):429–440. doi:10.1586/14760584.4.3.429

Watanabe S, Watanabe T, Noda T, Takada A, Feldmann H, Jasenosky LD, Kawaoka Y (2004) Production of novel Ebola virus-like particles from cDNAs: an alternative to Ebola Virus generation by reverse genetics. J Virol 78(2):999–1005

Watanabe S, Noda T, Kawaoka Y (2006) Functional mapping of the nucleoprotein of Ebola virus. J Virol 80(8):3743–3751. doi:10.1128/jvi.80.8.3743-3751.2006

Watanabe S, Noda T, Halfmann P, Jasenosky L, Kawaoka Y (2007) Ebola virus (EBOV) VP24 inhibits transcription and replication of the EBOV genome. J Infect Dis 196(Suppl 2):S284–290. doi:10.1086/520582

Watt A, Moukambi F, Banadyga L, Groseth A, Callison J, Herwig A, Ebihara H, Feldmann H, Hoenen T (2014) A novel life cycle modeling system for Ebola virus shows a genome length-dependent role of VP24 in virus infectivity. J Virol 88(18):10511–10524. doi:10.1128/jvi.01272-14

Weik M, Modrof J, Klenk HD, Becker S, Mühlberger E (2002) Ebola virus VP30-mediated transcription is regulated by RNA secondary structure formation. J Virol 76(17):8532–8539

Weis BL, Schleiff E, Zerges W (2013) Protein targeting to subcellular organelles via MRNA localization. Biochim Biophys Acta 1833(2):260–273

Welsch S, Kolesnikova L, Krähling V, Riches JD, Becker S, Briggs JA (2010) Electron tomography reveals the steps in filovirus budding. PLoS Pathog 6(4):e1000875. doi:10.1371/journal.ppat.1000875

Wenigenrath J, Kolesnikova L, Hoenen T, Mittler E, Becker S (2010) Establishment and application of an infectious virus-like particle system for Marburg virus. J Gen Virol 91(Pt 5): 1325–1334. doi:10.1099/vir.0.018226-0

Whelan SP, Barr JN, Wertz GW (2004) Transcription and replication of nonsegmented negative-strand RNA viruses. Curr Top Microbiol Immunol 283:61–119

Yamayoshi S, Noda T, Ebihara H, Goto H, Morikawa Y, Lukashevich IS, Neumann G, Feldmann H, Kawaoka Y (2008) Ebola virus matrix protein VP40 uses the COPII transport system for its intracellular transport. Cell Host Microbe 3(3):168–177. doi:10.1016/j.chom.2008.02.001

Yasuda J, Nakao M, Kawaoka Y, Shida H (2003) Nedd4 regulates egress of Ebola virus-like particles from host cells. J Virol 77(18):9987–9992

Filovirus Structural Biology: The Molecules in the Machine

Robert N. Kirchdoerfer, Hal Wasserman, Gaya K. Amarasinghe and Erica Ollmann Saphire

Abstract In this chapter, we describe what is known thus far about the structures and functions of the handful of proteins encoded by filovirus genomes. Amongst the fascinating findings of the last decade is the plurality of functions and structures that these polypeptides can adopt. Many of the encoded proteins can play multiple, distinct roles in the virus life cycle, although the mechanisms by which these functions are determined and controlled remain mostly veiled. Further, some filovirus proteins are multistructural: adopting different oligomeric assemblies and sometimes, different tertiary structures to achieve their separate, and equally essential functions. Structures, and the functions they dictate, are described for components of the nucleocapsid, the matrix, and the surface and secreted glycoproteins.

Contents

R.N. Kirchdoerfer · H. Wasserman
Department of Immunology and Microbiology, The Scripps Research Institute, La Jolla, CA 92037, USA

G.K. Amarasinghe (✉)
Department of Pathology and Immunology, Washington University School of Medicine, St. Louis, MO 63110, USA
e-mail: gamarasinghe@wustl.edu

E.O. Saphire (✉)
Department of Immunology and Microbiology, The Scripps Research Institute, The Skaggs Institute for Chemical Biology, La Jolla, CA 92037, USA
e-mail: erica@scripps.edu

Current Topics in Microbiology and Immunology (2017) 411:381–418
DOI 10.1007/82_2017_16

Published Online: 10 August 2017

1 Introduction

In this chapter, we describe what is currently known about the forms and form-driven functions of the proteins encoded by filoviruses. To begin with, we will clarify the potentially confusing terminology used to name the viruses and the proteins they encode.

The mononegaviral filovirus family (*Mononegavirales*: *Filoviridae*) includes three known genera: *Ebolavirus*, *Marburgvirus,* and *Cuevavirus* (Kuhn et al. 2014). The genus *Ebolavirus* has five members: Ebola virus, Sudan virus, Taï Forest virus, Bundibugyo virus, and Reston virus. The similar terms "Ebola virus," "ebolavirus or ebolaviruses" (used as the vernacular for the five members of the genus "*Ebolavirus*") understandably cause some confusion. "Ebola virus" (two words, capitalized E) refers to the specific and single virus of that name, while "ebolavirus" or "ebolaviruses" (lower case, one word) refers to one or more viruses in the *Ebolavirus* genus. The category "ebolaviruses" thus also includes Sudan virus, Bundibugyo virus, Reston virus, and Taï Forest viruses. The genus *Marburgvirus* has two members, termed Marburg virus and its variant Ravn virus. The genus *Cuevavirus* currently has only one member, Lloviu virus (Kuhn et al. 2014; Negredo et al. 2011).

All filoviruses contain seven genes. They are listed 3′ to 5′ in the negative-sense RNA genome, *NP, VP35, VP40, GP, VP30, VP24,* and *L*. In marburgviruses, these seven genes encode seven proteins of the same names: NP (the nucleoprotein), VP35 (a polymerase cofactor), VP40 (the matrix protein), GP (the surface glycoprotein), VP30 (also a polymerase cofactor), VP24 (a nucleocapsid protein), and L (the RNA-dependent RNA polymerase).

For the ebolaviruses and the cuevavirus, however, there are more than seven proteins, as not one, but several proteins result from the single *GP* gene: GP, sGP, ssGP, and Δ-peptide (Mehedi et al. 2011; Sanchez et al. 1996). In ebolaviruses, the major gene product is not the viral surface GP, but instead a secreted protein termed sGP. sGP constitutes 80% or more of the transcripts of the GP gene, is encoded in a single open reading frame, forms a dimer and is secreted abundantly from infected cells. The role of sGP is unclear, but it may serve as an immune decoy and to restore endothelial barrier function (Mohan et al. 2012; Wahl-Jensen et al. 2005).

In contrast, the viral surface GP results from a transcriptional editing event and a shift in open reading frame. Its *N*-terminal 295 amino acids are identical to those of sGP, but the shift in reading frame confers an alternate *C*-terminal region, which includes a transmembrane domain. GP is thus a membrane-anchored protein. It forms a trimer on the surface of the virus and is responsible for attachment and entry of new host cells. It is interesting that the glycoprotein required for propagation of the virus through its life cycle is only encoded after some transcriptional gymnastics. The third transcript produces a minor product whose function is unknown. This product, ssGP, results from the third possible reading frame of GP (Mehedi et al. 2011). Δ-peptide is a *C*-terminal glycosylated peptide released from sGP upon furin cleavage in the producer cell.

Here we describe what is known thus far about the structures and functions of the set of proteins encoded by filovirus genomes. Amongst the fascinating findings of the last decade is the plurality of functions and structures that these polypeptides can adopt. Many of the encoded proteins can play multiple, distinct roles in the virus life cycle, although the mechanisms by which these functions are determined and controlled remain mostly veiled. Further, some filovirus proteins are multi-structural: adopting different oligomeric assemblies and sometimes, different tertiary structures to achieve their separate, and equally essential functions.

Although we describe many findings here, much remains to be discovered. For some proteins, only a portion has been experimentally visualized. For others, we may have visualized the monomer, but do not yet know precisely how its biologically functional assembly or assemblies are built or function. In this chapter, we first describe the proteins in the nucleocapsid complex, followed by the matrix protein VP40, the glycoproteins, and finally, the polymerase, which has eluded high-resolution analysis.

2 NP

The nucleoproteins (NP) of filoviruses are 692–749 amino acids long and are encoded in the first gene of the negative-sense genome (Sanchez 1993). NP forms long, linear oligomers, encapsidating the viral RNA genome, and the NP-RNA complex is the functional template for the viral polymerase to carry out transcription and replication. NP does not interact directly with the polymerase (L). Instead, interactions are mediated by the polymerase cofactors VP30 and VP35 which are required to carry out RNA synthesis (Becker et al. 1998; Groseth et al. 2009). The *N*-terminal 450 amino acids of Ebola and Marburg virus NP (Bharat et al. 2011; Noda et al. 2010) are sufficient for RNA binding and oligomerization. This *N*-terminal region is likely to be homologous to the nucleoproteins of other non-segmented negative-sense RNA viruses (Bharat et al. 2011), in which the RNA binds in a cleft between two subdomains (Green et al. 2006).

Electron microscopy of full-length, recombinant NP oligomers shows long ropey filaments with a generally helical path (Mavrakis et al. 2002). Removal of the *C*-terminal region by genetic truncation causes NP oligomers to instead form into rigid, spiral cylinders similar to those of the measles virus nucleoprotein (Schoehn et al. 2004). In cell culture, narrow NP cylinders gather into large inclusions. Co-expression of VP24 and VP35 with NP results in somewhat wider cylinders whose density and diameter are indistinguishable from the nucleocapsids observed in filovirus-infected cells (Huang et al. 2002; Watanabe et al. 2006). Several studies using cryoelectron tomography of filovirus virions reveals the nucleocapsid to assemble into a left-handed helix, and that the nucleocapsid of Marburg virus has a diameter somewhat greater than that of the ebolaviruses (Beniac et al. 2012; Bharat et al. 2012; Bharat et al. 2011; Noda et al. 2005) (Fig. 1). It was estimated from these studies that NP binds to 6–12 nucleotides of RNA, and that VP24 and VP35 bind to alternating NP monomers (protomers) (Bharat et al. 2012; Bharat et al. 2011). NP has also been found to interact with the viral matrix protein VP40 to facilitate incorporation of the nucleocapsid into progeny virions (Noda et al. 2006).

Recent structural studies have provided additional insights into the role of NP and its interactions with RNA and with VP35. Studies from a variety of other mononegaviruses suggest that the nucleoprotein (termed NP for filoviruses or N for other mononegaviruses) readily associates with both host and viral RNA, and that the interaction with RNA is controlled by viral partners such as the phosphoprotein P (for other mononegaviruses) or VP35 (for filoviruses). Kirchdoerfer et al. (2015) and Leung et al. (2015), recently discovered that the *N*-terminal peptide of VP35 maintains NP in an RNA-free state, and determined the structures of the NP-VP35 peptide complex (Fig. 2a, b). The resulting structures are a reasonable approximation of the RNA-free form of filoviral NP and provide important insights into how VP35 association prevents binding of cellular RNA before delivery to the nascent nucleocapsid, where it may specifically bind the viral genome. The *N*-terminal domain of NP forms a two-lobed structure and is responsible for nucleocapsid assembly and binding of RNA. The *N*-terminal peptide of VP35 (residues 20–45 in Ebola virus) binds in a highly conserved pocket on the side of NP, roughly where the two lobes of the ordered Ebola NP *N*-terminal domain meet. Examination of the specific residues involved suggests that this interaction is conserved across the family *Filoviridae*.

Further analysis is needed to understand if NP undergoes a conformational change upon RNA binding, or uses another mechanism to close about the

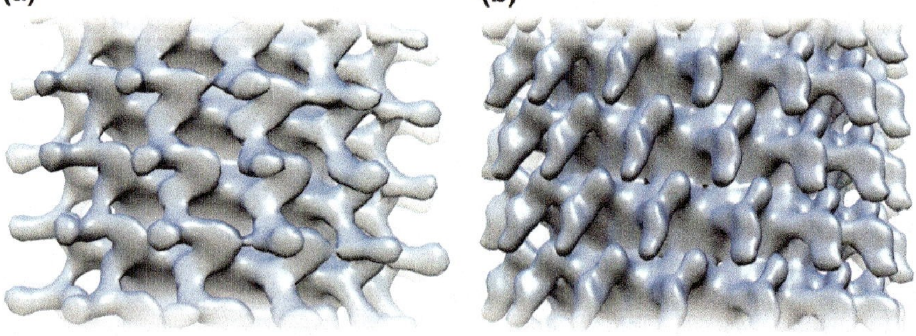

Fig. 1 Electron tomographic reconstructions of filovirus nucleocapsids. **a** Ebola virus (EMD-2043) and **b** Marburg virus (EMD-1986) nucleocapsids form left-handed helices with protrusions corresponding to the bound VP24 and VP35 proteins

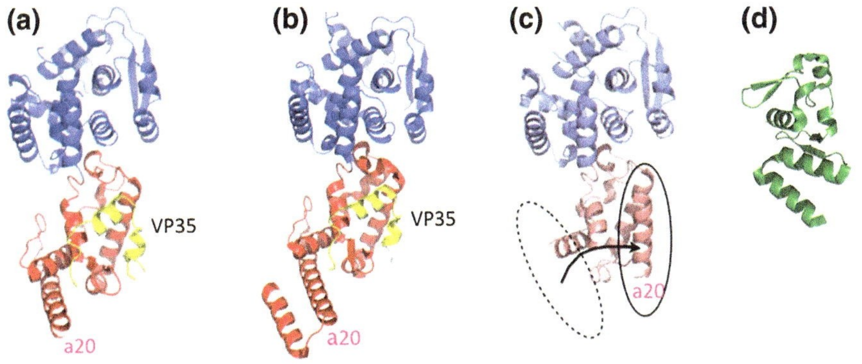

Fig. 2 Crystal structures of the NP ordered domains. (**a** and **b**) 2.4 Å (Kirchdoerfer et al. 2015) and 3.7 Å (Leung et al. 2015) crystal structures of the *N*-terminal domain of NP held in an RNA-free, monomeric state by the VP35 *N*-terminal peptide (*yellow*). The two lobes of the NP *N*-terminal domain are colored *red* and *blue*, and RNA is expected to bind between them. **c** 1.8 Å crystal structure (Dong et al. 2015) of the unbound NP *N*-terminal domain, in which helix α20 occupies the hydrophobic patch on the side of NP in place of the VP35 chaperoning peptide. **d** 1.75 Å structure of the 100-residue *C*-terminal domain of Ebola virus NP (Dziubanska et al. 2014)

RNA strand. Data are currently inconsistent whether the monomeric core of NP, itself, is sufficient to bind RNA, probably because different versions of the monomer have been analyzed—containing some or none of the additional *N*- and *C*-terminal arms that mediate NP oligomerization.

Dong et al. (2015) reported the structure of Ebola virus NP core alone (residues 19–405). This structure contains a fold similar to that observed in the VP35/NP complexes. However, instead of VP35 binding into the hydrophobic patch on monomeric NP, the *C*-terminal end of the NP *N*-terminal domain construct (helix α20) flipped up to occupy VP35's binding site (Fig. 2c). The functional significance of this interaction is currently unknown.

The *C*-terminal half of NP is flexible and acidic, scaffolds the other proteins in the nucleocapsid complex, and culminates in a ~100-residue ordered domain. Structures of this domain are available for Ebola virus (Fig. 2d), Taï Forest virus and Bundibugyo virus, and are consistent in fold (Dziubanska et al. 2014).

3 VP30

Viral protein 30 (VP30) plays an essential, though not fully understood, role in the filovirus life cycle. VP30 is unique to filoviruses, and unlike NP, VP35, and L, has no corresponding protein in the nucleocapsid of other mononegaviruses. VP30 is also essential: infectious recombinant Marburg and ebolaviruses cannot be produced from cDNAs that lack the VP30 gene, although the specific roles of and requirements for VP30 in Marburg and Ebola may differ (Enterlein et al. 2006). Many molecular functions have been ascribed to VP30. One key role in Ebola virus is as a transcriptional activator, allowing the viral polymerase to read through an RNA hairpin in the 5' untranslated region of NP mRNA (Weik et al. 2002). However, this function is not required for transcription in Marburg virus (Mühlberger et al. 1999), and it remains unclear why VP30 is essential for the rescue of Marburg viruses. VP30 has also been proposed to promote polymerase transcription re-initiation (Martinez et al. 2008).

Ebolavirus and marburgvirus VP30 proteins are 281–289 amino acids in length, while the recently discovered Lloviu virus VP30 contains a longer *N*-terminal region and is 328 amino acids in length. The *N*-terminal region of Ebola virus VP30 has been found to be phosphorylated (Modrof et al. 2002), to bind zinc and RNA (John et al. 2007), and to contribute to oligomerization of the protein (Hartlieb et al. 2003), but the structure of that essential domain remains unknown. The structure of the Ebola virus *C*-terminal region, however, has been determined by X-ray crystallography and shows a dimer of globular domains, in which a *C*-terminal helix reaches over to its neighbor to mediate dimerization (Clifton et al. 2014; Hartlieb et al. 2007) (Fig. 3a). The *C*-terminal domain has a structure similar to that of the pneumovirus M2-1 protein, which is also a viral transcriptional activator (Blondot et al. 2012), though there is no sequence homology, and VP30 and M2-1 play different specific roles in transcriptional activation of their respective viruses. VP30

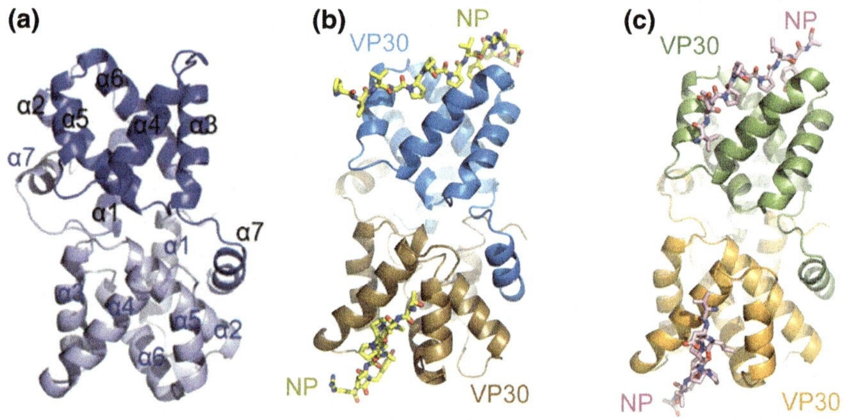

Fig. 3 VP30 *C*-terminal Domain. **a** The VP30 *C*-terminal domain forms a dimer and the unbound dimer structure has been determined from Ebola virus (Hartlieb et al. 2007) and Reston virus (Clifton et al. 2014) (pictured). Dimerization is assisted by α-helix 7 extending to contact the neighboring protomer. One protomer is light blue and the other dark blue and secondary structural elements are numbered. (**b** and **c**) Crystal structures of the Ebola virus (*blue* and *brown*) and Marburg virus (*green* and *orange*) VP30-NP complexes, which illustrate a proline-rich NP peptide (*yellow* for Ebola virus and *pink* for Marburg virus) binding into a groove on each copy of the VP30 dimer (Kirchdoerfer et al. 2016). PDB: 5T3T, 5T3W, 5VAO, 5VAP

allows polymerase read through of a hairpin, while M2-1 has an anti-termination function (Blondot et al. 2012).

Both the VP30 *N*- and *C*-terminal regions individually associate with NP inclusions in transfected cells (Hartlieb et al. 2007; Modrof et al. 2002). Interactions between NP and VP30 are thought to regulate the replication and transcription switch. Recent structural studies have identified a polypeptide region within the flexible *C*-terminal half of NP that binds the VP30 *C*-terminal domain (Kirchdoerfer et al. 2016). Crystal structures of the VP30-NP complexes made by both Ebola virus and Marburg virus illustrate that a short proline-rich sequence from NP binds into a conserved shallow groove on the VP30 *C*-terminal domain (Fig. 3b, c). Functional analysis suggests that this interaction is important for regulating viral RNA synthesis, and that binding of VP30 to this NP peptide is critical for mini-genome reporter activity and for synthesis of RNA (Kirchdoerfer et al. 2016). Structure-guided mutagenesis identified residues critical for the interaction, with some mutants enhancing the NP-VP30 interaction and others diminishing or abrogating binding (Kirchdoerfer et al. 2016).

4 VP35

VP35 is encoded by the genome's second gene and is the equivalent of viral phosphoproteins (P) found in other mononegaviruses. Despite this functional similarity, VP35 has not been found to be strongly phosphorylated in infected cells (Elliott et al. 1985). However, VP35 is similar to the other viruses' P proteins in its essential role in virus transcription and replication. VP35 acts as a polymerase cofactor, interacting with both the polymerase (L) and the viral nucleoprotein (NP). In addition, VP35 acts to blunt the host innate immune response by interfering with sensing of dsRNA (Cardenas et al. 2006; Feng et al. 2007; Luthra et al. 2013; Prins et al. 2009) and by RNA interference (Fabozzi et al. 2011).

VP35 ranges in length from 320 to 341 amino acids. The protein can be separated into several regions. The *N*-terminal peptide holds NP in a monomeric, RNA-free state as previously described (Fig. 2). An internal region, residues 80–120 in Ebola virus numbering, forms a coiled coil and is responsible for the oligomerization of the VP35 protein (Moller et al. 2005). The *C*-terminal RNA-binding domain, 217–340 in Ebola virus numbering, forms a four-helix bundle linked to a fifth short helix and a three-stranded mixed β-sheet (Leung et al. 2009) (Fig. 4a). The remaining, intervening regions of the protein are unlikely to form stable secondary structure, as judged by hydrogen–deuterium exchange mass spectrometry (H/DXMS) (Kimberlin et al. 2010) and have thus far prevented visualization of the complete molecule. This domain organization is similar to the P proteins of other mononegaviruses, as they also have internal oligomerization domains and folded *C*-terminal domains connected by flexible linkers (Gerard et al. 2009). Of note, however, is the larger size of the filovirus *C*-terminal domain, as well as its additional functionality in innate immune antagonism. Also like the P

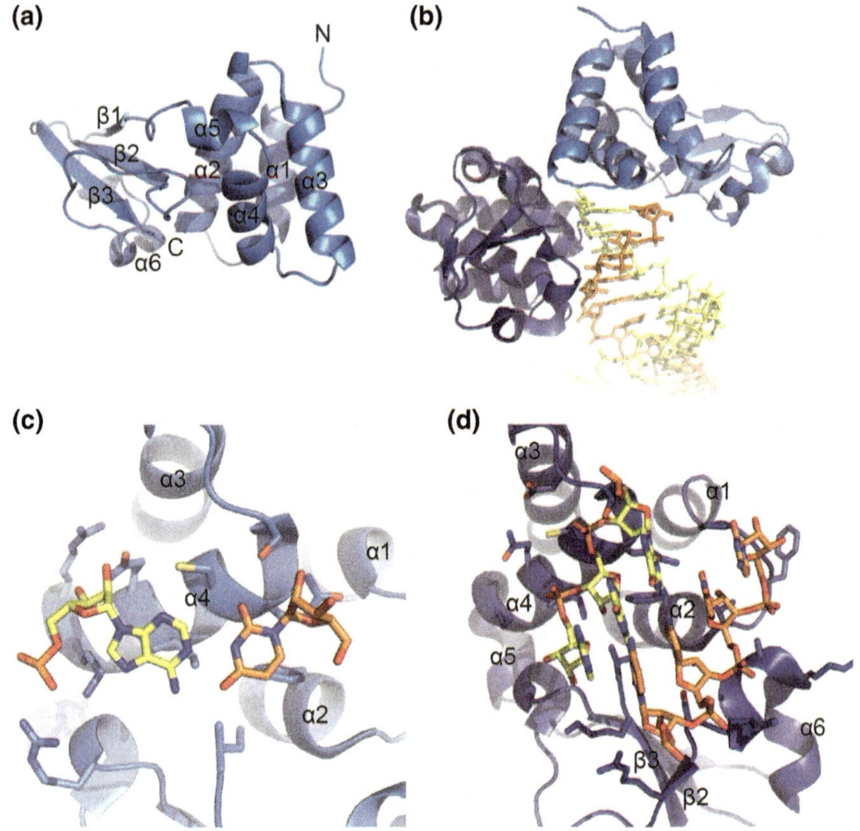

Fig. 4 VP35 *C*-terminal, RNA-binding domain. **a** The filovirus VP35 *C*-terminal domain forms a compact domain (PDB: 3KS4). **b** These domains from ebolaviruses form an asymmetric dimer (*blue* and *purple*) to bind RNA (PDB: 3KS8). (**c** and **d**) The different protomers in the dimer use different residues, although roughly contained in similar faces of the protein to contact either the end or the backbone of dsRNA

proteins of other mononegaviruses, the *C*-terminal domain of VP35 is responsible for the protein's interaction with the NP-RNA complex.

Structures of filovirus VP35 RNA-binding domains have been determined by X-ray crystallography for Ebola virus, Reston virus, and Marburg virus in both double stranded RNA (dsRNA)-bound and unliganded states (Figs. 4 and 5). Interestingly, all structures of the VP35 RNA-binding domain in complex with dsRNA for any ebolavirus (whether Ebola or Reston), illustrate how VP35 binds to the dsRNA as a striking, asymmetric dimer (Figs. 4b and 5a). In this dimer, one of the copies of VP35 uses a hydrophobic surface to cap the termini of the RNA strands while the second copy of VP35 contacts the phosphate backbone (Kimberlin et al. 2010; Leung et al. 2010) (Figs. 4c and d). Three mutations were found to interfere with innate immune antagonism. Of these, R305A and K309A

(Ebola virus numbering), both disrupt interactions with the phosphate backbone, whereas R312A disrupts formation of the dimer interface. Structures of the Marburg virus VP35 RNA-binding domain in complex with dsRNA illustrate that the fold of the RNA-binding domain is similar. However, all visualized interactions of Marburg virus VP35 are with the sugar-phosphate backbone (Fig. 5b, c), in a manner similar to the second copy in the asymmetric dimer in the ebolavirus VP35 structures (Bale et al. 2012b; Ramanan et al. 2012). Despite the observation of different binding modes in the crystals, both ebolavirus and Marburg virus VP35 RNA binding domains probably both cap and coat dsRNA in solution (Bale et al. 2012).

VP35 must be an oligomer to function in the virus life cycle (Moller et al. 2005), with oligomerization mediated by the central domain. Two separate X-ray crystal structures of the oligomerization domain from Marburg virus VP35 illustrate a trimeric coiled coil (Bruhn et al. 2016) (Fig. 6). In these structures, Marburg virus VP35 forms an assembly typical of other trimeric, coiled coils with all "knob" residues leucines, isoleucines, or valines (Bruhn et al. 2016). Two prolines (78 and 109) flank this coiled coil and are conserved among marburgviruses and most ebolaviruses. Beyond these prolines, however, very little sequence identity exists in the coiled coil domain (18% identity vs. 33% for the entire VP35) (Bruhn et al. 2016). The predicted coiled coil of the ebolaviruses is also 12 residues longer than that of Marburg virus (Geisbert and Jahrling 1995).

Although no high-resolution structure is yet available for an ebolavirus VP35 oligomerization domain, light scattering experiments suggest that the particle size is consistent with a tetramer instead of a trimer (Bruhn et al. 2016; Edwards et al. 2016). Distinct oligomeric assemblies between Marburg and Ebola virus VP35 would not be inconsistent with some of the observed functional differences in VP35 of the different viruses (Messaoudi et al. 2015) or the demonstrated inability of these proteins to be exchanged in minigenome experiments (Mühlberger et al. 1998,

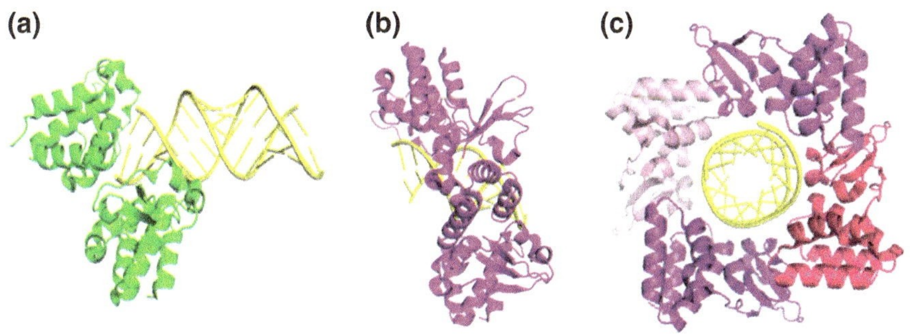

Fig. 5 VP35-dsRNA complexes. **a** Two copies of the ebolavirus VP35 RNA-binding domain (*green*) form an asymmetric dimer to cap the end of a dsRNA oligo (*yellow*). Pictured here is the complex from Reston virus with an 18 bp dsRNA (PDB:3KS8). (**b** and **c**) In crystal structures, Marburg virus VP35 (*purple*) instead spirals about the backbone of the dsRNA (*yellow*). Pictured are side **b** and end **c** views (PDB:4GHA)

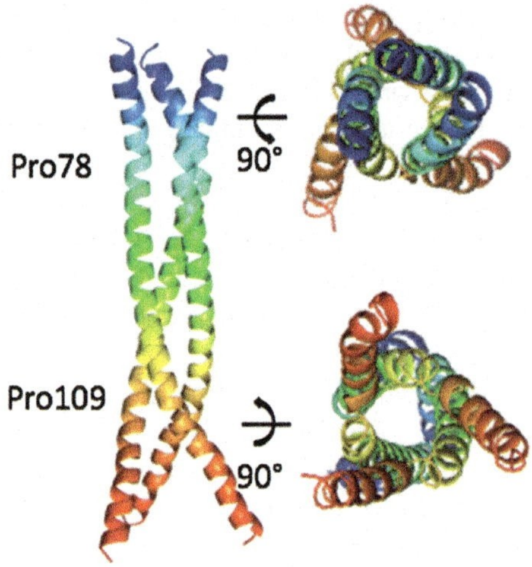

Fig. 6 VP35 oligomerization domain. The central oligomerization domain of Marburg virus is a triple-stranded coiled coil. Here, each polypeptide is colored *blue* to *red*, *N*- to *C* terminus. Prolines 78 and 109 flank the central coiled coil (PDB:5TOI) (Bruhn et al. 2016)

1999). A further difference is that oligomerization of VP35 is required for potent interferon antagonism by ebolaviruses (Bale et al. 2012b; Edwards et al. 2016), but not for maximal antagonism by Marburg virus. In general, Marburg virus VP35 is a less potent interferon antagonist (Bale et al. 2012b; Edwards et al. 2016).

5 VP24

The VP24 protein is important for assembly and function of the viral ribonucleoprotein complex and plays other roles as well (Hoenen et al. 2010b; Noda et al. 2007a; Watanabe et al. 2007). VP24 (233–241 amino acids in length) adopts a pyramidal α/β fold, $\sim$70 × 30 × 30 Å in dimension, with two neighboring cavities, one hydrophobic and the other hydrophilic (Zhang et al. 2012) (Fig. 7a). VP24 of the ebolaviruses inhibits interferon signaling by binding to karyopherin α proteins (Reid et al. 2006). Ebolavirus VP24 interacts with the *C*-terminal portion of karyopherin α5 by binding at a site that overlaps with that of a nonclassical nuclear localization signal. By binding karyopherin α5, VP24 prevents it from achieving nuclear transport of tyrosine phosphorylated STAT1 (Xu et al. 2014) (Fig. 7b).

VP24 of Marburg virus targets Keap1, a repressor of antioxidant transcriptional signaling (Edwards et al. 2014; Page et al. 2014). In Marburg virus VP24, but not ebolavirus VP24, residues 201–217 form a pair of long antiparallel beta strands that jut out from the VP24 core (Zhang et al. 2014) (Fig. 7c). These beta strands mediate dimerization of Marburg virus VP24, and a dimer/monomer equilibrium of Marburg virus VP24 may regulate its roles in ribonucleoprotein function and Keap1 interactions (Johnson et al. 2016a).

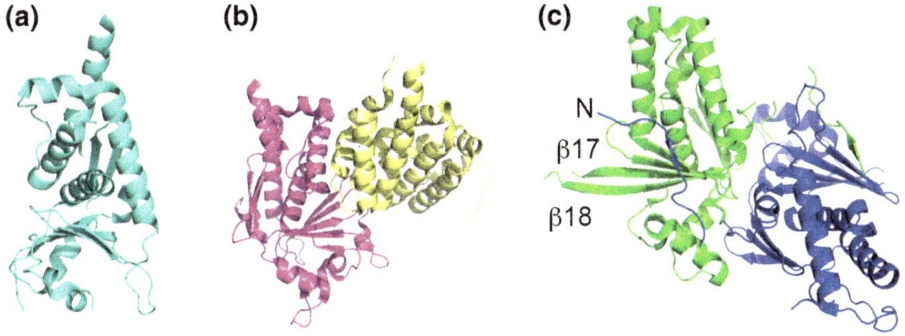

Fig. 7 VP24. **a** VP24 has a pyramidal shape. Illustrated here is Sudan virus VP24 (Zhang et al. 2012) (PDB:3VHF), but the fold is similar for Reston virus and Ebola virus as well. **b** Ebola virus VP24 (*pink*) binds the *C*-terminal portion of karyopherin α5 (*yellow*) to prevent nuclear translation of STAT1(Xu et al. 2014) (PDB:4U2X). **c** Marburg virus VP24 contains a projecting beta hairpin (β17-β18) and forms a dimer, with dimerization assisted by the *N*-terminal strand (Johnson et al. 2016a; Zhang et al. 2014). Here the two monomers are *green* and *blue*, respectively (PDBs: 4OR8 and 4M0Q)

Notably, for VP24, as in VP35, we observe different types of interactions with host factors for Ebola virus and Marburg virus. VP35 has multiple domains, each with separate roles to play, but governed by oligomerization at center. In the course of the virus life cycle, VP24 (at least for Marburg virus) may adopt distinct oligomeric assemblies to accomplish multiple functions. This introduces a theme of multi-form, multi-function proteins—a theme most evident in the viral protein VP40.

6 VP40

Viral Protein 40 (VP40) is the main structural component of the filovirion matrix (Elliott et al. 1985), and the equivalent of the matrix (M) proteins found in other mononegaviruses. VP40 polymerizes to form protein filaments out of which the virion matrix is constructed and from which the characteristic filamentous shape of the virion is achieved (Beniac et al. 2012; Bornholdt et al. 2013). The VP40 matrix also recruits the nucleocapsid to the virion (Bharat et al. 2012; Bharat et al. 2011; Hoenen et al. 2010a; Noda et al. 2007b) and has essential interactions with the glycoprotein and the cell membrane during budding. Expression of VP40 alone is sufficient to bud filamentous virus-like particles (Beniac et al. 2012; Jasenosky et al. 2001; Johnson et al. 2006; Noda et al. 2002; Panchal et al. 2003; Timmins et al. 2001). VP40 also interacts with the viral surface protein GP. Indeed, coexpression of VP40 alters the distribution pattern of GP, leading to GP accumulation at VP40-positive basolateral clusters from which budding occurs (Kolesnikova 2007). VP40 has also been shown to be critical for transcriptional control (Hoenen et al. 2010b), and VP40 for Marburg virus has been shown to be immunosuppressive

(Oda et al. 2015; Valmas and Basler 2011; Valmas et al. 2010). These multiple functions are believed to be controlled by the ability of the VP40 polypeptide to adopt different structures.

VP40 is encoded in the third gene of the negative-sense genome (Sanchez 1993). Ebola virus VP40 is 326 amino acids in length, while Marburg virus VP40 is 303 amino acids. A VP40 protomer is divided into two compact domains: an *N*-terminal domain (NTD) circa residues 1–194, and a *C*-terminal domain (CTD) circa residues 195 to C terminus (in Ebola virus numbering). The NTD is built around a central β-sandwich and has several α-helices on its surface farthest from the CTD (Dessen et al. 2000). Residues 1–43 and 320–326 are disordered. This simple two-domain structure supports a surprising variety of interfaces, alternative forms, and corresponding functions.

In some forms of the protomer, the CTD is held against the NTD by a limited number of hydrophilic and hydrophobic contacts (Dessen et al. 2000). In other forms, the CTD has been sprung and floats at a disordered position relative to the NTD, with the residues circa 193–200 becoming a flexible linker (Bornholdt et al. 2013; Gomis-Ruth et al. 2003; Scianimanico et al. 2000). This form of the protomer is not fully known, as the flexible linker and the sprung CTD are not visible in crystal structures. A second structural variation is that in some forms, residues 44–68 cling to the surface of the NTD farthest from the CTD (Dessen et al. 2000), while in other forms, these residues are absent from the crystal structures, having become part of the disordered *N*-tail (Bornholdt et al. 2013; Gomis-Ruth et al. 2003).

These alternative forms of the VP40 protomer underlie three distinct forms of oligomerization, each with its corresponding function:

First, a dimer, with the two protomers joined symmetrically NTD–NTD via a *dimer interface* located on the surface of the NTD farthest from the CTD (Bornholdt et al. 2013; Dessen et al. 2000; Timmins et al. 2003) (Fig. 8). Each

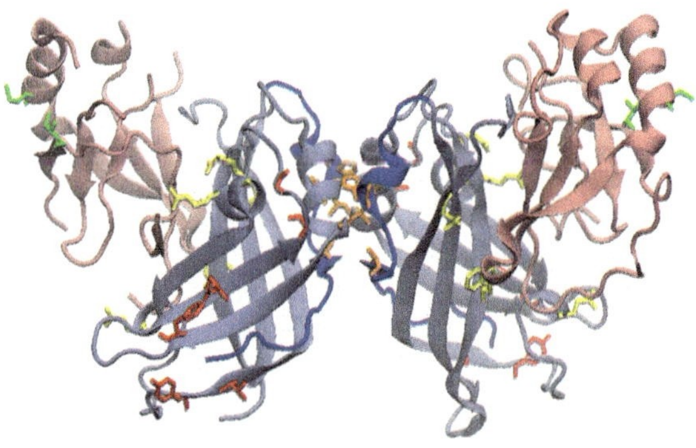

Fig. 8 VP40 dimer. The cartoon representation is *bright blue* for residues 44–68, *light blue* for the rest of the NTD, *pink* for the CTD (PDB:4LDB). Both CTDs are unsprung. At center, both instances of residues 44–68 are unsprung (*blue*), and it is here that the dimer interface is found (*orange*). In this form, the oligomerization interface (*yellow*) is hidden under the CTD, and the octamer interface (*red*) is hidden under residues 44–68 (*blue*)

protomer is maximally ordered, i.e., both the CTD and residues 44–68 cling to the NTD. This form of VP40, sometimes described as a butterfly shape, is essential for trafficking through the cell. Indeed, in late-stage infection, VP40 localizes to the cell membrane (Adu-Gyamfi et al. 2012; Bornholdt et al. 2013; Hoenen et al. 2005; Nanbo et al. 2013; Panchal et al. 2003; Timmins et al. 2001); mutants that cannot dimerize experience reduced localization, suggesting that the dimer recruits assistance from the cell's transportation network (Bornholdt et al. 2013). VP40 also interacts with the viral nucleoprotein and VP35, escorting nucleocapsids to the cell surface and incorporating them into virions (Geisbert and Jahrling 1995; Hoenen et al. 2010a; Johnson et al. 2006; Noda et al. 2006; Noda et al. 2007b).

Next, a hexamer is formed by linearly connecting three dimers (Bornholdt et al. 2013). To make this assembly possible, the four central protomers spring their CTDs; this displacement of the CTD reveals a new *oligomerization interface* on the NTD, by means of which the three dimers join NTD–NTD. A *CTD-to-CTD interface* then allows hexamers to polymerize, forming protein filaments out of

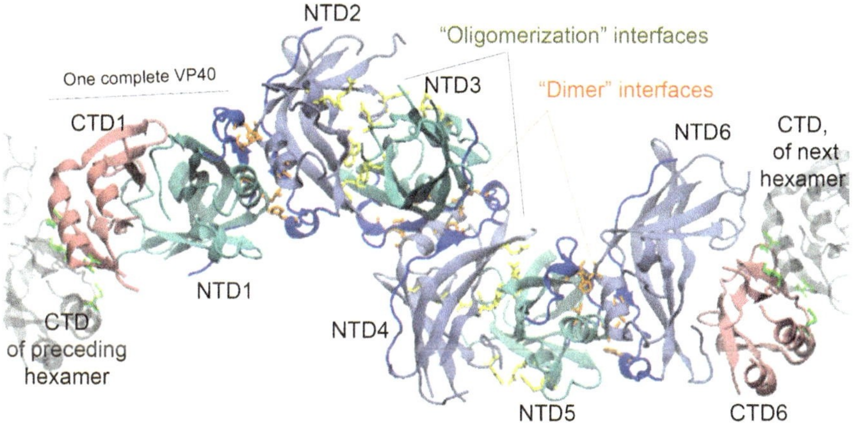

Fig. 9 VP40 hexamer (PDB:4LDD). A zig-zagging hexameric structure of VP40 is made by six NTDs at center (alternating *cyan* and *blue-gray*), with each end capped by one visible CTD (*pink*). The CTDs that belong to the NTDs at center do remain attached by their polypeptide bonds, but have "sprung" away and have become disordered in solvent channels in the crystals. Each hexameric building block, with the visible section described as (CTD-NTD)-$(NTD)_4$-(NTD-CTD), repeats in crystals by interactions of the CTD at each end with the CTD at the end of another hexameric building block. The CTD–CTD interface is hydrophobic, flat and "slippery", potentially allowing flexibility in the extended filaments built from these hexamers. In this figure, the central hexamer is composed of three dimers, each centered on its dimer interface (involving residues illustrated in *orange*). Note that the extreme-left and extreme-right copies of VP40 are similar to those of Fig. 8, while the central four protomers have sprung their CTDs, so only the NTDs are visible. This springing of the central CTDs exposes the instances of the oligomerization interface (*yellow*) to either side of the central dimer, joining the central dimer to the left dimer and right dimer. At extreme left and extreme right, the CTD-to-CTD interface (*green*) allows for polymerization with other hexamers (*gray*). For clarity, the following bonds are not drawn: the non-exposed instances of the oligomerization interface and the octameric interface (*red* in Fig. 8) which is still hidden in this structure under residues 44–68 (*royal bue*).

which the virion matrix is constructed (Bornholdt et al. 2013) (Fig. 9). Interestingly, VP40 hexamers were observed at sites of budding from the plasma membrane (Adu-Gyamfi et al. 2012; Adu-Gyamfi et al. 2013) prior to resolution of the hexameric crystal structure.

The CTD also includes residues that interact with phosphatidylserine in the cell membrane; this triggers hexamerization (Adu-Gyamfi et al. 2012, 2013, 2015; Bornholdt et al. 2013; Scianimanico et al. 2000; Soni et al. 2013) and leads to budding (Adu-Gyamfi et al. 2013, 2015; Dessen et al. 2000; Harty 2009; Jasenosky et al. 2001; Panchal et al. 2003; Ruigrok et al. 2000; Scianimanico et al. 2000; Soni and Stahelin 2014). VP40 is also sufficient to induce phosphatidylserine exposure on the outer plasma membrane leaflet at sites of budding (Adu-Gyamfi et al. 2015), which helps explain Ebola virus's phosphatidylserine-dependent entry (Kondratowicz et al. 2011), and makes a clear link between a matrix protein's use of a lipid for budding and then subsequent entry. Determining the functional dynamics of these VP40/lipid interactions remains an open problem. Dependency of budding on NTD residues has moreover been reported for a loop region (Adu-Gyamfi et al. 2014), for residues circa Pro 53 (Yamayoshi and Kawaoka 2007), and for motifs in residues 7–13 that interact with host proteins (Harty 2009; Harty et al. 2000; Jasenosky et al. 2001; Licata et al. 2003; Panchal et al. 2003; Timmins et al. 2001). Additionally, the lipid phosphatidylinositol 4,5-bisphosphate ($PI(4,5)P_2$) in the plasma membrane apparently serves as an important anchor for VP40: $PI(4,5)P_2$ triggers VP40 hexamerization (Johnson et al. 2016b), while VP40 clusters the $PI(4,5)P_2$ lipid (Gc et al. 2016).

The third form is a symmetrical octameric ring (Gomis-Ruth et al. 2003; Timmins et al. 2003). In this form, all eight CTDs are sprung, as well as all instances of residues 44–68 (Fig. 10). Every second protomer–protomer association forming the ring is NTD–NTD via the *oligomerization interface* described above. The remaining protomer–protomer associations are NTD–NTD via a new *octamer interface* that is revealed when residues 44–68 are peeled away. The NTD alone is thus sufficient to form the octamer; indeed, deleting the CTD forces octamerization (Bornholdt et al. 2013; Gomis-Ruth et al. 2003; Ruigrok et al. 2000; Timmins et al. 2003). Other ways of encouraging octamerization include deleting residues 320–326 (Scianimanico et al. 2000), mutating I307R or W95A (Bornholdt et al. 2013) or P283/286L (Panchal et al. 2003), and applying disruptive agents such as urea (Scianimanico et al. 2000); this suggests a general rule that springing all CTDs leads to octamerization. The sprung CTDs float above and below the ring's lumen (Scianimanico et al. 2000); their function thereafter is unknown.

RNA binds to the ring's lumen at the NTD–NTD cleft of the octamer interface (Gomis-Ruth et al. 2003) (Fig. 10). This is striking because no other form of VP40 binds RNA. Binding of RNA or an RNA-protein complex could be the natural trigger for formation of the octamer (Gomis-Ruth et al. 2003; Radzimanowski et al. 2014; Timmins et al. 2003). A three-nucleotide UGA repeat was visualized bound

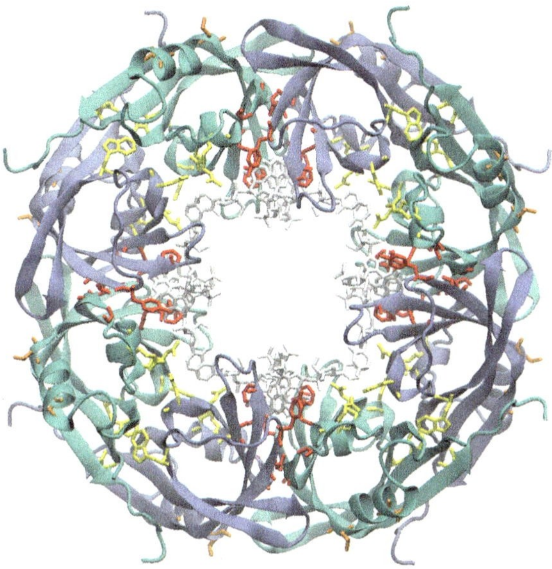

Fig. 10 VP40 octamer (PDB:1H2C). NTDs alternate *light blue* and *cyan*. In crystal structures, all eight CTDs are deleted. All instances of residues 44–68 are sprung; this disrupts the dimer interface (remnants *orange*) and exposes the octamer interface (*red*). Going around the ring, the octamer interface (*red*) alternates with the oligomerization interface (*yellow*). RNA (*white*) binds in the cleft of the octamer interface

to each of the eight copies of VP40 in a crystal structure of the octamer (Gomis-Ruth et al. 2003)—the nature and identity of additional nucleotides in the bound RNA are not yet known.

Octamers do not localize to the cell membrane (Bornholdt et al. 2013; Hoenen et al. 2010; Panchal et al. 2003; Timmins et al. 2001) and are not found in virions (Gomis-Ruth et al. 2003). They are instead specific to the infected cell and thus represent a non-structural form of this viral structural protein. Instead of assembling the viral matrix, this form of VP40 is thought to regulate viral transcription (Bornholdt et al. 2013; Hoenen et al. 2010a, b). The octamer is essential to the virus life cycle (Hoenen et al. 2005), presumably to achieve the incompletely understood regulatory function.

For Marburg virus VP40, octameric rings have a propensity to stack into rod-like structures (Timmins et al. 2003). The octameric ring form of Marburg virus VP40 may be linked to its immunosuppression function: the NTD of Marburg virus VP40 is necessary and sufficient to achieve antagonism of interferon β-stimulated enhancement of reporter gene expression, and the NTD alone is thought to form the octameric ring (Oda et al. 2015).

VP40 is thus a multi-form, multi-function protein, called a *transformer protein* (Bornholdt et al. 2013; Radzimanowski et al. 2014; Wasserman and Saphire 2016). VP40 has multiple structural states with the power to express in one form, then transform to another. There is, in particular, a convincing narrative that VP40 molecules express as dimers, localize to the vicinity of the cell membrane, and there, triggered by contact with lipids, transform into hexamers to build and bud virions (Adu-Gyamfi et al. 2012, 2013, 2015; Bornholdt et al. 2013). Other sources have found VP40 to form monomers (Dessen et al. 2000; Panchal et al. 2003;

Radzimanowski et al. 2014; Ruigrok et al. 2000; Scianimanico et al. 2000) or hexameric rings (Harty 2009; Radzimanowski et al. 2014; Ruigrok et al. 2000; Scianimanico et al. 2000; Timmins et al. 2003), or that the regulatory function is not specific to the octamer (Hoenen et al. 2010). Some issues are still subject to debate, but in many cases, these differences reflect the evolution of our understanding of VP40 in recent years.

Ebola virus and Marburg virus VP40 are 34% identical (49% homologous) in amino acid sequence, with differences concentrated in the *C*-terminal domains. Not surprisingly, the *N*-terminal domains are similar in structure (2.4 Å root-mean-square deviation (r.m.s.d.) among Cα atoms), and the NTD–NTD dimer interfaces they assemble are critical for matrix assembly and budding (Fig. 11a). The *C*-terminal domains, however, are only 16% identical in sequence and differ in structure (5.6 Å r.m.s.d. among Cα atoms). The Marburg virus CTD is more loosely folded than that of Ebola virus, key residues such as 208–221 trace a different path, and the basic patch of the CTD is expansive, larger and flatter than that of Ebola virus (Fig. 11b), which involves more projecting, flexible and disordered loop structures. In support of these structural differences, Marburg virus VP40 has been shown to associate with membranes in a nonspecific manner, although lipid binding is dependent upon the anionic charge density of the membrane (Wijesinghe and Stahelin 2015). Marburg virus VP40 is displaced from the plasma membrane inner leaflet when plasma membrane anionic charge is neutralized (Wijesinghe and Stahelin 2015), a classical feature of nonspecific electrostatic interactions. In contrast, Ebola virus VP40, which is thought to undergo selective interactions with phosphatidylserine (Adu-Gyamfi et al. 2015), was not displaced when the plasma membrane inner leaflet anionic charge was neutralized (Adu-Gyamfi et al. 2015; Johnson et al. 2016).

7 GP

Substantial progress has been made in the last decade in understanding the structures and functions of two major products of the filovirus *GP* gene: the viral surface glycoprotein GP and the secreted glycoprotein sGP. GP is 676–681 amino acid long, contains a transmembrane domain and forms a trimer on the surface of virions and infected cells (Feldmann et al. 1994). The GP trimer can also be cleaved from cell surfaces and released as an ectodomain (Dolnik et al. 2004). The sGP protein is 364 amino acids long, does not have any transmembrane domain, and instead forms a dimer that is secreted abundantly from infected cells (Sanchez et al. 1996; Sanchez et al. 1998).

The viral surface GP precursor is cleaved by furin in the producer cell to yield two distinct subunits: GP1 and GP2 (Volchkov et al. 1998). Of these, GP1 is responsible for binding receptor(s), while GP2 drives the conformational changes required for fusion of virus and host membranes. GP1 and GP2 remain associated

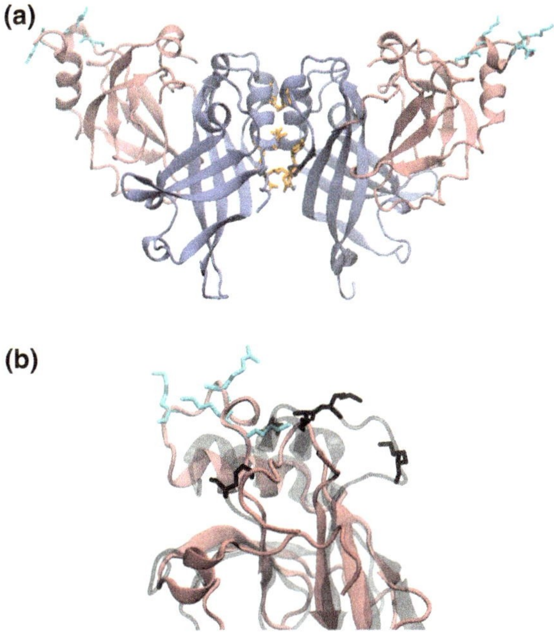

Fig. 11 Marburg virus VP40 and the basic patch. **a** Marburg virus VP40 dimer (PDB:5B0V). The cartoon representations are light *blue* for the NTD, *pink* for the CTD. Interfaces are suggested via colored bonds for selected residues: dimer interface, T40/N42/Y43/L49/T105, *orange*; basic patch, K210/K211/R215/K218, *cyan*. The basic patch is a primary site of lipid binding. Additional basic patch residues not visible here because disordered: K264/K265/R266. **b** Rotated detail of the Marburg virus VP40 CTD, illustrating the extensive, flat basic patch (*left*). Juxtaposed in translucent gray is the Ebola virus VP40 CTD (PDB:4LDB), aligned using MultiSeq/STAMP, with its less flat and more distributed basic batch. Ebola virus VP40 basic patch residues seen here are K221/K270/K274/K275/K279, *black*; not visible because they are disordered, K224/K225

by a disulfide bond and three copies of the GP1-GP2 complex form the trimer known as $GP_{1,2}$, or here, simply referred to as "GP."

In the fusion process, a dramatic conformational change in thought to occur in which this GP complex springs from a metastable prefusion arrangement of GP1 and GP2 intertwined into a more stable, six-helix bundle of GP2 alone (with GP1 released or moved out of the way). The six-helix bundle structures of GP2 were first revealed for Ebola virus in 1998 (Malashkevich et al. 1999; Weissenhorn et al. 1998a, b) (Fig. 12a, b), and for Marburg virus in 2012 (Koellhoffer et al. 2012) (Fig. 12c). In these structures, the central feature of the post-fusion GP2 is a long, triple-stranded coiled coil formed by the three copies of heptad repeat 1, flanked on the outside by the shorter helices of heptad repeat 2.

The prefusion conformation of Ebola virus GP was first determined in 2008, crystallized in complex with a human survivor antibody termed KZ52 (Lee et al. 2008) (Fig. 13). This structure revealed that the viral surface GP forms a bowl or "chalice" shape, with the three GP1 receptor-binding subunits angling outward,

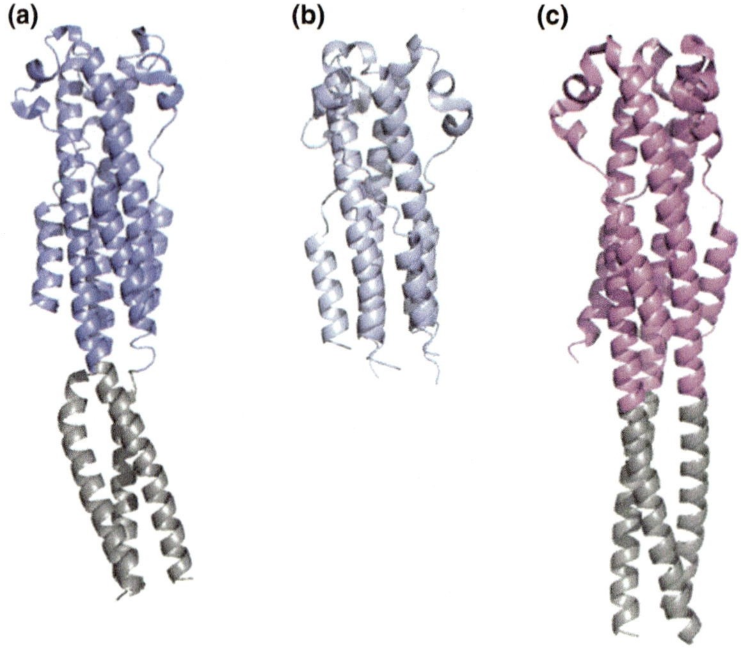

Fig. 12 Filovirus GP2 in its post-fusion structure. (**a** and **b**) Crystal structures of Ebola virus GP2 in its six-helix bundle conformation (Malashkevich et al. 1999; Weissenhorn et al. 1998a). (PDB: 1EBO (*darker blue*) and 2EBO (*lighter blue*), respectively). **c** Crystal structure of Marburg virus GP2 (Koellhoffer et al. 2012) (PDB: 4G2K). In **a** and **c**, GCN4 sequences that were introduced to support trimerization are colored gray

bound together at the bottom by the three GP2 fusion subunits that encircle and cradle them at the base. The trimer interfaces are made primarily of GP2-GP2 interactions at the center (Lee et al. 2008), with the additional packing of the internal fusion loop of each GP2 around the outside of the trimer into the neighboring monomer's GP1. During cellular entry, this fusion loop must unwind from the trimer and penetrate the target cell membrane.

Subsequent structures of GP from Sudan virus, using an isolate of variant Gulu (Dias et al. 2011) (Fig. 14) and isolate Boniface (Bale et al. 2012a), illustrate a similar trimeric organization, but different electrostatics at the trimeric interfaces. Sudan virus GP is much more acidic at this interaction site than Ebola virus (EBOV). Also in the Sudan virus (SUDV) structures, the *N*-terminal half of the fusion loop adopts a different, lower hanging conformation than that observed in EBOV—this region is only 33% identical in sequence between SUDV and EBOV (Dias et al. 2011). The upper, *C*-terminal portion, however, is more conserved in sequence and structure and is anchored to the neighboring monomer's GP1 by Arg 89 in both viruses (Dias et al. 2011). The Sudan virus structures revealed the

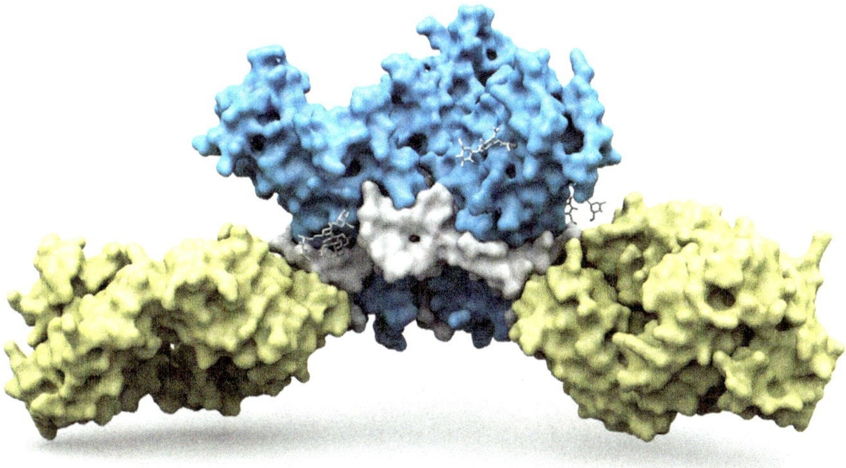

Fig. 13 Ebola virus GP. 3.4 Å Crystal structure of Ebola virus GP in its prefusion conformation, bound by the human survivor antibody KZ52 (Lee et al. 2008) (PDB:3CSY). The three copies of GP1 are colored *blue* and the three copies of GP2 colored *white*. Fab KZ52 is *yellow*. KZ52 binds the base of GP, bridging GP1 (*blue*) to GP2 (*white*). The stalk region of GP2 is disordered in this structure

GP1-GP2 disulfide bond common to filovirus GPs, C53 of GP1 and C609 of GP2, anchor those subunits together at the bottom of the trimer.

Each of the GP1 and GP2 subunits contains several subdomains. In GP1 is a base region that forms a spool about which GP2 is wound (Fig. 15a). Atop the base is the receptor-binding "head". The head is topped and masked by a subdomain termed the "glycan cap" that has several *N*-linked glycosylation sites (Lee et al. 2008). Upward and outward of the glycan cap is a flexible, heavily *N*- and *O*-glycosylated, mucin-like domain. The structure of the mucin-like domain has been studied by tomography (Tran et al. 2014), small-angle X-ray scattering (SAXS) (Hashiguchi et al. 2015) and most recently by cryoelectron microscopy (Beniac and Booth 2017). In solution, the mucin-containing GP ectodomain has a radius of gyration, R_G, about 20 Å larger than the mucin-deleted GP. The mucin-like domains are likely to be quite flexible in solution. Indeed, a smaller volume is visible by tomography and microscopy (Beniac et al. 2012; Tran et al. 2014; Beniac and Booth 2017), in which flexibility could be constrained by neighboring copies of GP on the viral surface.

The GP2 subunit contains an *N*-terminal peptide (released from GP1 by furin cleavage), an internal fusion loop, and two heptad repeats that flank a CX_6CC switch region (Lee et al. 2008). The first heptad repeat, HR1, which forms a single long helix in the post-fusion, six-helix bundle (Fig. 15b), is broken into four segments that wrap about the base of GP1 in the prefusion GP1-GP2 complex (Bale

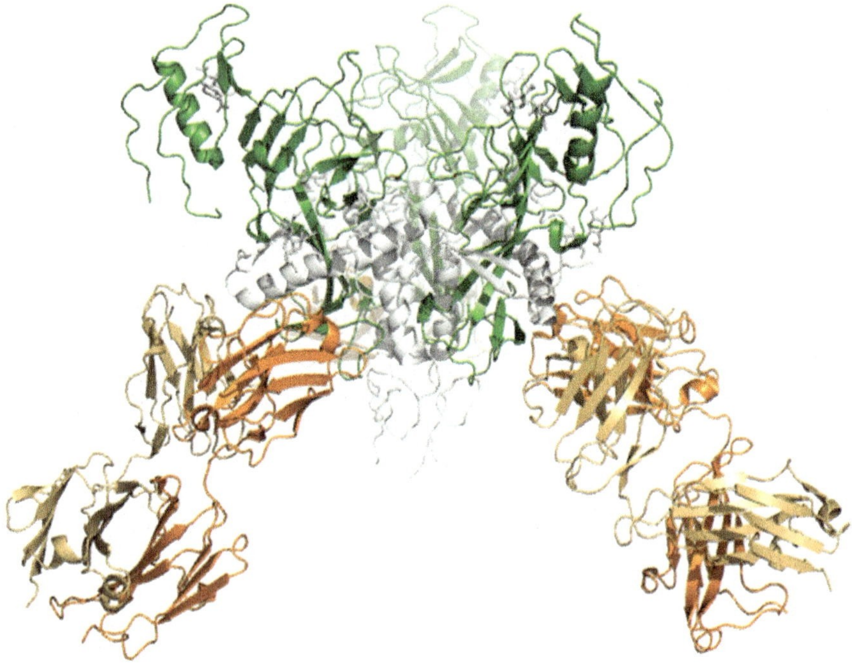

Fig. 14 Sudan virus GP. 3.35 Å crystal structure of Sudan virus GP in complex with Fab 16F6 (Dias et al. 2011) (PDB:3S88). GP1 is colored *green* and GP2 *white*. The heavy chain of the antibody is *orange* and the *light chain* colored *light orange*. 16F6 binds roughly the same location as KZ52, but is Sudan specific. The CX_6CC, switch region between the two heptad repeats of GP2 is visible at the bottom between the antibody fragments

et al. 2012a; Lee et al. 2008) (Fig. 15a). In that assembly, five β strands of the GP1 base combine with a GP2 β to form a semicircular spool about which the first heptad repeat is wound. On the other side, those same five GP1 strands combine with the two antiparallel β strands of the GP2 fusion loop to form a continuous, twisted β sheet. For GP2 to rearrange into its post-fusion six-helix bundle, both the continuous GP1-GP2 β sheet interactions and the GP1 spool-GP2 HR1 interactions must be broken (Bale et al. 2012a; Lee et al. 2008).

Below the GP base are the other portions of GP2: the chain reversal or switch region, followed by a stalk and the transmembrane domain. The chain reversal region contains a conserved CX_6CC sequence, which forms a 360° loop before continuing downward to the membrane. The final Cys is C609, which disulfide bonds to GP1 C53 (Fig. 15d). The first two are C601 and C608, which form an intra-GP2 disulfide bond (Fig. 15d). This bond is observed in both pre-fusion and post-fusion structures, and so it probably remains intact during the fusion process.

GP2 must unwind from around GP1 to adopt its post-fusion structure. It is not known if the GP1-GP2 disulfide must be reduced for this to occur. Given the proximity of that C53-C609 GP1-GP2 disulfide to the C601-C608 intra-GP2

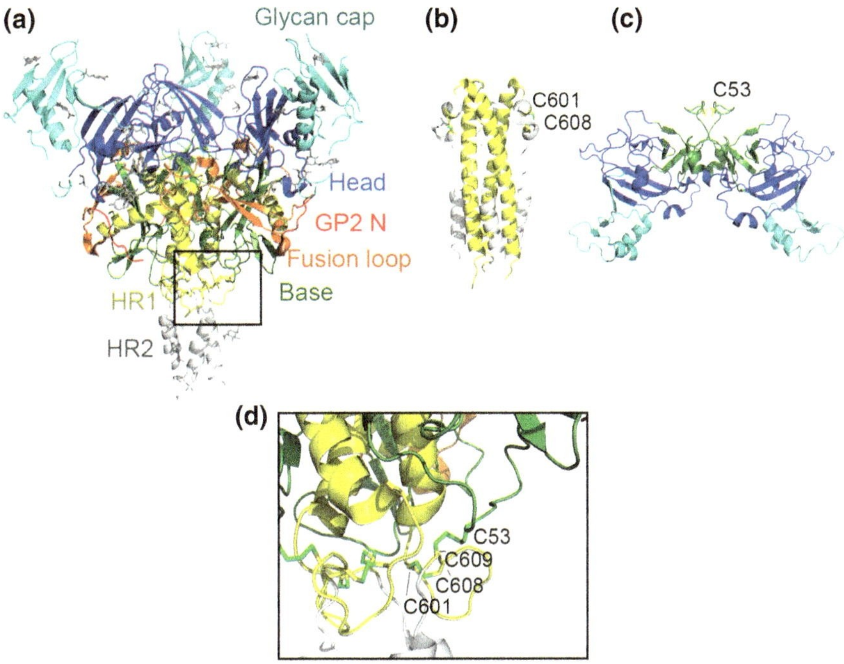

Fig. 15 Prefusion GP, post-fusion GP and sGP compared. **a** 2.2 Å Crystal structure of prefusion, unbound Ebola virus GP with residues colored by subdomain. The base region of GP1 is *dark green*, the head blue and the glycan cap colored *cyan*. The mucin-like domain (not pictured) is attached to the glycan cap. The *N*-terminal peptide of GP2 released by furin cleavage is *red*, the internal fusion loop *orange*, the first heptad repeat and CX_6CC switch region *yellow* and the second heptad repeat is *white*. Visible glycans are illustrated in ball-and-stick (Zhao et al. 2016) (PDB:5JQ3). **b** The crystal structure of the post-fusion six-helix bundle conformation of GP2 contains the first heptad repeat and CX_6CC region (*both yellow*) as well as the second heptad repeat (*white*). In this structure, HR1 forms a single, long central helix, while in the prefusion structure, it is wound about the GP1 base. HR2 packs on the outside of the three HR1 helices. The intra-GP2 disulfide bond C610-C608 is visible between the heptad repeats (Weissenhorn et al. 1998) (PDB:1EBO). **c** 5.5 Å structure of sGP illustrating dimerization mediated by a C53-C53 disulfide bond and interactions of residues that correspond to the base and head regions of GP1. The C306-C306 disulfide bond is not visible (Pallesen et al. 2016) (PDB:5KEM). **d** Zoomed-in view of the CX_6CC region of GP2 illustrating the C53-C609 GP1-GP2 disulfide, as well as the C601-C608 intra-GP2 disulfide

disulfide, which appears to remain intact, it is possible that neither disulfide is reduced during fusion. Further studies will reveal if GP1 simply rotates out of the way, or if instead, continual enzymatic processing of GP1 removes its steric hindrance to GP2 rearrangement (Bale et al. 2012a), or if another host factor is involved. Below the base is the GP2 stalk, which contains the second heptad repeat of GP2 (HR2), the membrane-proximal external region, the transmembrane domain and a very short cytoplasmic region (Zhao et al. 2016).

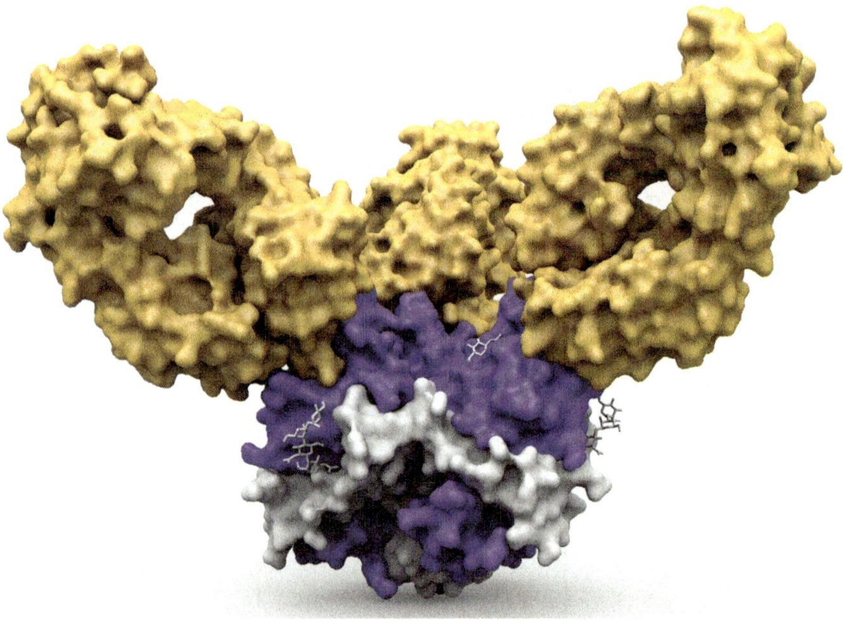

Fig. 16 Marburg virus GP. The 3.6 Å crystal structure of Marburg virus GP in complex with the human survivor antibody MR78 is illustrated with GP1 in purple, GP2 in *white* and the antibody Fab fragment in yellow (Hashiguchi et al. 2015) (PDB:3X2D). Visible glycans are illustrated in ball-and-stick. Residues corresponding to the glycan cap were removed by enzymatic cleavage to improve diffraction, but are thought to occupy a different position from where they are in Ebola virus. In Ebola virus, the glycan cap binds into and masks the receptor-binding site. They do not mask the same site in Marburg virus, and this antibody binds equally well to cleaved and uncleaved Marburg virus GP but only to cleaved Ebola virus GP (Flyak et al. 2015; Hashiguchi et al. 2015)

The structure of Marburg virus GP was determined in 2015, in complex with the antibody of a human survivor of Marburg virus infection (Fig. 16). Marburg virus and Ebola virus share only 28% sequence identity overall (but 85% within the receptor-binding site). The differences in sequence in GP result in several differences in structure. While Ebola virus entry requires cleavage by cathepsin B (Chandran 2005; Martinez et al. 2010; Schornberg et al. 2006), Marburg virus entry is independent of cathepsin B (Gnirss et al. 2012; Misasi et al. 2012). Further, the intra-GP1 disulfide bond formed by C121 and C147 in ebolavirus GPs does not exist in Marburg virus. Instead, these two cysteines are replaced by L106 and H131. As a result, the polypeptides that contain these residues differ in structure and flexibility. In ebolaviruses, the polypeptide bearing C147 (residues 145–150) turns inward toward the trimer center to disulfide bond with C121. In Marburg virus, the equivalent polypeptide turns outward to solvent, away from the trimer center. A second difference lies at the base of the cathepsin cleavage loop. In Marburg virus, these residues pack against the outside of the GP2 fusion loop and form a clear alpha helix. In ebolaviruses, the equivalent residues form a loop rather than a helix (Bale et al. 2012a; Dias et al. 2011; Lee et al. 2008; Zhao et al. 2016).

In Ebola virus, there is a large tunnel between monomers, with the entrance "plugged" by the two Phe residues in a loop containing an Asp-Phe-Phe sequence, also known as the "DFF" lid (Zhao et al. 2016). This interaction is likely important for Ebola virus stability as displacement of the DFF lid by the small molecules toremifene and ibuprofen destabilizes the GP structure (Zhao et al. 2016). In Marburg virus, however, the DFF lid is replaced instead by a helical structure (Hashiguchi et al. 2015).

After entry in the endosome, host cathepsin enzymes cleave filovirus GPs. This cleavage event strips the mucin-like domain and glycan cap from GP leaving the receptor-binding head well exposed (Bornholdt et al. 2016; Wang et al. 2016). This receptor-binding head forms a wave crest-and-trough morphology with a hydrophilic crest atop a hydrophobic wave trough. The essential domain C of the receptor Niemann Pick C1 (NPC1) binds into that GP trough (Fig. 17) using two protruding loops (Wang et al. 2016). The first loop contacts one side of the trough, while the second loop binds into the trough itself. In the second loop are residues F503, F504, and Y506, which make the majority of the contact there to GP (Wang et al. 2016).

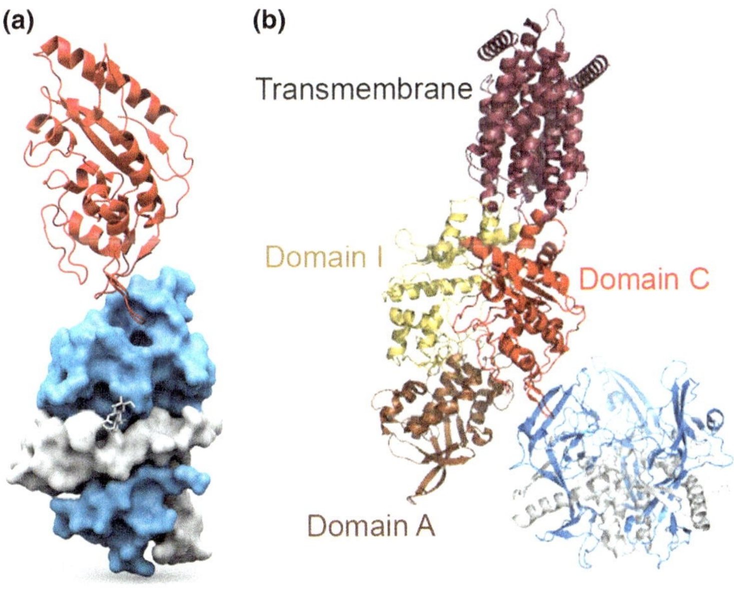

Fig. 17 NPC1 in complex with Ebola virus GP. **a** The 2.3 Å crystal structure of cleaved GP in complex with domain C of NPC1 is illustrated with a surface representation for one GP monomer (GP1 *blue* and GP2 *white*) and a ribbon model for domain C in *red* (Wang et al. 2016) (PDB:5F1B). An extended loop of NPC1 is visible binding down into the recessed, hydrophobic trough that becomes exposed upon cathepsin cleavage of GP. **b** A lower resolution, 6.6 Å cryo-EM reconstruction of intact NPC1 in complex with a cleaved GP trimer. Domains A and I and the thirteen transmembrane coils of NPC1 are also visible (Gong et al. 2016) (PDB: 5JNX)

Notably, F503 is a key determinant of filovirus tropism (Ndungo 2016; Ng 2015). An antibody against this receptor-binding site, MR78, also uses a Phe and Tyr residue to interact with the site in a similar fashion (Hashiguchi et al. 2015), as do the native residues of the Ebola virus glycan cap.

Conformational adjustments do occur upon NPC1 domain C binding, such as release of the short 3_{10} helix in the β3-α1 loop of GP1 from its contact with the internal fusion loop of GP2 (Wang et al. 2016). However, the NPC1–GP structures have not "sprung" into the post-fusion six-helix bundle (Gong et al. 2016; Wang et al. 2016). Indeed, it is not yet clear exactly what does spring GP from its prefusion to post-fusion structures. Low pH, cleavage and binding of NPC1 domain C are all essential for Ebola virus infection (Carette et al. 2011; Chandran 2005; Côté et al. 2011; Miller et al. 2012), but these three things together do not drive conformational change in GP (Bale 2011; Bornholdt et al. 2016; Wang et al. 2016).

Many structures of filovirus GP have been determined in complex with antibodies, revealing sites targeted by antibody elicited by immunization or natural infection. Some sites involve residues of GP2 wrapped about GP1, such as the "base" and the internal fusion loop. The "stalk" epitopes contain only GP2. Other sites involve only residues in GP1, such as the glycan cap, the receptor-binding site or "head" subdomain, and the mucin-like domain.

The "base" is a GP1/GP2 containing epitope at the bottom of the GP core that is recognized by the Ebola virus-specific human antibody KZ52 (Lee et al. 2008) (Fig. 13), the Sudan virus-specific antibody 16F6 (Dias et al. 2011) (Fig. 14), the Ebola-specific antibodies 2G4 and 4G7 (Murin et al. 2014; Pallesen et al. 2016; Tran et al. 2016) (Fig. 18b), mAb 114 (Misasi et al. 2016) (Fig. 18a) and others. More of the antibody contact at this site is to GP2 than GP1, yet GP1 is critical as it maintains GP2 in its prefusion conformation. Antibodies against the base likely neutralize by mechanically interfering with conformational changes required for infection.

Antibodies against the base are typically virus species specific. A major structural difference that may cause the virus specificity is the GP2 *N*-terminal peptide, which differs in sequence, structure and flexibility between different ebolaviruses (Bale et al. 2012a). In Ebola virus, its sequence is EAIVNAQPK, and it is hydrogen bonded to the GP core and forms a critical part of the KZ52 epitope, bound between the antibody heavy and light chains (Lee et al. 2008). In Sudan virus, the GP2 *N*-terminal peptide sequence is different (QVNTRATGK), disordered, and not bound by 16F6, the one anti-Sudan virus antibody visualized against this site (Bale et al. 2012a; Dias et al. 2011). Instead, 16F6 binds the underlying core of GP beneath the peptide. Studies by deuterium exchange mass spectrometry suggest that the GP2 *N*-terminal peptide may be fundamentally more mobile in SUDV than in EBOV, probably because of the presence of a Gly (EBOV) at position 509 versus a Pro (SUDV), where the peptide anchors to the core (Bale et al. 2012a). Further, EBOV contains an Asn at position 506 and a Gln at position 508, which hydrogen bond to the core. A Gln508Arg point mutation allows GP to escape recognition by multiple base-binding antibodies (Audet et al. 2014), presumably by breaking the hydrogen

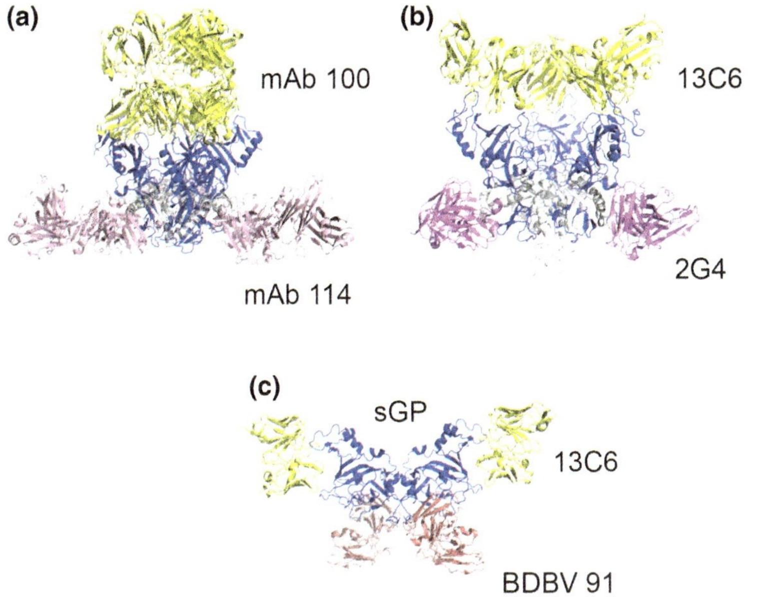

Fig. 18 GP/sGP cross-reactive antibodies. **a** 6.7 Å crystal structure of a ternary complex of mAb 100 against the apex of GP1 (*yellow*) and mAb 114 against the base with Ebola virus GP (GP1 *blue* and GP2 *white* (PDB:5FHC) (Misasi et al. 2016). **b** 4.3Å cryo-EM structure of the ternary complex of antibodies 13C6 (*yellow*) against the glycan cap, and 2G4 (*pink*) against the base with Ebola virus GP (PDB:5KEL) (Pallesen et al. 2016). Only the variable portions of each Fab fragment could be built. **c** 5.5 Å cryo-EM structures of the complex of antibodies 13C6 (*yellow*) and BDBV 91 (*pink*) with sGP (*blue*) (PDB:5KEM) (Pallesen et al. 2016). Only the variable portions of the Fab fragments could be built

bonds necessary to maintain the GP2 peptide in the right position for antibody binding.

The internal fusion loop is another GP2/GP1-containing epitope that requires the assembled trimer for binding. Below the fusion loop and the base is the stalk region, which is formed by heptad repeat 2 of GP2. Both the fusion loop and stalk are relatively conserved among filoviruses, and antibodies against these sites have been identified in immunized animals (Furuyama et al. 2016; Keck et al. 2016) and in human survivors of Ebola virus and Bundibugyo virus infection (Bornholdt et al. 2016; Flyak et al. 2016).

Tomography of the viral surface suggests that one GP spike exists per 250 nm^2 on a virus-like particle surface, with an average center-to-center spacing between GP spikes of 15 ± 4 nm (Tran et al. 2016). Hence, it is likely possible for an antibody against an epitope like the base or the stalk to bridge neighboring spikes (Tran et al. 2016).

In the GP1 subunit, major epitopes include the glycan cap, the receptor-binding site and the mucin-like domain. Many antibodies against GP1 also bind sGP, such as 13C6 (Murin et al. 2014; Pallesen et al. 2016; Tran et al. 2016) (Fig. 18b, c) and

Q314 (Zhang et al. 2016) against the glycan cap region, and mAb 100 (Misasi et al. 2016; Zhang et al. 2016) (Fig. 18a), Q206, Q411 (Misasi et al. 2016; Zhang et al. 2016), and FVM04 (Howell et al. 2016) which anchor both the glycan cap and the crest of the receptor-binding site. Many anti-GP1 antibodies bind the abundant sGP preferentially.

Curiously, binding of FVM04 at the apex of GP is enhanced by mutations at the base of GP—distal from the FVM04 epitope (Howell et al. 2016). This suggests that mutations and perhaps antibody binding at the base alter the GP structure.

Several competition groups of antibodies have also been identified in the mucin-like domain. These antibodies tend to be more weakly neutralizing, but offer some protection in mouse models (Wilson et al. 2000). One of these has an unusual structure: mAb 13F6-1-2 uses an extremely rare Vλx light chain to recognize GP residues 405–413 in a broad, flat groove that reaches atypically across both heavy and light chains, rather than between them (Lee et al. 2008b) (Fig. 19a). Antibody 14G7, in contrast, binds a tandem β turn conformation formed by GP residues 482–491 in a pocket between heavy and light chains (Olal et al. 2012) (Fig. 19b). The β turn conformation effectively doubles contact surface with the antibody combining regions.

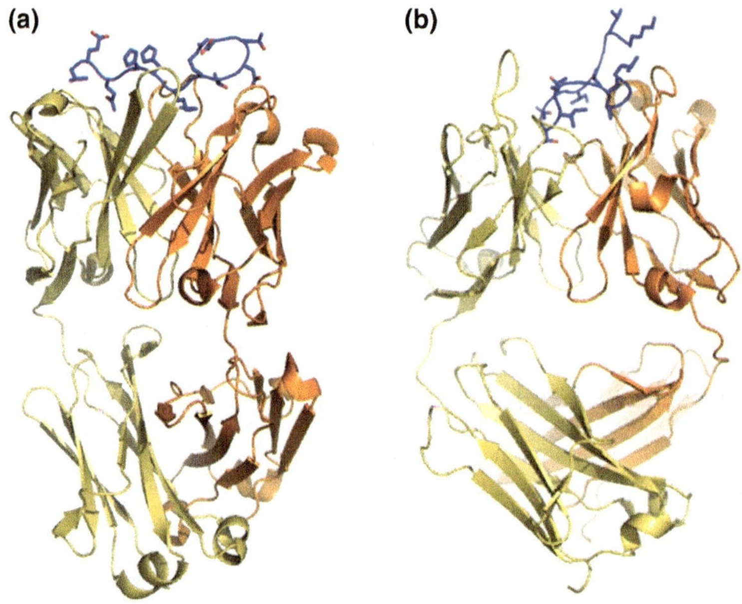

Fig. 19 Antibodies against the mucin-like domain. **a** 2.0 Å crystal structure of antibody 13F6-1-2 in complex with its GP mucin-like domain linear epitope, residues 405–413 (*blue*). The GP peptides lies in a broad, flat groove across both heavy (*orange*) and light chains (*yellow*), (Lee et al. 2008) (PDB: 2QHR). **b** 2.8 Å structure of antibody 4G7 recognizes GP residue 482–491 in a tandem β turn conformation between heavy (*orange*) and light chains (*yellow*), (Olal et al. 2012) (PDB: 2Y6S)

Notable differences between the mucin-like domains of Marburg virus and Ebola virus are relevant for antibody recognition. In ebolaviruses, furin cleaves *C*-terminal of the mucin-like domain so that the domain remains attached solely to the GP1 subunit. In Marburg virus, however, furin cleaves within the mucin-like domain, so that some of it remains attached to GP1 and another potion is attached to GP2. This additional GP2 anchor in Marburg virus results in an alternate position of the mucin-like domain, lower on the GP spike. This lower position is thought to cover the sides of the GP core, while leaving the upper receptor-binding site more exposed. In contrast, in Ebola virus, the solely GP1-anchored mucin-like domain is positioned upwards, leaving the base of the GP more exposed (Hashiguchi et al. 2015). Consistent with alternate positions of the mucin-like domain, multiple antibodies have been noted against the hydrophobic receptor-binding trough elicited during Marburg virus infection (Flyak et al. 2015; Hashiguchi et al. 2015). This site is masked in ebolaviruses by the glycan cap. Further, the alternate furin cleavage site in Marburg virus, which leaves a portion of the mucin-like domain attached to GP2 results in a Marburg-specific, GP2-containing antibody epitope termed the "wing" (Fusco et al. 2015).

The structure of the secreted dimeric sGP was recently revealed by cryo-EM (Pallesen et al. 2016). Each sGP monomer contains residues that roughly correspond to the base, head and glycan cap of GP1. As a result, both GP and sGP are recognized by antibodies that target epitopes like the glycan cap. In sGP, the two monomers are angled outward from each other. As a result, two copies of the anti-glycan cap antibody 13C6 face each other when bound to sGP, but are parallel when bound to GP (Pallesen et al. 2016) (Fig. 18b, c).

Dimerization of sGP is achieved by a hydrophobic patch formed largely by residues that would correspond to the base subdomain of GP1 and some of the head, as well as by C53-C53 and C306-C306 inter-chain disulfide bonds. The C306-C306 disulfide bond has been detected biochemically but was not visible in the cryo-EM structure (Barrientos et al. 2004; Falzarano et al. 2006; Pallesen et al. 2016) (Fig. 15c). C306 is unique to sGP and is the result of a shift in reading frame from that which encodes GP (Sanchez et al. 1996). C53 is shared by both GP and sGP, but in GP, C53 of GP1 anchors to C609 of GP2 (Fig. 15d). The fold of the core of sGP remains similar to that of GP1, explaining why so many antibodies cross-react between the two proteins. Other antibodies that recognize quaternary epitopes specific for sGP have also been noted (Pallesen et al. 2016).

8 L

Due to difficulties in producing sufficient amounts of material for in vitro characterizations, very little is known about the structure of the filovirus RNA-dependent RNA polymerase, L. The L protein is the largest gene and protein produced by the virus, varying from 2196 to 2331 amino acids in length. The protein is expressed as a single polypeptide carrying all of the enzymatic activities for RNA synthesis.

Much of what has been hypothesized about filovirus L proteins is based on the homology amongst other non-segmented negative strand virus L proteins. These L proteins each contain six conserved regions, but only a handful of highly conserved residues (Fig. 20) (Poch et al. 1990). Studies of vesicular stomatitis Indiana virus L have indicated that regions I–IV form the core polymerase domain (Rahmeh et al. 2010). Region III contains sequence motifs A–D found in all viral RNA-dependent polymerases (Poch et al. 1990). These sequence motifs contain the essential catalytic residues for RNA synthesis. Region V contains the polyribonucleotide transferase activity (PRNTase), which modifies nascent 5′ triphosphorylated mRNA transcripts to create GpppG products through a covalent intermediate with a histidine in a conserved "HR" motif (Li et al. 2008; Ogino et al. 2010). This nascent transcript is then G^7 and 2'O methylated to form the mRNA cap structure through the methyltransferase (MTase) activity of region VI (Galloway et al. 2008).

The structure of L remains the largest mystery in filovirus structural biology, but much remains to be learned for the remaining proteins as well. We do not yet have a high-resolution view of the assembled nucleocapsid, nor do we understand all the interactions of its component proteins that achieve replication and transcription, or how those complexes switch between those separate roles. We have not yet visualized a complete VP35 nor a complete VP30. We do not yet entirely understand what drives changes in function for multifunctional proteins like VP35 and VP24: are there structural adjustments important to change from a role in the nucleoprotein assemblies versus a role in immunosuppression? Or, are the distinct functions simply driven by interaction with separate binding partners at separate stages of the viral life cycle or in separate subcellular locations? We have not yet illuminated the intermediate steps by which GP undergoes conformational change. Finally, structural differences that remain to be determined in many of the proteins, whether subtle or sizeable, may dictate why filoviruses differ in virulence in human and in animal models.

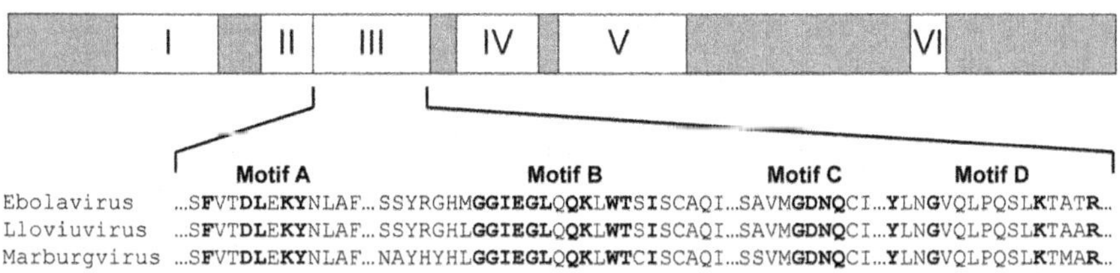

Fig. 20 Sequence alignment of filovirus L proteins. Region III of the L polymerase protein contains sequence motifs similar among non-segmented negative-sense RNA viruses (*bold*)

References

Adu-Gyamfi E, Digman MA, Gratton E, Stahelin RV (2012) Investigation of Ebola VP40 assembly and oligomerization in live cells using number and brightness analysis. Biophys J 102:2517–2525

Adu-Gyamfi E, Soni SP, Xue Y, Digman MA, Gratton E, Stahelin RV (2013) The Ebola virus matrix protein penetrates into the plasma membrane: A key step in VP40 oligomerization and viral egress. J Biol Chem 288:5779–5789

Adu-Gyamfi E, Soni SP, Jee CS, Digman MA, Gratton E, Stahelin RV (2014) A loop region in the *N*-terminal domain of Ebola virus VP40 is important in viral assembly, budding, and egress. Viruses 6:3837–3854

Adu-Gyamfi E, Johnson KA, Fraser ME, Scott JL, Soni SP, Jones KR, Digman MA, Gratton E, Tessier CR, Stahelin RV (2015) Host Cell Plasma Membrane Phosphatidylserine Regulates the Assembly and Budding of Ebola Virus. J Virol 89:9440–9453

Audet J, Wong G, Wang H, Lu G, Gao GF, Kobinger G, Qiu X (2014) Molecular characterization of the monoclonal antibodies composing ZMAb: a protective cocktail against Ebola virus. Scientific Reports 4:6881

Bale S, Halfmann P, Fusco ML, Kawaoka Y, Saphire EO (2011) Ebola virus glycoprotein needs an additional trigger, beyond proteolytic priming for membrane fusion. PLoS Negl Trop Dis 5: e1395

Bale S, Dias JM, Fusco ML, Hashiguchi T, Wong AC, Liu T, Keuhne AI, Li S, Woods VL Jr, Chandran K, Dye JM, Saphire EO (2012a) Structural basis for differential neutralization of ebolaviruses. Viruses 4:447–470

Bale S, Julien JP, Bornholdt ZA, Kimberlin CR, Halfmann P, Zandonatti MA, Kunert J, Kroon GJ, Kawaoka Y, Macrae IJ, Wilson IA, Saphire EO (2012b) Marburg Virus VP35 Can Both Fully Coat the Backbone and Cap the Ends of dsRNA for Interferon Antagonism. PLoS Pathog 8: e1002916

Bankamp B, Horikami SM, Thompson PD, Huber M, Billeter M, Moyer SA (1996) Domains of the measles virus N protein required for binding to P protein and self-assembly. Virology 216:272–277

Barrientos LG, Martin AM, Rollin PE, Sanchez A (2004) Disulfide bond assignment of the Ebola virus secreted glycoprotein SGP. Biochem Biophys Res Commun 323:696–702

Becker S, Rinne C, Hofsass U, Klenk HD, Muhlberger E (1998) Interactions of Marburg virus nucleocapsid proteins. Virology 249:406–417

Beniac DR and Booth TF (2017) Structure of the Ebola virus glycoprotein spike within the virion envelope at 11 Å resolution. Scientific Reports. 7:46374

Beniac DR, Melito PL, Devarennes SL, Hiebert SL, Rabb MJ, Lamboo LL, Jones SM, Booth TF (2012) The organisation of Ebola virus reveals a capacity for extensive, modular polyploidy. PLoS ONE 7:e29608

Bharat TA, Riches JD, Kolesnikova L, Welsch S, Krahling V, Davey N, Parsy ML, Becker S, Briggs JA (2011) Cryo-electron tomography of Marburg virus particles and their morphogenesis within infected cells. PLoS Biol 9:e1001196

Bharat TA, Noda T, Riches JD, Kraehling V, Kolesnikova L, Becker S, Kawaoka Y, Briggs JA (2012) Structural dissection of Ebola virus and its assembly determinants using cryo-electron tomography. Proc Natl Acad Sci USA 109:4275–4280

Blondot ML, Dubosclard V, Fix J, Lassoued S, Aumont-Nicaise M, Bontems F, Eleouet JF, Sizun C (2012) Structure and functional analysis of the RNA- and viral phosphoprotein-binding domain of respiratory syncytial virus M2-1 protein. PLoS Pathog 8: e1002734

Bornholdt ZA, Noda T, Abelson DM, Halfmann P, Wood MR, Kawaoka Y, Saphire EO (2013) Structural rearrangement of ebola virus VP40 begets multiple functions in the virus life cycle. Cell 154:763–774

Bornholdt ZA, Ndungo E, Fusco ML, Bale S, Flyak AI, Crowe JE Jr, Chandran K, Saphire EO (2016a) Host-Primed Ebola Virus GP Exposes a Hydrophobic NPC1 Receptor-Binding Pocket, Revealing a Target for Broadly Neutralizing Antibodies. MBio 7:e02154–02115

Bornholdt ZA, Turner HL, Murin CD, Li W, Sok D, Souders CA, Piper AE, Goff A, Shamblin JD, Wollen SE, Sprague TR, Fusco ML, Pommert KB, Cavacini LA, Smith HL, Klempner M, Reimann KA, Krauland E, Gerngross TU, Wittrup KD, Saphire EO, Burton DR, Glass PJ, Ward AB, Walker LM (2016b) Isolation of potent neutralizing antibodies from a survivor of the 2014 Ebola virus outbreak. Science 351:1078–1083

Bruhn JF, Kirchdoerfer RN, Urata SM, Li S, Tickle IJ, Bricogne G, Saphire EO (2016) Crystal structure of the Marburg virus VP35 oligomerization domain. J Virol 91:e01085-16

Cardenas WB, Loo YM, Gale M Jr, Hartman AL, Kimberlin CR, Martinez-Sobrido L, Saphire EO, Basler CF (2006) Ebola virus VP35 protein binds double-stranded RNA and inhibits alpha/beta interferon production induced by RIG-I signaling. J Virol 80:5168–5178

Carette JE, Raaben M, Wong AC, Herbert AS, Obernosterer G, Mulherkar N, Kuehne AI, Kranzusch PJ, Griffin AM, Ruthel G, Dal Cin P, Dye JM, Whelan SP, Chandran K, Brummelkamp TR (2011) Ebola virus entry requires the cholesterol transporter Niemann-Pick C1. Nature 477:340–343

Chandran K, Sullivan NJ, Felbor U, Whelan SP, Cunningham JM (2005) Endosomal proteolysis of the Ebola virus glycoprotein is necessary for infection. Science 308:1643–1645

Clifton MC, Kirchdoerfer RN, Atkins K, Abendroth J, Raymond A, Grice R, Barnes S, Moen S, Lorimer D, Edwards TE, Myler PJ, Saphire EO (2014) Structure of the Reston ebolavirus VP30 *C*-terminal domain. Acta Crystallogr F Struct Biol Commun 70:457–460

Côté M, Misasi J, Ren T, Bruchez A, Lee K, Filone CM, Hensley L, Li Q, Ory D, Chandran K, Cunningham J (2011) Small molecule inhibitors reveal Niemann-Pick C1 is essential for Ebola virus infection. Nature 477:344–348

Dessen A, Volchkov V, Dolnik O, Klenk HD, Weissenhorn W (2000) Crystal structure of the matrix protein VP40 from Ebola virus. EMBO J 19:4228–4236

Dias JM, Kuehne AI, Abelson DM, Bale S, Wong AC, Halfmann P, Muhammad MA, Fusco ML, Zak SE, Kang E, Kawaoka Y, Chandran K, Dye JM, Saphire EO (2011) A shared structural solution for neutralizing ebolaviruses. Nat Struct Mol Biol 18:1424–1427

Dolnik O, Volchkova V, Garten W, Carbonnelle C, Becker S, Kahnt J, Stroher U, Klenk HD, Volchkov V (2004) Ectodomain shedding of the glycoprotein GP of Ebola virus. EMBO J 23:2175–2184

Dong S, Yang P, Li G, Liu B, Wang W, Liu X, Xia B, Yang C, Lou Z, Guo Y, Rao Z (2015) Insight into the Ebola virus nucleocapsid assembly mechanism: crystal structure of Ebola virus nucleoprotein core domain at 1.8 a resolution. Protein Cell 6:351–362

Dziubanska PJ, Derewenda U, Ellena JF, Engel DA, Derewenda ZS (2014) The structure of the *C*-terminal domain of the Zaire ebolavirus nucleoprotein. Acta Crystallogr. 70:2420–2429

Edwards MR, Johnson B, Mire CE, Xu W, Shabman RS, Speller LN, Leung DW, Geisbert TW, Amarasinghe GK, Basler CF (2014) The Marburg virus VP24 protein interacts with Keap1 to activate the cytoprotective antioxidant response pathway. Cell Rep 6:1017–1025

Edwards MR, Liu G, Mire CE, Sureshchandra S, Luthra P, Yen B, Shabman RS, Leung DW, Messaoudi I, Geisbert TW, Amarasinghe GK, Basler CF (2016) Differential Regulation of Interferon Responses by Ebola and Marburg Virus VP35 Proteins. Cell Rep 14:1632–1640

Elliott LH, Kiley MP, McCormick JB (1985) Descriptive analysis of Ebola virus proteins. Virology 147:169–176

Enterlein S, Volchkov V, Weik M, Kolesnikova L, Volchkova V, Klenk HD, Muhlberger E (2006) Rescue of recombinant Marburg virus from cDNA is dependent on nucleocapsid protein VP30. J Virol 80:1038–1043

Fabozzi G, Nabel CS, Dolan MA, Sullivan NJ (2011) Ebolavirus proteins suppress the effects of small interfering RNA by direct interaction with the mammalian RNA interference pathway. J Virol 85:2512–2523

Falzarano D, Krokhin O, Wahl-Jensen V, Seebach J, Wolf K, Schnittler HJ, Feldmann H (2006) Structure-function analysis of the soluble glycoprotein, sGP, of Ebola virus. ChemBioChem 7:1605–1611

Feldmann H, Nichol ST, Klenk HD, Peters CJ, Sanchez A (1994) Characterization of filoviruses based on differences in structure and antigenicity of the virion glycoprotein. Virology 199:469–473

Feng Z, Cerveny M, Yan Z, He B (2007) The VP35 protein of Ebola virus inhibits the antiviral effect mediated by double-stranded RNA-dependent protein kinase PKR. J Virol 81:182–192

Flyak AI, Ilinykh PA, Murin CD, Garron T, Shen X, Fusco ML, Hashiguchi T, Bornholdt ZA, Slaughter JC, Sapparapu G, Klages C, Ksiazek TG, Ward AB, Saphire EO, Bukreyev A, Crowe JE Jr (2015) Mechanism of human antibody-mediated neutralization of Marburg virus. Cell 160:893–903

Flyak AI, Shen X, Murin CD, Turner HL, David JA, Fusco ML, Lampley R, Kose N, Ilinykh PA, Kuzmina N, Branchizio A, King H, Brown L, Bryan C, Davidson E, Doranz BJ, Slaughter JC, Sapparapu G, Klages C, Ksiazek TG, Saphire EO, Ward AB, Bukreyev A, Crowe JE Jr (2016) Cross-reactive and potent neutralizing antibody responses in human survivors of natural ebolavirus infection. Cell 164:392–405

Furuyama W, Marzi A, Nanbo A, Haddock E, Maruyama J, Miyamoto H, Igarashi M, Yoshida R, Noyori O, Feldmann H, Takada A (2016) Discovery of an antibody for pan-ebolavirus therapy. Scientific reports 6:20514

Fusco ML, Hashiguchi T, Cassan R, Biggins JE, Murin CD, Warfield KL, Li S, Holtsberg FW, Shulenin S, Vu H, Olinger GG, Kim do H, Whaley KJ, Zeitlin L, Ward AB, Nykiforuk C, Aman MJ, Berry JD, Saphire EO (2015) Protective mAbs and cross-reactive mAbs Raised by immunization with engineered marburg virus GPs. PLoS Pathog 11:e1005016

Galloway SE, Richardson PE, Wertz GW (2008) Analysis of a structural homology model of the 2'-O-ribose methyltransferase domain within the vesicular stomatitis virus L protein. Virology 382:69–82

Gc JB, Gerstman BS, Stahelin RV, Chapagain PP (2016) Ebola virus VP40 protein VP40 hexamer enhances clustering of PI(4,5)P2 lipids in the plasma membrane. Physical Chemistry Chemical Physics 18:28409–28417

Geisbert TW, Jahrling PB (1995) Differentiation of filoviruses by electron microscopy. Virus Res 39:129–150

Gerard FC, Ribeiro Ede A Jr, Leyrat C, Ivanov I, Blondel D, Longhi S, Ruigrok RW, Jamin M (2009) Modular organization of rabies virus phosphoprotein. J Mol Biol 388:978–996

Gnirss K, Kühl A, Karsten C, Glowacka I, Bertram S, Kaup F, Hofmann H, Pöhlmann S (2012) Cathepsins B and L activate Ebola but not Marburg virus glycoproteins for efficient entry into cell lines and macrophages independent of TMPRSS2 expression. Virology 424:3–10

Gomis-Ruth FX, Dessen A, Timmins J, Bracher A, Kolesnikowa L, Becker S, Klenk HD, Weissenhorn W (2003) The matrix protein VP40 from Ebola virus octamerizes into pore-like structures with specific RNA binding properties. Structure 11:423–433

Gong X, Qian H, Zhou X, Wu J, Wan T, Cao P, Huang W, Zhao X, Wang X, Wang P, Shi Y, Gao GF, Zhou Q, Yan N (2016) Structural Insights into the Niemann-Pick C1 (NPC1)-Mediated Cholesterol Transfer and Ebola Infection. Cell 165:1467–1478

Green TJ, Zhang X, Wertz GW, Luo M (2006) Structure of the vesicular stomatitis virus nucleoprotein-RNA complex. Science 313:357–360

Groseth A, Charton JE, Sauerborn M, Feldmann F, Jones SM, Hoenen T, Feldmann H (2009) The Ebola virus ribonucleoprotein complex: a novel VP30-L interaction identified. Virus Res 140:8–14

Hartlieb B, Modrof J, Muhlberger E, Klenk HD, Becker S (2003) Oligomerization of Ebola virus VP30 is essential for viral transcription and can be inhibited by a synthetic peptide. J Biol Chem 278:41830–41836

Hartlieb B, Muziol T, Weissenhorn W, Becker S (2007) Crystal structure of the *C*-terminal domain of Ebola virus VP30 reveals a role in transcription and nucleocapsid association. Proc Natl Acad Sci USA 104:624–629

Harty RN (2009) No exit: targeting the budding process to inhibit filovirus replication. Antiviral Res 81:189–197

Harty RN, Brown ME, Wang G, Huibregtse J, Hayes FP (2000) A PPxY motif within the VP40 protein of Ebola virus interacts physically and functionally with a ubiquitin ligase: implications for filovirus budding. Proc Natl Acad Sci USA 97:13871–13876

Hashiguchi T, Fusco ML, Bornholdt ZA, Lee JE, Flyak AI, Matsuoka R, Kohda D, Yanagi Y, Hammel M, Crowe JE Jr, Saphire EO (2015) Structural basis for Marburg virus neutralization by a cross-reactive human antibody. Cell 160:904–912

Hoenen T, Volchkov V, Kolesnikova L, Mittler E, Timmins J, Ottmann M, Reynard O, Becker S, Weissenhorn W (2005) VP40 octamers are essential for Ebola virus replication. J Virol 79:1898–1905

Hoenen T, Biedenkopf N, Zielecki F, Jung S, Groseth A, Feldmann H, Becker S (2010a) Oligomerization of Ebola virus VP40 is essential for particle morphogenesis and regulation of viral transcription. J Virol 84:7053–7063

Hoenen T, Jung S, Herwig A, Groseth A, Becker S (2010b) Both matrix proteins of Ebola virus contribute to the regulation of viral genome replication and transcription. Virology 403:56–66

Howell KA, Qiu X, Brannan JM, Bryan C, Davidson E, Holtsberg FW, Wec AZ, Shulenin S, Biggins JE, Douglas R, Enterlein SG, Turner HL, Pallesen J, Murin CD, He S, Kroeker A, Vu H, Herbert AS, Fusco ML, Nyakatura EK, Lai JR, Keck ZY, Foung SK, Saphire EO, Zeitlin L, Ward AB, Chandran K, Doranz BJ, Kobinger GP, Dye JM, Aman MJ (2016) Antibody treatment of ebola and sudan virus infection via a uniquely exposed epitope within the glycoprotein receptor-binding site. Cell Rep 15:1514–1526

Huang Y, Xu L, Sun Y, Nabel G (2002) The assembly of ebola virus nucleocapsid requires virion-associated proteins 35 and 24 and posttranslational modification of nucleoprotein. Mol Cell 10:307

Jasenosky LD, Neumann G, Lukashevich I, Kawaoka Y (2001) Ebola virus vp40-induced particle formation and association with the lipid bilayer. J Virol 75:5205–5214

John SP, Wang T, Steffen S, Longhi S, Schmaljohn CS, Jonsson CB (2007) Ebola virus VP30 is an RNA binding protein. J Virol 81:8967–8976

Johnson RF, Bell P, Harty RN (2006) Effect of Ebola virus proteins GP, NP and VP35 on VP40 VLP morphology. Virol J 3:31

Johnson B, Li J, Adhikari J, Edwards MR, Zhang H, Schwarz T, Leung DW, Basler CF, Gross ML, Amarasinghe GK (2016a) Dimerization controls marburg virus VP24-dependent modulation of host antioxidative stress responses. J Mol Biol 428:3483–3494

Johnson KA, Taghon GJ, Scott JL, Stahelin RV (2016b) The Ebola Virus matrix protein, VP40, requires phosphatidylinositol 4,5-bisphosphate (PI(4,5)P2) for extensive oligomerization at the plasma membrane and viral egress. Sci Rep 6:19125

Keck ZY, Enterlein SG, Howell KA, Vu H, Shulenin S, Warfield KL, Froude JW, Araghi N, Douglas R, Biggins J, Lear-Rooney CM, Wirchnianski AS, Lau P, Wang Y, Herbert AS, Dye JM, Glass PJ, Holtsberg FW, Foung SK, Aman MJ (2016) Macaque monoclonal antibodies targeting novel conserved epitopes within filovirus glycoprotein. J Virol 90:279–291

Kimberlin CR, Bornholdt ZA, Li S, Woods VL Jr, MacRae IJ, Saphire EO (2010) Ebolavirus VP35 uses a bimodal strategy to bind dsRNA for innate immune suppression. Proc Natl Acad Sci USA 107:314–319

Kirchdoerfer RN, Abelson DM, Li S, Wood MR, Saphire EO (2015) Assembly of the Ebola virus nucleoprotein from a chaperoned VP35 complex. Cell Rep 12:140–149

Kirchdoerfer RN, Moyer CL, Abelson DM, Saphire EO (2016) The Ebola virus VP30-NP interaction is a regulator of viral RNA synthesis. PLoS Pathog 12:e1005937

Koellhoffer JF, Malashkevich VN, Harrison JS, Toro R, Bhosle RC, Chandran K, Almo SC, Lai JR (2012) Crystal structure of the Marburg virus GP2 core domain in its postfusion conformation. Biochem 51:7665–7675

Kolesnikova L, Ryabchikova E, Shestopalov A, Becker S (2007) Basolateral budding of Marburg virus: VP40 retargets viral glycoprotein GP to the basolateral surface. J Infect Dis 196 S2: S232–S236

Kondratowicz AS, Lennemann NJ, Sinn PL, Davey RA, Hunt CL, Moller-Tank S, Meyerholz DK, Rennert P, Mullins RF, Brindley M, Sandersfeld LM, Quinn K, Weller M, McCray PB Jr, Chiorini J, Maury W (2011) T-cell immunoglobulin and mucin domain 1 (TIM-1) is a receptor for Zaire Ebolavirus and Lake Victoria Marburgvirus. Proc Natl Acad Sci USA 108:8426–8431

Kuhn JH, Andersen KG, Bao Y, Bavari S, Becker S, Bennett RS, Bergman NH, Blinkova O, Bradfute S, Brister JR, Bukreyev A, Chandran K, Chepurnov AA, Davey RA, Dietzgen RG, Doggett NA, Dolnik O, Dye JM, Enterlein S, Fenimore PW, Formenty P, Freiberg AN, Garry RF, Garza NL, Gire SK, Gonzalez JP, Griffiths A, Happi CT, Hensley LE, Herbert AS, Hevey MC, Hoenen T, Honko AN, Ignatyev GM, Jahrling PB, Johnson JC, Johnson KM, Kindrachuk J, Klenk HD, Kobinger G, Kochel TJ, Lackemeyer MG, Lackner DF, Leroy EM, Lever MS, Muhlberger E, Netesov SV, Olinger GG, Omilabu SA, Palacios G, Panchal RG, Park DJ, Patterson JL, Paweska JT, Peters CJ, Pettitt J, Pitt L, Radoshitzky SR, Ryabchikova EI, Saphire EO, Sabeti PC, Sealfon R, Shestopalov AM, Smither SJ, Sullivan NJ, Swanepoel R, Takada A, Towner JS, van der Groen G, Volchkov VE, Volchkova VA, Wahl-Jensen V, Warren TK, Warfield KL, Weidmann M, Nichol ST (2014) Filovirus RefSeq entries: evaluation and selection of filovirus type variants, type sequences, and names. Viruses 6:3663–3682

Lee JE, Fusco ML, Hessell AJ, Oswald WB, Burton DR, Saphire EO (2008a) Structure of the Ebola virus glycoprotein bound to an antibody from a human survivor. Nature 454:177–182

Lee JE, Kuehne A, Abelson DM, Fusco ML, Hart MK, Saphire EO (2008b) Complex of a protective antibody with its Ebola virus GP peptide epitope: unusual features of a V lambda x light chain. J Mol Biol 375:202–216

Leung DW, Ginder ND, Fulton DB, Nix J, Basler CF, Honzatko RB, Amarasinghe GK (2009) Structure of the Ebola VP35 interferon inhibitory domain. Proc Natl Acad Sci USA 106:411–416

Leung DW, Prins KC, Borek DM, Farahbakhsh M, Tufariello JM, Ramanan P, Nix JC, Helgeson LA, Otwinowski Z, Honzatko RB, Basler CF, Amarasinghe GK (2010) Structural basis for dsRNA recognition and interferon antagonism by Ebola VP35. Nat Struct Mol Biol 17:165–172

Leung DW, Borek D, Luthra P, Binning JM, Anantpadma M, Liu G, Harvey IB, Su Z, Endlich-Frazier A, Pan J, Shabman RS, Chiu W, Davey RA, Otwinowski Z, Basler CF, Amarasinghe GK (2015) An intrinsically disordered peptide from Ebola virus VP35 controls viral RNA synthesis by modulating nucleoprotein-RNA interactions. Cell Rep 11:376–389

Li J, Rahmeh A, Morelli M, Whelan SP (2008) A conserved motif in region v of the large polymerase proteins of nonsegmented negative-sense RNA viruses that is essential for mRNA capping. J Virol 82:775–784

Licata JM, Simpson-Holley M, Wright NT, Han Z, Paragas J, Harty RN (2003) Overlapping motifs (PTAP and PPEY) within the Ebola virus VP40 protein function independently as late budding domains: involvement of host proteins TSG101 and VPS-4. J Virol 77:1812–1819

Luthra P, Ramanan P, Mire CE, Weisend C, Tsuda Y, Yen B, Liu G, Leung DW, Geisbert TW, Ebihara H, Amarasinghe GK, Basler CF (2013) Mutual antagonism between the Ebola virus VP35 protein and the RIG-I activator PACT determines infection outcome. Cell Host Microbe 14:74–84

Malashkevich VN, Schneider BJ, McNally ML, Milhollen MA, Pang JX, Kim PS (1999) Core structure of the envelope glycoprotein GP2 from Ebola virus at 1.9-A resolution. Proc Natl Acad Sci USA 96:2662–2667

Martinez MJ, Biedenkopf N, Volchkova V, Hartlieb B, Alazard-Dany N, Reynard O, Becker S, Volchkov V (2008) Role of Ebola virus VP30 in transcription reinitiation. J Virol 82:12569–12573

Martinez O, Johnson J, Manicassamy B, Rong L, Olinger GG, Hensley LE, Basler CF (2010) Zaire Ebola virus entry into human dendritic cells is insensitive to cathepsin L inhibition. Cell Microbiol 12:148–157

Mavrakis M, Kolesnikova L, Schoehn G, Becker S, Ruigrok RW (2002) Morphology of Marburg virus NP-RNA. Virology 296:300–307

Mehedi M, Falzarano D, Seebach J, Hu X, Carpenter MS, Schnittler HJ, Feldmann H (2011) A new Ebola virus nonstructural glycoprotein expressed through RNA editing. J Virol 85:5406–5414

Messaoudi I, Amarasinghe GK, Basler CF (2015) Filovirus pathogenesis and immune evasion: insights from Ebola virus and Marburg virus. Nat Rev Microbiol 13:663–676

Miller EH, Obernosterer G, Raaben M, Herbert AS, Deffieu MS, Krishnan A, Ndungo E, Sandesara RG, Carette JE, Kuehne AI, Ruthel G, Pfeffer SR, Dye JM, Whelan SP, Brummelkamp TR, Chandran K (2012) Ebola virus entry requires the host-programmed recognition of an intracellular receptor. EMBO J 31:1947–1960

Misasi J, Chandran K, Yang JY, Considine B, Filone CM, Cote M, Sullivan N, Fabozzi G, Hensley L, Cunningham J (2012) Filoviruses require endosomal cysteine proteases for entry but exhibit distinct protease preferences. J Virol 86:3284–3292

Misasi J, Gilman MS, Kanekiyo M, Gui M, Cagigi A, Mulangu S, Corti D, Ledgerwood JE, Lanzavecchia A, Cunningham J, Muyembe-Tamfun JJ, Baxa U, Graham BS, Xiang Y, Sullivan NJ, McLellan JS (2016) Structural and molecular basis for Ebola virus neutralization by protective human antibodies. Science 351:1343–1346

Modrof J, Muhlberger E, Klenk HD, Becker S (2002) Phosphorylation of VP30 impairs ebola virus transcription. J Biol Chem 277:33099–33104

Mohan GS, Li W, Ye L, Compans RW, Yang C (2012) Antigenic subversion: a novel mechanism of host immune evasion by Ebola virus. PLoS Pathog 8:e1003065

Moller P, Pariente N, Klenk HD, Becker S (2005) Homo-oligomerization of Marburgvirus VP35 is essential for its function in replication and transcription. J Virol 79:14876–14886

Mühlberger E, Lotfering B, Klenk HD, Becker S (1998) Three of the four nucleocapsid proteins of Marburg virus, NP, VP35, and L, are sufficient to mediate replication and transcription of Marburg virus-specific monocistronic minigenomes. J Virol 72:8756–8764

Mühlberger E, Weik M, Volchkov VE, Klenk HD, Becker S (1999) Comparison of the transcription and replication strategies of Marburg virus and Ebola virus by using artificial replication systems. J Virol 73:2333–2342

Murin CD, Fusco ML, Bornholdt ZA, Qiu X, Olinger GG, Zeitlin L, Kobinger GP, Ward AB, Saphire EO (2014) Structures of protective antibodies reveal sites of vulnerability on Ebola virus. Proc Natl Acad Sci USA 111:17182–17187

Nanbo A, Watanabe S, Halfmann P, Kawaoka Y (2013) The spatio-temporal distribution dynamics of Ebola virus proteins and RNA in infected cells. Scientific Reports 3:1206

Ndungo E, Herbert AS, Raaben M, Obernosterer G, Biswas R, Miller EH, Wirchnianski AS, Carette JE, Brummelkamp TR, Whelan SP, Dye JM, Chandran K (2016). A single residue in Ebola virus receptor NPC1 influences cellular host range in reptiles. mSphere 1:e00007–16

Negredo A, Palacios G, Vazquez-Moron S, Gonzalez F, Dopazo H, Molero F, Juste J, Quetglas J, Savji N, de la Cruz Martinez M, Herrera JE, Pizarro M, Hutchison SK, Echevarria JE, Lipkin WI, Tenorio A (2011) Discovery of an ebolavirus-like filovirus in europe. PLoS Pathog 7:e1002304

Ng M, Ndungo E, Kaczmarek ME, Herbert AS, Binger T, Kuehne AI, Jangra RK, Hawkins JA, Gifford RJ, Biswas R, Demogines A, James RM, Yu M, Brummelkamp TR, Drosten C, Wang LF, Kuhn JH, Muller MA, Dye JM, Sawyer SL, Chandran K (2015) Filovirus receptor NPC1 contributes to species-specific patterns of ebolavirus susceptibility in bats. Elife 4: Le11785

Noda T, Sagara H, Suzuki E, Takada A, Kida H, Kawaoka Y (2002) Ebola virus VP40 drives the formation of virus-like filamentous particles along with GP. J Virol 76:4855–4865

Noda T, Aoyama K, Sagara H, Kida H, Kawaoka Y (2005) Nucleocapsid-like structures of Ebola virus reconstructed using electron tomography. J Vet Med Sci 67:325–328

Noda T, Ebihara H, Muramoto Y, Fujii K, Takada A, Sagara H, Kim JH, Kida H, Feldmann H, Kawaoka Y (2006) Assembly and budding of Ebolavirus. PLoS Pathog 2:e99

Noda T, Halfmann P, Sagara H, Kawaoka Y (2007a) Regions in Ebola virus VP24 that are important for nucleocapsid formation. J Infect Dis 196(Suppl 2):S247–250

Noda T, Watanabe S, Sagara H, Kawaoka Y (2007b) Mapping of the VP40-binding regions of the nucleoprotein of Ebola virus. J Virol 81:3554–3562

Noda T, Hagiwara K, Sagara H, Kawaoka Y (2010) Characterization of the Ebola virus nucleoprotein-RNA complex. J Gen Virol 91:1478–1483

Oda S, Noda T, Wijesinghe KJ, Halfmann P, Bornholdt ZA, Abelson DM, Armbrust T, Stahelin RV, Kawaoka Y, Saphire EO (2015) Crystal structure of marburg virus VP40 reveals a broad, basic patch for matrix assembly and a requirement of the *N*-terminal domain for immunosuppression. J Virol 90:1839–1848

Ogino T, Yadav SP, Banerjee AK (2010) Histidine-mediated RNA transfer to GDP for unique mRNA capping by vesicular stomatitis virus RNA polymerase. Proc Natl Acad Sci USA 107:3463–3468

Olal D, Kuehne AI, Bale S, Halfmann P, Hashiguchi T, Fusco ML, Lee JE, King LB, Kawaoka Y, Dye JM Jr, Saphire EO (2012) Structure of an antibody in complex with its mucin domain linear epitope that is protective against Ebola virus. J Virol 86:2809–2816

Page A, Volchkova VA, Reid SP, Mateo M, Bagnaud-Baule A, Nemirov K, Shurtleff AC, Lawrence P, Reynard O, Ottmann M, Lotteau V, Biswal SS, Thimmulappa RK, Bavari S, Volchkov VE (2014) Marburgvirus hijacks nrf2-dependent pathway by targeting nrf2-negative regulator keap1. Cell Rep 6:1026–1036

Pallesen J, Murin CD, de Val N, Cottrell CA, Hastie KM, Turner HL, Fusco ML, Flyak AI, Zeitlin L, Crowe JE Jr, Andersen KG, Saphire EO, Ward AB (2016) Structures of Ebola virus GP and sGP in complex with therapeutic antibodies. Nat Microbiol 1:16128

Panchal RG, Ruthel G, Kenny TA, Kallstrom GH, Lane D, Badie SS, Li L, Bavari S, Aman MJ (2003) In vivo oligomerization and raft localization of Ebola virus protein VP40 during vesicular budding. Proc Natl Acad Sci USA 100:15936–15941

Poch O, Blumberg BM, Bougueleret L, Tordo N (1990) Sequence comparison of five polymerases (*L* proteins) of unsegmented negative-strand RNA viruses: theoretical assignment of functional domains. J Gen Virol 71(Pt 5):1153–1162

Prins KC, Cardenas WB, Basler CF (2009) Ebola virus protein VP35 impairs the function of interferon regulatory factor-activating kinases IKKepsilon and TBK-1. J Virol 83:3069–3077

Radzimanowski J, Effantin G, Weissenhorn W (2014) Conformational plasticity of the Ebola virus matrix protein. Protein Sci 23:1519–1527

Rahmeh AA, Schenk AD, Danek EI, Kranzusch PJ, Liang B, Walz T, Whelan SP (2010) Molecular architecture of the vesicular stomatitis virus RNA polymerase. Proc Natl Acad Sci U S A 107:20075–20080

Ramanan P, Edwards MR, Shabman RS, Leung DW, Endlich-Frazier AC, Borek DM, Otwinowski Z, Liu G, Huh J, Basler CF, Amarasinghe GK (2012) Structural basis for Marburg virus VP35-mediated immune evasion mechanisms. Proc Natl Acad Sci USA 109:20661–20666

Reid SP, Leung LW, Hartman AL, Martinez O, Shaw ML, Carbonnelle C, Volchkov VE, Nichol ST, Basler CF (2006) Ebola virus VP24 binds karyopherin alpha1 and blocks STAT1 nuclear accumulation. J Virol 80:5156–5167

Ruigrok RW, Schoehn G, Dessen A, Forest E, Volchkov V, Dolnik O, Klenk HD, Weissenhorn W (2000) Structural characterization and membrane binding properties of the matrix protein VP40 of Ebola virus. J Mol Biol 300:103–112

Sanchez A, Kiley MP, Holloway BP, McCormick JB, Auperin DD (1993) Sequence analysis of the Ebola virus genome: organization, genetic elements, and comparison with the genome of Marburg virus. Virus Res. 29:215–240

Sanchez A, Trappier SG, Mahy BW, Peters CJ, Nichol ST (1996) The virion glycoproteins of Ebola viruses are encoded in two reading frames and are expressed through transcriptional editing. Proc Natl Acad Sci USA 93:3602–3607

Sanchez A, Yang ZY, Xu L, Nabel GJ, Crews T, Peters CJ (1998) Biochemical analysis of the secreted and virion glycoproteins of Ebola virus. J Virol 72:6442–6447

Schoehn G, Mavrakis M, Albertini A, Wade R, Hoenger A, Ruigrok RW (2004) The 12 a structure of trypsin-treated measles virus N-RNA. J Mol Biol 339:301–312
Schornberg K, Matsuyama S, Kabsch K, Delos S, Bouton A, White J (2006) Role of endosomal cathepsins in entry mediated by the Ebola virus glycoprotein. J Virol 80:4174–4178
Scianimanico S, Schoehn G, Timmins J, Ruigrok RH, Klenk HD, Weissenhorn W (2000) Membrane association induces a conformational change in the Ebola virus matrix protein. EMBO J 19:6732–6741
Soni SP, Stahelin RV (2014) The Ebola virus matrix protein VP40 selectively induces vesiculation from phosphatidylserine-enriched membranes. J Biol Chem 289:33590–33597
Soni SP, Adu-Gyamfi E, Yong SS, Jee CS, Stahelin RV (2013) The Ebola virus matrix protein deeply penetrates the plasma membrane: an important step in viral egress. Biophys J 104:1940–1949
Timmins J, Scianimanico S, Schoehn G, Weissenhorn W (2001) Vesicular release of ebola virus matrix protein vp40. Virology 283:1–6
Timmins J, Schoehn G, Kohlhaas C, Klenk HD, Ruigrok RW, Weissenhorn W (2003) Oligomerization and polymerization of the filovirus matrix protein VP40. Virology 312:359–368
Tran EE, Simmons JA, Bartesaghi A, Shoemaker CJ, Nelson E, White JM, Subramaniam S (2014) Spatial localization of the Ebola virus glycoprotein mucin-like domain determined by cryo-electron tomography. J Virol 88:10958–10962
Tran EE, Nelson EA, Bonagiri P, Simmons JA, Shoemaker CJ, Schmaljohn CS, Kobinger GP, Zeitlin L, Subramaniam S, White JM (2016) Mapping of ebolavirus neutralization by monoclonal antibodies in the ZMapp cocktail using cryo-electron tomography and studies of cellular entry. J Virol 90:7618–7627
Valmas C, Basler CF (2011) Marburg virus VP40 antagonizes interferon signaling in a species-specific manner. J Virol 85:4309–4317
Valmas C, Grosch MN, Schumann M, Olejnik J, Martinez O, Best SM, Krahling V, Basler CF, Muhlberger E (2010) Marburg virus evades interferon responses by a mechanism distinct from ebola virus. PLoS Pathog 6:e1000721
Volchkov VE, Feldmann H, Volchkova VA, Klenk HD (1998) Processing of the Ebola virus glycoprotein by the proprotein convertase furin. Proc Natl Acad Sci USA 95:5762–5767
Wahl-Jensen VM, Afanasieva TA, Seebach J, Stroher U, Feldmann H, Schnittler HJ (2005) Effects of Ebola virus glycoproteins on endothelial cell activation and barrier function. J Virol 79:10442–10450
Wang H, Shi Y, Song J, Qi J, Lu G, Yan J, Gao GF (2016) Ebola viral glycoprotein bound to its endosomal receptor niemann-pick C1. Cell 164:258–268
Wasserman H, Saphire EO (2016) More than Meets the Eye: Hidden Structures in the Proteome. Annu Rev Virol 3:373–386
Watanabe S, Noda T, Kawaoka Y (2006) Functional mapping of the nucleoprotein of Ebola virus. J Virol 80:3743–3751
Watanabe S, Noda T, Halfmann P, Jasenosky L, Kawaoka Y (2007) Ebola virus (EBOV) VP24 inhibits transcription and replication of the EBOV genome. J Infect Dis 196(Suppl 2):S284–290
Weik M, Modrof J, Klenk HD, Becker S, Muhlberger E (2002) Ebola virus VP30-mediated transcription is regulated by RNA secondary structure formation. J Virol 76:8532–8539
Weissenhorn W, Calder LJ, Wharton SA, Skehel JJ, Wiley DC (1998a) The central structural feature of the membrane fusion protein subunit from the Ebola virus glycoprotein is a long triple-stranded coiled coil. Proc Natl Acad Sci USA 95:6032–6036
Weissenhorn W, Carfi A, Lee KH, Skehel JJ, Wiley DC (1998b) Crystal structure of the Ebola virus membrane fusion subunit, GP2, from the envelope glycoprotein ectodomain. Mol Cell 2:605–616
Wijesinghe KJ, Stahelin RV (2015) Investigation of the lipid binding properties of the Marburg virus matrix protein VP40. J Virol 90:3074–3085

Wilson JA, Hevey M, Bakken R, Guest S, Bray M, Schmaljohn AL, Hart MK (2000) Epitopes involved in antibody-mediated protection from Ebola virus. Science 287:1664–1666

Xu W, Edwards MR, Borek DM, Feagins AR, Mittal A, Alinger JB, Berry KN, Yen B, Hamilton J, Brett TJ, Pappu RV, Leung DW, Basler CF, Amarasinghe GK (2014) Ebola virus VP24 targets a unique NLS binding site on karyopherin alpha 5 to selectively compete with nuclear import of phosphorylated STAT1. Cell Host Microbe 16:187–200

Yamayoshi S, Kawaoka Y (2007) Mapping of a region of Ebola virus VP40 that is important in the production of virus-like particles. J Infect Dis 196(Suppl 2):S291–295

Zhang AP, Bornholdt ZA, Liu T, Abelson DM, Lee DE, Li S, Woods VL Jr, Saphire EO (2012) The ebola virus interferon antagonist VP24 directly binds STAT1 and has a novel. Pyramidal Fold. PLoS Pathog 8:e1002550

Zhang AP, Bornholdt ZA, Abelson DM, Saphire EO (2014) Crystal structure of Marburg virus VP24. J Virol 88:5859–5863

Zhang Q, Gui M, Niu X, He S, Wang R, Feng Y, Kroeker A, Zuo Y, Wang H, Wang Y, Li J, Li C, Shi Y, Shi X, Gao GF, Xiang Y, Qiu X, Chen L, Zhang L (2016) Potent neutralizing monoclonal antibodies against Ebola virus infection. Sci Rep 6:25856

Zhao Y, Ren J, Harlos K, Jones DM, Zeltina A, Bowden TA, Padilla-Parra S, Fry EE, Stuart DI (2016) Toremifene interacts with and destabilizes the Ebola virus glycoprotein. Nature 535:169–172

Part IV
Tools and Taxonomy

Reverse Genetics of Filoviruses

Thomas Hoenen, Janine Brandt, Yíngyún Caì, Jens H. Kuhn and Courtney Finch

Abstract Reverse genetics systems are used for the generation of recombinant viruses. For filoviruses, this technology has been available for more than 15 years and has been used to investigate questions regarding the molecular biology, pathogenicity, and host adaptation determinants of these viruses. Further, reporter-expressing, recombinant viruses are increasingly used as tools for screening for and characterization of candidate medical countermeasures. Thus, reverse genetics systems represent powerful research tools. Here we provide an overview of available reverse genetics systems for the generation of recombinant filoviruses, potential applications, and the achievements that have been made using these systems.

Contents

T. Hoenen (✉) · J. Brandt
Friedrich-Loeffler-Institut, Südufer 10, 17493 Greifswald—Insel Riems, Germany
e-mail: thomas.hoenen@fli.de

Y. Caì · J.H. Kuhn (✉) · C. Finch
Integrated Research Facility at Fort Detrick (IRF-Frederick), Division of Clinical Research (DCR), National Institute of Allergy and Infectious Diseases (NIAID), National Institutes of Health (NIH), B-8200 Research Plaza, Fort Detrick, Frederick, MD 21702, USA
e-mail: kuhnjens@mail.nih.gov

Current Topics in Microbiology and Immunology (2017) 411:421–446
DOI 10.1007/82_2017_55

Published Online: 17 September 2017

1 Introduction

About 50 years have passed since the discovery of filoviruses during an outbreak of a then unknown, severe viral hemorrhagic fever amongst workers of a German biomedical company, which used kidney cells from grivets imported from Uganda for the purposes of vaccine production. Marburgviruses were identified as the agents responsible for this outbreak, and about a decade later ebolaviruses were identified as closely related viruses responsible for concurrent outbreaks of severe hemorrhagic fevers in Sudan (now South Sudan) and Zaire (now Democratic Republic of the Congo). Three decades after these discoveries, reverse genetics systems for filoviruses became available, which helped to greatly increase our knowledge regarding their molecular biology and pathophysiology. Here, after a short introduction into the molecular biology of filoviruses, we describe the development of reverse genetics systems for filoviruses and provide examples for the discoveries made possible by these systems.

1.1 Marburg Virus

Marburg virus (MARV) and the more distantly related Ravn virus (RAVV) are the two members of the genus *Marburgvirus* (Kuhn 2017). In contrast to RAVV, MARV is well characterized on the molecular level (Brauburger et al. 2012, 2015). MARV has a negative-sense, single-stranded, nonsegmented, RNA genome. This genome is roughly 19 kb in length with the following gene order: 3′-leader-*NP*-*VP35*-*VP40*-*GP*-*VP30*-*VP24*-*L*-trailer-5′ (Fig. 1a) (Feldmann et al. 1992). Each of the seven genes is flanked by 3′ and 5′ untranslated regions containing transcriptional initiation and termination signals, and each gene is either separated from downstream genes by intergenic regions or, in one case, overlaps (Feldmann et al. 1992).

The seven genes encode seven structural proteins (Fig. 1b): nucleoprotein (NP), polymerase cofactor (VP35), matrix protein (VP40), glycoprotein ($GP_{1,2}$), viral protein 30 (VP30), nucleocapsid-associated protein (VP24), and the RNA-dependent RNA polymerase (L), respectively. NP, VP35, VP30, L, and VP24 comprise the MARV ribonucleoprotein complex (RNP) (Schmidt and Mühlberger 2016).

NP coats and stabilizes the MARV genome and antigenome (Becker et al. 1998; Bharat et al. 2011), and mediates the contact between the polymerase complex and the genome or antigenome (Becker et al. 1998; Mühlberger et al. 1998; Mavrakis et al. 2002; Kolesnikova et al. 2000). NP interacts with VP35, while VP35 also interacts with L (Mühlberger et al. 1999; Becker et al. 1998; Möller et al. 2005). NP also recruits cellular endosomal sorting components to the cell membrane, facilitating both transport of viral components to the membrane and budding (Dolnik et al. 2010, 2014, 2015a). VP30 is a poorly understood component of the RNP. It is

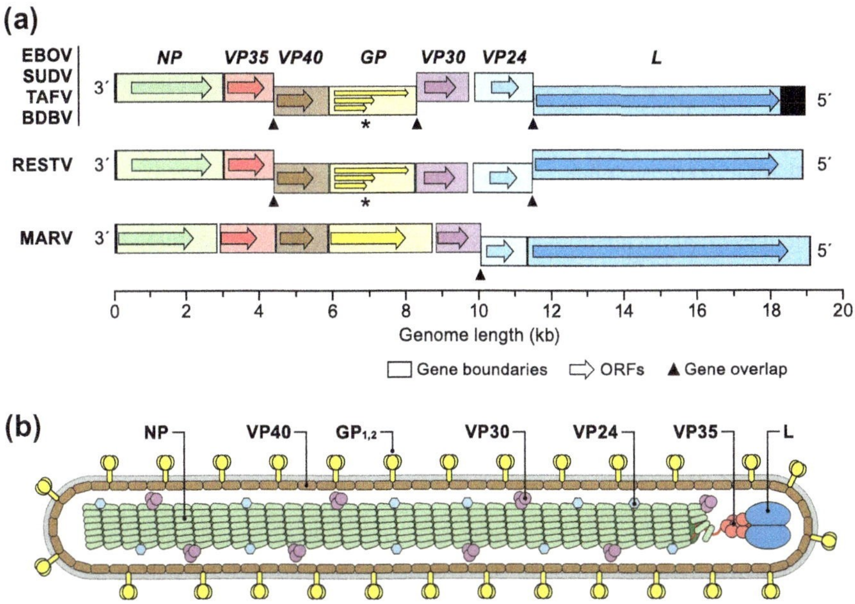

Fig. 1 Filoviruses. **a** Genome, drawn to scale, showing all genes (labeled in *italics*), open reading frames (*colored arrows*), and gene overlaps (*triangles*). **b** Virion, drawn not to scale, showing all structural proteins (labeled *unitalicized*)

not essential for the formation of the RNP and is not required in the context of minigenome assays (Mühlberger et al. 1998; Becker et al. 1998). Nevertheless, the presence of VP30 is essential for the rescue of infectious MARV from full-length genomes; furthermore, VP30 down-regulation leads to a reduction in MARV protein production (Enterlein et al. 2006; Fowler et al. 2005). Even less clear is the function of VP24, which appears less closely associated with the RNP than NP, L, VP30, and VP35. The most recent data indicate that VP24 may also be involved in the regulation of RNA synthesis (Wenigenrath et al. 2010) and in the release of infectious MARV particles, but not virion budding itself (Bamberg et al. 2005). VP40, the only MARV protein other than $GP_{1,2}$ not directly associated with the RNP, is a matrix protein that mediates virion budding (Kolesnikova et al. 2007, 2012; Wijesinghe and Stahelin 2015).

During transcription, which occurs via a stop–start mechanism, the viral polymerase recognizes the termination and initiation signals of each gene as it moves along the genome, transcribing accordingly. This movement leads to the sequential transcription of all seven genes (from *NP* to *L*). As the genes are transcribed, all nascent mRNAs are capped and polyadenylated. L caps the mRNAs immediately after transcription initiation, whereas the poly(A) tails are generated via a stuttering mechanism subsequent to transcription termination (Mühlberger et al. 1996). An inherent characteristic of sequential transcription is a concentration gradient among the expressed proteins. Because L falls off the genome occasionally, genes at the 5′

end of the genome are transcribed less frequently than those at the 3′ end, leading to a much higher concentration of, for instance, NP compared to that of L (Mühlberger et al. 1996). During replication, L ignores the transcription initiation and termination signals, thereby synthesizing a full-length antigenome that serves as the template for the viral polymerase to generate full-length progeny genomes (Mühlberger 2007; Mühlberger et al. 1998; Brauburger et al. 2015).

Although the 3′ leader and 5′ trailer regions of the MARV genome are noncoding, these regions are critical for virus replication and transcription because they contain promoters. For example, the replication promoter is comprised of two regions. The first region is located in the 3′ leader region, whereas the second region is part of the 3′ untranslated region (negative-sense orientation) of the *NP* gene. The transcription promoter is also located in the leader region (Enterlein et al. 2009). Similar to the 3′ leader region, the 5′ trailer region contains the complementary replication promoter, which serves to replicate the MARV antigenome (Schmidt and Mühlberger 2016).

1.2 Ebolaviruses

The genus *Ebolavirus* comprises five genetically distinct viruses: Ebola virus (EBOV) and Sudan virus (SUDV), which were co-discovered in 1976 and became the founding members of the genus; Bundibugyo virus (BDBV); Reston virus (RESTV); and Taï Forest virus (TAFV) (Kuhn 2017). EBOV along with SUDV, BDBV, and MARV are known to cause severe viral hemorrhagic fevers in humans with case-fatality rates averaging 41% (Kuhn 2015). Only one (nonfatal) human case of TAFV infection has been described thus far (Formenty et al. 1999; Le Guenno et al. 1999). The Asian filovirus RESTV can infect humans but appears to be nonpathogenic (Miranda et al. 1999; Miranda and Miranda 2011; Albariño et al. 2017).

The genetic diversity in the nucleotide sequences between the various ebolaviruses ranges, based on nucleotides, from 32 to 42%, whereas the genetic difference between ebolaviruses and marburgviruses is approximately 57%. However, the molecular differences among variants and isolates of individual ebolaviruses are very low and indicate a high degree of genetic stability for each virus. This low genetic diversity suggests strong ecological adaptation. The molecular disparity between genetic variants of EBOV is around 2.7%, whereas that of SUDV and RESTV variants is 5.2 and 4.5%, respectively (Carroll et al. 2013). In contrast, the two described marburgviruses (MARV and RAVV) are highly distinct from each other, but genomes of their variants and isolates have little variation (Towner et al. 2009, 2006).

Filoviruses form pleomorphic particles that vary in length. The average unit lengths of the virions in association with peak infectivity range from ≈880 nm for MARV to ≈980 nm for EBOV (Beniac et al. 2012; Bharat et al. 2012). The various lengths of the EBOV and MARV particles result from subtle differences in the structure of their nucleocapsid. Compared to the MARV nucleocapsid, the EBOV nucleocapsid has fewer nucleoprotein (NP) subunits per helix rotation, but has also more helical turns (Martin et al. 2017).

Ebolavirus and marburgvirus genomes are similar in length and gene arrangement. However, marburgvirus genomes have only one gene overlap between the *VP30* and the *VP24* genes, whereas ebolavirus genomes contain up to three overlaps (between the *VP35* and *VP40*, *GP* and *VP30* [absent in RESTV], and *VP24* and *L* genes). These overlaps are 18–20 bases in length and are limited primarily to conserved sequences that serve as transcription signals (Sanchez et al. 1993; Bukreyev et al. 1995). Furthermore, the lengths of the 5′ trailer regions vary among filoviruses. The longest trailer region of 676 bases is found in EBOV, whereas RESTV has the shortest trailer region with just 25 bases (Volchkov et al. 1999).

With respect to their lifecycles and the functions of their individual proteins, ebolaviruses strongly resemble marburgviruses. One striking difference is the organization of their *GP* gene. In all filoviruses, the *GP* gene encodes the surface glycoprotein $GP_{1,2}$, which is the only viral transmembrane surface protein. $GP_{1,2}$ forms virion-surface spikes involved in a receptor-mediated virion entry into host cells (Bukreyev et al. 1993; Feldmann et al. 1991; Sanchez et al. 1993). In contrast to marburgviruses, whose *GP* gene contains only a single open reading frame (ORF), ebolavirus $GP_{1,2}$ is encoded by two partially overlapping ORFs. In the center of the *GP* gene is a translational stop codon that prevents the synthesis of the full-length glycoprotein (Sanchez et al. 1996; Volchkov et al. 1995). During transcription, three different sets of mRNAs can be produced through an editing mechanism in the *GP* gene caused by a poly-U repeat, the so-called editing site, upstream of the stop codon (Volchkov et al. 1995; Mehedi et al. 2013; Sanchez et al. 1996). Primarily, the mRNA is not subject to editing during transcription, and the translation runs until the stop codon of the first ORF is encountered. This termination results in the synthesis of the precursor soluble glycoprotein (pre-sGP). The nonstructural pre-sGP is post-translationally cleaved into mature sGP and Δ-peptide, and both cleavage products are secreted from the infected cell (Volchkova et al. 1999; Radoshitzky et al. 2011). If, however, a non-templated adenosine residue is inserted co-transcriptionally at the editing site, a frameshift takes place, and the two ORFs become linked. Translation then continues until the stop codon of the second ORF is reached, resulting in the $GP_{1,2}$ precursor protein, preGP (Volchkov et al. 2001; Sanchez et al. 1996). preGP is post-translationally cleaved by furin-like proteases into the disulfide bond-linked subunits GP_1 and GP_2. These two subunits are functionally analogous to marburgvirus GP_1 and GP_2 and mediate receptor binding and membrane fusion, respectively (Volchkov et al. 1998; Feldmann et al. 1994; Jeffers et al. 2002). Co-transcriptional insertion of two adenosine residues at the editing site results in the expression of a small nonstructural glycoprotein, the small soluble glycoprotein (ssGP), which is also secreted from the infected cell (Mehedi et al. 2011; Volchkov et al. 1998). The exact role of ssGP, sGP, and Δ-peptide in the lifecycle of ebolaviruses remains incompletely understood.

2 Filovirus Reverse Genetics

One definition for reverse genetics is research from genotype to phenotype rather than from phenotype to genotype (Griffiths et al. 2000). Thus, reverse genetics relies on knowledge of a gene sequence of interest and the ability to express that gene in a model system. With respect to viruses, reverse genetics typically refers to the cloning of a complete virus genome and the generation or "rescue" of a functional virus from that clone. Other definitions of reverse genetics are also commonly used, e.g., the generation of genomes or genome analogs (such as minigenomes) from cDNA (Hoenen et al. 2011); however, within this chapter, we will only discuss reverse genetics as is refers to the production of recombinant viruses from a cDNA clone. For positive-sense RNA viruses, the concept of reverse genetics is relatively simple. The viral RNA (vRNA) essentially serves as an mRNA that can be directly read by ribosomes. Consequently, the "naked" vRNA is infectious; introduction of the vRNA into a host cell leads to immediate expression of viral proteins, followed by virion assembly and egress. Thus, to establish a reverse genetics system, a cDNA copy of the virus genome is cloned into an expression plasmid that, upon transfection into a cell, serves as the template to synthesize naked vRNA. The first such viral reverse genetics system was created for poliovirus in 1981 (Racaniello and Baltimore 1981).

Reverse genetics systems for negative-sense RNA viruses are more challenging to create because the viral proteins required for transcription must be provided in trans to achieve rescue of the full-length clone. In addition, when full-length genomic (rather than the antigenomic) sense cDNA is transfected, the vRNA must be transcribed into mRNAs and antigenomes. Thus, reverse genetics systems for such viruses typically consist of a transcription plasmid encoding the viral antigenome and several expression plasmids encoding the viral proteins required for replication and transcription. Upon transfection of all plasmids into a host cell, viral transcription and replication are initiated. The first reverse genetics system to generate negative-sense RNA viruses was developed in 1994 for rabies virus (Schnell et al. 1994). This seminal work laid the foundation for reverse genetics systems for all negative-sense RNA viruses, including ebolaviruses and marburgviruses. The initial rabies virus system relied on an antigenomic cDNA clone under the control of the Escherichia phage T7 RNA polymerase (T7) promoter. To express rabies virus in mammalian cells, the rabies virus plasmid was transfected into cells previously infected with a T7-expressing recombinant vaccinia virus. In addition, plasmids expressing N (analog to filoviral NP), P (analog to filoviral VP35), and L had to be transfected into the same cells (Schnell et al. 1994). Filovirus rescue is even more complex as rescue additionally requires the presence of VP30 (Volchkov et al. 2001; Neumann et al. 2002; Towner et al. 2005; Enterlein et al. 2006; Krähling et al. 2010; Dolnik et al. 2014; Albariño et al. 2013).

2.1 *Marburg Virus Reverse Genetics*

2.1.1 Components and Concepts

MARV NP, VP35, VP30, and L are required to rescue full-length MARV from a cDNA clone. The VP30 requirement for rescue first described by Enterlein et al. was surprising, as VP30 is not required for MARV minigenome replication and can even be replaced by EBOV VP30 to achieve rescue of full-length MARV (Enterlein et al. 2006; Krähling et al. 2010; Dolnik et al. 2014; Albariño et al. 2013). With the exception of the original MARV reverse genetics system described by Enterlein et al., which used a T7 promoter for helper gene expression, MARV *NP*, *VP35*, *VP30*, and *L* genes are typically cloned under the control of an RNA polymerase II (RNA Pol II) promoter, such as the cytomegalovirus (CMV) promoter or the chicken beta actin (CAG) promoter. Along with these helper plasmids, a plasmid encoding a cDNA copy of the MARV antigenome is transfected (Albariño et al. 2013). The full-length antigenome is cloned under the control of a T7 promoter (Enterlein et al. 2006; Krähling et al. 2010; Dolnik et al. 2014; Albariño et al. 2013). T7 is delivered either via a helper plasmid in the same manner as the ribonucleocapsid genes or is stably expressed by a genetically modified mammalian cell line (Enterlein et al. 2006; Krähling et al. 2010; Dolnik et al. 2014; Albariño et al. 2013).

The goal of any viral reverse genetics system is to generate infectious virus from cDNA. Thus, while the initial round of viral replication is generated from plasmid transfection, all subsequent rounds are a product of viral infection of new cells. Figure 2 compares a natural filovirus replication cycle (Fig. 2a) to that of a filovirus reverse genetics system (Fig. 2b).

The primary difference between the natural MARV lifecycle and a launch from a reverse genetics system is how "virus" is introduced into the cell and how initial genome transcription and replication occur. During natural MARV infection, MARV particles enter the host cell via receptor-mediated endocytosis. As the endosome matures and acidifies, $GP_{1,2}$ facilitates the fusion of the virion and endosomal membranes. This fusion results in the release of the RNP into the cytoplasm (Bhattacharyya et al. 2012; Miller and Chandran 2012). At this point, viral mRNAs are transcribed from the MARV genome by the viral polymerase (L), resulting in the synthesis of all MARV proteins, and positive-sense MARV antigenomes are transcribed from the genome as templates for negative-sense progeny vRNA genomes. Ultimately, new virions form at the cell membrane at the site of filopodia (Kolesnikova et al. 2007). In contrast, a reverse genetics system involves the formation and transfection of plasmid-carrying liposomes formed with the aid of a commercially available transfection reagent. The liposomes enter the cell, and their contents are released into the cytoplasm. Before the viral polymerase complex can begin transcription and replication, the RNP genes and the viral genome must be transcribed. Once that transcription has occurred, NP must coat the until-then free antigenomic RNA to form the RNP (a crucial step that does not occur in the

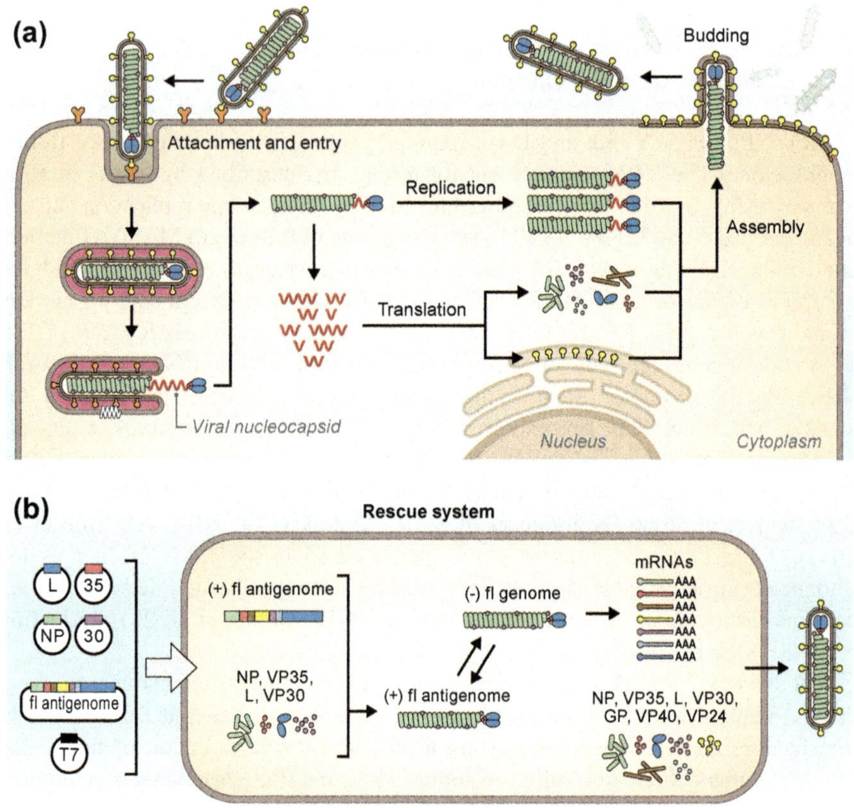

Fig. 2 Filovirus lifecycle. **a** Natural lifecycle showing virion attachment to and entry into a host cell, uncoating, replication and transcription, protein production, and virion egress. **b** Nonnatural initiation of the lifecycle using reverse genetics. Plasmids expressing the filoviral ribonucleocapsid proteins NP, VP35, VP30, and L under T7 promoters (or alternatively eukaryotic promoters—not shown), a plasmid expressing T7, and a plasmid encoding the full-length (fl) filovirus antigenome under a T7 promoter are transfected into a target cell. Coating of the naked filovirus genome and formation of the ribonucleocapsid complexes are the rate-limiting steps

natural lifecycle during which vRNA is always encapsidated, and which therefore is referred to as artificial or illegitimate encapsidation (Conzelmann 2004)). From this point forward, subsequent rounds of transcription and replication proceed in the same manner as in a naturally infected cell (Schmidt and Mühlberger 2016).

Currently, only three MARV reverse genetics systems have been reported (Table 1). These systems are based on MARV isolates from both human patients and frugivorous bats (Enterlein et al. 2006; Krähling et al. 2010; Albariño et al. 2013). Their derivatives have been used to answer key scientific questions:

Table 1 Available Marburg virus reverse genetic systems originating from three different laboratories. Each entry refers to the original publication (subsequent publications on each system do exist and are discussed in the main text)

Virus used for reverse genetics[a]	Rescue cell line	Used plasmids encoding	Reference
Marburg virus/H.sapiens-tc/KEN/1980/Mt. Elgon-Musoke	Recombinant baby hamster kidney cell line BSRT7/5 (derived from BHK-21), which expresses T7	• MARV antigenome (T7 promoter) • MARV NP (T7 promoter) • MARV VP35 (T7 promoter) • MARV VP30 (T7 promoter) • MARV L (T7 promoter) • T7	(Enterlein et al. 2009)
Marburg virus/H.sapiens-tc/KEN/1980/Mt. Elgon-Musoke	Vero cells	• MARV antigenome[b] (T7 promoter) • MARV NP (CAG promoter) • MARV VP35 (CAG promoter) • MARV VP30 (CAG promoter) • MARV L (CAG promoter) • T7	(Krähling et al. 2010)
Marburg virus/R.aegyptiacus-tc/UGA/2007/Kitaka-371Bat	BSRT7/5 and BHK-21 cell lines	• MARV antigenome or genome (T7 promoter) • Codon-optimized MARV NP (CAG promoter) • Codon-optimized MARV VP35 (CAG promoter) • Codon-optimized MARV VP30 (CAG promoter) • Codon-optimized MARV L (CAG promoter) • Codon-optimized T7 (with BHK-21 cell lines)	(Albariño et al. 2013)

[a]For nomenclature see (Kuhn et al. 2013, 2014); [b]The authors do not clearly indicate the nature of the genome

2.1.2 Site-Directed Mutagenesis

Important research has been performed using MARV mutants created via site-directed mutagenesis of cDNA clones. This research includes findings regarding host specificity, evasion of host innate immunity, interaction of MARV with the cellular host factor tumor susceptibility gene 101 (Tsg101), MARV transcription and replication, and the role of the GP_2 cytoplasmic domain in the MARV lifecycle (Enterlein et al. 2009; Dolnik et al. 2014; Albariño et al. 2015b; Koehler et al. 2015; Mittler et al. 2013).

Wild-type MARV does not infect rodents. A single MARV VP40 mutation, D184N, had previously been found to arise during MARV adaptation to guinea pigs (*Cavia porcellus*) by serial passaging (Lofts et al. 2007). To further understand this mutation, a wild-type MARV cDNA clone was altered to introduce the D184N mutation into VP40 (Koehler et al. 2015). This study demonstrated that, while the VP40 mutation did not alter viral protein production or interferon (IFN) antagonism, the mutation enhanced MARV replication in human and guinea pig cell lines in addition to causing a slight increase in VP40-mediated budding. Further research using minigenome and virus-like particle (VLP) systems confirmed these findings and revealed that D148 N enhanced recruitment of NP to virion budding sites (Koehler et al. 2015).

Albariño et al. sought to disrupt VP35's ability to bind dsRNA through the generation of two mutant viruses each bearing mutations in the *VP35* gene (Albariño et al. 2015b). MARV encoding VP35 R294A and MARV encoding VP35 R301A were reduced in their ability to inhibit the IFN response in comparison to wild-type MARV. In IFN-producing cell lines, the R301A mutation inhibited MARV replication to a higher extent compared to the R294A mutation, whereas growth kinetics of both mutant viruses and wild-type control virus were similar in IFN-deficient cells (Albariño et al. 2015b).

To examine putative endosomal sorting motifs important for MARV budding, Dolnik et al. mutated one of these motifs encoded by the *NP* gene. Specifically, the motif PSAP, located in the C-terminal domain of NP, was modified to AAAA. The mutant MARV was unable to recruit Tsg101 (Dolnik et al. 2014), a part of the endosomal sorting complexes required for transport (ESCRT) machinery known to bind MARV NP PSAP (Slagsvold et al. 2006; Bieniasz 2006). The mutant's inability to recruit Tsg101 was associated with a reduction in virion budding and consequently viral spread. Additionally, with the loss of the Tsg101-NP interaction, nucleocapsids accumulated within the cytoplasm of the host cell. Further analysis revealed that Tsg101 is involved in the transport of nucleocapsids within the cell. In the absence of the Tsg101-NP interaction, the nucleocapsids followed a shorter trajectory, and some nucleocapsids were immobile (Dolnik et al. 2014).

Mutations were also introduced into the MARV genome via site-directed mutagenesis to improve understanding of the bipartite MARV replication promoter (Enterlein et al. 2009). A MARV mutant, in which the second promoter element was destroyed, could not be rescued, indicating that the second promoter element is essential for replication activity (Enterlein et al. 2009).

Mittler et al. sought to elucidate the function of the MARV GP_2 cytoplasmic tail domain and GP_2 acylation by establishing mutant MARVs encoding a series of GP_2 cytoplasmic tail variants (Mittler et al. 2013). Three viruses could be rescued: a GP_2 cytoplasmic domain deletion mutant, a GP_2 cytoplasmic domain deletion mutant with mutated acylation sites at amino acid positions 671 and 673, and a GP_2 mutant with an intact cytoplasmic domain but mutated acylation sites at positions 671 and 673. Deletion of the GP_2 cytoplasmic domain resulted in viruses causing smaller sized plaques in plaque assays compared to those observed with wild-type MARV. Additionally, the percentage of infected cells after just one cycle of replication was much lower compared to a wild-type control, indicating a viral cell-entry defect. Deleting the GP_2 cytoplasmic domain also increased the concentration of MARV overall protein content within host cells and in the cell-culture supernatant (Mittler et al. 2013). Surprisingly, loss of GP_2 acylation alone did not significantly alter MARV infectivity. However, the MARV mutant combining both the loss of acylation and the deletion of the cytoplasmic domain was significantly less infectious compared to wild-type MARV or the cytoplasmic domain deletion mutant (Mittler et al. 2013).

2.1.3 Reporter Viruses

Beyond simply generating MARV mutants bearing functionally altered genes, reverse genetics was used to create reporter gene-encoding viruses. MARV mutants encoding optical reporter genes have been instrumental in research areas such as intracellular virus localization, rapid drug screening assays, and MARV isolate comparisons (Schmidt et al. 2011; Albariño et al. 2013; Dolnik et al. 2014).

Schmidt et al. published the first reporter-encoding MARV in 2011. The researchers cloned a sequence encoding enhanced green fluorescent protein (eGFP) flanked by MARV transcription initiation and termination signals into the *VP35-VP40* intergenic region. Although the viral replication rate of the eGFP-encoding MARV was lower than that of wild-type virus, eGFP was expressed and viral spread was easily detectable by live-cell imaging (Schmidt et al. 2011).

Albariño et al. created an Egyptian rousette (*Rousettus aegyptiacus*) MARV isolate encoding eGFP (Albariño et al. 2013). In this case, the *eGFP* gene was cloned into the *NP-VP35* intergenic region. The mutant virus caused only a roughly log-fold reduction in peak viral titer compared to that observed with wild-type virus in baby hamster (*Mesocricetus auratus*) kidney (BHK-21) cells. In macrophages, however, replication of eGFP-encoding virus was significantly impaired and induced distinct differences in the intrinsic immune response compared to wild-type virus (Albariño et al. 2013). Thus, while a foreign gene can be expressed by a replicating MARV, this additional gene cassette may alter modulation of the host innate immune response.

Optical reporter-encoding MARVs have been instrumental in examining MARV nucleocapsid transport and budding (Dolnik et al. 2014; Schudt et al. 2013).

Using a MARV encoding VP40 N-terminally fused to red fluorescent protein (RFP) as an additional transcriptional unit upstream of the *VP40* gene, fluorescently tagged actin, tubulin, and myosin 10 (an actin-dependent motor protein), and a MARV VP30-eGFP fusion expression plasmid, live-cell imaging tracked colocalization of these proteins. Imaging revealed colocalization of RFP-VP40 at the cell membrane; colocalization of actin and the nucleocapsid; recruitment of RFP-VP40-associated nucleocapsids at filopodia; colocalization of RFP-VP40 and myosin 10; and actin-dependent nucleocapsid transport from the point of viral replication to the site of MARV particle budding (Schudt et al. 2013). These observations confirmed previous hypotheses of VP40 recruiting nucleocapsids to the site of budding and of co-transportation of VP40 and the nucleocapsids by myosin 10 along actin filaments, with budding preferentially occurring at filopodia (Kolesnikova et al. 2007; Berg and Cheney 2002; Welsch et al. 2010; Schudt et al. 2013). Similarly, a gene encoding a VP30-RFP fusion protein was cloned downstream of the *VP35* gene into the *VP35-VP40* intergenic region. Live-cell imaging revealed colocalization of the VP30-RFP fusion with Tsg101 in inclusion bodies and cotransport of VP30-RFP with Tsg101. This experiment confirmed the critical role of Tsg101 in nucleocapsid transport (Dolnik et al. 2014).

2.2 *Ebolavirus Reverse Genetics*

2.2.1 Rescue Procedure

Similar to MARV, a number of research groups have successfully generated reverse genetics systems for ebolaviruses and used them to study ebolavirus molecular biology. In fact, the generation of recombinant EBOV from cDNA preceded that of recombinant MARV and appears to be easier to achieve. The reason for easier generation of recombinant EBOV is not well understood, since the overall strategy for rescue is identical for viruses of both filovirus genera: expression of a viral antigenome from plasmid DNA (by T7-driven initial transcription) is followed by illegitimate encapsidation of the "naked" antigenome and replication and transcription by the RNP proteins NP, VP35, VP30, and L. This transcription results in the production of all viral proteins, and ultimately the formation of recombinant viruses.

Details of rescue protocols vary somewhat between different groups and studies, particularly with respect to the cell lines used for rescue. Good transfection efficacy and high susceptibility and efficient propagation of recombinant viruses are required. Grivet (*Chlorocebus aethiops*) Vero cells (or their subclones) are used most often, which although somewhat more difficult to transfect very efficiently replicate filoviruses (Hoenen et al. 2012), in addition to baby hamster (*Mesocricetus auratus*) kidney BSR-T7 cells (Volchkov et al. 2001). Some groups also have used mixtures of 293T and Vero cells, trying to combine the advantages of high transfectability of human 293T cells with the high propagation capacity of Vero cells

(Towner et al. 2005). Recently, human Huh-7 cells have been used for rescue, and less mutations were observed in the genomes of rescued viruses compared to viruses rescued using Vero cells (Tsuda et al. 2015).

Another recent optimization of the rescue procedure has been the inclusion of a hammerhead ribozyme at the 5′ end of the antigenome to ensure the creation of an authentic 5′ antigenomic terminus. This modification led to a modest increase in rescue efficacy in some, but not all, cell lines tested (Tsuda et al. 2015). Finally, codon optimization of the support plasmids (i.e., the expression plasmids for the RNP proteins and T7) was used to improve rescue efficacy (Albariño et al. 2015a). An overview of some of the ebolavirus reverse genetics systems is given in Table 2. However, these systems and the rescue protocols are improved continuously even within research groups, and therefore numerous variations exist. Current detailed rescue protocols for recombinant filoviruses are provided in (Hoenen and Feldmann 2017; Groseth 2017).

2.2.2 Molecular Biology

The purpose of the majority of studies using recombinant ebolaviruses was to better understand the molecular biology of these viruses. Indeed, the very first studies aimed at a better understanding of the molecular biology of the *GP* gene and its products. Using EBOVs with mutations in the *GP* gene editing site, RNA editing was shown to play a role in modulating $GP_{1,2}$-mediated cytotoxicity (Volchkov et al. 2001). Viruses with preGP furin cleavage knock-out mutations remained functional, thereby demonstrating that preGP cleavage into GP_1 and GP_2 subunits is dispensable in vitro (Neumann et al. 2002) and in vivo (Neumann et al. 2007). Similarly, the importance of individual amino acid residues for $GP_{1,2}$ function has been investigated using reverse genetics (Mpanju et al. 2006). Finally, the role of $GP_{1,2\Delta TM}$ shedding has been studied with recombinant viruses, and it has been suggested that shedding regulates the amount of cell-expressed $GP_{1,2}$, and thus as an additional mechanism to control $GP_{1,2}$-mediated cytotoxicity, as well as $GP_{1,2}$ incorporation into virions (Dolnik et al. 2015b).

Reverse genetics also has been used to show an importance of RNA binding by VP40 octamers for the EBOV lifecycle, even though the exact mechanism for this phenomenon remains unclear (Hoenen et al. 2005). Further, the main function of EBOV VP40, i.e., virus budding, has been investigated using reverse genetics. Here recombinant viruses encoding VP40 variants with two knocked-out late domain motifs (which were deemed essential for virion budding based on data from VLP experiments) were shown to be only mildly attenuated and not strongly impaired in budding, suggesting that either additional late budding motifs or other budding mechanisms must exist (Neumann et al. 2005).

A third viral protein that has been rather extensively studied by reverse genetics is VP30. Initially, it was shown that VP30 is involved in transcription reinitiation (Martínez et al. 2008). Later studies with recombinant viruses investigated the role of VP30 in ebolavirus transcription, particularly the importance of dynamic

Table 2 Ebolavirus reverse genetic systems from various laboratories. Indicated are the virus isolate used as basis for the full-length clone, modifications (intentional or unintentional) introduced into the sequence of that isolate during cloning of the basic full-length clone as well as their consequences (i.e., silent or non-silent changes in open reading frames, changes in noncoding regions (NCR), knock-out (KO) or addition of restriction sites), the rescue strategy, and special features

Virus used for reverse genetics[a]	Genetic modification compared to databank sequence	Rescue strategy	Special features	Reference
Ebola virus/H. sapiens-tc/COD/1976/Yambuku-Mayinga	C6182A (*GP* silent, *Sac*I KO); C6185T (*GP* silent, *Sac*I KO); A18227T (NCR)	BSR T7/5 cells, stable expression of T7, pT or pGEM helper plasmids		(Volchkov et al. 2001)
Ebola virus/H. sapiens-tc/COD/1976/Yambuku-Mayinga	A insertion 9744/9745 (NCR); A18226T (NCR); A insertion 18495/18496 (NCR)	Vero cells, plasmid-derived T7, pCAGGS helper plasmids		(Neumann et al. 2002)
Ebola virus/H. sapiens-tc/COD/1976/Yambuku-Mayinga	G2833C (NCR); A3803G (NCR) A4187G (NCR); T11193C (NCR); G14187A (NCR); G18138A (L.M2186I); T deletion at 18138; AGC insertion 2851/2852 (NCR, *Bsi*WI insertion)	50/50 mix of Vero and 293T cells, plasmid-derived T7, pCAGGS helper plasmids	Contains introduced *Bsi*WI site for introduction of an additional transcriptional unit	(Towner et al. 2005)
Reston virus/M. fascicularis-tc/USA/1989/Philippines89-Pennsylvania	C814G (*NP* silent, *Xho*I KO); G7252A (*GP* silent); T8582C (VP30 silent, *Xma*I insertion); T8996A (*VP30* silent, *Sac*I KO); C17546T (*L* silent, *Kpn*I KO)	Vero cells, plasmid-derived T7, pCAGGS helper plasmids	Optimized for easy exchange of genes; low-copy plasmid for full-length genome	(Groseth et al. 2012)

(continued)

Table 2 (continued)

Virus used for reverse genetics[a]	Genetic modification compared to databank sequence	Rescue strategy	Special features	Reference
Ebola virus/H. sapiens-tc/COD/1976/Yambuku-Mayinga	C2149G (*NP* silent, *Sph*I KO); A11043G (*VP24* silent, *Bam*HI KO); C13194G (*L* silent; *Xho*I KO); C15639G (*L* silent; *Xma*I KO)	Vero cells, plasmid-derived T7, pCAGGS helper plasmids	Optimized for easy exchange of genes; low-copy plasmid for full-length genome	(Hoenen et al. 2012)
Ebola virus/H. sapiens-tc/COD/1976/Yambuku-Mayinga	none	Vero, Huh7, BHK T7, Cos7, 293 cells, plasmid-derived T7, pCAGGS helper plasmids	Hammerhead ribozyme at 5′ genome end; low-copy plasmid for full-length genome	(Tsuda et al. 2015)
Ebola virus/H.sapiens-wt/LBR/2014/Makona-201403007	T8104C (NCR)	Huh7 cells, plasmid-derived T7, pCAGGS helper plasmids	Codon-optimized pCAGGS helper plasmids (based on Mayinga isolate)	(Albariño et al. 2015a)
Ebola virus/H.sapiens-tc/GIN/2014/Makona-C07	none	Huh7 cells, plasmid-derived T7, pCAGGS helper plasmids	Low-copy plasmid for full-length genome	(Dietzel et al. 2017)

[a]For nomenclature see (Kuhn et al. 2013, 2014)

phosphorylation at a serine residue cluster involving S29 for the regulation of transcription, replication, and nucleocapsid transport (Martinez et al. 2011; Biedenkopf et al. 2016).

Although most recombinant ebolavirus molecular biology studies used viruses with point mutations, some work has also been done using reporter-expressing viruses. These viruses express fluorescent or luminescent reporter proteins either from additional transcriptional units or as fusions with viral proteins. The expression of viral protein/fluorescent reporter fusions, of course, lends itself to detailed spatial and spatiotemporal analyses of viral protein localization in infected cells using microscopic approaches. For instance, a recombinant EBOV expressing a fusion of the fluorescent protein mCherry and L allowed studying the localization of L in cells over the first 24 h of infection, ultimately leading to the conclusion that viral inclusion bodies are sites of EBOV genome replication (Hoenen et al. 2012). However, also reporter proteins expressed individually by recombinant viruses can be used for answering questions regarding the molecular biology of ebolaviruses. For example, a recombinant EBOV expressing luciferase from an additional transcriptional unit was used to study the role of VP24 in down-regulating virus genome replication and transcription, showing that this function is much less prominent during virus infection than in studies using VP24 overexpressed from a plasmid (Watt et al. 2014).

2.2.3 Pathogenicity Determinants and Host Adaptation

The second major field in which recombinant ebolaviruses have been used is that of understanding pathogenicity and host adaptation determinants. In particular, researchers are interested in understanding the molecular mechanisms that dictate pathogenicity and virulence. The *GP* gene has been proposed for a long time to be the decisive factor in pathogenicity. However, results of reverse genetics studies using full-length clones of both EBOV and RESTV as well as chimeric EBOV/RESTV with their *GP* genes swapped indicated that the *GP* gene contributes but is not solely responsible for pathogenicity, at least in a laboratory mouse model (Groseth et al. 2012). A potential role of sGP for pathogenicity also has been investigated, but so far, studies have not reached a consensus on sGP's importance (Hoenen et al. 2015; Volchkova et al. 2015). In contrast, a role of the IFN antagonism function of EBOV VP35 has clearly been demonstrated by several groups and in several animal models. Mutating a single amino acid of VP35 to render the protein unable to inhibit IRF-3 activation completely abolishes the lethality of the recombinant virus in otherwise highly lethal laboratory mouse or guinea pig models (Hartman et al. 2008a, b; Prins et al. 2010).

In most small animal models of ebolavirus infection, an adaptation of the viruses by sequential passaging is typically required to achieve pathogenicity. The underlying mutations have been the subject of intense studies. For mouse-adaptation, eight mutations have been observed, and by introducing these mutations either individually or in various combinations into recombinant viruses, Ebihara et al.

showed that VP24 and NP are necessary and sufficient for EBOV adaptation to laboratory mice (Ebihara et al. 2006). Interestingly, VP24 has also been shown to be crucial for adaptation to guinea pigs (Mateo et al. 2011). However, somewhat surprisingly, results of this study suggest that the role of VP24 in adaption might not be due to the VP24's function as an IFN signaling inhibitor, but rather due to VP24's role in nucleocapsid assembly and condensation.

Reverse genetics has also been used to study host adaptation in the context of human infection. Dietzel et al. investigated mutations that occurred during the prolonged Ebola virus disease outbreak due to EBOV infection in Western Africa from 2013 to 2016 within extensive human-to-human transmission chains. Using recombinant viruses, the authors characterized mutations in NP, $GP_{1,2}$, and L that appear to modulate protein function and result in faster virus growth in tissue culture (Dietzel et al. 2017). Similarly, Albariño et al. investigated mutations in VP30 and L that confer slight advantages in replication and transcription, and showed that recombinant viruses carrying these mutations outcompete wild-type viruses (Albariño et al. 2016).

2.2.4 Using Recombinant Viruses for Medical Countermeasure Evaluation

A third field in which recombinant viruses are used increasingly is the screening and testing of candidate medical countermeasures. Reporter-expressing filoviruses are typically used for this purpose, as they are more convenient to use in mid- or high-throughput screens compared to wild-type viruses that can only be quantified via cytopathic effect-based assays (Towner et al. 2005). In most cases, fluorescent reporters such as eGFP are used, although luciferase-expressing viruses have also been developed. With these luciferase-expressing viruses, the readout is possible after 2 days, compared to 3–5 days for eGPF-expressing viruses used at similar MOIs or 1–2 weeks for viruses without a reporter (Hoenen et al. 2013).

Different strategies are used for reporter expression. The first strategy is to express the reporter from an additional transcriptional unit inserted between two existing genes (Towner et al. 2005). As this strategy distorts the transcriptional gradient, attenuation of recombinant viruses can be a consequence. Indeed, this attenuation was observed in vitro (albeit not in all tested cell lines) and in vivo, with insertions proximal to the 3′ end of the genome having a stronger attenuating effect than insertions distal to it (Ebihara et al. 2007). Nevertheless, these viruses are now commonly used and have been proven invaluable for identifying and characterizing antivirals (Panchal et al. 2010, 2012; Madrid et al. 2013; Warren et al. 2010; Johansen et al. 2013; Welch et al. 2016). Similarly, eGFP-expressing ebolaviruses carrying $GP_{1,2}$s from various filoviruses (i.e., BDBV, SUDV, Lloviu virus, and MARV) have been used to characterize filovirus-neutralizing antibodies (Ilinykh et al. 2016).

While introducing additional transcriptional units for generation of reporter-expressing viruses remains the most commonly used strategy, alternative approaches have also been pursued to circumvent disturbing the transcriptional gradient. One rather intriguing strategy is to express the reporter as a fusion with a viral protein but separated by a self-cleaving P2A peptide. For instance, a ZsGreen-expressing EBOV was created by introducing a *ZsGreen-P2A-VP40* ORF in place of the original *VP40* open reading frame. The resulting virus expressed high levels of ZsGreen but was somewhat attenuated vitro (Albariño et al. 2015a). Similarly, the above-mentioned recombinant virus expressing an L-mCherry fusion protein was readily rescuable and was not noticeably attenuated in Vero cells (Hoenen et al. 2012). However, no in vivo data with this virus have yet been published, and whereas the mCherry-derived fluorescence allows tracking of L in infected cells using live-cell imaging technologies, overall signal strengths are lower than those observed with viruses expressing a reporter from an additional transcriptional unit. The applications for this L-mCherry virus, therefore, lie in investigating the molecular biology of EBOV L rather than high-throughput screening approaches for candidate medical countermeasures.

3 Conclusions

Reverse genetics systems for the generation of recombinant filoviruses have been and continue to be invaluable research tools in filovirology. Use of reverse genetics requires a maximum-containment (BSL-4) laboratory, as opposed to lifecycle modeling ("surrogate") systems such as minigenomes, which are important tools to study the filovirus lifecycle at lower containment (Hoenen et al. 2011). However, by their very nature, recombinant viruses are free of the artificial aspects that differentiate lifecycle modeling systems from infectious viruses. Therefore, reverse genetic systems for the generation of recombinant filoviruses are important to validate findings from studies performed at lower containment levels. Further, with these systems, researchers also can ask scientific questions in more authentic and complex contexts (including in vivo, thus enabling studies on an organism level).

Recombinant viruses have extensively been used to understand the molecular biology of ebolaviruses and to dissect host adaptation and pathogenicity factors. Recombinant reporter-expressing viruses are replacing wild-type viruses as tools for in vitro characterization of candidate medical countermeasures. However, in vivo applications of reporter-expressing viruses are still exceedingly scarce, owing to the attenuation of these viruses in vivo. Therefore, developing non-attenuated reporter-expressing viruses will be an important achievement for the study of filoviruses. Further, reverse genetics systems currently only exist for the generation of EBOV and RESTV. Expansion of this spectrum to include other ebolaviruses will be important to better understand the differences and commonalities between these different viruses.

With the publication of the first MARV reverse genetics system in 2006 (Enterlein et al. 2006), complex MARV molecular biology became possible. Hypotheses regarding viral replication, virus-host interaction, and viral pathogenesis could be posed and tested. Phenomena once merely observable can now be explained by exploring mechanisms-of-action in controlled experimental settings. Therefore, the advent of this system marked a new frontier in MARV research. Despite this progress, numerous obstacles have yet to be overcome. MARV reverse genetics systems are not yet widely distributed, and rescue of MARV from these systems is not necessarily foolproof or highly efficient. Furthermore, MARV reverse genetics only exist for two distinct MARV isolates, and the repertoire of reporter-encoding MARVs should be extended to yield viruses that are not impaired in growth or virulence.

Acknowledgements We thank Laura Bollinger and Jiro Wada (IRF-Frederick) for critically editing this chapter and for figure preparation, respectively.

Disclaimer: The views and conclusions contained in this document are those of the author and should not be interpreted as necessarily representing the official policies, either expressed or implied, of the US Department of Health and Human Services or of the institutions and companies affiliated with the author. This work was supported in part through Battelle Memorial Institute's prime contract with the US National Institute of Allergy and Infectious Diseases (NIAID) under Contract No. HHSN272200700016I. Subcontractors to Battelle Memorial Institute who performed this work are Y.C., J.H.K., and C.F., employees of Tunnell Government Services, Inc.

References

Albariño CG, Uebelhoer LS, Vincent JP, Khristova ML, Chakrabarti AK, McElroy A, Nichol ST, Towner JS (2013) Development of a reverse genetics system to generate recombinant Marburg virus derived from a bat isolate. Virology 446(1–2):230–237. doi:10.1016/j.virol.2013.07.038

Albariño CG, Wiggleton Guerrero L, Lo MK, Nichol ST, Towner JS (2015a) Development of a reverse genetics system to generate a recombinant Ebola virus Makona expressing a green fluorescent protein. Virology 484:259–264. doi:10.1016/j.virol.2015.06.013

Albariño CG, Wiggleton Guerrero L, Spengler JR, Uebelhoer LS, Chakrabarti AK, Nichol ST, Towner JS (2015b) Recombinant Marburg viruses containing mutations in the IID region of VP35 prevent inhibition of host immune responses. Virology 476:85–91. doi:10.1016/j.virol.2014.12.002

Albariño CG, Guerrero LW, Chakrabarti AK, Kainulainen MH, Whitmer SLM, Welch SR, Nichol ST (2016) Virus fitness differences observed between two naturally occurring isolates of Ebola virus Makona variant using a reverse genetics approach. Virology 496:237–243. doi:10.1016/j.virol.2016.06.011

Albariño CG, Wiggleton Guerrero L, Jenks HM, Chakrabarti AK, Ksiazek TG, Rollin PE, Nichol ST (2017) Insights into Reston virus spillovers and adaption from virus whole genome sequences. PLoS ONE 12(5):e0178224. doi:10.1371/journal.pone.0178224

Bamberg S, Kolesnikova L, Möller P, Klenk H-D, Becker S (2005) VP24 of Marburg virus influences formation of infectious particles. J Virol 79(21):13421–13433. doi:10.1128/JVI.79.21.13421-13433.2005

Becker S, Rinne C, Hofsäss U, Klenk H-D, Mühlberger E (1998) Interactions of Marburg virus nucleocapsid proteins. Virology 249(2):406–417. doi:10.1006/viro.1998.9328

Beniac DR, Melito PL, Devarennes SL, Hiebert SL, Rabb MJ, Lamboo LL, Jones SM, Booth TF (2012) The organisation of Ebola virus reveals a capacity for extensive, modular polyploidy. PLoS ONE 7(1):e29608. doi:10.1371/journal.pone.0029608

Berg JS, Cheney RE (2002) Myosin-X is an unconventional myosin that undergoes intrafilopodial motility. Nat Cell Biol 4(3):246–250. doi:10.1038/ncb762

Bharat TA, Riches JD, Kolesnikova L, Welsch S, Krähling V, Davey N, Parsy M-L, Becker S, Briggs JA (2011) Cryo-electron tomography of Marburg virus particles and their morphogenesis within infected cells. PLoS Biol 9(11):e1001196. doi:10.1371/journal.pbio.1001196

Bharat TA, Noda T, Riches JD, Kraehling V, Kolesnikova L, Becker S, Kawaoka Y, Briggs JAG (2012) Structural dissection of Ebola virus and its assembly determinants using cryo-electron tomography. Proc Natl Acad Sci U S A 109(11):4275–4280. doi:10.1073/pnas.1120453109

Bhattacharyya S, Mulherkar N, Chandran K (2012) Endocytic pathways involved in filovirus entry: advances, implications and future directions. Viruses 4(12):3647–3664

Biedenkopf N, Lier C, Becker S (2016) Dynamic phosphorylation of VP30 is essential for Ebola virus life cycle. J Virol 90(10):4914–4925. doi:10.1128/JVI.03257-15

Bieniasz PD (2006) Late budding domains and host proteins in enveloped virus release. Virology 344(1):55–63. doi:10.1016/j.virol.2005.09.044

Brauburger K, Hume AJ, Mühlberger E, Olejnik J (2012) Forty-five years of Marburg virus research. Viruses 4(10):1878–1927. doi:10.3390/v4101878

Brauburger K, Deflubé LR, Mühlberger E (2015) Filovirus transcription and replication. Biology and pathogenesis of rhabdo- and filoviruses. World Scientific Publishing, Singapore

Bukreyev A, Volchkov VE, Blinov VM, Netesov SV (1993) The GP-protein of Marburg virus contains the region similar to the 'immunosuppressive domain' of oncogenic retrovirus P15E proteins. FEBS Lett 323(1–2):183–187

Bukreyev AA, Belanov EF, Blinov VM, Netesov SV (1995) Complete nucleotide sequences of Marburg virus genes 5 and 6 encoding VP30 and VP24 proteins. Biochem Mol Biol Int 35 (3):605–613

Carroll SA, Towner JS, Sealy TK, McMullan LK, Khristova ML, Burt FJ, Swanepoel R, Rollin PE, Nichol ST (2013) Molecular evolution of viruses of the family *Filoviridae* based on 97 whole-genome sequences. J Virol 87(5):2608–2616. doi:10.1128/JVI.03118-12

Conzelmann KK (2004) Reverse genetics of *Mononegavirales*. Curr Top Microbiol Immunol 283:1–41

Dietzel E, Schudt G, Krähling V, Matrosovich M, Becker S (2017) Functional characterization of adaptive mutations during the West African Ebola virus outbreak. J Virol 91(2):e01913–e01916. doi:10.1128/JVI.01913-16

Dolnik O, Kolesnikova L, Stevermann L, Becker S (2010) Tsg101 is recruited by a late domain of the nucleocapsid protein to support budding of Marburg virus-like particles. J Virol 84 (15):7847–7856. doi:10.1128/JVI.00476-10

Dolnik O, Kolesnikova L, Welsch S, Strecker T, Schudt G, Becker S (2014) Interaction with Tsg101 is necessary for the efficient transport and release of nucleocapsids in Marburg virus-infected cells. PLoS Pathog 10(10):e1004463. doi:10.1371/journal.ppat.1004463

Dolnik O, Stevermann L, Kolesnikova L, Becker S (2015a) Marburg virus inclusions: a virus induced microcompartment and interface to multivesicular bodies and the late endosomal compartment. Eur J Cell Biol 94(7–9):323–331. doi:10.1016/j.ejcb.2015.05.006

Dolnik O, Volchkova VA, Escudero-Perez B, Lawrence P, Klenk H-D, Volchkov VE (2015b) Shedding of Ebola virus surface glycoprotein is a mechanism of self-regulation of cellular cytotoxicity and has a direct effect on virus infectivity. J Infect Dis 212(Suppl 2):S322–S328. doi:10.1093/infdis/jiv268

Ebihara H, Takada A, Kobasa D, Jones S, Neumann G, Theriault S, Bray M, Feldmann H, Kawaoka Y (2006) Molecular determinants of Ebola virus virulence in mice. PLoS Pathog 2 (7):e73. doi:10.1371/journal.ppat.0020073

Ebihara H, Theriault S, Neumann G, Alimonti JB, Geisbert JB, Hensley LE, Groseth A, Jones SM, Geisbert TW, Kawaoka Y, Feldmann H (2007) In vitro and in vivo characterization of recombinant Ebola viruses expressing enhanced green fluorescent protein. J Infect Dis 196 (Suppl 2):S313–S322. doi:10.1086/520590

Enterlein S, Volchkov V, Weik M, Kolesnikova L, Volchkova V, Klenk H-D, Mühlberger E (2006) Rescue of recombinant Marburg virus from cDNA is dependent on nucleocapsid protein VP30. J Virol 80(2):1038–1043. doi:10.1128/JVI.80.2.1038-1043.2006

Enterlein S, Schmidt KM, Schümann M, Conrad D, Krähling V, Olejnik J, Mühlberger E (2009) The Marburg virus 3' noncoding region structurally and functionally differs from that of Ebola virus. J Virol 83(9):4508–4519. doi:10.1128/JVI.02429-08

Feldmann H, Will C, Schikore M, Slenczka W, Klenk H-D (1991) Glycosylation and oligomerization of the spike protein of Marburg virus. Virology 182(1):353–356

Feldmann H, Mühlberger E, Randolf A, Will C, Kiley MP, Sanchez A, Klenk H-D (1992) Marburg virus, a filovirus: messenger RNAs, gene order, and regulatory elements of the replication cycle. Virus Res 24(1):1–19

Feldmann H, Nichol ST, Klenk H-D, Peters CJ, Sanchez A (1994) Characterization of filoviruses based on differences in structure and antigenicity of the virion glycoprotein. Virology 199 (2):469–473. doi:10.1006/viro.1994.1147

Formenty P, Hatz C, Le Guenno B, Stoll A, Rogenmoser P, Widmer A (1999) Human infection due to Ebola virus, subtype Cote d'Ivoire: clinical and biologic presentation. J Infect Dis 179 (Suppl 1):S48–S53. doi:10.1086/514285

Fowler T, Bamberg S, Möller P, Klenk H-D, Meyer TF, Becker S, Rudel T (2005) Inhibition of Marburg virus protein expression and viral release by RNA interference. J Gen Virol 86(Pt 4):1181–1188. doi:10.1099/vir.0.80622-0

Griffiths AJH, Miller JH, Suzuki DT, Lewontin RC, Gelbart WM (2000) An introduction to genetic analysis, Reverse Genetics, 7th edn. W. H. Freeman, New York

Groseth A (2017) Generation of recombinant Ebola viruses using reverse genetics. Methods Mol Biol 1628:177–188. doi:10.1007/978-1-4939-7116-9_13

Groseth A, Marzi A, Hoenen T, Herwig A, Gardner D, Becker S, Ebihara H, Feldmann H (2012) The Ebola virus glycoprotein contributes to but is not sufficient for virulence in vivo. PLoS Pathog 8(8):e1002847. doi:10.1371/journal.ppat.1002847

Hartman AL, Bird BH, Towner JS, Antoniadou Z-A, Zaki SR, Nichol ST (2008a) Inhibition of IRF-3 activation by VP35 is critical for the high level of virulence of Ebola virus. J Virol 82 (6):2699–2704. doi:10.1128/JVI.02344-07

Hartman AL, Ling L, Nichol ST, Hibberd ML (2008b) Whole-genome expression profiling reveals that inhibition of host innate immune response pathways by Ebola virus can be reversed by a single amino acid change in the VP35 protein. J Virol 82(11):5348–5358. doi:10.1128/JVI.00215-08

Hoenen T, Feldmann H (2017) Reverse genetics systems for filoviruses. Methods Mol Biol 1602:159–170. doi:10.1007/978-1-4939-6964-7_11

Hoenen T, Volchkov V, Kolesnikova L, Mittler E, Timmins J, Ottmann M, Reynard O, Becker S, Weissenhorn W (2005) VP40 octamers are essential for Ebola virus replication. J Virol 79 (3):1898–1905. doi:10.1128/JVI.79.3.1898-1905.2005

Hoenen T, Groseth A, de Kok-Mercado F, Kuhn JH, Wahl-Jensen V (2011) Minigenomes, transcription and replication competent virus-like particles and beyond: reverse genetics systems for filoviruses and other negative stranded hemorrhagic fever viruses. Antiviral Res 91 (2):195–208. doi:10.1016/j.antiviral.2011.06.003

Hoenen T, Shabman RS, Groseth A, Herwig A, Weber M, Schudt G, Dolnik O, Basler CF, Becker S, Feldmann H (2012) Inclusion bodies are a site of Ebolavirus replication. J Virol 86 (21):11779–11788. doi:10.1128/JVI.01525-12

Hoenen T, Groseth A, Callison J, Takada A, Feldmann H (2013) A novel Ebola virus expressing luciferase allows for rapid and quantitative testing of antivirals. Antiviral Res 99(3):207–213. doi:10.1016/j.antiviral.2013.05.017

Hoenen T, Marzi A, Scott DP, Feldmann F, Callison J, Safronetz D, Ebihara H, Feldmann H (2015) Soluble glycoprotein is not required for Ebola virus virulence in guinea pigs. J Infect Dis 212(Suppl 2):S242–S246. doi:10.1093/infdis/jiv111

Ilinykh PA, Shen X, Flyak AI, Kuzmina N, Ksiazek TG, Crowe JE Jr, Bukreyev A (2016) Chimeric filoviruses for identification and characterization of monoclonal antibodies. J Virol 90(8):3890–3901. doi:10.1128/JVI.00101-16

Jeffers SA, Sanders DA, Sanchez A (2002) Covalent modifications of the Ebola virus glycoprotein. J Virol 76(24):12463–12472

Johansen LM, Brannan JM, Delos SE, Shoemaker CJ, Stossel A, Lear C, Hoffstrom BG, Dewald LE, Schornberg KL, Scully C, Lehar J, Hensley LE, White JM, Olinger GG (2013) FDA-approved selective estrogen receptor modulators inhibit Ebola virus infection. Sci Transl Med 5(190):190ra179. doi:10.1126/scitranslmed.3005471

Koehler A, Kolesnikova L, Welzel U, Schudt G, Herwig A, Becker S (2015) A single amino acid change in the Marburg virus matrix protein VP40 provides a replicative advantage in a species-specific manner. J Virol 90(3):1444–1454. doi:10.1128/JVI.02670-15

Kolesnikova L, Mühlberger E, Ryabchikova E, Becker S (2000) Ultrastructural organization of recombinant Marburg virus nucleoprotein: comparison with Marburg virus inclusions. J Virol 74(8):3899–3904

Kolesnikova L, Bohil AB, Cheney RE, Becker S (2007) Budding of Marburgvirus is associated with filopodia. Cell Microbiol 9(4):939–951. doi:10.1111/j.1462-5822.2006.00842.x

Kolesnikova L, Mittler E, Schudt G, Shams-Eldin H, Becker S (2012) Phosphorylation of Marburg virus matrix protein VP40 triggers assembly of nucleocapsids with the viral envelope at the plasma membrane. Cell Microbiol 14(2):182–197. doi:10.1111/j.1462-5822.2011.01709.x

Krähling V, Dolnik O, Kolesnikova L, Schmidt-Chanasit J, Jordan I, Sandig V, Günther S, Becker S (2010) Establishment of fruit bat cells (*Rousettus aegyptiacus*) as a model system for the investigation of filoviral infection. PLoS Negl Trop Dis 4(8):e802. doi:10.1371/journal.pntd.0000802

Kuhn JH (2015) Ebolavirus and marburgvirus infections. In: Kasper DL, Fauci AS, Hauser SL, Longo DL, Jameson JL, Loscalzo J (eds) Harrison's principles of internal medicine, vol 2, 19th edn. McGraw-Hill Education, Columbus, USA, pp 1323–1329

Kuhn JH (2017) Guide to the correct use of filoviral nomenclature. Curr Top Microbiol Immunol. doi:10.1007/82_2017_7

Kuhn JH, Bao Y, Bavari S, Becker S, Bradfute S, Brister JR, Bukreyev AA, Chandran K, Davey RA, Dolnik O, Dye JM, Enterlein S, Hensley LE, Honko AN, Jahrling PB, Johnson KM, Kobinger G, Leroy EM, Lever MS, Mühlberger E, Netesov SV, Olinger GG, Palacios G, Patterson JL, Paweska JT, Pitt L, Radoshitzky SR, Saphire EO, Smither SJ, Swanepoel R, Towner JS, van der Groen G, Volchkov VE, Wahl-Jensen V, Warren TK, Weidmann M, Nichol ST (2013) Virus nomenclature below the species level: a standardized nomenclature for natural variants of viruses assigned to the family *Filoviridae*. Arch Virol 158 (1):301–311. doi:10.1007/s00705-012-1454-0

Kuhn JH, Andersen KG, Bào Y, Bavari S, Becker S, Bennett RS, Bergman NH, Blinkova O, Bradfute S, Brister JR, Bukreyev A, Chandran K, Chepurnov AA, Davey RA, Dietzgen RG, Doggett NA, Dolnik O, Dye JM, Enterlein S, Fenimore PW, Formenty P, Freiberg AN, Garry RF, Garza NL, Gire SK, Gonzalez J-P, Griffiths A, Happi CT, Hensley LE, Herbert AS, Hevey MC, Hoenen T, Honko AN, Ignatyev GM, Jahrling PB, Johnson JC, Johnson KM, Kindrachuk J, Klenk H-D, Kobinger G, Kochel TJ, Lackemeyer MG, Lackner DF, Leroy EM, Lever MS, Mühlberger E, Netesov SV, Olinger GG, Omilabu SA, Palacios G, Panchal RG, Park DJ, Patterson JL, Paweska JT, Peters CJ, Pettitt J, Pitt L, Radoshitzky SR, Ryabchikova EI, Saphire EO, Sabeti PC, Sealfon R, Shestopalov AM, Smither SJ, Sullivan NJ, Swanepoel R, Takada A, Towner JS, van der Groen G, Volchkov VE, Volchkova VA, Wahl-Jensen V, Warren TK, Warfield KL, Weidmann M, Nichol ST (2014) Filovirus RefSeq entries: evaluation and selection of filovirus type variants, type sequences, and names. Viruses 6(9):3663–3682. doi:10.3390/v6093663

Le Guenno B, Formenty P, Boesch C (1999) Ebola virus outbreaks in the Ivory Coast and Liberia, 1994–1995. Curr Top Microbiol Immunol 235:77–84

Lofts LL, Ibrahim MS, Negley DL, Hevey MC, Schmaljohn AL (2007) Genomic differences between guinea pig lethal and nonlethal Marburg virus variants. J Infect Dis 196(Suppl 2): S305–S312. doi:10.1086/520585

Madrid PB, Chopra S, Manger ID, Gilfillan L, Keepers TR, Shurtleff AC, Green CE, Iyer LV, Dilks HH, Davey RA, Kolokoltsov AA, Carrion R Jr, Patterson JL, Bavari S, Panchal RG, Warren TK, Wells JB, Moos WH, Burke RL, Tanga MJ (2013) A systematic screen of FDA-approved drugs for inhibitors of biological threat agents. PLoS ONE 8(4):e60579. doi:10.1371/journal.pone.0060579

Martin B, Canard B, Decroly E (2017) Filovirus proteins for antiviral drug discovery: structure/function bases of the replication cycle. Antiviral Res 141:48–61. doi:10.1016/j.antiviral.2017.02.004

Martinez MJ, Volchkova VA, Raoul H, Alazard-Dany N, Reynard O, Volchkov VE (2011) Role of VP30 phosphorylation in the Ebola virus replication cycle. J Infect Dis 204(Suppl 3):S934–S940. doi:10.1093/infdis/jir320

Martínez MJ, Biedenkopf N, Volchkova V, Hartlieb B, Alazard-Dany N, Reynard O, Becker S, Volchkov V (2008) Role of Ebola virus VP30 in transcription reinitiation. J Virol 82(24):12569–12573. doi:10.1128/JVI.01395-08

Mateo M, Carbonnelle C, Reynard O, Kolesnikova L, Nemirov K, Page A, Volchkova VA, Volchkov VE (2011) VP24 is a molecular determinant of Ebola virus virulence in guinea pigs. J Infect Dis 204(Suppl 3):S1011–S1020. doi:10.1093/infdis/jir338

Mavrakis M, Kolesnikova L, Schoehn G, Becker S, Ruigrok RWH (2002) Morphology of Marburg virus NP-RNA. Virology 296(2):300–307. doi:10.1006/viro.2002.1433

Mehedi M, Falzarano D, Seebach J, Hu X, Carpenter MS, Schnittler H-J, Feldmann H (2011) A new Ebola virus nonstructural glycoprotein expressed through RNA editing. J Virol 85(11):5406–5414. doi:10.1128/JVI.02190-10

Mehedi M, Hoenen T, Robertson S, Ricklefs S, Dolan MA, Taylor T, Falzarano D, Ebihara H, Porcella SF, Feldmann H (2013) Ebola virus RNA editing depends on the primary editing site sequence and an upstream secondary structure. PLoS Pathog 9(10):e1003677. doi:10.1371/journal.ppat.1003677

Miller EH, Chandran K (2012) Filovirus entry into cells—new insights. Curr Opin Virol 2(2):206–214. doi:10.1016/j.coviro.2012.02.015

Miranda ME, Miranda NL (2011) Reston ebolavirus in humans and animals in the Philippines: a review. J Infect Dis 204(Suppl 3):S757–S760. doi:10.1093/infdis/jir296

Miranda ME, Ksiazek TG, Retuya TJ, Khan AS, Sanchez A, Fulhorst CF, Rollin PE, Calaor AB, Manalo DL, Roces MC, Dayrit MM, Peters CJ (1999) Epidemiology of Ebola (subtype Reston) virus in the Philippines, 1996. J Infect Dis 179(Suppl 1):S115–S119. doi:10.1086/514314

Mittler E, Kolesnikova L, Herwig A, Dolnik O, Becker S (2013) Assembly of the Marburg virus envelope. Cell Microbiol 15(2):270–284. doi:10.1111/cmi.12076

Möller P, Pariente N, Klenk H-D, Becker S (2005) Homo-oligomerization of Marburgvirus VP35 is essential for its function in replication and transcription. J Virol 79(23):14876–14886. doi:10.1128/JVI.79.23.14876-14886.2005

Mpanju OM, Towner JS, Dover JE, Nichol ST, Wilson CA (2006) Identification of two amino acid residues on Ebola virus glycoprotein 1 critical for cell entry. Virus Res 121(2):205–214. doi:10.1016/j.virusres.2006.06.002

Mühlberger E (2007) Filovirus replication and transcription. Future Virol 2(2):205–215. doi:10.2217/17460794.2.2.205

Mühlberger E, Trommer S, Funke C, Volchkov V, Klenk H-D, Becker S (1996) Termini of all mRNA species of Marburg virus: sequence and secondary structure. Virology 223(2):376–380. doi:10.1006/viro.1996.0490

Mühlberger E, Lötfering B, Klenk H-D, Becker S (1998) Three of the four nucleocapsid proteins of Marburg virus, NP, VP35, and L, are sufficient to mediate replication and transcription of Marburg virus-specific monocistronic minigenomes. J Virol 72(11):8756–8764

Mühlberger E, Weik M, Volchkov VE, Klenk H-D, Becker S (1999) Comparison of the transcription and replication strategies of Marburg virus and Ebola virus by using artificial replication systems. J Virol 73(3):2333–2342

Neumann G, Feldmann H, Watanabe S, Lukashevich I, Kawaoka Y (2002) Reverse genetics demonstrates that proteolytic processing of the Ebola virus glycoprotein is not essential for replication in cell culture. J Virol 76(1):406–410

Neumann G, Ebihara H, Takada A, Noda T, Kobasa D, Jasenosky LD, Watanabe S, Kim JH, Feldmann H, Kawaoka Y (2005) Ebola virus VP40 late domains are not essential for viral replication in cell culture. J Virol 79(16):10300–10307. doi:10.1128/JVI.79.16.10300-10307. 2005

Neumann G, Geisbert TW, Ebihara H, Geisbert JB, Daddario-DiCaprio KM, Feldmann H, Kawaoka Y (2007) Proteolytic processing of the Ebola virus glycoprotein is not critical for Ebola virus replication in nonhuman primates. J Virol 81(6):2995–2998. doi:10.1128/JVI. 02486-06

Panchal RG, Kota KP, Spurgers KB, Ruthel G, Tran JP, Boltz RC, Bavari S (2010) Development of high-content imaging assays for lethal viral pathogens. J Biomol Screen 15(7):755–765. doi:10.1177/1087057110374357

Panchal RG, Reid SP, Tran JP, Bergeron AA, Wells J, Kota KP, Aman J, Bavari S (2012) Identification of an antioxidant small-molecule with broad-spectrum antiviral activity. Antiviral Res 93(1):23–29. doi:10.1016/j.antiviral.2011.10.011

Prins KC, Delpeut S, Leung DW, Reynard O, Volchkova VA, Reid SP, Ramanan P, Cárdenas WB, Amarasinghe GK, Volchkov VE, Basler CF (2010) Mutations abrogating VP35 interaction with double-stranded RNA render Ebola virus avirulent in guinea pigs. J Virol 84 (6):3004–3015. doi:10.1128/JVI.02459-09

Racaniello VR, Baltimore D (1981) Cloned poliovirus complementary DNA is infectious in mammalian cells. Science 214(4523):916–919

Radoshitzky SR, Warfield KL, Chi X, Dong L, Kota K, Bradfute SB, Gearhart JD, Retterer C, Kranzusch PJ, Misasi JN, Hogenbirk MA, Wahl-Jensen V, Volchkov VE, Cunningham JM, Jahrling PB, Aman MJ, Bavari S, Farzan M, Kuhn JH (2011) Ebolavirus Δ-peptide immunoadhesins inhibit marburgvirus and ebolavirus cell entry. J Virol 85(17):8502–8513. doi:10.1128/JVI.02600-10

Sanchez A, Kiley MP, Holloway BP, Auperin DD (1993) Sequence analysis of the Ebola virus genome: organization, genetic elements, and comparison with the genome of Marburg virus. Virus Res 29(3):215–240

Sanchez A, Trappier SG, Mahy BWJ, Peters CJ, Nichol ST (1996) The virion glycoproteins of Ebola viruses are encoded in two reading frames and are expressed through transcriptional editing. Proc Natl Acad Sci U S A 93(8):3602–3607

Schmidt KM, Mühlberger E (2016) Marburg virus reverse genetics systems. Viruses 8(6):178. doi:10.3390/v8060178

Schmidt KM, Schümann M, Olejnik J, Krähling V, Mühlberger E (2011) Recombinant Marburg virus expressing eGFP allows rapid screening of virus growth and real-time visualization of virus spread. J Infect Dis 204(Suppl 3):S861–S870. doi:10.1093/infdis/jir308

Schnell MJ, Mebatsion T, Conzelmann K-K (1994) Infectious rabies viruses from cloned cDNA. EMBO J 13(18):4195–4203

Schudt G, Kolesnikova L, Dolnik O, Sodeik B, Becker S (2013) Live-cell imaging of Marburg virus-infected cells uncovers actin-dependent transport of nucleocapsids over long distances. Proc Natl Acad Sci U S A 110(35):14402–14407. doi:10.1073/pnas.1307681110

Slagsvold T, Pattni K, Malerød L, Stenmark H (2006) Endosomal and non-endosomal functions of ESCRT proteins. Trends Cell Biol 16(6):317–326. doi:10.1016/j.tcb.2006.04.004

Towner JS, Paragas J, Dover JE, Gupta M, Goldsmith CS, Huggins JW, Nichol ST (2005) Generation of eGFP expressing recombinant Zaire Ebola virus for analysis of early pathogenesis events and high-throughput antiviral drug screening. Virology 332(1):20–27. doi:10.1016/j.virol.2004.10.048

Towner JS, Khristova ML, Sealy TK, Vincent MJ, Erickson BR, Bawiec DA, Hartman AL, Comer JA, Zaki SR, Ströher U, Gomes da Silva F, del Castillo F, Rollin PE, Ksiazek TG, Nichol ST (2006) Marburgvirus genomics and association with a large hemorrhagic fever outbreak in Angola. J Virol 80(13):6497–6516. doi:10.1128/JVI.00069-06

Towner JS, Amman BR, Sealy TK, Carroll SA, Comer JA, Kemp A, Swanepoel R, Paddock CD, Balinandi S, Khristova ML, Formenty PBH, Albarino CG, Miller DM, Reed ZD, Kayiwa JT, Mills JN, Cannon DL, Greer PW, Byaruhanga E, Farnon EC, Atimnedi P, Okware S, Katongole-Mbidde E, Downing R, Tappero JW, Zaki SR, Ksiazek TG, Nichol ST, Rollin PE (2009) Isolation of genetically diverse Marburg viruses from Egyptian fruit bats. PLoS Pathog 5(7):e1000536. doi:10.1371/journal.ppat.1000536

Tsuda Y, Hoenen T, Banadyga L, Weisend C, Ricklefs SM, Porcella SF, Ebihara H (2015) An improved reverse genetics system to overcome cell-type-dependent Ebola virus genome plasticity. J Infect Dis 212(Suppl 2):S129–S137. doi:10.1093/infdis/jiu681

Volchkov VE, Becker S, Volchkova VA, Ternovoj VA, Kotov AN, Netesov SV, Klenk H-D (1995) GP mRNA of Ebola virus is edited by the Ebola virus polymerase and by T7 and vaccinia virus polymerases. Virology 214(2):421–430

Volchkov VE, Feldmann H, Volchkova VA, Klenk H-D (1998) Processing of the Ebola virus glycoprotein by the proprotein convertase furin. Proc Natl Acad Sci U S A 95(10):5762–5767

Volchkov VE, Volchkova VA, Chepurnov AA, Blinov VM, Dolnik O, Netesov SV, Feldmann H (1999) Characterization of the L gene and 5' trailer region of Ebola virus. J Gen Virol 80(Pt 2):355–362. doi:10.1099/0022-1317-80-2-355

Volchkov VE, Volchkova VA, Mühlberger E, Kolesnikova LV, Weik M, Dolnik O, Klenk H-D (2001) Recovery of infectious Ebola virus from complementary DNA: RNA editing of the GP gene and viral cytotoxicity. Science 291(5510):1965–1969. doi:10.1126/science.1057269

Volchkova VA, Klenk H-D, Volchkov VE (1999) Delta-peptide is the carboxy-terminal cleavage fragment of the nonstructural small glycoprotein sGP of Ebola virus. Virology 265(1):164–171. doi:10.1006/viro.1999.0034

Volchkova VA, Dolnik O, Martinez MJ, Reynard O, Volchkov VE (2015) RNA editing of the GP gene of Ebola virus is an important pathogenicity factor. J Infect Dis 212(Suppl 2):S226–S233. doi:10.1093/infdis/jiv309

Warren TK, Warfield KL, Wells J, Enterlein S, Smith M, Ruthel G, Yunus AS, Kinch MS, Goldblatt M, Aman MJ, Bavari S (2010) Antiviral activity of a small-molecule inhibitor of filovirus infection. Antimicrob Agents Chemother 54(5):2152–2159. doi:10.1128/AAC.01315-09

Watt A, Moukambi F, Banadyga L, Groseth A, Callison J, Herwig A, Ebihara H, Feldmann H, Hoenen T (2014) A novel life cycle modeling system for Ebola virus shows a genome length-dependent role of VP24 in virus infectivity. J Virol 88(18):10511–10524. doi:10.1128/JVI.01272-14

Welch SR, Guerrero LW, Chakrabarti AK, McMullan LK, Flint M, Bluemling GR, Painter GR, Nichol ST, Spiropoulou CF, Albariño CG (2016) Lassa and Ebola virus inhibitors identified using minigenome and recombinant virus reporter systems. Antiviral Res 136:9–18. doi:10.1016/j.antiviral.2016.10.007

Welsch S, Kolesnikova L, Krähling V, Riches JD, Becker S, Briggs JA (2010) Electron tomography reveals the steps in filovirus budding. PLoS Pathog 6(4):e1000875. doi:10.1371/journal.ppat.1000875

Wenigenrath J, Kolesnikova L, Hoenen T, Mittler E, Becker S (2010) Establishment and application of an infectious virus-like particle system for Marburg virus. J Gen Virol 91(Pt 5):1325–1334. doi:10.1099/vir.0.018226-0

Wijesinghe KJ, Stahelin RV (2015) Investigation of the lipid binding properties of the Marburg virus matrix protein VP40. J Virol 90(6):3074–3085. doi:10.1128/JVI.02607-15

Guide to the Correct Use of Filoviral Nomenclature

Jens H. Kuhn

Abstract The International Committee on Taxonomy of Viruses (ICTV) currently recognizes three genera and seven species as part of the mononegaviral family *Filoviridae*. Eight distinct filoviruses (Bundibugyo virus, Ebola virus, Lloviu virus, Marburg virus, Ravn virus, Reston virus, Sudan virus, and Taï Forest virus) have been assigned to these seven species. This chapter briefly summarizes the status quo of filovirus classification and focuses on the importance of differentiating between filoviral species and filoviruses and the correct use of taxonomic and vernacular filovirus names and abbreviations in written and oral discourse.

Contents

J.H. Kuhn (✉)
Integrated Research Facility at Fort Detrick (IRF-Frederick), Division of Clinical Research (DCR), National Institute of Allergy and Infectious Diseases (NIAID) National Institutes of Health (NIH), B-8200 Research Plaza, Fort Detrick, Frederick, MD 21702, USA
e-mail: kuhnjens@mail.nih.gov

Current Topics in Microbiology and Immunology (2017) 411:447–460
DOI 10.1007/82_2017_7

Published Online: 27 June 2017

1 Introduction

Virus taxonomy consists of two subdisciplines, classification and nomenclature.

1.1 Classification

Virus classification is the continuous attempt to sort viruses into categories that are called taxa (singular: taxon). In practice, virus classification is a way to gather viruses into progressively more inclusive groups based on common genomic, phylogenetic, and phenotypic properties. Taxa are hierarchically organized into ranks. In virology, these ranks are called, in descending order, order, family, subfamily, genus, and species (King et al. 2011a). Importantly, each lower taxon is included in the next higher taxon rank: a virus that is a member of a particular species is automatically a member of the higher ranked taxa that include that species. This organization is akin to other biological classification schemes, for instance that of zoology. If a particular animal is a primate (an animal belonging to the order Primates) and if the order Primates is included in the class Mammalia, then that animal is both a primate and a mammal because all primates are mammals ("inclusion principle").

Assigning viruses to hierarchical taxa ideally occurs based on objectively measured evolutionary relationships. The goal of virus classification is therefore the clarification that, for instance,

- a virus 1 is not the same thing as a virus 2, requiring both of them to be assigned to different lowest taxa (species);
- but that both viruses share a common ancestor and therefore should be assigned to the same higher taxon (genus); and
- that both viruses are more closely related to each other than either of them are to a virus 3, which therefore needs to be assigned to a different lower (species) and higher (genus) taxon.

This clarification has meaning beyond purely evolutionary interests: viruses that are closely related to each other share more properties with each other than either of them shares with a more distantly related virus. Therefore, successful diagnostics or medical countermeasures developed against virus 1 are more likely to work on virus 2 than on virus 3. Empirical treatment, as already possible for bacterial diseases, may therefore become possible for viral infections if virus taxonomy is "correct," i.e., truly reflects the evolutionary relationship between a relatively uncharacterized virus and a better characterized virus.

1.2 Nomenclature

Virus nomenclature is the process of assigning names to the individual concept of the mind (taxon) and its physical members (viruses) to allow precise written and oral communication because names are easier to remember than long lists of alphanumerical codes. Therefore, nomenclature efforts for the 3 viruses could result in

- naming viruses 1, 2, and 3 "rabies virus," "vesicular stomatitis Indiana virus"; and "Lassa virus," respectively;
- assigning these viruses to the species "*Rabies lyssavirus*" in the genus "*Lyssavirus,*" "*Indiana vesiculovirus*" in the genus "*Vesiculovirus,*" and "*Lassa mammarenavirus*" in the genus "*Mammarenavirus,*" respectively;
- including the genera "*Lyssavirus*" and "*Vesiculovirus*" together in the family "*Rhabdoviridae*"; and
- including the genus "*Mammarenavirus*" in a different family called "*Arenaviridae*".

1.3 Taxonomic Process

The International Committee on Taxonomy of Viruses (ICTV), the only committee of the Virology Division of the International Union of Microbiological Societies (IUMS), is responsible for all matters of virus taxonomy. The ICTV's activities are regulated by Statutes [(King et al. 2011b); latest update at http://www.ictvonline.org/statutes.asp], and the Rules of virus classification and nomenclature are laid out in a Code, the International Code of Virus Classification and Nomenclature [ICVCN (King et al. 2011a); latest update at https://talk.ictvonline.org/information/w/ictv-information/383/ictv-code]. Importantly, the vast majority of virology and microbiology specialty journals, including those by Springer Publishing, require authors to follow official ICTV taxonomy (see Springer "Instructions for Authors"). This requirement means that applying ICTV nomenclature in written discourse is not optional but quasi-mandated, although unfortunately not strictly enough enforced.

The ICTV bases all its decisions on scientific debate and available (published) data. At the heart of all discussions and decisions lays the input of individual experts in the field (through the submission of taxonomic proposals or public comment on such proposals) and the consultation of ICTV Study Groups (listed at http://www.ictvonline.org/studygroups.asp?se=5). ICTV Study Groups typically consist of several experts on the viruses belonging to a particular viral order or family and serve as mediators between the needs and concerns of their respective virologist constituency on the one hand and the Code requirements and concerns of the ICTV on the other.

Taxonomic process begins with the submission to the ICTV of a taxonomic proposal ("TaxoProp") to change current virus classification and/or nomenclature and/or the Code via https://talk.ictvonline.org/files/taxonomy-proposal-templates/. Such TaxoProps can be written and submitted by anybody. The TaxoProp is then evaluated for adherence to the Code, scientific accuracy, and taxonomic merit by the respective ICTV Study Group. Finally, the ICTV Executive Committee and the IUMS Virology Division make a decision based on ICTV Study Group input and public comments whether to accept/ratify, reject, or delay the proposal. Importantly, absence of ICTV Study Group input and/or public comments may have to be regarded as input by the ICTV to ensure taxonomic progress ("absence of dissent is consent"), and the ICTV also can (but very rarely does) overrule Study Group recommendations. During every given year, the then-official ICTV taxonomy is posted at http://www.ictvonline.org/virusTaxonomy.asp, and annual update articles are published in Springer's *Archives of Virology* [latest update: (Adams et al. 2017)]. Importantly, this official taxonomy only lists taxa but not their member viruses, i.e., the ICTV Master Species List cannot be used to look up the names of particular viruses or their abbreviations.

2 Current Filovirus Taxonomy

2.1 *Filovirus Classification*

The family *Filoviridae*, the members of which are collectively referred to as "filoviruses", has been established around the prototype filovirus, Marburg virus, which was discovered and named that way in 1967 (Siegert et al. 1967). Molecular studies revealed that Marburg virus shares several properties with viruses of the current families *Pneumoviridae* and *Rhabdoviridae* (e.g., a monopartite nonsegmented, linear, negative-sense single-stranded RNA genome and the general gene order 3′-*N*-*P*-*M*-*G*-*L*-5′). However, Marburg virus also differs from pneumoviruses and rhabdoviruses considerably (e.g., encoding the additional protein VP24 for which there is no homolog in other virus groups). The commonalities and differences indicated that Marburg virus ought to be classified in a new family, *Filoviridae* (Kiley et al. 1982), which however needed be included together with *Pneumoviridae* and *Rhabdoviridae* in a higher taxon, i.e., the order today called *Mononegavirales* (Pringle 1991). Since 1967, seven additional distinct viruses have been discovered that share more properties with Marburg virus than with any other virus in that order. Hence, these viruses were also classified in the family *Filoviridae* [reviewed in (Kuhn et al. 2011; Bukreyev et al. 2014)]. Phylogenetic analyses based on coding-complete filovirus genome sequences, filovirus core RNA-dependent RNA polymerase module amino acid sequences, glycoprotein gene nucleotide or deduced amino acid sequences, and pairwise genome

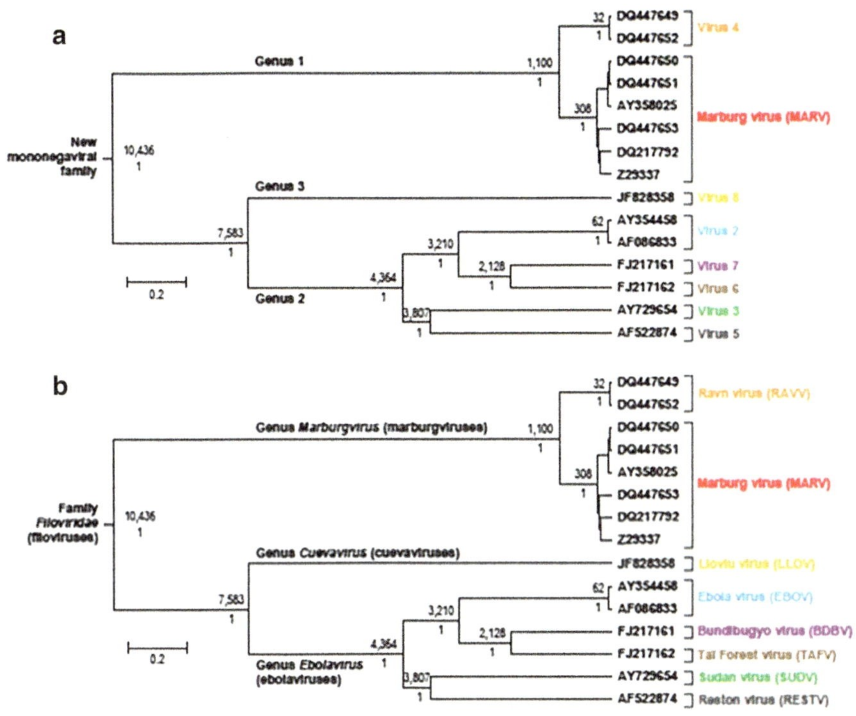

Fig. 1 Official ICTV taxonomy of the family *Filoviridae* (Bukreyev et al. 2014; Kuhn et al. 2011; Carroll et al. 2013). Graphical presentation of a Bayesian coalescent analysis using representative cuevavirus, marburgvirus, and ebolavirus genomes (based on data published by Serena Carroll/CDC (Carroll et al. 2013)). **a** Evolutionary relationships (classification) of newly discovered filoviruses relative to Marburg virus. **b** Same analysis with all clades and viruses named based on current nomenclature consensus

comparisons using PASC or DEmARC methodologies all confirmed the existence of seven distinct filovirus lineages (i.e., species) that cluster into three clades (i.e., genera) in one mononegaviral family (Fig. 1a) (Bao et al. 2012; Lauber and Gorbalenya 2012; Carroll et al. 2013; Kuhn et al. 2011). This classification of distinct filoviruses into several species of more than one genus was *de facto* established by the ICTV via the ICTV *Filoviridae* Study Group chaired by Netesov in 2000 (Netesov et al. 2000) and upheld by the successor Study Groups chaired by Feldmann (2005) (Feldmann et al. 2005) and Kuhn (present) (Kuhn et al. 2011). References to the eight distinct filoviruses as “subtypes” or “strains” of either Marburg virus or Ebola virus can unfortunately still be found in manuscripts published by scientists who have not reviewed or understood filovirus phylogeny in depth. However, this classification is uncontroversial among evolutionary filovirologists and therefore will not be discussed here any further.

Table 1 Official ICTV taxonomy of the family *Filoviridae* (Bukreyev et al. 2014; Kuhn et al. 2011)

Order *Mononegavirales*
Family *Filoviridae*
Genus *Marburgvirus*
Species *Marburg marburgvirus*
Virus: Marburg virus (MARV)
Virus: Ravn virus (RAVV)
Genus *Ebolavirus*
Species *Bundibugyo ebolavirus*
Virus: Bundibugyo virus (BDBV)
Species *Reston ebolavirus*
Virus: Reston virus (RESTV)
Species *Sudan ebolavirus*
Virus: Sudan virus (SUDV)
Species *Taï Forest ebolavirus*
Virus: Taï Forest virus (TAFV)
Species *Zaire ebolavirus*
Virus: Ebola virus (EBOV)
Genus *Cuevavirus*
Species *Lloviu cuevavirus*
Virus: Lloviu virus (LLOV)

2.2 *Filovirus Nomenclature*

In contrast to classification, nomenclature is unfortunately almost always contentious as names appear tightly tied to emotions (Payne 2016; Kuhn and Wahl-Jensen 2010). Ideally, names for any biological entity (e.g., genes, proteins, organisms, viruses) remain stable as long as possible to ensure precise and unambiguous scientific communication. On the other hand, new discoveries or newly emerging sensitivities often challenge old or well-established names. Ultimately, names should honor those who first created them as much as possible and should be unique, memorable, devised in a way not to cause confusion, be adaptable to new developments, and eligible for electronic database usage by following a universal format. Combined, these requirements restrain the freedom in naming. The most current nomenclature of the family *Filoviridae* is outlined in Table 1 and Fig. 1b.

3 Application of Filovirus Nomenclature

3.1 *Taxa Versus Taxon Members*

3.1.1 Concept

During manuscript writing, it is important to clearly distinguish between viral taxa and their members. Taxa are concepts of the human mind, i.e. not things, whereas taxon members (e.g., viruses, animals, plants) are physical objects and not concepts of the human mind (Calisher and Mahy 2003; Drebot et al. 2002; Kuhn and Jahrling 2010; Van Regenmortel 2003, 2006, 2007, 2016). A taxon, for instance a species, can be regarded as the perfect average of all its variable members, perfection in conceptual space (Kuhn and Jahrling 2010). In zoology, one can for instance refer to *Homo sapiens*, the species that has been created for all humans that have ever lived, that currently live, or that will live in the future. Included in the concept of *Homo sapiens* is that its members are upright-walking bipedal mammals. This idea, the species, is not challenged by the occasional observation that individual members of this species are not able to walk upright in a bipedal manner—the average of all members indeed "walks" upright in conceptual space. Keeping this concept in mind, it becomes instantly clear that one cannot talk to or study *Homo sapiens*, but that one can only talk to or study individual representatives of the species. These representatives, the members of the species *Homo sapiens*, are called "humans". The definition of the species *Homo sapiens* is based on studying as many humans as possible to approximate the "perfect" human in conceptual space. However, because one can never study all living humans, let alone all humans that have lived or have not yet lived, the species definition will always remain in flux. The smaller the sampling space, the less perfect the species definition.

Many virologists understand the differentiation between taxa and their members on the macroscopic level. For instance, in manuscript method sections, virologists note that "rhesus monkeys (*Macaca mulatta*)" have been infected with a particular virus and then continue referring to "rhesus monkeys" (the physical object) rather than "*Macaca mulatta*" (the concept) throughout the remainder of the manuscript. These virologists have only infected a few representatives of the species—and not all rhesus monkeys that have ever lived, currently live, or will live tomorrow. Likewise, the results obtained with several rhesus monkeys cannot necessarily be extrapolated to all rhesus monkeys. Although the same concept applies to virology, most virologists unfortunately do not yet apply the correct separation of taxa and their members to viruses. In the case of filoviruses, for instance, Ebola virus is the member of the species *Zaire ebolavirus*. It follows that one can centrifuge Ebola virus and one can get infected with Ebola virus, or one can use Ebola virus to infect rhesus monkeys, because Ebola virus is a physical object. But one cannot infect rhesus monkeys with *Zaire ebolavirus* or be infected by *Zaire ebolavirus* because *Zaire ebolavirus* is not a physical object. The separation of object and concept allows us to have species for animals (and possibly also viruses) that do not exist anymore. Contrary to common parlance, the zoological species *Tyrannosaurus rex* is not extinct. Instead, the reader has an immediate idea about the animal member of that species despite the fact that there are no living members of that species anymore. The concept *T. rex* is "alive," whereas animal representatives of that concept are not.

3.1.2 Orthography

In all biological taxonomies, taxon names are written with a capital first letter. In virology, taxon names of all ranks are italicized and thereby are instantly recognizable as taxa (e.g., "*Mononegavirales*: *Filoviridae*: *Ebolavirus*: *Zaire ebolavirus*"), whereas names of ranks higher than genus are not italicized in non-virological taxonomies (e.g., "Primates: Hominidae: *Homo*: *Homo sapiens*" in zoology). The names of members of taxa are not italicized and are always written in lower case except if a name component is a proper noun [e.g., "rabies virus" is the member of the species "*Rabies lyssavirus*" and "Sudan virus" is the member of the species "*Sudan ebolavirus*" ("Sudan virus" is capitalized because "Sudan" is a proper noun)]. In contrast to taxon names, virus names may be abbreviated (e.g., Sudan virus: "SUDV").

Virological taxa of different ranks above species can be easily differentiated by ICVCN-proscribed name suffixes. These are "*-virales*", "*-viridae*", "*-virinae*", and "*-virus*" for order, family, subfamily, and genus names, respectively. Thus, the name "*Mononegavirales*" immediately implies an order, whereas the names "*Filoviridae*" and "*Ebolavirus*" immediately imply a family and a genus, respectively. Higher taxa almost always have more than one virus member. The collective members (tangible things) of a higher taxon (mental construct) are named by keeping the word stem of the taxon name, removing the italics, placing the word into lower case, and adding the suffix "-viruses". Importantly, collective members of taxa above species are never to be abbreviated (there is no abbreviation for "poxviruses", there is none for "arenaviruses", and hence there is none for "ebolaviruses" or "marburgviruses"). The correct application of filoviral above-species taxon orthography is depicted in Tables 2, 3 and 4.

Taxon member-specific adjectives are formed by using the singular of the member name ("The polymerase of ebolaviruses"→"The ebolavirus polymerase") or by replacing the suffix "-viruses" with "-viral" (→"The ebolaviral polymerase").

From the deliberations above, it logically follows that

- the sentence "The hantavirus/hantaviral glycoprotein binds a protein receptor" would mean that all 62 currently classified members of the genus *Hantavirus*, i.e. all hantaviruses, have a glycoprotein that binds a protein receptor; whereas "Hantaan virus glycoprotein binds a protein receptor" would mean that (possibly

Table 2 Orthography of viral order and order member names

Order name (capitalized, italicized, specific suffix "*-virales*", never abbreviated)	Name of the collective members of the order (word stem of the order name appended to the specific suffix "-viruses", not capitalized, not italicized, never abbreviated)
Caudovirales	caudoviruses
Nidovirales	nidoviruses
Picornavirales	picornaviruses
→*Mononegavirales*	→mononegaviruses

Table 3 Orthography of viral family and family member names

Family name (capitalized, italicized, specific suffix "*-viridae*", never abbreviated)	Name of the collective members of the family (word stem of the family name appended to the specific suffix "-viruses", not capitalized, not italicized, never abbreviated)
Arenaviridae	arenaviruses
Paramyxoviridae	paramyxoviruses
Poxviridae	poxviruses
→*Filoviridae*	→filoviruses

Table 4 Orthography of viral genus and genus member names

Genus name (capitalized, italicized, specific suffix "*-virus*", never abbreviated)	Name of the collective members of the genus (word stem of the genus name appended to the specific suffix "-viruses", not capitalized, not italicized, never abbreviated)
Flavivirus	flaviviruses
Hantavirus	hantaviruses
Henipavirus	henipaviruses
→*Cuevavirus*	→cuevaviruses
→*Ebolavirus*	→ebolaviruses
→*Marburgvirus*	→marburgviruses

only) this one particular hantavirus, namely Hantaan virus, binds a protein receptor; and therefore that

- the sentence "The ebolavirus/ebolavirus glycoprotein binds NPC1" means that all 5 currently classified members of the genus *Ebolavirus*, i.e. all ebolaviruses (BDBV + EBOV + RESTV + SUDV + TAFV), have a glycoprotein that binds NPC1, whereas "Bundibugyo virus binds NPC1" would mean that (possibly only) one particular ebolavirus, namely Bundibugyo virus, binds NPC1.

The taxonomic inclusion principle implies that

- if dengue virus 1 (DENV-1) and yellow fever virus (YFV) are members of the genus *Flavivirus*, then dengue virus 1 is a flavivirus and yellow fever virus is also a flavivirus; and therefore
- if Ebola virus (EBOV) and Bundibugyo virus (BDBV) are members of the genus *Ebolavirus*, then Ebola virus is an ebolavirus and Bundibugyo virus (BDBV) is also an ebolavirus; and
- if Marburg virus (MARV) and Ravn virus (RAVV) are members of the genus *Marburgvirus*, then Marburg virus is a marburgvirus and Ravn virus is also a marburgvirus; and
- because the genera *Ebolavirus* and *Marburgvirus* are included in the family *Filoviridae*, Ebola virus and Marburg virus are (two) filoviruses, and ebolaviruses and marburgviruses are (seven) filoviruses.

Table 5 Nomenclature cheat sheet for filovirus manuscript writing

Potential problem	Potential solution
Search your draft manuscript for the terms "*Bundibugyo ebolavirus*", "*Lloviu cuevavirus*", "*Marburg marburgvirus*", "*Reston ebolavirus,*" "*Sudan ebolavirus,*" "*Taï Forest ebolavirus,*" and "*Zaire ebolavirus*"	These terms are current species names (Bukreyev et al. 2014). Species are not objects (Kuhn and Jahrling 2010). Hence, these terms should only appear very rarely in your manuscript. Typically, these terms ought to be listed in the method section, linked to their member viruses, which are usually the subject of scientific papers. Therefore, replace these terms throughout the paper with virus names: "Bundibugyo virus," "Lloviu virus," "Marburg virus" or "Ravn virus," "Reston virus," "Sudan virus," "Taï Forest virus," or "Ebola virus," respectively (Bukreyev et al. 2014)
Search your draft manuscript for the terms "*Lake Victoria marburgvirus*" and "*Cote d'Ivoire ebolavirus*"/"*Côte d'Ivoire ebolavirus*"/"*Ivory Coast ebolavirus*"	These species names are no longer official. Replace with "*Marburg marburgvirus*" or "*Taï Forest ebolavirus,*" respectively (Bukreyev et al. 2014)
Search your draft manuscript for the terms "Bundibugyo ebolavirus," "Lloviu cuevavirus," "Marburg marburgvirus," "Reston ebolavirus," "Sudan ebolavirus," "Taï Forest ebolavirus"/"Cote d'Ivoire ebolavirus"/"Côte d'Ivoire ebolavirus"/ "Ivory Coast ebolavirus," or "Zaire ebolavirus"	These terms do not exist anymore. They either need to be italicized (if referring to species) or, more likely, need to be replaced with current virus names (see above) (Bukreyev et al. 2014)
Search your draft manuscript for the terms "BEBOV"/"UEBOV", "MBGV", "Marburg virus Ravn," "REBOV", "SEBOV", "CIEBOV"/"ICEBOV", or "ZEBOV"	These terms are outdated. Replace with "BDBV", "MARV", "RAVV", "RESTV", "SUDV", "TAFV", or "EBOV", respectively (Bukreyev et al. 2014; Kuhn et al. 2010)
Search your draft manuscript for the abbreviation "EBOV"	Do you refer to one virus, namely "Ebola virus (species *Zaire ebolavirus*)"? In that case the abbreviation is correctly applied (Kuhn et al. 2010). Do you refer to several viruses of the genus *Ebolavirus*? Then replace with "ebolavirus(es)/ebolaviral" because the collective members of higher taxa are not abbreviated
Search your draft manuscript for the abbreviation "MARV"	Do you refer to one virus, namely "Marburg virus (species *Marburg marburgvirus*)"? In this case, the abbreviation is correctly applied (Kuhn et al. 2010). Do you refer to several viruses of the genus *Marburgvirus*? Then replace with "marburgvirus(es)/marburgviral" because the collective members of higher taxa are not abbreviated

(continued)

Table 5 (continued)

Potential problem	Potential solution
Search your draft manuscript for the terms "Cuevavirus", "Ebolavirus" and "Marburgvirus"	These terms do not exist. Either italicize (if referring to genera) or place in lower case plural (when referring to all members of these genera): "cuevaviruses", "ebolaviruses", or "marburgviruses"
Search your draft manuscript for the terms "ebolavirus"/"ebolaviral" and "marburgvirus"/"marburgviral"	These terms refer to groups of viruses, namely all members of the genera *Ebolavirus* and *Marburgvirus*, respectively. If you want to refer to one particular virus, replace with, for instance, "Ebola virus" or "Marburg virus"
Search your draft manuscript for the term "species"	Remember that a species is not a thing, but an idea (Kuhn and Jahrling 2010). Species can therefore not be "used" in any way; they can also not be discovered or go extinct. Rephrase your sentences accordingly, for instance: (a) replace "three species of monkeys were infected with Ebola virus" with "monkeys of three species were infected with Ebola virus"; (b) replace "rhesus monkeys infected with the *Zaire ebolavirus* species" with "rhesus monkeys infected with Ebola virus"; (c) "Ebola virus is thought to infect three species of fruit bats" with "Ebola virus is thought to infect fruit bats of three species"; or (d) replace "Five ebolavirus species have been discovered" with "Five ebolaviruses have been discovered" or "Five ebolavirus species have been established"
Search your draft manuscript for the term "member" and the verb "include(d)"	If in taxonomic context, ensure that "include(d)" is only used for taxa (species are included in genera, genera are included in families and so on—but species are not members of genera and genera are not members of families). Likewise, ensure that "member" is only used for viruses (physical objects): a virus is a member of a species, a genus, a family and so on, but it is not "included" in any of them (Van Regenmortel 2003, 2006, 2007, 2016)
Search your draft manuscript for the term "strain"	If referring to a filovirus, use the word "strain" only if you refer to a laboratory mouse- or guinea pig-adapted virus that causes disease in those rodents. Otherwise, replace with "variant" or "isolate" ["Kikwit", "Makona", "Yambuku" are EBOV variants; "Kikwit-9510621", "Makona-C05", and "Yambuku-Mayinga" are examples for isolates of each of these variants (Kuhn et al. 2013a, b)].

Applied to all ranks, these principles mean that if the genera *Cuevavirus*, *Ebolavirus*, and *Marburgvirus* are included in the family *Filoviridae*, and measles virus is a member of the genus *Morbillivirus* in the family *Paramyxoviridae*, and the families *Filoviridae* and *Paramyxoviridae* are included in the order *Mononegavirales*, then

- (a) measles virus and Ebola virus are mononegaviruses; (b) ebolaviruses and morbilliviruses are mononegaviruses; and (c) paramyxoviruses and filoviruses are mononegaviruses, but
- (d) measles virus is not a filovirus and (e) Ebola virus is not a morbillivirus.

4 Conclusion

Filovirus taxon and virus nomenclature, if correctly applied as outlined here, mirrors the nomenclature of all other viruses. Correct application of this nomenclature is not only important to avoid confusion in scientific discourse but also helps to properly populate electronic databases such as GenBank. Uniformly applied standards in databases ensure that search queries result in the maximum of appropriate retrieved hits: if all Ebola virus sequence entries are labeled with the correct "Ebola virus", a search for "Ebola virus" will retrieve all entries; however, if 30% of the entries are labeled incorrectly with "*Zaire ebolavirus*", then a search for "Ebola virus" will retrieve only 70% of all desired entries. The differentiation between taxa and taxon members via application of correct orthography and term usage (see Table 5 for a cheat sheet for filovirus manuscript authors) is therefore important rather than merely a philosophical exercise.

Acknowledgements I would like to thank Laura Bollinger (IRF-Frederick) for editing the manuscript and Victoria Wahl-Jensen (National Biodefense Analysis and Countermeasures Center, Frederick, MD, USA) for critical comments.

Disclaimer The views and conclusions contained in this document are those of the author and should not be interpreted as necessarily representing the official policies, either expressed or implied, of the US Department of Health and Human Services or of the institutions and companies affiliated with the author. This work was supported in part through Battelle Memorial Institute's prime contract with the US National Institute of Allergy and Infectious Diseases (NIAID) under Contract No. HHSN272200700016I. A subcontractor to Battelle Memorial Institute who performed this work is: J.H.K., an employee of Tunnell Government Services, Inc.

References

Adams MJ, Lefkowitz EJ, King AMQ, Harrach B, Harrison RL, Knowles NJ, Kropinski AM, Krupovic M, Kuhn JH, Mushegian AR, Nibert M, Sabanadzovic S, Sanfaçon H, Siddell SG, Simmonds P, Varsani A, Zerbini FM, Gorbalenya AE, Davison AJ (2017) Changes to taxonomy and the International Code of Virus Classification and Nomenclature ratified by the International Committee on Taxonomy of Viruses (2017). Arch Virol 162. doi:10.1007/s00705-017-3358-5

Bao Y, Chetvernin V, Tatusova T (2012) PAirwise Sequence Comparison (PASC) and its application in the classification of filoviruses. Viruses 4(8):1318–1327. doi:10.3390/v4081318

Bukreyev AA, Chandran K, Dolnik O, Dye JM, Ebihara H, Leroy EM, Mühlberger E, Netesov SV, Patterson JL, Paweska JT, Saphire EO, Smither SJ, Takada A, Towner JS, Volchkov VE, Warren TK, Kuhn JH (2014) Discussions and decisions of the 2012–2014 International Committee on Taxonomy of Viruses (ICTV) *Filoviridae* Study Group, January 2012–June 2013. Arch Virol 159(4):821–830. doi:10.1007/s00705-013-1846-9

Calisher CH, Mahy BW (2003) Taxonomy: get it right or leave it alone. Am J Trop Med Hyg 68 (5):505–506

Carroll SA, Towner JS, Sealy TK, McMullan LK, Khristova ML, Burt FJ, Swanepoel R, Rollin PE, Nichol ST (2013) Molecular evolution of viruses of the family *Filoviridae* based on 97 whole-genome sequences. J Virol 87(5):2608–2616. doi:10.1128/JVI.03118-12

Drebot MA, Henchal E, Hjelle B, LeDuc JW, Repik PM, Roehrig JT, Schmaljohn CS, Shope RE, Tesh RB, Weaver SC, Calisher CH (2002) American Committee on Arthropod-borne Viruses Subcommittee on inter relationships among catalogued arboviruses, improved clarity of meaning from the use of both formal species names and common (vernacular) virus names in virological literature. Arch Virol 147(12):2465–2472. doi:10.1007/s00705-002-0938-8

Feldmann H, Geisbert TW, Jahrling PB, Klenk H-D, Netesov SV, Peters CJ, Sanchez A, Swanepoel R, Volchkov VE (2005) Family *Filoviridae*. In: Fauquet CM, Mayo MA, Maniloff J, Desselberger U, Ball LA (eds) Virus taxonomy—eighth report of the International Committee on Taxonomy of Viruses. Elsevier/Academic Press, San Diego, USA, pp 645–653

Kiley MP, Bowen ETW, Eddy GA, Isaäcson M, Johnson KM, McCormick JB, Murphy FA, Pattyn SR, Peters D, Prozesky OW, Regnery RL, Simpson DIH, Slenczka W, Sureau P, van der Groen G, Webb PA, Wulff H (1982) Filoviridae: a taxonomic home for Marburg and Ebola viruses? Intervirology 18(1–2):24–32

King AMQ, Adams MJ, Carstens EB, Lefkowitz EJ (2011a) The international code of virus classification and nomenclature. In: King AMQ, Adams MJ, Carstens EB, Lefkowitz EJ (eds) Virus taxonomy—ninth report of the International Committee on Taxonomy of Viruses. Elsevier/Academic Press, London, United Kingdom, pp 1273–1277

King AMQ, Adams MJ, Carstens EB, Lefkowitz EJ (2011b) The statutes of the ICTV. In: King AMQ, Adams MJ, Carstens EB, Lefkowitz EJ (eds) Virus taxonomy—ninth report of the International Committee on Taxonomy of Viruses. Elsevier/Academic Press, London, United Kingdom, pp 1269–1272

Kuhn JH, Jahrling PB (2010) Clarification and guidance on the proper usage of virus and virus species names. Arch Virol 155(4):445–453. doi:10.1007/s00705-010-0600-9

Kuhn JH, Wahl-Jensen V (2010) Being obsessive-compulsive about terminology and nomenclature is not a vice, but a virtue. Bionomina 1(1):11–14. doi:10.11646/bionomina.1.1.2

Kuhn JH, Becker S, Ebihara H, Geisbert TW, Johnson KM, Kawaoka Y, Lipkin WI, Negredo AI, Netesov SV, Nichol ST, Palacios G, Peters CJ, Tenorio A, Volchkov VE, Jahrling PB (2010) Proposal for a revised taxonomy of the family *Filoviridae*: classification, names of taxa and viruses, and virus abbreviations. Arch Virol 155(12):2083–2103. doi:10.1007/s00705-010-0814-x

Kuhn JH, Becker S, Ebihara H, Geisbert TW, Jahrling PB, Kawaoka Y, Netesov SV, Nichol ST, Peters CJ, Volchkov VE, Ksiazek TG (2011) Family *Filoviridae*. In: King AMQ, Adams MJ,

Carstens EB, Lefkowitz EJ (eds) Virus taxonomy—ninth report of the International Committee on Taxonomy of Viruses. Elsevier/Academic Press, London, United Kingdom, pp 665–671

Kuhn JH, Bao Y, Bavari S, Becker S, Bradfute S, Brister JR, Bukreyev AA, Cai Y, Chandran K, Davey RA, Dolnik O, Dye JM, Enterlein S, Gonzalez J–P, Formenty P, Freiberg AN, Hensley LE, Honko AN, Ignatyev GM, Jahrling PB, Johnson KM, Klenk H–D, Kobinger G, Lackemeyer MG, Leroy EM, Lever MS, Lofts LL, Mühlberger E, Netesov SV, Olinger GG, Palacios G, Patterson JL, Paweska JT, Pitt L, Radoshitzky SR, Ryabchikova EI, Saphire EO, Shestopalov AM, Smither SJ, Sullivan NJ, Swanepoel R, Takada A, Towner JS, van der Groen G, Volchkov VE, Wahl-Jensen V, Warren TK, Warfield KL, Weidmann M, Nichol ST (2013a) Virus nomenclature below the species level: a standardized nomenclature for laboratory animal-adapted strains and variants of viruses assigned to the family *Filoviridae*. Arch Virol 158(6):1425–1432. doi:10.1007/s00705-012-1594-2

Kuhn JH, Bao Y, Bavari S, Becker S, Bradfute S, Brister JR, Bukreyev AA, Chandran K, Davey RA, Dolnik O, Dye JM, Enterlein S, Hensley LE, Honko AN, Jahrling PB, Johnson KM, Kobinger G, Leroy EM, Lever MS, Mühlberger E, Netesov SV, Olinger GG, Palacios G, Patterson JL, Paweska JT, Pitt L, Radoshitzky SR, Saphire EO, Smither SJ, Swanepoel R, Towner JS, van der Groen G, Volchkov VE, Wahl-Jensen V, Warren TK, Weidmann M, Nichol ST (2013b) Virus nomenclature below the species level: a standardized nomenclature for natural variants of viruses assigned to the family *Filoviridae*. Arch Virol 158 (1):301–311. doi:10.1007/s00705-012-1454-0

Lauber C, Gorbalenya AE (2012) Genetics-based classification of filoviruses calls for expanded sampling of genomic sequences. Viruses 4(9):1425–1437. doi:10.3390/v4091425

Netesov SV, Feldmann H, Jahrling PB, Klenk H-D, Sanchez A (2000) Family *Filoviridae*. In: van Regenmortel MHV, Fauquet CM, Bishop DHL et al (eds) Virus taxonomy—seventh report of the International Committee on Taxonomy of Viruses. Academic Press, San Diego, USA, pp 539–548

Payne A (2016) Why do taxonomists write the meanest obituaries? The open nature of the science of classification virtually guarantees fights. Nautilus http://nautil.us/issue/35/boundaries/why-do-taxonomists-write-the-meanest-obituaries

Pringle CR (1991) The order *Mononegavirales*. Arch Virol 117(1–2):137–140

Siegert R, Shu H-L, Slenczka W, Peters D, Müller G (1967) Zur Ätiologie einer unbekannten, von Affen ausgegangenen menschlichen Infektionskrankheit. Dtsch Med Wochenschr 92 (51):2341–2343

Van Regenmortel MHV (2003) Viruses are real, virus species are man-made, taxonomic constructions. Arch Virol 148(12):2481–2488. doi:10.1007/s00705-003-0246-y

Van Regenmortel MHV (2006) Virologists, taxonomy and the demands of logic. Arch Virol 151 (7):1251–1255. doi:10.1007/s00705-006-0786-z

van Regenmortel MHV (2007) Virus species and virus identification: past and current controversies. Infect Genet Evol 7(1):133–144. doi:10.1016/j.meegid.2006.04.002

Van Regenmortel MHV (2016) Classes, taxa and categories in hierarchical virus classification: a review of current debates on definitions and names of virus species. Bionomina 10(1):1–21. doi: 10.11646/bionomina.10.1.1

CPSIA information can be obtained
at www.ICGtesting.com
Printed in the USA
BVOW07*0200180118
505652BV00001B/14/P